**AVIAN BIOLOGY**
**Volume IX**

# CONTRIBUTORS

Alan H. Brush

Donald S. Farner

H.-G. Hartwig

John C. Wingfield

# AVIAN BIOLOGY
# Volume IX

EDITED BY

*DONALD S. FARNER*

*JAMES R. KING*

*KENNETH C. PARKES*

*Carnegie Museum of Natural History*
*Pittsburgh, Pennsylvania*

ACADEMIC PRESS
**Harcourt Brace & Company, Publishers**
London San Diego New York
Boston Sydney Tokyo Toronto

ACADEMIC PRESS LIMITED
24/28 Oval Road,
London NW1 7DX

*United States Edition published by*
ACADEMIC PRESS, INC.
San Diego, CA 92101



This book is printed on acid-free paper

**A catalogue record for this book is available from the British Library**
ISBN 0–12–249409–1

Typeset by Fakenham Photosetting Limited, Fakenham, Norfolk
Printed in Great Britain by TJ Press (Padstow) Ltd, Padstow, Cornwall

*These volumes are dedicated to the memory of*

A. J. "JOCK" MARSHALL

(1911–1967)

*whose journey among men was too short by half*

# CONTENTS

## Chapter 3. Endocrinology of Reproduction in Wild Species

*John C. Wingfield and Donald S. Farner*

## CONTRIBUTORS

Numbers in parentheses indicate the pages on which the authors' contributions begin.

ALAN H. BRUSH (121), Department of Physiology and Neurobiology, University of Connecticut, Storrs, Connecticut 06268, USA

DONALD S. FARNER (163), formerly Department of Zoology, University of Washington, Seattle, Washington 98195, USA

H.-G. HARTWIG (1), Department of Anatomy II, Heinrich-Heine-Universität Düsseldorf, D-40225, Düsseldorf, FRG

JOHN C. WINGFIELD (163), Department of Zoology, University of Washington, Seattle, Washington 98195, USA

# GENERAL PREFACE

The birds are the best known of the large and adaptively diversified classes of animals. Between 9,000 and 10,000 species are currently recognized (depending in part on taxonomic taste). A few genuinely unknown species are discovered almost every year, and the list is also being expanded as we learn that taxa formerly considered "species" actually constitute two or more reproductively isolated forms that must be called sibling species. Nevertheless, our inventory of living species and knowledge of their distribution is much more nearly complete for birds than for any other class of animals. It is noteworthy that our relatively advanced knowledge of birds is attributable to a very substantial degree to a large group of dedicated and skilled amateur ornithologists.

Because of the abundance of empirical information on distribution, habitat, life cycles, breeding habits, etc., it has been relatively easier to use birds instead of other animals in the study of the general aspects of ethology, ecology, population biology, evolutionary biology, physiological ecology, and other fields of biology of contemporary interest. Model systems based on birds have had a prominent role in the development of these fields. The function of this multivolume treatise in relation to the place of birds in biological science is therefore envisioned as two-fold: to present a reasonable assessment of selected aspects of avian biology for those having this field as their primary interest, and to contribute to the broader fields of biology in which investigations using birds are of substantial significance.

Three decades have passed since the publication of A. J. Marshall's "Biology and Comparative Physiology of Birds", but progress in the fields included in this treatise has made most of the older chapters obsolete. Avian biology has shared in the so-called information explosion. The number of serial publications devoted mainly to avian biology has increased by about 20% per decade since 1940, and the spiral has been amplified by the parallel increase in page production and by the spread of publication into ancillary journals. By 1964, there were about 215 exclusively ornithological journals and about 245 additional serials publishing appreciable amounts of information on avian biology.*

These data reflect only the quantitative acceleration in the output of information in recent time. The qualitative changes have been much more impressive. Avifaunas that were scarcely known except as lists of species a decade ago have become accessible to investigation because of improved transportation and facilities in many parts of the world. New instrumentation has allowed the development of new fields of study and has extended the scope of old ones. Obvious examples include the use of radar in visualizing migration, of telemetry in studying the physiology of flying birds, and of spectography in analyzing bird sounds. The development of mathematical modeling, for instance, in evolutionary biology and population ecology has supplied new perspectives for old problems and a new arena for the examination of empirical data. New techniques for investigations at the molecular level, including DNA, have profoundly affected the fields of systematics, population studies, mating systems, and other disciplines. All of these developments—social, practical, and theoretical—have profoundly affected many aspects of avian biology in the last two decades. "Avian Biology" was conceived as an appropriate medium for another inventory of information, hypotheses, and new questions.

Marshall's "Biology and Comparative Physiology of Birds" was the first treatise in the English language that regarded ornithology as consisting of more than anatomy, taxonomy, zoology, and life history. This viewpoint was in part a product of the times; but it also reflected Marshall's own holistic philosophy and his understanding that "life history" had come to include the entire spectrum of physiological, demographic, and behavioral adaptation. This treatise is the direct descendant of Marshall's initiative. We have attempted to preserve the view that ornithology belongs to anyone who studies birds, whether it be on the level of molecules, individuals, or populations. To emphasize our intentions we have called the work "Avian Biology".

It has been proclaimed by various oracles that sciences based on taxonomic units (such as insects, birds, or mammals) are obsolete, and that the forefront of biology is process oriented rather than taxon oriented. This narrow vision of biology derives from an understandable but nevertheless myopic philosophy of reductionism and from the hyperspecialization that characterizes so much of science today. It fails to notice that lateral synthesis as well as vertical analysis are inseparable partners in the search for biological principles. Avian biologists of both stripes have together contributed a disproportionately large share of the information and thought that have produced contemporary principles in zoogeography, systematics,

* Baldwin, P. A., and Oehlert, D. E. (1964). "Studies in Biological Literature and Communications. No. 4. The Status of Ornithological Literature." Biological Abstracts, Inc., Philadelphia.

ethology, demography, comparative physiology, and other fields too numerous to mention. The record speaks for itself.

In part, this progress results from the attributes of birds themselves. They are active and visible during the daytime; they have diversified into virtually all major habitats and modes of life; they are small enough to be studied in useful numbers but not so small that observation is difficult; and, not least, they are esthetically attractive. In short, they are relatively easy to study. For this reason we find in avian biology an alliance of specialists and generalists who regard birds as the best natural vehicle for the exploration of process and pattern in the biological realm. It is an alliance that seems still to be increasing in vigor and scope.

In the early planning stages of the treatise we established certain working rules that we have been able to follow with rather uneven success:

(1) "Avian Biology" is the conceptual descendant of Marshall's earlier treatise, but is more than simply a revision of it. We have obviously been able to cover more subjects, in greater depth, in nine volumes than Marshall could in two volumes. The only topic covered by Marshall and omitted in the present series is avian embryology, which, under a new banner of developmental biology, has expanded and specialized to the extent that a significant review of recent advances would be a treatise in itself. On the other hand, several topics not included in Marshall's work have been fully covered in "Avian Biology".
(2) Since we expect the volumes to be useful for reference purposes as well as for the instruction of advanced students, we have asked authors to summarize established facts and principles as well as to review recent advances.
(3) We have attempted to arrange a balanced account of avian biology as it has stood in the past two decades. We have not only retained chapters outlining modern concepts of structure and function in birds, as is traditional, but have also encouraged contributions representing multidisciplinary approaches and synthesis of new points of view.
(4) We have attempted to avoid a parochial view of avian biology by seeking diversity among authors with respect to nationality, age, and ornithological heritage.
(5) As a corollary of the preceding point, we have not intentionally emphasized any single school of thought, nor have we sought to dictate the treatment given to controversial subjects. Our single concession to conceptual conformity is in taxonomic and nomenclatural usage, as explained by Kenneth Parkes in the Note on Taxonomy.

We began our work with a careful plan for a logical topical development throughout all volumes. Only its dim vestiges remain. Owing to belated defections by a few authors and conflicting commitments by others we have been obliged to sacrifice logical sequence in order to retain authors whom we regarded as the best for the task. In short, we gave first priority to the maintenance of general quality, trusting that each reader would supply logical cohesion by selecting chapters that are germane to his individual interests.

DONALD S. FARNER
JAMES R. KING

## AN APPRECIATION OF DONALD S. FARNER AND JAMES R. KING

The creators and editors of "Avian Biology", Donald S. Farner and James R. King, died on 18 May 1988 and 7 April 1991, respectively. Farner and King originally were mentor and student, and subsequently became one of the most effective sets of editorial collaborators in modern biology. Their impressive accomplishments and diverse honors in research, teaching, and professional service have already been described in professional obituaries (*Auk* **106**, 710–713, 1989; *Auk* **109**, 643–645, 1992) and I offer here only a brief outline of their careers, preferring to focus on the production of "Avian Biology".

Donald Sankney Farner was born in 1915, acquired his BS degree from Hamline University (1937), and his MA (1939) and PhD (1941) degrees from the University of Wisconsin. After military service in World War II, he joined the faculty of the University of Kansas in 1945. In 1947, he moved to Washington State University where he was first Associate Professor of Zoology (1947–1952) and then Professor of Zoophysiology (1952–1965). He also served there as Dean of the Graduate School (1961–1964). He moved to the University of Washington in 1965, where he was Professor of Zoology (1965–1985) and Chairman of Zoology (1966–1981). He retired from the faculty in 1985, although he remained professionally active. For most of his career, Farner's research centered in the related areas of photoregulatory biology, endocrinology, and reproductive physiology. His laboratory was remarkably productive, producing over 260 publications, and was particularly notable for his collaborative efforts with postdoctoral scholars and visiting senior scientists. Don was also a prominent and progressive biopolitician, playing major roles in both national and international scientific organizations (e.g. American Ornithologists' Union, International Ornithological Congress, American Society of Zoologists, International Union for Biological Sciences).

James Roger King was born in 1927 and acquired his BA degree from San Jose State College (1951). He received his MA (1953) and PhD (1957) degrees from Washington State University, where he was one of Farner's earliest and most successful PhD students. King's first faculty position was at the University of Utah as Assistant Professor of Experimental Biology (1957–1960). He was then convinced to return to Washington State University where he served as Assistant Professor (1960–1962), Associate Professor (1962–1967), and Professor of Zoophysiology (1967–1991), as well as Chairman of the Department of Zoology (1972–1978). In his research, King made pioneering contributions to several areas of environmental physiology—including chronobiology, energetics, biophysical ecology, and nutritional ecology. The value of these efforts was magnified by King's impressive skills for identifying key questions, designing rigorous experiments to address them, and placing the results in a context appropriate for understanding organisms within their natural environments.

What attributes made Farner and King such remarkably effective editors? Both men, of course, commanded the respect of the biologists who authored chapters because they remained leaders in their research areas throughout their careers, publishing a stream of valuable and skillfully written papers. They also had substantial editorial experience prior to embarking on this series. For example, Don edited *The Auk* (1960–1962) and Jim edited *The Condor* (1965–1968). They combined this experience and scientific prominence while avoiding the egocentric trap of imposing a narrow style of writing on authors who conformed to the general format for the series and used clear, correct, and reasonably concise English. This is a set of traits likely to attract prominent scientists as authors.

Perhaps more important than these editorial attributes was the strength and breadth of King and Farner's view of avian biology and scientific processes in general. They initiated "Avian Biology" with the clear conviction that modern ornithology should consist of a holistic view of the organism and not unduly emphasize historically important disciplines such as taxonomy, anatomy, or faunistics. Although they certainly valued the continuing contributions possible from these areas, they also were convinced that it would be both valuable and scientifically invigorating to emphasize a broader focus within ornithology. At least partially, this reflects the breadth of their backgrounds in biology. Farner's earliest publications, for example, dealt with diverse taxa (e.g., insects, fish, amphibians) and topics (e.g., taxonomy, natural history, and population biology). King's earliest research dealt with natural history and faunistics, and throughout his career he published occasionally on a variety of non-physiological topics (e.g., bird song, migration). Both men therefore were impressive for their far-reaching knowledge and the coverage of topics in "Avian Biology" was similarly extensive, encompassing areas as diverse as behavior, ecology, morphology, paleontology, and systematics. King and Farner also were purposefully broad in the authors they recruited for the series. Edited volumes too often reflect national biases or an editor's narrow range of acquaintances. Farner and King assiduously avoided this hazard. They were well-disposed to do so, having very extensive national and international contacts. Between the two of them, they maintained an impressive knowledge of the personalities, skills, and activities of a substantial fraction of the research-active ornithologists in the world. Their single overriding basis for identifying authors was scientific merit; they recruited the individual likely to make the most valuable contribution and were unusually successful at ignoring complicating criteria such as age or geographic location. Consequently, authors from 12 countries contributed to "Avian Biology" and included both senior and very prominent scientists as well as some who were Assistant Professors when they prepared their chapters.

Finally, both Don and Jim were strong adherents to the tenets of intellectual freedom. As academic or scientific administrators, they were forceful and effective defenders of the scholarly tradition of free inquiry and expression. For an editor, academic freedom commonly means publishing views with which one personally disagrees and could readily suppress by either overt or subtle means. Well aware of the pernicious effects of failing to meet this ethical challenge, Farner and King were notably successful in ensuring expression of the diversity of viewpoints so necessary for scientific progress.

Due in large part to the high professional standards and the intellectual acumen of Don Farner and Jim King, "Avian Biology" has been one of the most successful synthetic compendia dealing with any group of organisms. Similarly, ornithology has progressed substantially in the 21 years since this series began and stands today as one of the most vigorous of the taxonomically defined scientific disciplines. Through efforts such as the production of "Avian Biology" and their many other professional contributions, Donald S. Farner and James R. King significantly invigorated ornithology and facilitated its evolution as a discipline.

Glenn E. Walsberg
Department of Zoology
Arizona State University
Tempe, AZ 85287–1501

## PREFACE TO VOLUME IX

This is the final volume of the "Avian Biology" series. When the first volume appeared in 1971, no set number of volumes was anticipated—as long as the editors found authors willing to contribute their particular expertise in providing appropriate chapters, and as long as Academic Press was satisfied with the series, we would issue new volumes. Much has changed in the ensuing decades. We have lost the two brilliant avian biologists, Donald S. Farner and James R. King, who had the original concept for the series and co-edited the first eight volumes. New methods of research, particularly at molecular levels, have had a major impact on several aspects of biology in general and ornithology in particular. Few if any of the aspects of avian biology covered in earlier volumes have remained static, yet in only one field—the fossil record of birds—did we publish an updated account. There will no doubt be similar treatises in the future that will summarize the advances in our knowledge, but nevertheless I believe that the 59 chapters of "Avian Biology" will continue to be of great significance to students and others for years to come, as references to the sum total of our knowledge of birds at the time of publication.

Dr Walsberg's tribute to Drs Farner and King in this volume reflects their stature as scientists and colleagues. I want to add my own gratitude for their having invited me to participate in this complex undertaking. I am sure they would have joined me also in thanking Jean E. Thomson Black and Andrew Richford, our editors at Academic Press, for their help and their exceptional patience with us and with the chapter authors.

Kenneth C. Parkes

# NOTE ON TAXONOMY

Early in the planning stages of "Avian Biology" it became apparent that it would be desirable to have the manuscript read by a taxonomist, whose responsibility it would be to monitor uniformity of usage in classification and nomenclature. Other multiauthored compendia have been criticized by reviewers for use of obsolete scientific names and for lack of concordance from chapter to chapter. As neither of the other editors was a taxonomist, they invited me to perform this service as Taxonomic Editor. This was my principal contribution to the first five volumes of the series; as a full editor from Volume VI onwards, I have continued to monitor the classification and nomenclature along with other editorial duties.

A brief discussion of the ground rules that we tried to follow is in order. Insofar as possible, the classification of birds down to the family level in Volumes I through VII followed that presented by Dr Storer in Chapter 1, Volume I. Shortly after the appearance of Volume VII, the American Ornithologists' Union published the sixth edition of its "Check-list of North American Birds". Having been a member of the Union's Committee on Classification and Nomenclature that compiled this edition, I was able to anticipate some of the changes effected in scientific and English names, and incorporate them in Volumes VI and VII, where they were presented in an Addendum to the original "Note on Taxonomy". In Volume VIII and the present, final volume, of "Avian Biology", nomenclature follows the usage of the AOU "Check-list" for those bird species covered therein.

This is not the place to debate the classification and nomenclature of the AOU "Check-list", or to explain the rationale for each of the changes from previous usage. Suffice it to say that these aspects of ornithology, like others reflected in the volumes of "Avian Biology", are not static, nor should they be. Although the original goal had indeed been for uniformity throughout, I felt that it was more important for the nomenclature of the later volumes to accord with that standardly adopted by all American ornithological journals than to conform rigidly to what appeared in earlier volumes of this series. Changes have been relatively few, and nomenclature remains consistent within a volume; two authors (Dr Brodkorb in

Volume I and Dr Olson in Volume VIII) use their chapters to present their own ideas on classification, although these depart substantially from the classification followed elsewhere in the series.

In general, the scientific names of non-North American taxa have been those used by the Peters "Check-list"; exceptions include those orders and families covered in the earliest Peters volumes for which more recent classifications have become widely accepted. To supplement the Peters list I have relied on several standard regional references, using my own taxonomic judgment in cases of conflicting usage.

At this writing (summer 1992), the taxonomy, taxonomy-based scientific nomenclature, and English nomenclature of the birds of the world are in a state of turmoil. Conflicting classifications of parts of the Class Aves appear regularly in the major ornithological journals. It would be both premature and confusing for the final volume of "Avian Biology" to depart from the conventions employed in the previous volumes.

Standardization of English names for bird species is a difficult goal, as evidenced not only by the lack of concordance in the several lists of birds of the world published in the last several years, but also by the emotions aroused by recent British proposals to change the "official" English names of birds of the Western Palearctic. Some writers even question the validity of the goal of standardization. Within a multiauthored work such as "Avian Biology", however, it is mandatory insofar as possible that every English name be associated with a single scientific name, and, conversely, that only one English name be used for a given species.

Reliance on a standard reference such as Peters has meant that certain species appear under scientific names quite different from those used in much of the previous ornithological literature. For example, the Zebra Finch, widely used as a laboratory species, was long known as *Taeniopyga castanotis*. In Volume 14 of the Peters "Check-list" (pp. 357–358, 1968), *Taeniopyga* is considered a subgenus of *Poephila*, and castanotis a subspecies of *P. guttata*. Thus the species name of the Zebra Finch becomes *Poephila guttata*. In such cases the more familiar scientific name was sometimes given in parentheses in earlier volumes of "Avian Biology".

Strict adherence to standard references also means that some birds appear under scientific names that, for either taxonomic or nomenclatorial reasons, would not be chosen by either the chapter author or the taxonomic editor. Laboratory workers not uncommonly perpetuate long-obsolete scientific names as used in the earliest literature of their specialty. Similarly, the standardized English name may not be the one most familiar to the chapter author. As a taxonomist, I naturally hold some opinions that differ from those of the authors of the Peters list and the other reference works used. I feel strongly, however, that a general treatise such as "Avian Biology" should not be used as a vehicle for taxonomic or nomenclatural innovation or for the furtherance of personal opinions, no matter how strongly held. I therefore apologize to those authors in whose chapters names have been altered for the sake of uniformity and offer as solace the fact that I have had my own objectivity strained several times by having to use names that do not reflect my own taxonomic judgment.

Within each chapter the first mention of a species of wild bird includes both the scientific name and the English name, or the scientific name alone. If the same species is mentioned by English name later in the same chapter, the scientific name is usually omitted. Scientific names are also usually omitted for domesticated or laboratory birds. The reader may make the assumption throughout that, unless otherwise indicated, the following statements apply:

(1) "Duck" or "domestic duck" refers to domesticated forms of *Anas platyrhynchos*; the breed is sometimes specified, as in "Pekin duck".
(2) "Goose" or "domestic goose" refers to domesticated forms of *Anser anser*.
(3) "Pigeon" or "domestic pigeon" or "homing pigeon" refers to domesticated forms of *Columba livia*.
(4) "Turkey" or "domestic turkey" refers to domesticated forms of *Meleagris gallopavo*.
(5) "Chicken", "domestic fowl", "domestic hen", "domestic rooster", "capon", etc., refer to domesticated forms of *Gallus gallus*; these are often collectively called "*Gallus domesticus*" in biological literature.
(6) "Japanese Quail" refers to laboratory strains of the genus *Coturnix*, the exact taxonomic status of which is uncertain. See Moreau and Wayre, *Ardea* **56**, 209–227, 1968.
(7) "Canary" or "domestic canary" refers to domesticated forms of *Serinus canaria*.
(8) "Guineafowl" or "Guineahen" refers to domesticated forms of *Numida meleagris*.
(9) "Ring Dove" refers to domesticated and laboratory strains of the genus *Streptopelia*, often called "Barbary Dove", and often incorrectly given specific status as *S. risoria*. Now thought to have descended from the African Collared Dove, *S. roseogrisea*, the Ring Dove of today *may* possibly be derived in part from *S. decaocto* of Eurasia; at the time of publication of Volume 3 of Peters's "Check-list of Birds of the World" (p. 92, 1937), *S. decaocto* was thought to be the direct ancestor of "*risoria*". See Goodwin, "Pigeons and Doves of the World" (3rd edn), pp. 117–119, 1983.

Name changes effected in this volume of "Avian Biology" have been made solely in order to conform with the usage of the 6th edition of the AOU "Check-list of North American Birds" and its Supplements as published in *The Auk*, in accordance with the policy mentioned in the "Note on Taxonomy" in Volume VIII (p. xx, 1985).

KENNETH C. PARKES

## CONTENTS OF OTHER VOLUMES

**Volume IV**



**Volume V**



**Volume VI**



**Volume VII**

## Volume VIII

# Chapter 1

# THE CENTRAL NERVOUS SYSTEM OF BIRDS: A STUDY OF FUNCTIONAL MORPHOLOGY

*H.-G. Hartwig*

*Department of Anatomy II*
*Heinrich-Heine-Universität Düsseldorf*
*D-40225 Düsseldorf, FRG*





Avian Biology, Vol. IX

ISBN 0-12-249409-1

## I. Introduction

The central nervous system (CNS) of birds has long been the subject of intensive research. The names of such well-known scientists as S. Ramón y Cajal, C. Golgi, C. U. Ariens Kappers, G. C. Huber, E. C. Crosby, N. Holmgren, H. Bergquist, H. Kuhlenbeck, to mention but a few, are closely associated with the major works on avian neurobiology. As early as 1934 Graf Haller von Hallerstein presented a detailed historical review of brain research in birds and other vertebrates. A few decades later there appeared Kuhlenbeck's (1968–1978) fundamental treatise and Pearson's (1972) extensive overview of findings on the functional anatomy of the avian CNS. The material in this chapter and the cited references are limited for reasons of length. The interested reader is therefore referred to Kuhlenbeck (1968–1978) and others for discussions of the sometimes contradictory results and concepts found in the older literature.

The present chapter is focused on more recent concepts of brain and spinal cord topography and functional morphology. Modern techniques, e.g., mapping neuronal connections with tracer substances or identifying neurotransmitters and neurohormones with immunocyto- and histochemical methods, have provided the neurobiologist with entirely new insights. Wherever possible I have followed the guidelines of the International Committee on Avian Anatomical Nomenclature (cf., Breazile, 1979). Additional information dealing primarily with descriptive anatomical details can be found in a number of textbooks which also supply coordinates for stereotaxic surgery (chicken: Van Tienhoven and Juhasz, 1962; pigeon: Karten and Hodos, 1967; Japanese Quail: Baylé *et al.*, 1974; Ring Dove: Vowels *et al.*, 1975; 3-day-old chick: Youngren and Phillips, 1978).

Among vertebrates the highest level of morphological and functional evolution is achieved by birds and mammals (for a general treatment of the evolution of the brain and intelligence, see Jerison, 1973). The behavior of birds includes highly specific responses to complex visual, auditory, tactile, and gustatory stimuli. It has been discovered that some species also respond to magnetic (review, Ossenkopp and Barbeito, 1978), and olfactory information (see page 72). The complexity of behavioral patterns is reflected in the complexity of the CNS. Next to mammals, birds have evolved the greatest index of cerebralization among vertebrates. For a review and discussion of the problems related to establishing a "cerebralization index", see Kuhlenbeck (1973) and Jerison (1973).

In birds the relative weights of brains (brain weight in relation to body weight) range from e.g., 1:25 in House Sparrow (*Passer domesticus*) and 1:91 in pigeons through lower values in domesticated birds (approx. 1:260 in mallards and 1:400 in chickens), down to relative weights of only 1:1200 in Ostriches (*Struthio camelus*). In this context it should be noted that the size of the vertebrate brain is influenced primarily by two different factors: (1) the absolute size of the body, and (2) the degree of cerebralization (Starck, 1982).

The absolute weights of brains in different species of birds have only rarely been measured on a broader comparative basis. Moreover, data are frequently incomplete. They often refer to only one or two individual specimens of a species without reporting such vital information as sex, body weight, and age. Gadow and Selenka (1891) summarize the following representative weights:

| | |
|---|---|
| *Corvus monedula* | 4.35 g |
| *Pica pica* (female) | 4.8 g |
| *Pica pica* (male) | 5.2 g |
| *Serrinus serinus* | 0.75 g |
| *Garrulus glandarius* | 4.15 g |
| *Upupa epops* | 5.6 g |
| *Charadrius* sp. | 2.2 g |
| *Anser anser* (domestic) | 12.4 g |
| *Perdix* sp. | 1.85 g |
| *Alauda* sp. | 0.8 g |
| *Anas crecca* | 4.1 g |
| *Gallus gallus* (female) | 3.5 g |
| *Gallus gallus* (male) | 3.8 g |
| *Vanellus vanellus* | 2.25 g |

In this context it is interesting to note that the brains of homing pigeons are significantly (5%) larger than those of fantails and strassers, a difference that is independent of body weight (Haase *et al.*, 1977).

Jerison (review, 1973) attempted to assess the evolution of the vertebrate brain by collecting "brain: body data" for a total of 198 species. He also took into careful consideration numerous errors that can be found among data of this sort in many compendia. Jerison's studies included 94 species of mammals, 52 of birds, 20 of reptiles, and 32 of bony fish. He created a system for the functional "mapping of brain: body data" into so-called minimum convex polygons. The minimum convex polygons of homeothermic vertebrates can readily be separated from those of poikilothermic vertebrates (Fig. 1). The data presented by Jerison (1973) suggest that the increase in the relative brain weights of vertebrates from the "lower" to the "higher" classes is in the order of a factor of 10. Finally it should be mentioned that relative brain weights in very small birds (e.g., hummingbirds) increase as do the relative weights of other organs of crucial function (e.g., heart). Organs that control central functions can apparently not be miniaturized to the same extent as can the somatic locomotor system.

Certain centers in the brains and spinal cords of some bird species have grown larger and more highly specialized in their fine structure during evolution. Functions controlled or regulated by these centers have become comparably complex. Examples of centers that have become extremely large or complex include the lumbosacral spinal medulla in Ostriches (cf. Kuhlenbeck, 1975) and the auditory system in certain species of song birds and owls (for a review of comparative

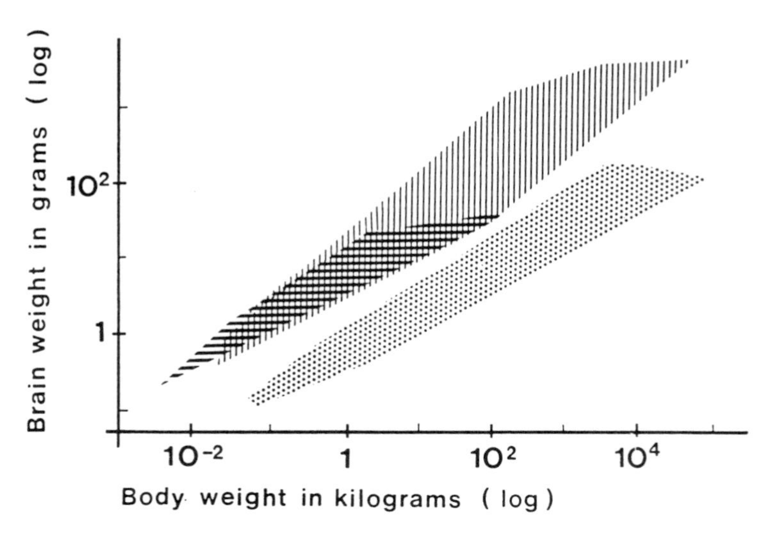


FIG. 1. Relationships of brain to body weights in terrestrial vertebrates (minimum convex polygons of Jerison). Upper polygon encloses two classes of homeothermic ("higher") vertebrates, lower polygon encloses three classes of poikilothermic ("lower") vertebrates (after Jerison, 1973; fig. 8.4).

studies on hearing in vertebrates, see Popper and Fay, 1980). Nevertheless, birds share with other vertebrates common features in the topographical arrangement of functional components of the CNS. General principles of the overall functional morphological pattern (Bauplan) of the bird CNS have been described in fine detail by Kuhlenbeck (1978). These principles are briefly summarized below.

## II. General Features of the Central Nervous System

The CNS develops from the ectodermal neural plate that closes to form the neural tube. The tissue originating from the neural tube becomes arranged in longitudinal strips or zones (primary zonal system, cf. Kuhlenbeck, 1978). These zones in turn become clearly differentiated into (1) dorsoventral, and (2) rostrocaudal regions that are later responsible for specific functions. In a rostrocaudal sequence the major subdivisions of the CNS are as follows: (1) the prosencephalon (telencephalon and diencephalon), (2) the deuterencephalon (mesencephalon and rhombencephalon including cerebellum), and (3) the spinal cord.

The spinal cord and the deuterencephalon develop from the unpaired roof and floor plates, and the paired alar and basal plates of the neural tube (Fig. 2).

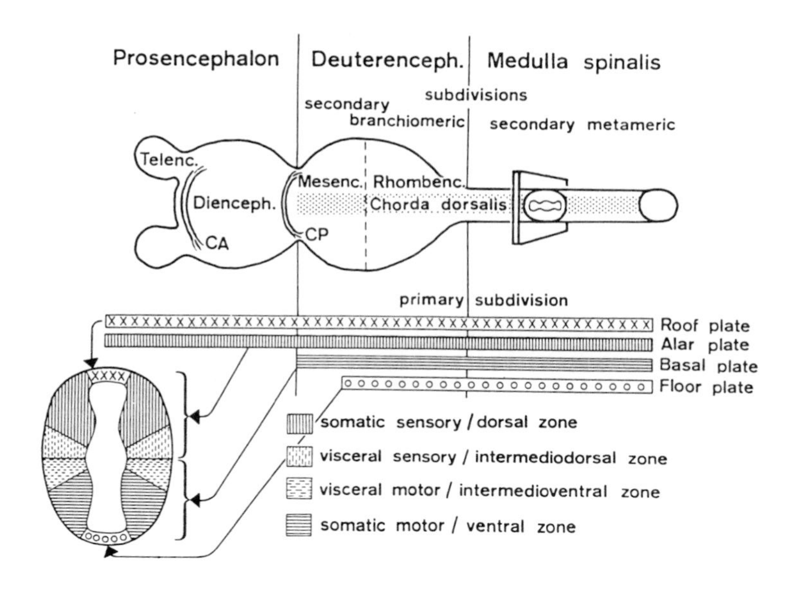


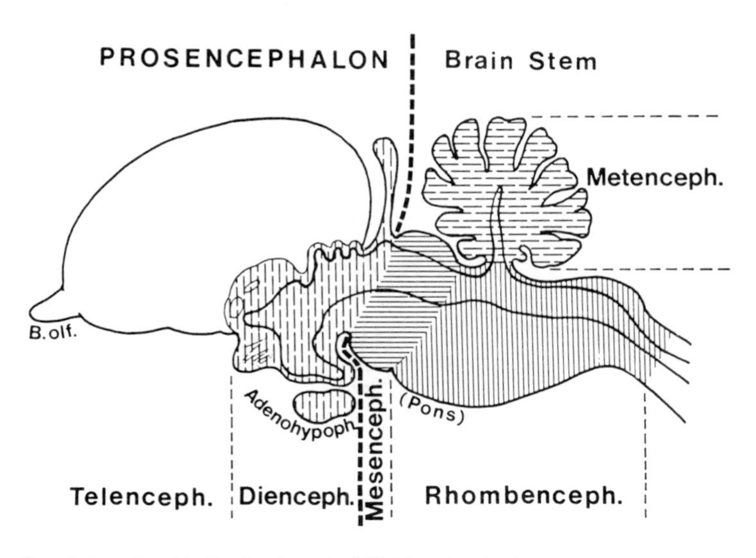


FIG. 2. Dorsal view of an idealized embryonic CNS (above) and a diagrammatic midsagittal view of the brain of an adult bird (below). Note differences between the general morphological patterns of the prosencephalon and the more caudal components of the CNS which develop in close relationship with the chorda dorsalis. Boundary between prosencephalon and meso-rhombencephalon (or deuterencephalon of Kuhlenbeck) is indicated in the adult brain by an interrupted line. In the embryonic brain this boundary area is indicated by the posterior commissure (CP). CA, anterior commissure as the borderline between diencephalon and telencephalon; B.olf., bulbus olfactorius.

Generally, only cells of the alar and basal plates are able to differentiate into neurons. It has been suggested that the floor and basal plates disappear at the level of the rostral end of the chorda dorsalis. This level marks the boundary between the mesencephalon and the prosencephalon. Apparently the prosencephalon is derived only from the alar and roof plates (for details, cf. Kuhlenbeck, 1978). In the course of phylogenetic development the material of the spinal cord and the deuterencephalon that is derived from the alar plates becomes further subdivided into dorsal and intermediodorsal zones. The material of the basal plates becomes that of the intermedioventral and ventral zones. In dorsoventral sequence these secondary zones (cf. Kuhlenbeck, 1978) are later concerned with the following functional divisions: somatic sensory, visceral sensory, visceral motor, and somatic motor areas (Fig. 2). In mammals, neurons of the thoracic intermedioventral zone form a distinct lateral column. Birds do not have such a column.

The roof plate of the rhombencephalon develops into an extensive but thin epithelial lamina covering the fourth ventricle (precursor or anlage of the choroid plexus). At the same time the alar plate is displaced laterally. Thus, the dorsoventral sequence of the secondary zones is transformed into a lateromedial sequence (Fig. 3).

At the level of the diencephalon the secondary zones seem to be derived exclusively from alar plate material. The following subdivisions can be distinguished in dorsoventral sequence: epithalamus, thalamus dorsalis, thalamus ventralis, and hypothalamus. At the level of the telencephalon the secondary zones, also originating only from the alar plate, consist of four distinct basal zones

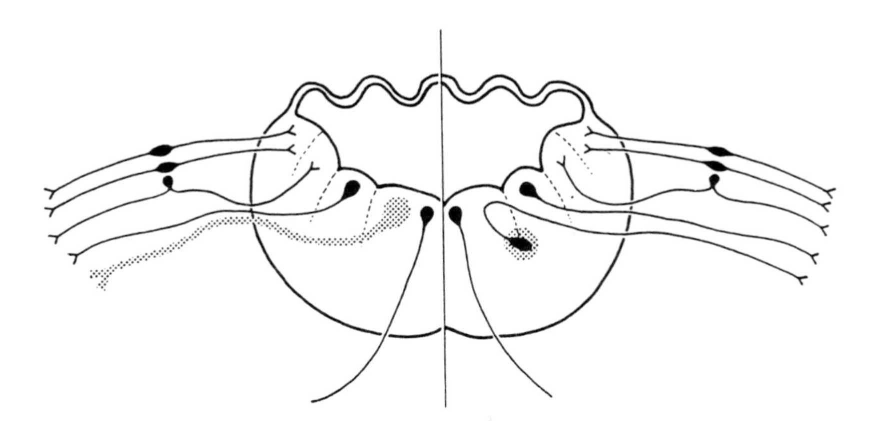

FIG. 3. Simplified diagram of an early developmental stage of the medulla oblongata in transverse section. On the left somatomotor (most medially located), branchiomotor (dotted structure), visceromotor, viscerosensory, and somatosensory areas are arranged in a mediolateral sequence. On the right, ventrolateral migration of branchiomotor perikarya is shown. Note axon of branchiomotor neuron passes back through original position of developing perikarya before leaving the brain stem.

and four dorsal or pallial zones (for a thorough review, see Kuhlenbeck, 1978). In birds and in mammals zone D1, which is originally pallial, becomes basal secondarily (see in this context, Starck, 1962). The avian telencephalon is characterized by a massive development of the basal zones (see p. 37). Bergquist (1957) provided a critical review of the historical development of ideas on the morphological and functional subdivisions of the telencephalon in birds and other vertebrates.

The system of secondary zones as described above has been found in all vertebrate species. This also holds true for all major ascending and descending channels for transmitting information (cf. Kuhlenbeck, 1978).

Somatosensory and viscerosensory neuronal pathways ascending from the spinal cord reach: (1) the tegmental reticular formations in the lower brain stem, (2) the cerebellum, (3) the tectum mesencephali, and (4) finally, via thalamic nuclei of the diencephalon, the telencephalon. In birds, however, it has been shown that somatosensory information, which arises from input channels located in the head and neck, may bypass thalamic relay nuclei and reach the telencephalic nucleus basalis directly (for details, see p. 45). Ascending pathways conducting somatosensory and viscerosensory information can be grouped into two distinct systems, the spinocerebellar and the spino-bulbo-tecto-thalamic systems. Somatosensory, viscerosensory and some special sensory (e.g., gustatory) information is also carried to these two systems by the cranial nerves of the deuterencephalon. Spino-bulbo-tecto-thalamic pathways form the so-called medial lemniscal system. The lateral lemniscal system conducts vestibular and auditory information to the mesencephalic, diencephalic, and telencephalic nuclei involved in higher integration. Optic sensory information reaches the telencephalon via primary optic centers in the mesencephalon (tectum mesencephali or optic tectum), and via thalamic nuclei ("lateral geniculate area"). The brain stem receives visual information via the nucleus of the basal optic root (or ectomammillary nucleus) and the suprachismatic area of the hypothalamus. In contrast to the sensory pathways mentioned above, those conducting olfactory information pass directly to the telencephalon without being relayed in the thalamus.

Descending neuronal pathways that terminate on motoneurons and the interneurons associated with them (final common pathway of Sherrington; for review, see Kuhlenbeck, 1978), as well as the visceromotor pathways in the spinal cord and the deuterencephalon, can be separated into the following main tracts: (1) the reticulospinal, (2) the vestibulospinal, (3) the tecto-tegmental (or tecto-tegmento-spinal, one major component being the rubrospinal pathway), and (4) the cerebello-tegmental tracts. In addition, birds have in common with other vertebrates the descending pretectal, the hypothalamic, and the telencephalic neuronal pathways. Ascending and descending pathways that interconnect the telencephalon with caudal centers of the CNS are gathered into the medial and lateral forebrain bundles. Convincing evidence for the existence of pathways that descend directly from cortical areas of the telencephalon to the spinal cord (cortico-spinal tract in mammals) has not yet been found. This may be so because the cortex cerebri of birds does not reach a level of development comparable to that of the mammalian cortex. For a detailed discussion of possible homologies with pyramidal tract fibers in Ungulata, see Zeier and Karten (1971).

In all vertebrates information is exchanged between the left and the right sides of the CNS through several main commissures. Seen in rostrocaudal sequence they are:

(1) the Commissura pallii (very small in birds);
(2) the Commissura anterior;
(3) the Commissura supraoptica dorsalis et ventralis;
(4) the Commissura habenularis;
(5) the Commissura posterior and Commissura tectalis;
(6) the Commissura tegmentalis, and;
(7) the Decussatio lemniscalis.

Finally, most pathways, with the apparent exception of parts of the ascending spinocerebellar system, cross or decussate the midline. For details on commissures and a discussion of the fiber systems that cross the midline, but whose function is still poorly understood, see Kuhlenbeck (1978). Although the subject has not yet been thoroughly investigated in birds, it can be assumed that avian communication channels and the sites to which they project are arranged in orderly and distinct patterns (somatotopic order) similar to those seen in mammals and other vertebrates. In birds the most detailed data in this field of research have been collected in studies on the retinotopic order of the visual information channel and the tonotopic order of the auditory system. Details are presented in the corresponding sections.

## III. Structural Components of the Central Nervous System

### A. NEURONS

Neurons stem from the neuroepithelial cells of the neural tube and from cells of the adjacent neural crest and the sensory placodes. Neuroblasts develop into a number of different types of neurons. These can be described according to the gross form of the cell body or other cytoarchitectonic features, such as the structural characteristics of the nucleus. Nissl staining and other neurohistological techniques reveal light microscopical differences in the makeup of the rough endoplasmic reticulum between types of neurons.

Several approaches for classifying the neurons in the CNS of homeothermic animals have emerged. For example, two types of multipolar neurons can be distinguished on the basis of the appearance of their axons following treatment with Golgi techniques (for review, see Kuhlenbeck, 1970). A Golgi-type-I neuron possesses an axon that extends to a target area rather distant from the location of its perikaryon. In most cases the axon bears a myelin sheath. The axon of the Golgi-type-II neuron remains more or less in the neighborhood of the perikaryon. Kuhlenbeck (1970) classified neurons into five types based on their overall structure: (1) unipolar, (2) pseudounipolar, (3) bipolar, (4) multipolar, and (5) neuroepithelial neurons.

The distribution of the different types of neurons in the CNS of birds was studied by earlier investigators (Ramón y Cajal, Golgi, Ariens Kappers and others; for review, see Kuhlenbeck, 1970). They used the silver impregnation techniques developed by Golgi (impregnation of the cell membrane) and those introduced by Cajal (impregnation of neuro-fibrillar material). More recent cytoarchitectonic studies based on various silver impregnation methods have been made on the spinal medulla of the chicken (Matsushita, 1968). Silver impregnation techniques have also been applied successfully in the analysis of the neuronal circuitry of highly differentiated regions of the CNS such as the optic tectum, the retina and the cerebellum. Differences in the staining characteristics of nucleic acids in the nucleus and the soma (Nissl staining) have become valuable criteria for distinguishing between different types of neurons at the level of the light microscope. Recently developed specific and highly sensitive histochemical, immunocytochemical and autoradiographic techniques have made possible detailed functional mapping of monoamine- or peptide hormone-containing or steroid hormone-binding neuronal systems in the CNS (monoamines: Ikeda and Goto, 1974; Yamamoto *et al.*, 1977; Dubé and Parent, 1981; peptide hormones: Goossens *et al.*, 1977; Bons *et al.*, 1978a, b; steroid hormone-binding neurons: Pfaff, 1976). Moreover, topographical arrangements of specifically activated neuronal circuits have been made visible by means of the $^{14}$C-deoxiglucose technique (see e.g., specifically activated auditory nuclei in the Guineafowl, Scheich *et al.*, 1979).

In general, the neurons of the CNS of birds closely resemble those of mammals not only in gross form but also in ultrastructure. Moreover, the results of immunocytochemical studies suggest that the molecular composition of neurofilaments in the domestic fowl and man are similar (Dahl and Bignami, 1977). Additional aspects of ultrastructure will be treated in sections IV–IX. There are major differences between mammals and birds with respect to types of neurons and circuitry in the telencephalon. Since birds did not evolve a cerebral cortex comparable to that of mammals, neuronal characteristics commonly observed in the mammalian cerebral cortex do not exist in the telencephalon of birds. Details of the comparative neurology of the telencephalon of all classes of vertebrates have been extensively dealt with in a treatise by Ebbesson and co-workers (Ebbesson, 1980).

## B. NEUROGLIA AND MESOGLIA

Neurons and their processes are intimately intermingled with the somata and processes of neuroglial and mesoglial cells (Hortega cells). There is general agreement among neurobiologists on the mechanical and metabolic functions of the glial components of the CNS. The number of glial cells in the CNS of mammals, and possibly of birds, apparently exceeds that of neurons (cf. Kuhlenbeck, 1970).

Silver impregnation techniques are not only used to study neurons but they have also long been valuable in investigating differences in the glial elements of the CNS. Earlier neuroscientists such as Ramón y Cajal, Rio del Hortega and others used them effectively; for review, see Kuhlenbeck (1970), Inoue (1971), King (1966); for a general historical review of investigations on glial elements, see Niessing *et al.* (1980). Recently Leonhardt *et al.* (1987) presented evidence based on immunocytochemical investigations showing that the organization of the neuroglia in the midsagittal plane of the CNS of vertebrates exhibits a characteristic morphological pattern distinct from that observed in lateral areas.

There are two major lineages of glial cells: neuroglial and mesoglial. Neuroglial elements are derived from the neuroepithelium, whereas mesoglial elements or Hortega cells are presumably of mesodermal origin (cf. Kuhlenbeck, 1970). Investigations based on silver impregnation techniques and the nuclear characteristics of glial cells as revealed by conventional staining have established three general types of neuroglial cells: (1) astrocytes, (2) oligodendrocytes, and (3) ependymal cells. For descriptions of these cell types in the chicken, see King (1966), Inoue (1970, 1971) and Inoue *et al.* (1971, 1973); in pigeons and parrots, see Stensaas and Stensaas (1968). Navascués *et al.* (1985) studied the proliferation of neuroglial cells from a glial precursor during the early development of the chick optic nerve. They found that proliferation coincides with the penetration of earliest optic fibers into the optic nerve anlage.

Astrocytes, which are characterized by an abundance of cytoplasmic fibrillar material, are thought to be involved in the exchange and transport of water, electrolytes, and metabolites between the blood, the cerebrospinal fluid, and neurons (review, Niessing *et al.*, 1980). Astrocytes often form perineuronal satellites in the gray matter. Their processes are frequently in contact with capillaries. Astrocytes can be subdivided into two major types, fibrous and protoplasmic astrocytes. In chickens, protoplasmic astrocytes have round or oval somata 8–10 μm in diameter with up to 10 or more processes radiating in all directions. A single protoplasmic astrocyte may cover an area 50–70 μm in diameter in a histological section. The surfaces of protoplasmic astrocytes, which are covered with granular projections in silver impregnated specimens, look spiny. Usually one or more processes make contact with a blood vessel. Protoplasmic astrocytes occur predominantly in gray matter areas (Inoue *et al.*, 1971).

The processes of fibrous astrocytes are fewer, thinner, and much straighter than those of protoplasmic astrocytes. These processes are not only longer and less branched but also have smoother surfaces. Fibrous astrocytes can be found predominantly in white matter areas tightly interposed between myelinated fibers. The processes of a single fibrous astrocyte in the chicken can extend through a sphere-like volume of tissue varying between 50 and 100 μm in diameter. Processes occasionally extend as far as 200 μm parallel to nerve fibers (Inoue *et al.*, 1971). There may be numerous transitional forms between protoplasmic and fibrous astrocytes. Fibrillous material is present in both types of astrocytes at light microscopical (Cajal's gold-sublimate technique) and at ultrastructural levels (for ultrastructural observations on glial elements of the chicken, see Lyser, 1972).

The oligodendrocytes of white matter areas are frequently aligned between myelinated nerve fibers. They are thought to produce myelin sheaths in the CNS. In contradistinction to astrocytes, they contain no cytoplasmic fibrillae. In gray matter areas oligodendrocytes can frequently be found in satellite position with respect to neuronal perikarya. In chickens oligodendrocytes in gray matter areas possess bodies that are almost spherical (6–8 μm diameter). Five to ten straight or tortuous primary processes project in all directions and extend through a sphere-like volume of tissue maximally 60 μm in diameter. Secondary or tertiary processes can usually be observed. The processes often possess nodular swellings and correspond in structural details to the type-I group of cells in the classification of Rio del Hortega (cf. Inoue *et al.*, 1971).

In white-matter areas oligodendrocytes have larger, round or oval cell bodies whose diameters range between 7 and 20 μm. The number of primary processes is at most seven. They are, however, remarkably tortuous and radiate in all directions, branching two or three times. In silver impregnated material the ends of the processes are ring-shaped or reticulotubular in structure. Presumably such endings encircle nerve fibers. Oligodendrocytes in white-matter areas conform with types II and III of Rio del Hortega (Inoue *et al.*, 1971). In addition to the above-described types of oligodendrocytes in the gray matter of chickens some oligodendrocytes appear with large somata and one to three thick processes the ends of which apparently surround axons. These oligodendrocytes resemble type IV in the Hortega classification (Inoue *et al.*, 1971). Type III and IV oligodendrocytes have not been observed in pigeons and parrots (Stensaas and Stensaas, 1968). Finally, Inoue *et al.* (1973) observed glial elements in adult chickens, pigeons, parakeets, and quail that resemble immature oligodendrocytes in young mammals (cat, rat) and in the embryonic chick. However, there was no evidence that these atypical cells develop further in the adult birds.

Among the glial elements the mesoglial, or microglial cells, also generally known as Hortega cells, are the least numerous (for review, see Kuhlenbeck, 1970; Niessing *et al.*, 1980). Like oligodendrocytes or astrocytes, they occur either as isolated cells or within clusters of cells in juxtaposition to neuronal perikarya (satellite position as seen in conventional neurohistological sections). Mesoglial cells have been called the morphological correlates of an intercellular substance in the vertebrate CNS. Hortega cells change their shapes, and most probably their functions, under various pathological conditions. Rod cells and compound granular corpuscles or rounded macrophages have been described. For a detailed review, consult Kuhlenbeck (1970). The Hortega cells in the chicken are morphologically similar to those of mammals, reptiles and amphibians (Inoue, 1971). Hortega cells in white-matter areas have processes that lie predominantly parallel to the course of the fibers. In gray-matter areas that are characterized by an isotropic distribution of neuronal and neuroglial processes, Hortega cells and their processes adhere to the isotropy of the neuronal structures. One example of this kind of area is the stratum griseum superficiale of the optic tectum. Hortega cells are smaller in young chickens than in adults (Stensaas and Stensaas, 1968). Their nuclei are elongated (3–4 μm in diameter) and have a chromatin-rich, dark nucleoplasm (Stensaas and Stensaas, 1968). Their processes are small, irregular in shape, and have spiny protrusions.

Ependymal cells retain their positions as the original neuroepithelium of the neural tube (for an extensive review, see Leonhardt, 1980). Ependymocytes are neuroglial elements that constitute a structural and functional boundary between the intraventricular cerebrospinal fluid (CSF) and the central nervous tissue. Fully developed ependymal cells resemble columnar, cuboidal or flat, ciliated epithelial cells. Cilia, however, are frequently lacking entirely or can be seen only in small patches (for details revealed in a SEM study in the brain of the adult pigeon, see Mestres and Rascher, 1987). The ependymal cells often have basal processes that penetrate up to 500 μm into the surrounding nervous tissue (Stensaas and Stensaas, 1968). Other processes are short and may be bifurcated. Adjacent ependymal cells have apical junctions that can be classified as either gap or tight junctions (for findings in the chicken, see Lyser, 1972).

The spinal cord contains a special type of ependymal astrocyte (Stensaas and Stensaas, 1968). These cells each send a long, smooth, process into the central nervous tissue. Multiple branches of these primary processes resemble the secondary processes of protoplasmic astrocytes. In general, these peculiar ependymal cells occur predominantly in areas of the vertebrate CNS in which the wall of the CNS does not exceed a certain thickness. If the wall is no thicker than approximately 500–700 μm, the processes of these cells may contact the pial surface. Horstmann (1954) described these relationships and coined the term tanycytes for these cells. In birds one finds large numbers of ependymal cells with long basal processes not only in the spinal cord but also in the telencephalon and the optic tectum. In addition to numerous morphological similarities between these cells and astrocytes, the two types of cells also resemble each other functionally. Horstmann and Fleischhauer concluded that astrocytes and tanycytes may be intermediate forms of the same cell type (for review, see Leonhardt, 1980).

In addition to the major types of glial cells described above, specialized glial elements can be found in the cerebellum (Bergmann cells) and in the retina (Müller cells). Kuhlenbeck (1970) regarded these as "somewhat related to both ependymal cells and astrocytes". Müller cells resemble tanycytes since they extend from the former pial surface (membrana limitans interna) to the former ventricular surface (membrana limitans externa) of the retina. Glial cells resembling Müller cells in that they contact both the inner surface and the outer surface of the CNS can also be found in the pineal glands of several species of birds. Lemmon (1985) raised monoclonal antibodies specific for glia in the chick CNS. It is interesting to note that one of the antibodies bound both to Müller cells in the retina and to the radial glia in the optic tectum of chickens. It is also interesting to note that endfeet of Müller cells (as well as newly forming outer segments) exhibit specific reactions with antibodies raised against chick transferrin in chicks from embryonic day 6 to an age of 3 weeks post hatching (Zeevalk and Hyndman, 1987).

### C. NEUROPIL

The fine structure of the neuropil can be thoroughly examined only with the electron microscope. Most electron microscope investigations in birds have dealt with (1) glial cells (cf. Lyser, 1972), (2) neurons and specialized, highly ordered areas of neuropil, such as the plexiform layers of the retina, and (3) synaptic structures in the cerebellar cortex and the hypothalamus (for references, see the corresponding sections of this treatise). In general, areas of neuropil in birds exhibit ultrastructural characteristics that are quite similar to those of mammalian neuropil.

### D. MENINGES

In birds the dura mater is tightly attached to the cranium and to the inner wall of the vertebral canal (Kuhlenbeck, 1973; for general aspects, see Bargmann *et al.*, 1980). Böhme (1973) reported that the leptomeninx of chickens always consists of a pia mater and an arachnoid. A neurothel is attached to the dura mater. The arachnoid tissue is present independently of the formation of a subarachnoid space. In chickens this space is consistently present at (1) the basal surface of the brain, in (2) the neighborhood of brain fissures, and (3) around large blood vessels (for details, see Böhme, 1973). The presence of "granulationes meningicae" in this species was documented by the same author (1972a, 1974). Folds of the dura mater in geese, pigeons, chickens and Ostriches appear on either side of the brain stem; these are directed towards the cerebro-cerebellar fissure and correspond to the tentorium cerebelli of mammals (Klintworth, 1968). Some contradictory opinions exist with respect to the presence of an epidural space in the vertebral canal. Dingler (1965) described such a space in the cervical region of the vertebral canal in swans, geese, and cormorants (exact species names not provided). She found that the vertebral canal in the cervical region of all the birds examined was much more narrow than in mammals. According to Kuhlenbeck (1973), birds do not possess a continuous epidural space at the level of the spinal cord. This question should be investigated by using modern techniques that avoid artifacts (e.g., fixation by freeze substitution and subsequent decalcification). Neither Kuhlenbeck nor Dingler reported having seen the neurothel, a sheet of tissue composed of four to five layers of epitheloid cells of mesodermal origin. It is found between the dura mater and the arachnoid. In a recent study using freeze-etching techniques, Van Rybroek and Low (1982) described the remarkable plasticity of intercellular junctions and overall cellular composition of the developing arachnoid membrane in White Leghorn chickens.

## E. BLOOD VESSELS

Blood to the avian brain is supplied primarily by the internal carotid arteries which divide into rostral and caudal cerebral branches (cf. Kuhlenbeck, 1973; review by Pearson, 1972). The spinal cord receives its blood supply through branches of paired segmental arteries from the dorsal aorta. These are often interconnected by transverse and longitudinal anastomoses. The longitudinal anastomoses form: (1) unpaired anterior spinal, and (2) paired dorsal spinal arteries, whereas the transverse anastomoses create a "vascorona" (Fig. 4). At the border between the

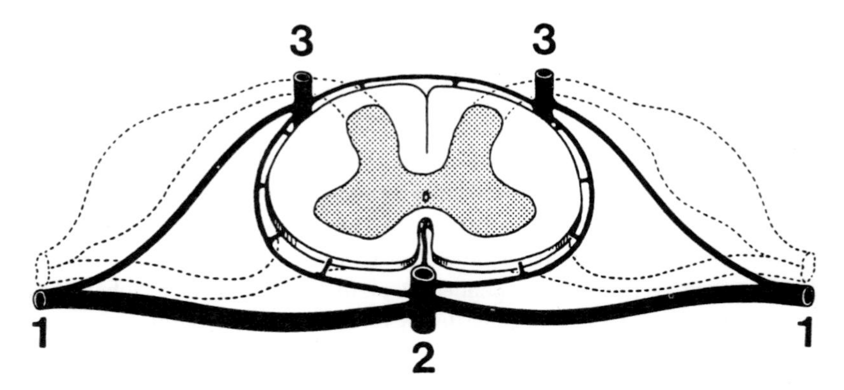


FIG. 4. General morphological pattern of the blood supply to the spinal cord. Note the paired segmental (radicular) arteries (1) providing blood to the unpaired ventral spinal artery (2), and the paired dorsal spinal arteries (3). The spinal arteries form a longitudinally anastomosing system of vessels. This longitudinally arranged network is interconnected by circular arteries that form a transversely anastomosing system of vessels ("vascorona"). This general pattern may vary considerably (predominantly reduction of certain segmental arteries).

medulla oblongata and the spinal cord one finds interconnections between the vascular provinces of the internal carotid arteries and those of the segmental arterial system. Birds lack the vertebral arteries so characteristic of mammals (for further details and review, see Kuhlenbeck, 1973). Studies have been done on the patterns of branching and distribution of the cerebral arteries including regional specializations in 21 of the usually 26–28 recognized orders of birds (Baumel, 1962, 1967; see also Wingstrand, 1951; Vitums *et al.*, 1965; for review and further references, see Pearson, 1972). Usually there is an anastomosis between the right and the left internal carotid artery. A rostral branch of the internal carotid artery gives rise to three major brain vessels at the level of the ventral parts of the cerebral hemispheres: (1) the cerebro-ethmoidal, (2) the medial cerebral, and (3) the posterior cerebral arteries. The extracranial external ophthalmic artery provides an important collateral route of circulation for the intracranial arteries (Vitums *et al.*, 1965). In 75% of the investigated specimens (adult *Columba livia*) the basilar artery, which is the sole supplier of blood to the myelencephalon, originates asymmetrically from the right internal carotid artery (Baumel, 1962).

The dura mater receives its blood supply mainly from extracranial blood vessels (external ophthalmic artery, occipital artery; for details of findings in the White-crowned Sparrow, *Zonotrichia leucophrys*, see Vitums *et al.*, 1965).

Blood from the cerebral veins drains into the general venous system via two pairs of ophthalmic veins in front of and behind the optic nerves, and via paired carotid veins which surround the internal carotid arteries. A variety of venous sinuses channel the blood to the jugular veins (for review, see Pearson, 1972). In many species of birds the ophthalmic veins that drain the blood from the eye form a "rete mirabilis ophthalmicum". This rete is part of a countercurrent system that can cool the blood of the internal carotid arteries, thus contributing to the maintenance of constant brain temperatures. Similar vascular relationships involved in the control of brain temperature can be found in certain species of mammals. For a review of the literature and findings in the pigeon, see Bernstein *et al.*, 1979; Kilgore *et al.* 1979; in Domestic Mallards, see Arad *et al.* (1987); in the Herring Gull, *Larus argentatus*, see Midtgard, 1984. It has been shown that the extrapulmonary gas exchange mediated by the rete mirabilis ophthalmicum in the pigeon raises the oxygen content and lowers the carbon dioxide content in the blood of the internal carotid artery (Bernstein *et al.*, 1984).

Capillaries form a more or less dense network in the CNS of birds. The hairpin-like capillary loops characteristic of marsupials and some reptiles do not occur in birds (Kuhlenbeck, 1973). Density and distribution patterns of brain capillaries vary according to the metabolic activities of the brain regions in which they are found. For a general discussion of brain function and the evolution of cerebral vascularization, see Scharrer (1962); for the histogenesis of telencephalic capillaries in the chicken and information about their permeability to peroxidase, see Delorme *et al.* (1970), and Schueler and Lierse (1970); for data on the innervation of cerebral blood vessels in the chicken, see Tagawa *et al.* (1979).

## IV. Basic Principles in the Topographical Arrangement of Functional Subunits

### A. SPINAL CORD

The spinal cord of a bird extends through the length of the cervical, thoracic, and lumbar vertebral canal. Both a cauda equina and a filum terminale, comparable to these structures in mammals, are lacking. The spinal cord is usually absent only in a small part of the tail (cf. Kuhlenbeck, 1975). One finds remarkable species-dependent variations in the number of spinal segments. The following are illustrative: House Sparrow 40, goose 53, swan 60 (from Dingler, 1965; exact species names not given), chicken 39 (Matsushita, 1968), pigeon 38, Ostrich 41 (Portmann, 1950; exact species names not given). Moreover, there are considerable species-dependent variations in the number of cervical, thoracic, and lumbosacral vertebral or spinal segments (i.e., 17 cervical vertebrae in the goose versus 25 in the swan (Dingler, 1965; exact species names not given). Portmann (1950) compiled the following table:

| | Segments | | | |
|---|---|---|---|---|
| | Cervical | Thoracic | Lumbosacral | Coccygeal |
| Ostrich | 15 | 8 | 19 | 9 |
| Pigeon | 12 | 8 | 12 | 6 |

Discrepancies in the data pertaining to the number of cervical and thoracic segments may be the result of differences in the interpretation of cervical ribs (cf. Kuhlenbeck, 1975).

The spinal medulla of birds exhibits two distinct swellings, the cervical and the lumbosacral enlargements. In birds that are able to fly, the cervical enlargement usually exceeds the lumbosacral swelling in size. In flightless birds this relationship is reversed (Kuhlenbeck, 1975). In apparently all birds the lumbosacral enlargement is characterized by a peculiar separation of the dorsal funiculi over a length of five to six segments (Portmann, 1950). The dorsal columns and some of the more ventral portions of the spinal cord are medially separated from each other (Fig. 5). The space that lies dorsal to the central canal between the separated parts of the cord is called the fossa or sinus rhomboidalis. It is filled with a cellular material of gelatinous consistency (gelatinous body). This conspicuous body has been the subject of numerous studies (for review, see Kuhlenbeck, 1975; Möller, 1978). According to more recent histochemical and ultrastructural observations the large cells of the body contain glycogen, which is stored predominantly in the form of β-particles (Welsch and Wächtler, 1969; Lyser, 1973; Paul, 1973; Agulleiro, 1975; Möller, 1978; for an extensive review, see DeGennaro, 1982). Consequently

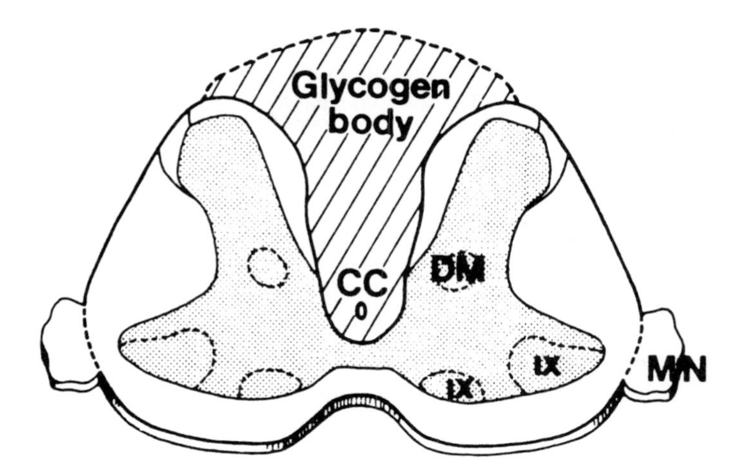


FIG. 5. Diagram of a cross-sectioned spinal cord in a pigeon at the level of the lumbosacral enlargement showing the glycogen body. CC, canalis centralis; MN, marginal nuclei (accessory lobes of Lachi in the older literature). Delineations of the dorsal magnocellular nucleus (DM) and motoneurons (lamina IX) in accordance with Leonard and Cohen (1975a). For details of the subdivision of gray-matter areas in the spinal cord, see Fig. 6.

the term "glycogen body" has replaced the older terms sinus rhomboidalis or gelatinous body. According to Welsch and Wächtler (1969) the glycogen body possesses the highest concentration of glycogen found in any vertebrate tissues. Storage of glycogen begins early in development (7–8 days of incubation in Japanese Quails, *Coturnix coturnix japonica*; DeGennaro and Benzo, 1987).

Cells of the glycogen body lie next to the nervous tissue without being separated from it by a basal lamina. However, in large birds (swan and goose; Dingler, 1965; exact species names not given) a few fibers of connective tissue may be found between the nervous tissue and the glycogen body. These fibers, which probably originate from the ligamenta denticulata, penetrate between the lateral funiculi and the tissue of the glycogen body, and cross the midline dorsal to the central canal. For further observations concerning the inter-relationships between the glycogen body and the meninges, see Kuhlenbeck (1975). In the pigeon the tissue of the glycogen body apparently extends from the pia mater dorsally down through the ependyma of the central canal (Welsch and Wächtler, 1969; cf. Fig. 5).

Möller (1978) removed small pieces of the glycogen body of 18-day chick embryos and maintained them in tissue culture. When the cells of the glycogen body, which are fully developed in 18-day chick embryos, were cultured in small numbers per tissue culture chamber, they lost their glycogen and altered their shapes dramatically. The cells that had lost their glycogen could be identified as astrocytes. This observation is in accord with ultrastructural findings on cells of the glycogen body that had been fixed *in situ* (for review, see Möller, 1978). Furthermore, by employing immunocytochemical techniques, Möller (1989) was able to demonstrate glial fibrillary acidic protein in the cells of the glycogen body of chickens and Zebra Finches. The astocytic nature of the glycogen-storing cells seems to be established by these results.

In chickens the cervical segments 14 and 15 have in cross sections a characteristic onion-shaped area that surrounds the ependyma of the central canal. The cells of this peculiar area contain large amounts of glycogen. Thus, this species appears to possess an additional brachial glycogen body (Sansone and Lebeda, 1976). Moreover, there is a ventrally located region of glycogen-rich ependyma that extends the entire length of the spinal cord and the deuterencephalon up to the level of the mesencephalic oculomotor complex (Sansone, 1977). Uehara and Ueshima (1982) studied the ontogenesis of the stores of glycogen in the chicken. By using tritiated steroid hormones, Reid *et al.* (1981) found autoradiographic evidence for the transient presence of steroid binding sites in the glycogen body of 10–18-day-old chick embryos. Thus the transient expression of steroid hormone receptors does not seem to be limited to developing neurons.

So far no convincing hypothesis has appeared with regard to the function of the glycogen body, which is apparently innervated by unmyelinated nerve fibers (for a review, see DeGennaro, 1982). Some of these fibers contain catecholamines (Paul, 1971, 1973). Paul demonstrated that prolonged cardiazol-induced convulsions bring about the liberation of some of the glycogen in the glycogen body of the chicken. Benzo *et al.* (1975) suggested that the glycogen body is metabolically linked to lipid synthesis and myelin formation in the CNS of bird embryos. For a critical review of the functional significance of such a connection, see DeGennaro (1982). The glycogen body is densely vascularized (Pessacq, 1967) For a discussion of the quantitative relationships between the glycogen-rich ependymal structures and the blood supply of the spinal cord, see Sarnat *et al.* (1975).

Areas of white matter in the periphery of the spinal cord and areas of central gray matter can be distinguished in transverse sections of the spinal cord. Left and right white-matter areas are separated dorsally and ventrally by the dorsal median septum and the ventral or median fissure, respectively. For a general discussion of the phylogenetic development of the latter structures from roof and floor plate material, see Kuhlenbeck, 1973, 1975. The white-matter areas contain long ascending and descending myelinated axons that form distinct bilateral bundles or funiculi. These are called, according to their positions, the dorsal, the lateral, and the ventral funiculi (Fig. 6). Distinct bundles of myelinated fibers cross the midline ventral to the central canal in the so-called commissura alba. In addition one finds a less well-developed, poorly myelinated dorsal commissural system (commissura grisea). The borderline between areas of white and of gray matter is not well defined in all regions. Bundles of myelinated fibers penetrate through gray-matter areas, whereas dendritic processes and sometimes neuronal perikarya may be found in white-matter areas. The latter are predominantly elements of the "intrinsic system" (see Section V.A).

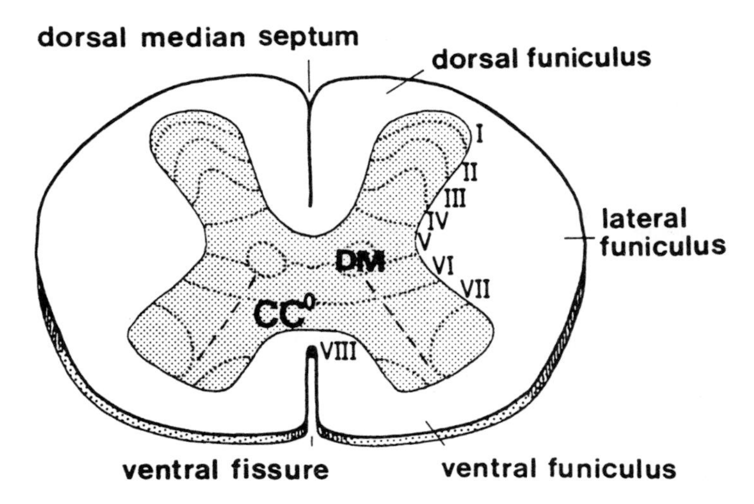


FIG. 6. Diagram of a cross-sectioned spinal cord in the pigeon at the level of the rostral pole of the cervical enlargement. Laminar patterning of gray-matter areas in accordance with Leonard and Cohen (1975a). CC, canalis centralis; DM, dorsal magnocellular column (presumably corresponding to Clarke's column in mammals; cf. Leonard and Cohen, 1975a,b).

Groups of neurons, the derivatives of alar and basal plates, are arranged in bilateral dorsal and ventral areas of gray matter. In cross sections these are termed the dorsal and ventral horns. The aggregations of perikarya and neuropil form continuous longitudinal columns, which can in principle be followed up to the level of the mesencephalon (see Figs 2 and 3).

The gray-matter areas of the spinal cord can be subdivided into more or less homogeneously composed nuclei or into a dorsoventrally oriented lamination pattern on the basis of cytoarchitectural features (for review, see Kuhlenbeck, 1975). These structural features are, e.g., (1) the shape, size and staining intensity of cell nuclei, and (2) perikaryal Nissl substance, (3) the size, number and orientation of primary dendritic processes, and (4) the density of neurons and glial elements. However, subdividing gray-matter areas on the basis of morphological analyses can be influenced by subjective and more or less uncontrollable factors inherent in the experiences and anticipations of the investigator (for critical remarks, see Hartwig and Wahren, 1982). Consequently, descriptions of the numbers and locations of nuclei, or laminae in the gray-matter areas of the spinal cord are sometimes contradictory. Cytoarchitectonic studies have examined the spinal cord in the chicken (Matsushita, 1968; Brinkman and Martin, 1973) and in the pigeon (Leonard and Cohen, 1975a). Leonard and Cohen (1975a) not only took into account cytoarchitectonic features, but they also considered functional experimental data. For a description of the distribution pattern of dorsal root fibers and their terminal fields, see Leonard and Cohen (1975b). The method applied by these authors seems to be a reliable one with which to distinguish functional and morphological components of gray-matter areas.

Leonard and Cohen (1975a) counted nine laminae of cells arranged in a dorsoventral sequence in the gray matter of the pigeon spinal cord. Particularly with respect to the dorsal column the pattern of cytoarchitectonic lamination in birds resembles rather closely that described for various species of mammals (Fig. 6).

In addition to the laminar organization mentioned above, there are distinct groups of neurons that form nuclei in the gray matter of the pigeon spinal cord (Leonard and Cohen, 1975a). The following nuclei are the most prominent:

(1) The dorsal magnocellular nucleus or column, which is located in the neck of the dorsal horn. It is not present in higher cervical segments. This aggregate of neurons can be assumed to correspond to Clarke's column in mammals and is thought to be the origin of the spinocerebellar pathway (Fig. 7).
(2) The column of Terni, which can be recognized in all thoracic segments. It lies immediately dorsal and dorsolateral to the central canal (Fig. 7). Cell clusters in this column are derivatives of the intermedioventral or visceromotor zone (cf. MacDonald and Cohen, 1970).
(3) Lateral marginal neurons. These form a distinct column in an area between or within the lateral funiculus and the pial surface in cervical and thoracic segments (nucleus marginalis minoris Hoffmann–Kölliker in older reports).

In lumbosacral segments laterally located marginal neurons create more ventrally located, distinct protuberances of the lateral funiculi (cf. Fig. 5; accessory lobes of Lachi or nuclei marginales majores Hofmann–Kölliker in the older literature). In an ultrastructural study DeGennaro and Benzo (1976) found that a large portion of the accessory lobes of Lachi in the chicken is made up of glycogen-rich cells that resemble those found in the glycogen body. Matsushita (1968), who confirmed the observations of earlier investigators in this species, found that the lateral marginal

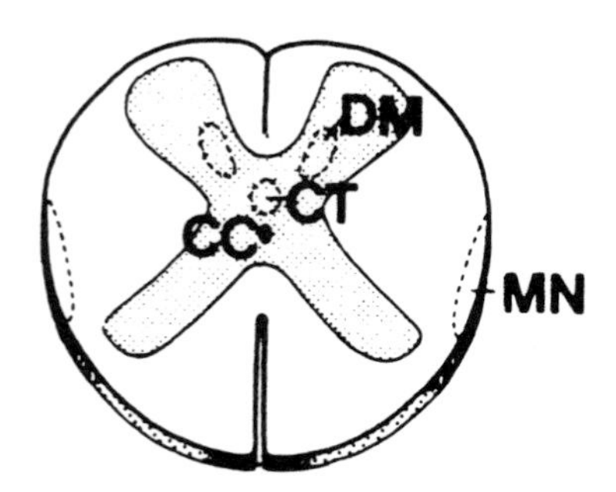


FIG. 7. Diagram of a cross-sectioned spinal cord in a pigeon at a thoracic level. Delineation of the column of Terni (CT) and the marginal nuclei (MN) in accordance with Leonard and Cohen (1975a,b). CC, canalis centralis; DM, dorsal magnocellular column.

neurons send axons into the anterior commissure system. The final destinations of these fibers, however, is not precisely known. Similar marginal nuclei have been implicated in mechanoreceptive functions in snakes and lampreys (for review of comparative anatomy of marginal nuclei, see Schroeder and Murray, 1987). In white Carneaux pigeons the relationships between the lateral longitudinal ligament and the ventral transverse ligament (specializations of the meninges) on the one hand and the accessory lobes on the other, suggest mechanoreceptor possibilities because tension on the ligaments could be transmitted to the lobes. However, experimental evidence supporting this suggestion is still lacking (Schroeder and Murray, 1987).

As in the structural organization in the mammalian spinal cord, layers I–IV in the spinal cord of the pigeon compose the head of the dorsal horn (Leonard and Cohen, 1975a). The dorsal magnocellular column is embedded within layer V. Layer VI covers the intermediate zone. It can be clearly distinguished only at the levels of the cervical and the lumbosacral enlargements. Layers VII–IX make up the ventral horn. The motoneurons are found in layer IX. The motoneurons innervating the hypaxonic nucleus seem to belong to two different groups. One is responsible for the muscles derived from the embryonic ventral mass; the other is responsible for the muscles derived from the embryonic dorsal mass. These results were obtained from leg muscles of chick embryos individually injected (horseradish peroxidase) at different stages of development (Hollyday, 1980). Motoneurons innervating dorsal wing muscles in 2–7-day-old chicks are located in a ventrolateral portion of the anterior column, whereas those projecting to ventral groups of muscles occupy dorsomedial positions within the anterior columns (Straznicky and Tay, 1983).

It is generally believed that efferent axons leave the spinal cord via the ventral roots whereas afferent axons enter the spinal cord via the dorsal roots (for a detailed discussion, see Coggeshall, 1980).

Afferent fibers originating from pseudounipolar cells in the spinal ganglia enter the spinal cord via the dorsal roots. These are arranged in a very specific way within the roots and become distributed in the cord according to very specific patterns. The dorsomedial portions of the dorsal roots contain thick fibers which, after entering the cord, bifurcate in a T-shaped fashion into ascending and descending collaterals (see Section V.A). The ascending collaterals contribute to the ipsilateral ascending dorsal funiculi. Other fibers in the dorsomedial portions of the dorsal roots may contact dorsal horn cells at the level of their entrance into the cord by means of additional axon collaterals. Fibers of the ventrolateral portions of the dorsal roots generally have smaller diameters. They contact dorsal horn cells but do not project into the dorsal funiculi. Ascending and descending axon collaterals of bifurcating afferent dorsal root fibers terminate on neurons located in the outermost part of the dorsal column, the substantia gelatinosa. Several investigators have reported that some efferent fibers may be found in the dorsal roots (for a critical review, see Kuhlenbeck, 1975). These earlier reports of efferent fibers in the dorsal roots have, however, not been substantiated by modern neurobiological techniques.

Efferent fibers leave the cord via the ventral roots. Chu-Wang and Oppenheim (1978b) counted the fibers in the ventral root and the corresponding motoneurons in segments 26–29 of the chick. They found more motoneurons and ventral root fibers in the chick embryo on days 5 and 6 of incubation than at any later period. At later stages of development (in their studies up to the 5th week after hatching) they found roughly 70% fewer motoneurons and ventral root fibers. On the first day after hatching they counted about 1,500 axons and corresponding motoneurons per ventral root. For further details, see also Chu-Wang and Oppenheim (1978a), Oppenheim *et al.* (1978), Oppenheim and Majors-Willard (1978). The figures for neurons reported by these authors match, on the whole, those reported a few years earlier by Hamburger (1975).

The fascinating phenomenon of neuronal cell death during ontogenesis was the subject of an extensive review by Kuhlenbeck in 1973. Since then numerous experimental investigations have been devoted to the study of spontaneous cell death in the various parts of the CNS. To list but a few of these studies: innervation of supernumerary hind limbs, Morris (1978); development of the segmental innervation of the fore limbs, Keynes and Stern (1984), Stirling and Summerbell (1977); development of afferent projections, Heaton (1977), McMillen-Carr and Simpson (1978a,b); response of glial cells to spontaneous death of neurons, Pannese (1978); development of spontaneous motility, Sedláček (1977, 1979), Sedláček and Doskocil (1978), Oppenheim (1975), Sohal (1976). None of these investigations was able to provide conclusive information about the functional significance or biological reason for the natural, spontaneous death of neurons. In chick embryos Catsicas *et al.* (1987b) showed that neuronal death observed during development of the isthmo-optic nucleus, which projects from the mesencephalon to the retina (for details, see p. 69), most probably reduces the imprecision of previously established connections.

Using a pharmacological approach, Zilles *et al.* (1981) treated a total of more than 200 chick embryos from day 10 to day 18 with α- or β-bungarotoxin or with *d*-tubocurarin. These substances block neural transmission based on acetylcholine. They subsequently analyzed the reactions of the motoneurons in the trochlear nucleus by means of quantitative histological methods, ultrastructural examination, and with retrograde axonal transport. The authors observed that spontaneous cell death no longer occurred. This was in agreement with earlier studies on the prevention of ontogenetic death of motoneurons in the spinal cord (for review, see Cowan, 1981). Application of drugs that block acetylcholine receptors causes the target organ to produce continually more receptors. The number of acetylcholine receptors in a peripheral target is apparently a vital point of reference for the neurons that will innervate that organ. The number of neurons that survive to maturity is determined by the number of acetylcholine receptors in the target organ. This assumption has been experimentally tested by Oppenheim (1984). He showed that certain drugs can prevent the ontogenetic loss of motoneurons in the chick embryo. These neurons persist until hatching. Thus, functional contact between neuron and target seems to be particularly important during the development of neuromuscular systems. It was shown in these experiments that there are also factors responsible for programmed neuronal death in spinal ganglia other than faulty peripheral connections (McMillan-Carr and Simpson, 1978a,b). In this context it is most interesting to note that during development of the brachial lateral motor column in the wingless mutant chick embryo, motor neuron number does not approach zero as muscle volume does. This suggests that most, but not all, motoneurons depend on limb bud-derived muscles for survival (Lanser and Fallon, 1987). For further pharmacological experiments in which programmed death of motoneurons was inhibited in chicks, see Douglas and Ribchester (1986).

Ontogenetically programmed death of neurons occurs in all parts of the central and most probably also the peripheral nervous system. Generally speaking, cell death occurs in other vertebrate classes and in invertebrates (for reviews, see Cowan, 1981; Cunningham, 1982) .

## B. MEDULLA OBLONGATA AND "PONS REGION"

The medulla oblongata is the rostral continuation of the spinal cord and the caudal-most part of the brain stem (Figs 8 and 9). The general morphology of the medulla oblongata and the rostrally adjacent "pons region" closely resembles that of the spinal cord. For critical remarks concerning the presence of a "pons" in birds, see p. 28. Consequently, the morphological junction between the medulla oblongata and the spinal cord was placed relatively arbitrarily in a transverse plane through the cranial-most rootlet of the first spinal nerve. At this level there are nuclei within the dorsal funiculi, which, seen in a sequence from medial to lateral, are called the nucleus gracilis, the nucleus cuneatus, and the nucleus cuneatus lateralis. The nucleus cuneatus and nucleus gracilis are important relay stations in the somatosensory system (cf. p. 42). The nucleus gracilis receives afferent fibers from the caudal regions of the body wall and from the legs. The nucleus cuneatus is associated with the cranial regions of the body wall, the neck and the wings. The nucleus cuneatus lateralis is a relay nucleus that sends fibers to the ipsilateral cerebellum, whereas axons from the nuclei gracilis and cuneatus cross the midline in the form of internal arcuate fibers. These fibers then ascend as medial lemnisci to the thalamus of the contralateral side (for a review of these connections, see Kuhlenbeck, 1975).

Tissue originating from the alar and basal plates extends rostrally through the brain stem. The cells that are derivatives of the alar plates are particularly prolifer-

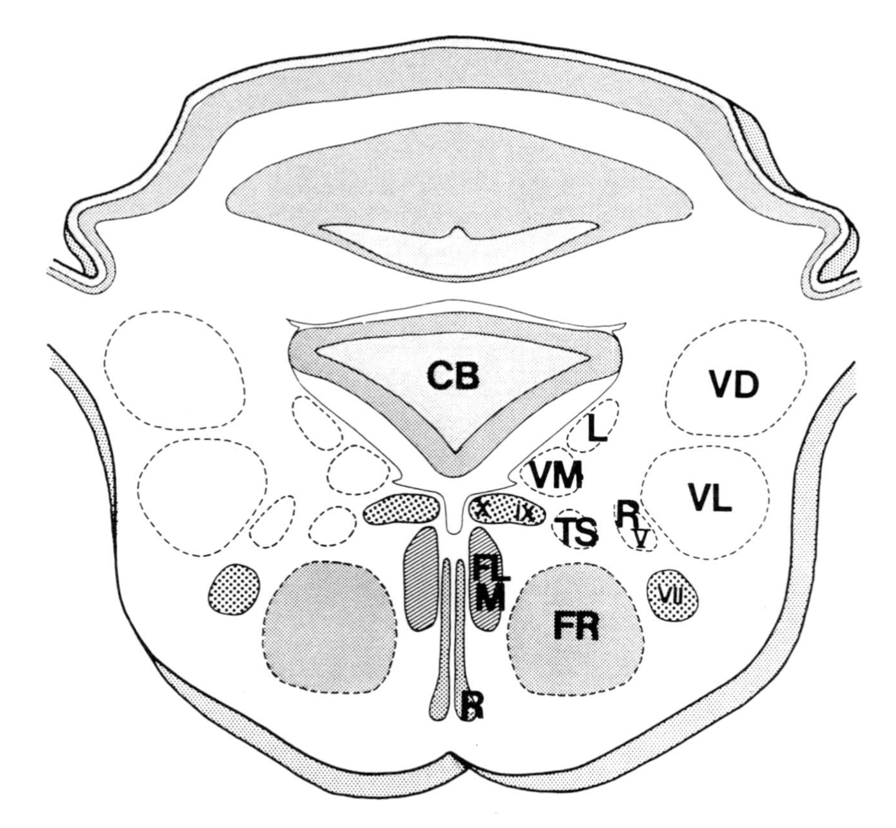


FIG. 8. Diagram of the medulla oblongata of a hatchling of Japanese quail at the level of the dorsal nuclei of the IXth and Xth nerves and the vestibular nuclei drawn from a transverse section. CB, cerebellum; FLM, fasciculus longitudinalis medialis; FR, formatio reticularis (including olivary structures); L, nucleus laminaris; R, raphe nuclei; $R_V$, area occupied by descending part or spinal root of the trigeminal nerve and its accompanying nucleus; TS, area of the solitary tract and its accompanying nucleus; VD, dorsal vestibular nucleus; VM, medial vestibular nucleus; VL, lateral vestibular nucleus; VII, area occupied by branchiomotor nuclei (here: nucleus of facial nerve); IX and X, area occupied by parasympathetic nuclei of cranial nerves (here glossopharyngeal and vagus nerves). Outer granular layer of cerebellum, which is still present in hatchlings, is not indicated.

ative at the level of the medulla oblongata. Proliferation proceeds primarily in lateral directions. The roof plate develops into the epithelial parts of the choroid plexus. The central canal of the spinal cord widens into the fourth ventricle. In the "pons region" a massive expansion of the alar plates overgrows the fourth ventricle to form the cerebellum. As a consequence, the ventral-to-dorsal sequence in the arrangement of the longitudinal columns of neurons in the spinal cord (somatomotor, visceromotor, visceral sensory and somatic sensory) becomes transposed into a medial-to-lateral sequence (Fig. 3).

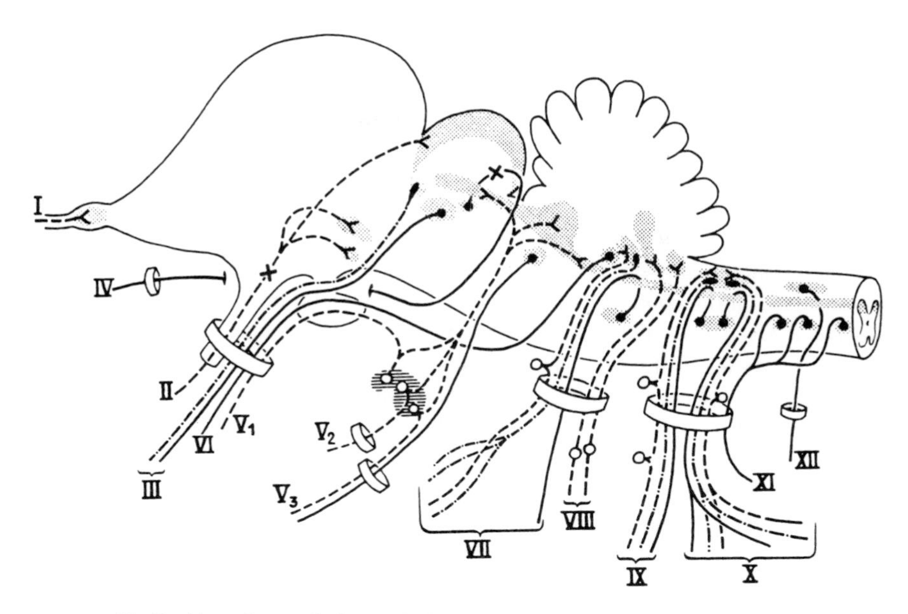


FIG. 9. Idealized lateral view of a brain of a bird. Shaded areas represent either afferent or efferent territories of cranial nerves (I–XII). Where peripheral cranial nerves exit through the bony skull is indicated by rings in the diagram. Where fibers cross the midline is indicated by an x. ——— somatomotor or branchiomotor fibers; –·–·–·–. visceromotor or parasympathetic fibers; –––– afferent fibers. Afferent and efferent perikarya are represented by open (○), respectively filled (●) circles (after Goller, 1972).

Moreover, these longitudinal columns are subdivided into branchiomeric subunits corresponding to the nuclei of cranial nerves. These subunits are comparable to the metameric subunits of the spinal cord that are associated with the spinal nerves (Kuhlenbeck, 1978). The separation of gray- and white-matter areas is not as clear in the brain stem as it is in the spinal cord. Systems of interneurons that integrate sensory-motor information and afferents of the autonomic system develop between gray- and white-matter areas. These systems become the reticular formation, an area that is characterized by a meshwork of fiber tracts between which lie numerous neurons (for details, see below). The differentiation of these systems of functionally specialized groups of interneurons correlates closely with the maturation of the functions of cranial nerves. In the course of functional specialization, nuclei of a given column may migrate from their original positions. Since functional specializations are species-specific, the position of the migrated nuclei may be correlated with different types of behavior or adaptations to different environmental demands on the individual species. For a general discussion of these relationships, see Kuhlenbeck (1975), Starck (1982), and for detailed findings in birds, see Stingelin (1965) and Dubbeldam (1968).

The following columns of neuronal aggregations can generally be distinguished at the level of the medulla oblongata, the "pons region", and, less clearly, in the mesencephalon: (1) somatic motor column, (2) visceral motor column, (3) visceral sensory column, and (4) somatic sensory column. Specialized motor neurons that supply the muscles derived from branchial arches migrate from the original position of the somatic motor columns in ventrolateral directions to form the column of branchial motor nuclei (N V, VII, IX, X, and XI).

*Somatic motor column.* The neurons in this column are similar to those in the motor nuclei of the spinal cord. They supply the skeletal muscles of the head that are not derived from the branchial mesoderm (intrinsic muscles of the tongue, extrinsic muscles of the eye). The hypoglossal nucleus is located in the medulla oblongata, the abducens nucleus in the "pons region", whereas the trochlear and oculomotor nuclei are located in the mesencephalon.

*Visceral motor column.* This column is made up of preganglionic parasympathetic neurons that innervate the cranial, cervical, thoracic and abdominal viscera. In some taxonomic groups of birds it is not possible to subdivide this column into specific nuclei (e.g., subunits of the salivatory nucleus) associated with individual cranial nerves. Ontogenetically this column develops in a position dorsolateral to the somatic motor column. For findings on ontogeny and spontaneous cell death in the dorsal motor nucleus of the vagus nerve in chick embryos, see Wright (1981). At the level of the midbrain the homolog of the mammalian Edinger–Westphal nucleus can be distinguished as a subunit of the oculomotor nucleus. Distinct components of the glossopharyngeal nucleus and the nucleus intercalatus can be differentiated within the medulla oblongata (Cohen *et al.*, 1970).

*Branchial motor nuclei.* Cytoarchitectonically the neurons of this column resemble those of the somatic motor column. However, they supply the muscles derived from branchial arches (V: muscles of mastication; VII: superficial facial muscles; IX and X: muscles of the pharynx and its derivatives; XI: some muscles of the neck). The neurons in the branchial motor nuclei originate in a matrix area of the ventricular wall located between the primordia of the somatic motor column and the primordia of the visceral motor column. In the course of ontogenesis these neurons migrate ventrolaterally. They are ultimately found at the border of, or within, the dorsal tegmentum of the brain stem. Later the axons of the branchial motor neurons reveal the course of migration (Fig. 3).

*Visceral sensory column.* In birds this column consists primarily of the nucleus of the tractus solitarius, which receives predominantly afferent fibers from the IXth and Xth nerves. Special visceral afferents (gustatory) generally seem to be limited to those of the IXth nerve from the oral cavity. The visceral sensory column occupies an intermediate position between the visceral motor and somatic sensory columns. For further details, especially concerning the gustatory system, see p. 47. Katz and Karten (1983) recently demonstrated the presence of organ-specific afferent projections to the nucleus of the solitary tract in the pigeon by means of tracer techniques.

*Somatic sensory column.* Neurons in this column receive afferents predominantly from cranial nerves V and VII. The column occupies a position lateral to the columns described above. In its caudal portions it is continuous with the dorsal column of the spinal cord. These two columns intermingle at cervical levels.

Afferent fibers of the Vth nerve, like other afferents, branch into long ascending and descending fibers. There are nuclei associated with bundles of these fibers (e.g., radix spinalis nervi trigemini). The descending fiber tract of the trigeminal nerve overlaps with Lissauer's tract at the level of the spinal cord. The VIIIth cranial nerve sends its afferent fibers to two functionally different complexes of nuclei belonging to a special somatic sensory system (auditory and vestibular nuclei). They are located in the extreme dorsolateral part of the medulla oblongata and the "pons region". The degree of development of these nuclei varies considerably from one species to the next (e.g., the auditory nuclei reach the highest level of differentiation in owls; Popper and Fay, 1980).

In addition to the growth and development that takes place within the abovementioned columns of afferent and efferent nuclei in the brain stem, there is also a considerable increase in the number of interneurons that develop ventral to the phylogenetically old columns. These neurons are intermingled with myelinated ascending and descending fiber tracts which give the area a reticular appearance. The area as a whole forms the tegmentum of the brain stem, and is known as the formatio reticularis. The reticular formation has been less thoroughly investigated in birds than in mammals. The following brief description of the reticular formation therefore applies to vertebrates in general (Kuhlenbeck, 1978, has provided details). The reticular formation is diffusely organized. Usually, however, its medial parts contain larger neurons than the more lateral parts and these give rise to efferent projections. The morphology of the medially located neurons closely resembles that of motor neurons in the somatic and branchial motor columns. These "motor neurons of the reticular formation" have long-distance projections that usually pass to the spinal cord, the diencephalon or to other neurons of the reticular formation. The lateral parts of the reticular formation contain smaller, predominantly "sensory" neurons which seem to project only over short distances, especially to the neurons of the medial part. Although the functions of the reticular formation in birds have not been thoroughly investigated, it seems likely that it is important for almost all functions controlled by the nervous system (cf. Brodal, 1981; for details of terminology of different reticular nuclei in birds, see Petrovicky, 1966).

A well-developed longitudinal fiber bundle in the brain stem, containing ascending as well as descending projections is the fasciculus longitudinalis medialis. This phylogenetically old fiber tract interconnects afferent and efferent nuclei of the brain stem. Some of its fibers can be followed down to the spinal cord.

Rostral to the medulla oblongata, at the level of the abducens nerve, the brain stem of birds exhibits a ventral flexure which can be considered to be the "pons region". In addition to the motor nucleus of the abducens nerve, the "pons region" contains afferent and efferent nuclei of the facial and trigeminal nerves. In birds the auditory system is relatively highly developed; large auditory nuclei can be found in the "pons region" (cf. p. 55). Birds do not have a true pons comparable to that of mammals since they do not possess a cortico-ponto-cerebellar system. In most mammals such a system is prominent. The pontine nuclei that project to the cerebellum are responsible for the gross anatomical features of the pons in mammals. These nuclei are apparently only very poorly developed in birds (see p. 51). The lack of pedunculi cerebri in birds can also be attributed to their lack of cortico-ponto-cerebellar fiber systems.

## C. CEREBELLUM

The cerebellum of birds is a well-developed laminated or cortical structure that forms a roof over the rostral part of the fourth ventricle. It has become a model organ for elucidating the organizational principles of the CNS as a whole. Numerous investigators of comparative anatomy, e.g., Cajal, Larsell, Jansen, Brodal, and Nieuwenhuys (for reviews, see Kuhlenbeck, 1975; Sarnat and Netzky, 1981) have conducted thorough studies of the cerebellum. Apparently in all vertebrate species investigated to date the cerebellum acts as a chronometric co-ordinator of motor functions. It governs the temporal sequencing of muscle fiber contractions in the synergistic and antagonistic muscles of the trunk and its appendages.

The cerebellum develops from material of the alar plates. This material migrates from the laterally located plates in a dorsomedial direction until the two sides unite in the midline to form the corpus of the cerebellum. The anterior aspect of the cerebellum is connected to the roof of the mesencephalon by the velum medullare anterius. The velum medullare posterius connects the posterior aspect of the cerebellum to the rostral extension of the spinal cord. The anterior velum contains important fiber projections (cf. p. 48), whereas the posterior velum consists mainly of a single layer of ependymal cells.

Two major functional subunits can be distinguished in the cerebellum of birds: (1) the archi- or vestibulocerebellum is the phylogenetically oldest part and is intimately connected with the vestibular system (cf. p. 49), and (2) the paleo- or spinocerebellum receives projections from the spinal cord and the trigeminal system. These projections carry information from cutaneous as well as from muscle and joint receptors. The paleocerebellum is intimately connected with several brain stem nuclei (see below). There are no apparent differences between the fine structural organization of the archi- and the paleocerebellum.

The superficial layer of the cerebellar cortex in all homeothermic vertebrates studied to date is called the stratum moleculare or molecular layer. This layer is composed of different kinds of interneurons and characteristic neuropil formations (synaptic contacts between systems of afferent fibers, the terminals of interneurons and the geometrically arranged dendritic trees of Purkinje cells; for details and a review of findings in birds, see Pearson, 1972; Jansen, 1972). The similarities between the fine structure of the mammalian and the bird cerebellum have recently been beautifully illustrated in findings on the pigeon and the chicken (cf. Paula-Barbosa and Sobrinho-Simoes, 1976; Freedman *et al.*, 1977; Vielvoye and Voogd, 1977).

Purkinje cell perikarya separate the molecular layer from the stratum granulosum or granular layer. The granular layer is composed of countless small neuronal perikarya which are the primary receptive elements in the cerebellar cortex. The axons of the Purkinje cells are heavily myelinated and project within a layer of white matter to cerebellar nuclei (cf. pp. 50–51).

Midsagittal sections of the cerebellum have such a characteristic appearance, created by the lobes with their layered cortex, that this part of the organ has been called the "vermis cerebelli" or the "arbor vitae".

During ontogenesis the medial corpus of the cerebellum and its lateral portions become subdivided by two major fissures running perpendicular to the neuroaxis into anterior, medial, and posterior lobes (for details on the ontogeny of the cerebellar fissures in the chick embryo, see Melian *et al.*, 1986). In all avian species thus far investigated the cerebellum is further subdivided by secondary and tertiary sulci into small folia or lobuli. Several earlier investigators introduced various terminologies to distinguish the individual lobuli of the cerebellum. For example, the primary folia have been listed, in a rostro-caudal sequence, as folia I–X; for a detailed review, see Pearson (1972). In general, the anterior lobe is subdivided into four and the posterior lobe into three lobuli or folia. The subdivisions of the medial lobe seem to be more variable (cf. Starck, 1982). The fourth ventricle extends into the white matter of the most medial part of the corpus of the cerebellum.

Three anatomically defined parts of the cerebellum deserve particular mention because they are so closely connected to specific functions: the right and the left flocculi, together with an intermediate nodulus, are intimately related to the vestibular nuclei and are more or less separated from the remaining parts of the cerebellum by the primary fissure. They are considered to be the constituent parts of the archicerebellum. Unfortunately the terminology for the majority of lobules and fissures is purely descriptive and does not include functional aspects.

Traditional neurohistological techniques do not reveal the zonal organization of the cerebellum. This first became evident in studies on the topography of fiber projections, on histochemical and electrophysiological features of different cerebellar compartments. The zonal organization of the cerebellum, which has been investigated predominantly in mammals, complements the somatotopic representation of the body on the surface of the cerebellar cortex but it differs from that representation in several respects. Feirabend *et al.* (1976) demonstrated that the cerebellum of the chicken can be divided bilaterally into six to nine longitudinal compartments based on differences in the caliber of Purkinje cell axons and on differences in afferent and efferent projections (for recent findings and review, see Feirabend and Voogd, 1986). These longitudinal compartments can also be found in the vestibular projections of the cerebellum in the chicken (cf. Wold, 1981). Two small pontine nuclei project to the cerebellum in birds (Brodal *et al.*, 1950; cf. Clarke, 1977). Birds, however, do not seem to have a system comparable to the cortico-ponto-cerebellar projections in mammals. Neither do they seem to have a neocerebellum. For details of possible "neocerebellar" functions in birds, see Goodman *et al.* (1964). The suggestion that birds in general do not have a cortico-ponto-cerebellar projection is supported by the results of tracing studies undertaken by Clarke (1977). This author showed that pigeons have medial and lateral nuclei in the pons region that probably belong to the reticular formation and not to a system of cortico-ponto-cerebellar projections. Ponto-cerebellar projections were re-examined in the chicken by means of the classical Fink–Heimer methods (silver impregnation of degenerating fiber systems and terminals) but not with the more sensitive tracer techniques (Freedman *et al.*, 1975). As in mammals, the reticular nuclei in the brain stem of birds may project to the cerebellum. In the parakeet such a projection has been shown to arise from catecholamine-containing nuclei in the reticular formation (locus coeruleus, dorsal tegmental nuclei, and disseminated tegmental perikarya) by Tohyama *et al.* (1974). Furthermore, catecholamine-containing fibers and terminals that form geometrically arranged net-like patterns in the granular and the molecular layers have been seen in the chicken cerebellum (Mugnaini and Dahl, 1975).

Afferent and efferent cerebellar projections follow three bilateral cerebellar peduncles (pedunculi cerebelli rostrales, intermedii et caudales). Since birds apparently lack a neocerebellum, the intermediate peduncles are only poorly developed. Details on the afferent (e.g., spinal, trigeminal, visual, vestibular, olivary) and the efferent (e.g., vestibular, rubral, olivary, thalamic) projections are discussed in Section V.

### D. MESENCEPHALON

The caudal border of the mesencephalon is delineated by the roots of the trochlear nerve. Axons of the trochlear nerve cross the midline before leaving the brain stem dorsally. Rostrally the posterior commissure forms a landmark between the periventricular derivatives of the brain stem and the prosencephalon. It should be noted, however, that derivatives of the mesencephalon that develop in lateral directions extend rostrally into originally prosencephalic territories.

Two cranial motor nerves, the fourth and the third, originate in the mesencephalon. The third cranial nerve contains a parasympathetic component that supplies the intrinsic muscles of the eye.

At the level of the mesencephalon one finds more neurons devoted to the control of higher integrative functions than in the rhombencephalon. Some of these neurons migrate dorsally to form the mesencephalic tectum. Others develop ventrally and form important parts of the tegmentum of the mesencephalon, which is continuous with the tegmentum of the "pons region" (cf. Fig. 10). Kitt and Brauth (1986a, b) investigated telencephalic projections arising from the locus coeruleus and subcoeruleus as well as from the tegmental nigral complex in the pigeon. Apparently there is a high degree of overlap between locus coeruleus and nigral-telencephalic terminal fields. In this respect birds differ considerably from mammals. In general, ascending tegmental projections reach the cerebral hemispheres via the medial and lateral forebrain bundles, the ansa lenticularis, the quintofrontal and the occipito-mesencephalic tracts. They exhibit sparse contralateral projections that cross the midline primarily via the dorsal supraoptic decussation.

With its primary visual and non-visual exteroceptive centers, the bird tectum represents a relatively large part of the brain and is bilaterally lobed. Because of

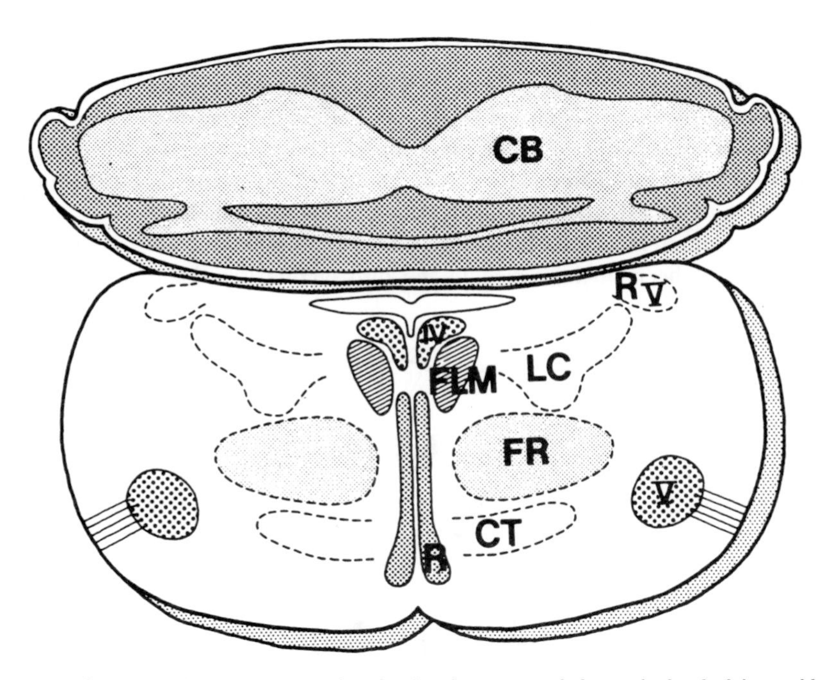


FIG. 10. Representative transverse section showing the mesencephalon at the level of the trochlear nucleus, drawn from a section of a hatchling Japanese quail brain. CB, cerebellum; CT, corpus trapecoideum; FR, reticular formation; LC, locus coeruleus; $R_V$, ascending root of the trigeminal nerve and its accompanying nuclei; FLM, fasciculus longitudinalis medialis; R, raphe nuclei; IV and V, motor nuclei of the trochlear and trigeminal nerves. Outer granular layer of hatchling cerebellum not indicated.

the exceptional size of bird eyes and the advanced development of the telencephalon, the cerebellum and the inner ear, the tectal hemispheres have become displaced from their original position dorsal to the aquaeduct. In birds these hemispheres come to lie ventrally, anteriorly and laterally, ultimately overlapping the lateral wall of the diencephalon (Fig. 11). The tectum of the mesencephalon forms a gross anatomical feature of the brain which has traditionally been called the lobus opticus or tectum opticum. This terminology does not indicate that visual function is only one of the multiple activities of this laminated structure. Consequently the misleading term "tectum opticum" should be avoided or replaced by the descriptive "tectum mesencephali" (cf. Breazile, 1979). The tectum mesencephali is an important relay station that integrates visual, acoustic, labyrinthine, trigeminal and ordinary exteroreceptive as well as other neuronal information. The outer neuronal laminae of the tectum are components of the visual system that receive retinal fibers. The inner neuronal laminae give rise primarily to efferent projections. Afferent and efferent projections are in turn arranged in laminar patterns. For a review and the results of tracing studies in which the retrograde transport of horseradish peroxidase was examined in the pigeon, see Reiner and Karten (1982). The part of the bird tectum that is devoted to auditory functions, the nucleus lateralis mesencephali or torus semicircularis, which corresponds to the inferior colliculus in mammals, is buried by that part of the tectum which corresponds to the mammalian superior colliculus. It is separated from the tectum mesencephali, in *sensu strictu*, by a small mesencephalic ventricle. This ventricle extends primarily as a lateral recess from the aqueduct into the mesencephalon.

Up to 16 different laminae consisting of fiber layers and layers of various types of neurons have been distinguished in the bird tectum (for detailed reviews, see

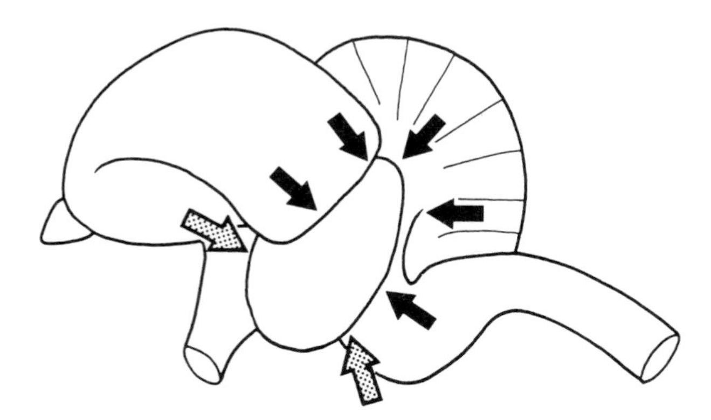

FIG. 11. Developing brain of a bird from a simplified lateral view. Stippled arrows indicate directions of massive growth of eye and inner ear and filled arrows indicate growth of telencephalon and cerebellum which result in a dislocation of the mesencephalic tectum toward a more lateral and basal location (after Starck, 1982).

Polyak, 1957; Kuhlenbeck, 1975). Details of the cytoarchitecture and the fiber projections of the tectum mesencephali are presented in Section V.

Species-specific differences in thickness and numbers of tectal laminae have been reported by several investigators. Unfortunately the number of thoroughly studied species is so limited that it is not possible to draw definite conclusions about species-specific patterns in laminar arrangements. Finally, it should be mentioned that the mesencephalon of birds, similar to that of all vertebrates, especially mammals, contains two important tegmental nuclei, the red nucleus or nucleus ruber and the nucleus tegmenti pedunculo-pontinus. These nuclei are involved in the supraspinal control of complex motor functions. The nucleus pedunculo-pontinus most probably represents the homolog of a part of the substantia nigra in mammals (Bertler *et al.*, 1964; Karten and Dubbeldam, 1973; Kitt and Brauth, 1982; for a recent histochemical demonstration of catecholamine-containing neurons in the corresponding region of the brain stem of the chicken, see Dubé and Parent, 1981).

## E. DIENCEPHALON

The avian diencephalon, which is the caudal part of the prosencephalon, contains the third ventricle. The complexity of its structural organization is by no means inferior to that of the mammalian diencephalon. It is located rostral to the mesencephalon, i.e., between the mesencephalon and the telencephalon. The dorso-caudal border of the diencephalon is formed by the posterior commissure, which contains only fibers that interconnect diencephalic, mesencephalic and rhombencephalic structures. More ventrally the borderline between the diencephalon and the mesencephalon becomes less clearly definable. From a phylogenetic point of view a layer of neuronal aggregates that forms the periventricular gray matter of the mesencephalon originates from the diencephalon. Because neuronal perikarya of both diencephalic and mesencephalic origin can be found at the same level in cross sections of the rostral mesencephalon, this part of the brain has been called the "synencephalon" (cf. Kuhlenbeck, 1975).

Rostrally the anterior commissure forms a landmark between the telencephalon and the diencephalon. In the region lateral to the lamina terminalis the border between telencephalic and diencephalic regions is difficult to delineate.

At the level of the diencephalon, which as part of the prosencephalon develops rostral to the chorda dorsalis, the classical longitudinal arrangement of columns of afferent or efferent nuclei is difficult to distinguish. However, the diencephalon may be subdivided in a dorsoventral sequence into the following regions which become distinct very early in development: (1) epithalamus, (2) thalamus proper or thalamus dorsalis, (3) thalamus ventralis, and (4) hypothalamus. In general the individual nuclei of each region have well-defined afferent and efferent projections. On the other hand, it is often rather difficult to delineate distinct boundaries between the individual nuclei within the regions. The functions of the major subdivisions of the diencephalon are briefly characterized below.

*Epithalamus.* Groups of nuclei belonging to the epithalamus, with the exception of the pineal organ, seem to possess characteristic features which have been well preserved in the course of vertebrate phylogeny. The bilateral habenular ganglia have multiple afferent and efferent connections with limbic structures. The main inputs pass via the stria medullaris thalami, the main outputs leave via the habenulo-peduncular tract. The habenular ganglia also receive important olfactory inputs. In chickens there is a certain structural asymmetry between the medial left and right habenular nuclei which has been shown to be sex-dependent (Gurusinghe and Ehrlich, 1985). The pineal organ develops from the epithalamic roof of the diencephalon into a gland-like structure that, in several bird species, has been shown to be an important component of endogenous time-keeping or circadian clock mechanisms of the central nervous system.

*Thalamus.* The thalamus proper or thalamus dorsalis is an important relay station and integration center for sensory inputs. The way in which it is subdivided into various thalamic nuclei is species dependent. However, both the direct olfactory projections to the telencephalon and the ascending projections arising in the quintofrontal tract from the principal sensory nucleus of the trigeminal nerve bypass the thalamic relay station. For a review, see Necker (1983a), and for further details on fiber systems that apparently bypass the thalamus dorsalis, see p. 45. Fibers that interconnect the thalamus dorsalis with the telencephalon form the medial and lateral forebrain bundles. The lateral bundle includes the thalamic radiation to the telencephalon (for review, see Kuhlenbeck, 1977).

*Thalamus ventralis.* The thalamus ventralis consists of the nucleus reticularis thalami and the zona incerta that is in part engaged in the control of complex motor functions. Furthermore, it has been suggested that the lateral geniculate nuclei also belong to the ventral thalamus (for review, see Kuhlenbeck, 1977). Only visual functions of the lateral geniculate nuclei have been thoroughly investigated in birds.

*Hypothalamus.* The optic chiasma divides the hypothalamus into preoptic and posterior regions. The hypothalamus has reciprocal connections with nearly all parts of the brain that are arranged in an orderly fashion into various laminae of fibers. The hypothalamus controls primarily autonomic functions via neuronal and neurohumoral signals. In many birds these functions are strongly influenced by daylength. In an elegant combination of different neurobiological techniques (cytoarchitectonic studies, immunocytochemistry, anterograde tracing) Cassone and Moore (1987) were able in House Sparrows to delineate the possible homolog of the mammalian suprachiasmatic nucleus. This nucleus receives a major input from the contralateral retina. The high degree of similarity between the structural characteristics and the pattern of messenger molecules observed in the avian and the mammalian suprachiasmatic nuclei is a fascinating example for the astounding similarity of the subcortical CNS in these two classes. In this context it should be noted that the hypothalamus of birds has repeatedly been shown to be directly photosensitive (for detailed review, see Oksche and Farner, 1974; Yokoyama *et al.*, 1978).

The diencephalon contains a considerable number of phylogenetically ancient commissures: (1) the anterior commissure, (2) the optic chiasma, (3) the supraoptic commissural fiber systems, (4) the infundibular, supramammillary and rostral tegmental commissures, and (5) the habenular commissures. For a review, see Kuhlenbeck (1977) and for recent findings on the anterior commissure and the supraoptic decussation, see Saleh and Ehrlich (1984), Ehrlich and Mills (1985), Ehrlich *et al.* (1988).

### F. TELENCEPHALON

The bird telencephalon has been studied from three primary lines of approach (for review, see Bergquist, 1957): (1) it has been examined from an ontogenetic and comparative point of view; (2) the cytoarchitecture of the telencephalon, i.e., the patterns in which its neurons, glial cells and fiber tracts are arranged has been studied and, (3) more recently its functional anatomy has been explored in an attempt to find areas that are functional equivalents of the cortex cerebri of mammals.

The numerous studies that could be grouped under (1) have led to complex terminologies and, at least in part, have produced contradictory results (for a critical review, see Kuhlenbeck, 1977). A number of areas in the avian telencephalon, which can be considered functional equivalents of cortical areas in mammals, have been described. Predominantly electrophysiological techniques and the analysis of stimulus-dependent differences in the metabolism of neurons have been employed to analyze the principles of functional anatomy (for review, see Benowitz, 1980).

Only a few investigations have been undertaken using cytoarchitectonic approaches. The techniques involved pose problems for several different reasons (for a review of technical problems, see Hartwig and Wahren, 1982). More recently Rehkämper *et al.* (1984, 1985) studied the areal pattern of the hyperstriatum ventrale in the domestic pigeon by using an automatic image analyzer, a technique that is well established in cytoarchitectonic studies of the cortex cerebri of mammals. This computerized technique eliminates unwanted subjective delineations of areal patterns. The results of this analysis correlate well with data on established areas in mammals that receive primary ascending projections. Other areal patterns correlate with telencephalic regions assumed to be associative in function. Furthermore, areas with as yet unknown functions have also been delineated (for details and critical review of contradictory terminology of telencephalic regions in the domestic pigeon, see Rehkämper *et al.*, 1984, 1985).

The bird telencephalon is a paired structure that develops as a bilateral outgrowth of the diencephalon. Only a small medial or "impar" portion, which is located anteriorly and ventrally of the interventricular foramen and includes the lamina terminalis, may be described as a direct derivative of the neural tube. Rostrally the paired hemispheres form the olfactory lobes. These are usually small in birds (Figs 12 and 13). The relative weights of the avian and the mammalian telencephalon are quite similar (for details and review, see Jerison, 1973; cf. Fig. 1). Nevertheless, the telencephalon of these two classes cannot be directly compared because in mammals it has evolved cortical structures whereas in birds cortical areas can be recognized only to a very limited extent. In consequence, brain structures closely related to the cortex cerebri of mammals (such as the corpus callosum, the internal capsule, the corticospinal, and cortico-ponto-cerebellar projections) are not present in birds.

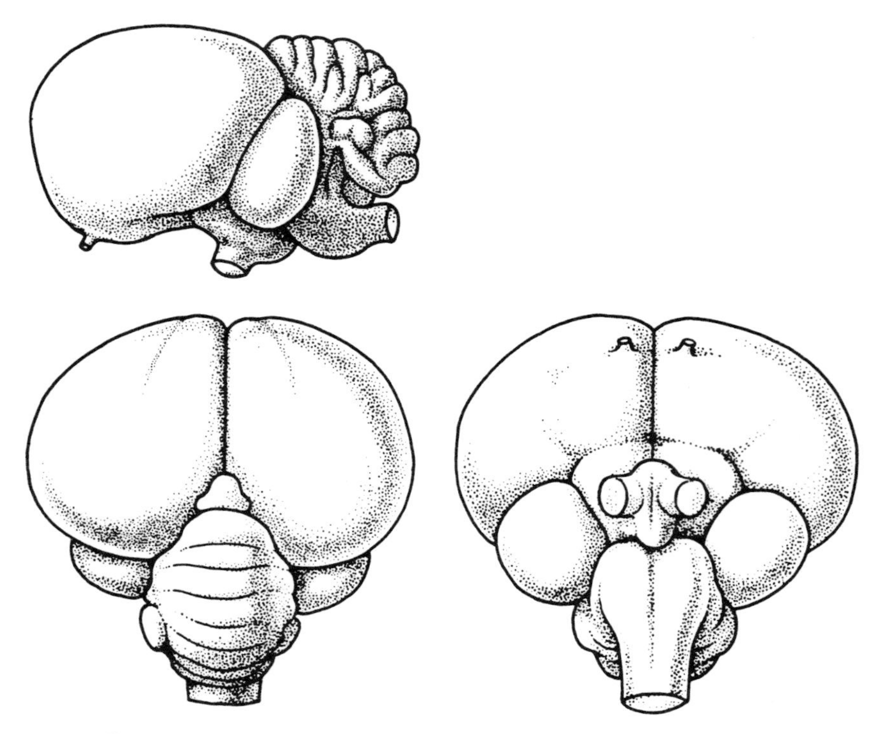

FIG. 12. Brain of Black Kite (*Milvus migrans*) from lateral, dorsal and ventral views. Please note massively developed telencephalic hemispheres with reduced olfactory lobes and strongly developed optic tectum (after Starck, 1982). Compare with Fig. 13.

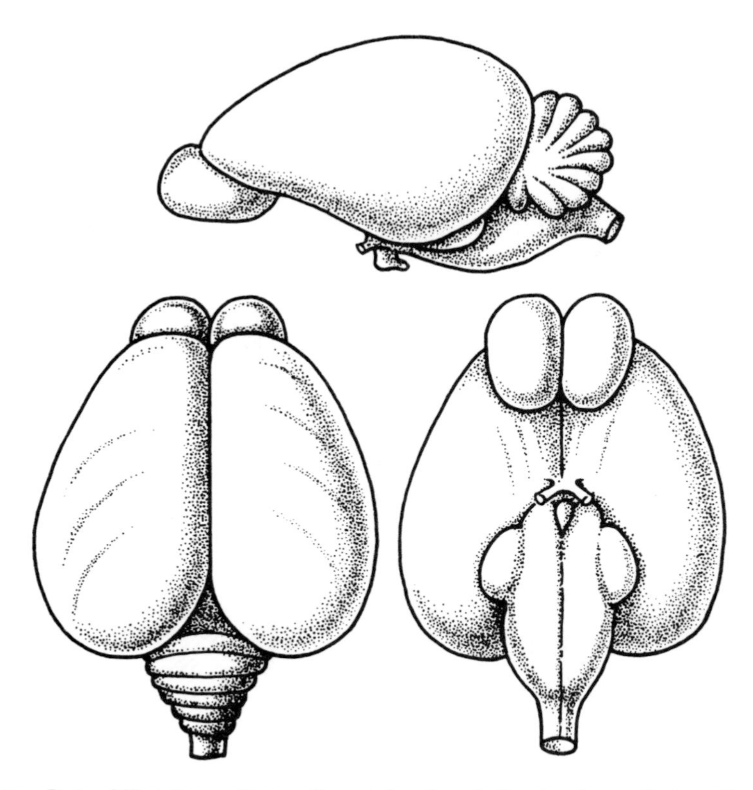

FIG. 13. Brain of Kiwi *Apteryx* (Apterygiformes) from lateral, dorsal and ventral views. Note large olfactory lobes and small mesencephalic tectum in this macrosomatic bird (after Starck, 1982).

In birds the lateral walls of the telencephalic hemispheres consist of hypertrophied areas of gray matter that, in the course of development, considerably reduce the size of the lateral ventricles. Gray-matter areas are interrupted by bundles of myelinated axons thus giving them a striated appearance (cf. Karten, 1969). Earlier investigators (e.g., Kappers, Herrick, Kuhlenbeck, Källén, Bergquist, Holmgren) used different approaches to compare the striatal components of birds with striatal structures in the telencephalon of other classes of vertebrates. One of the basic ideas of these earlier investigators led to the most widely used terminology for telencephalic gray-matter areas outside cortical regions (cf. Fig. 14). In fish the "striatal body" is made up of a primary structure, a paleostriatum, to which structures are added during phylogenetic development. In reptiles an archistriatum and a neostriatum may be distinguished in addition to the paleostriatal complex. Birds evolve a further striatal component, the hyperstriatum, which can be subdivided into hyperstriatum accessorium, hyperstriatum ventrale, and hyperstriatum dorsale (cf. Källén, 1962). Based on cytoarchitectonic details Stingelin (cf.

Stingelin and Senn, 1969) distinguished two different lines in the evolution of the avian telencephalon. One led to the dominance of the hyperstriatum accessorium, which ultimately forms the gross anatomical structure of the welt (Wulst). The other led, in addition to a well-developed Wulst, to the development of a prominent

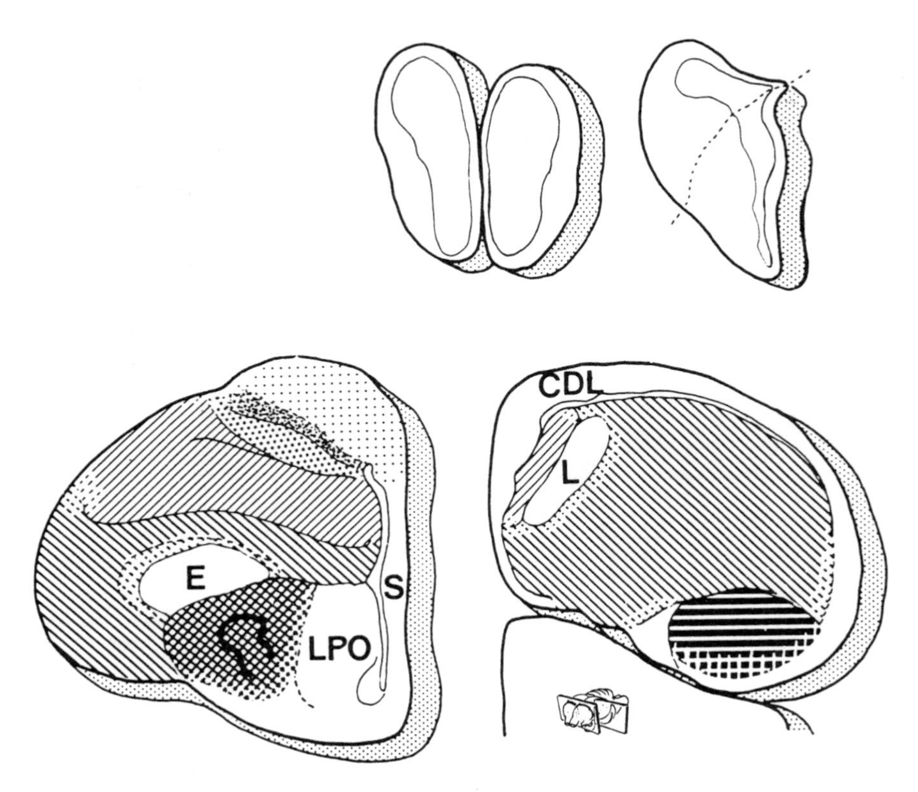


FIG. 14. Schematic transverse section through the telencephalon of a young chick according to Kuhlenbeck (upper left panel shows early developmental stage). The interrupted line in upper right panel separates pallial zones D and telencephalic basal zones B of Kuhlenbeck (1977). Lower panel shows cross sections through the brain of a pigeon (after Benowitz, 1980). Architectural subdivisions of telencephalic structures in this and in the following diagrams of the telencephalon are labeled with graphic patterns:

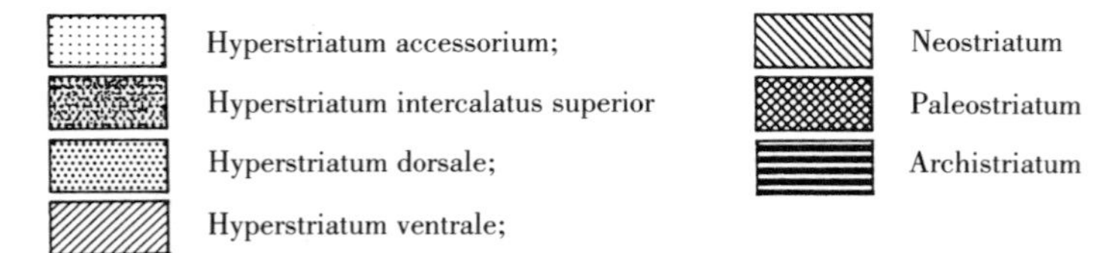


CDL, dorsal cortical layer; L, field L of Rose; LPO, lateral preoptic area; S, septum; E, ectostriatum.

hyperstriatum and neostriatum in the frontal region. The first line has been called the dorsal-front type and the second the basal-front type of evolution (Fig. 15).

Unfortunately the cytoarchitectonically distinct subdivisions of telencephalic gray-matter areas reflect only to a very limited degree functional entities. However, fairly recent results have been obtained by applying modern histochemical techniques and by tracing fiber projections. These techniques have facilitated the identification of functional compartments in the avian telencephalon and comparisons with cortical and subcortical gray-matter areas in mammals (Brauth *et al.*, 1978; for review, see Karten, 1969). Most important for the understanding of the functional organization of the avian telencephalon has been the discovery that certain areas within the neostriatum and the hyperstriatum are reciprocally con-

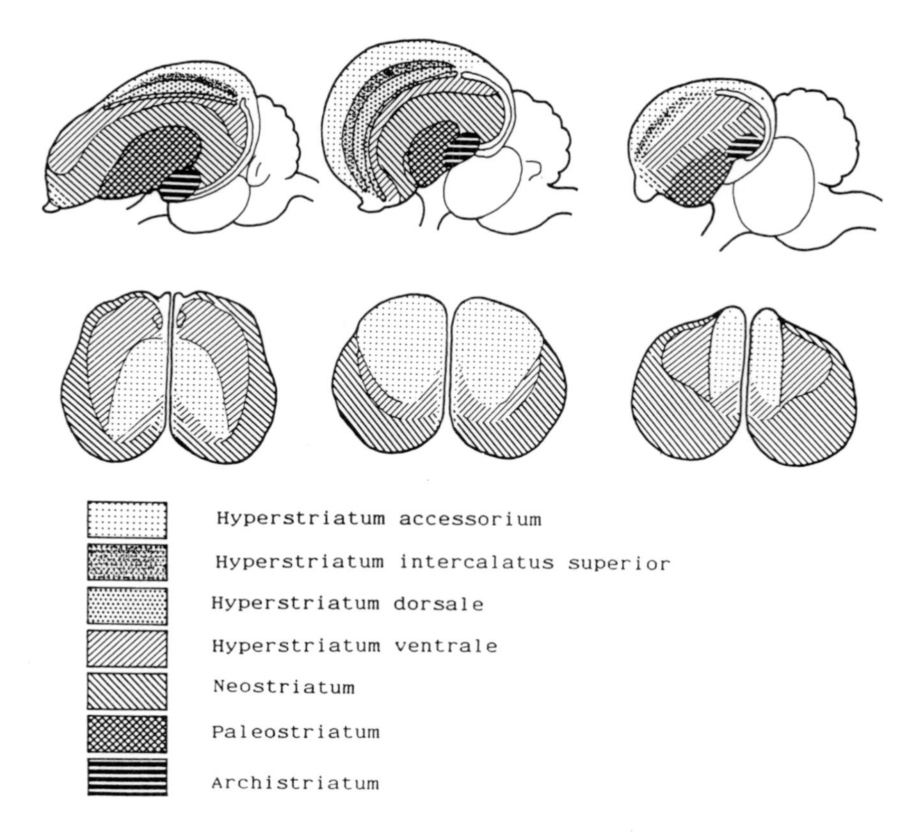


FIG. 15. Different types of telencephalic development after Stingelin (1958) shown in lateral (upper panels) and dorsal (lower panels) views. The architectural subunits presented in Fig. 14 have been projected to the surface of the brain. Left: parrot (basal-front type); middle: owl (dorsal-front type); right: pigeon (primitive type without specialized evolution).

nected with the dorsal thalamus. This pattern of fiber projections closely resembles thalamo-cortical and cortico-thalamic projections in mammals (for further details and review, see Benowitz, 1980; cf. Sections V–VII). After summarizing some of the above-mentioned findings, Karten (1969) concludes:

> Despite the divergent evolutionary lines of birds and mammals, the degree to which the brains of birds and mammals resemble each other in so many details except for the final pattern of alignment of interneuronal grouping is impressive . . . The differences between birds and mammals, however, may reflect different patterns of alignment of telencephalic neurons rather than a fundamental difference in identity between the representative neuronal populations themselves.

Equally important contributions to a better understanding of the structural and functional organization of the telencephalon were made by Ebbesson, who summarized his ideas in the so-called parcellation theory. The basic principles of the parcellation theory, however, had been introduced earlier (for details, see Ebbesson, 1980). Briefly, this theory suggests that in phylogeny and possibly also during ontogeny, the functional specialization of a given system of neurons is concurrent with proliferation and is then followed by the separation or parcellation of aggregates of neurons most of which have originated from a single primordium. These processes include complex modifications in the wiring diagram. In part this theory has been beautifully confirmed by autoradiographic results obtained in $^3$H-thymidine pulse-labeled brains of chick embryos (Tsai *et al.*, 1981a). These authors (Tsai *et al.*, 1981b) also showed that cell populations in anatomically distinct and well-separated visual areas in the telencephalon of an adult bird have a common locus of origin in the proliferating neuroepithelium.

## V. Somatosensory Reflex Systems

Somatosensory reflex systems in birds resemble in various aspects those of mammals. Differences may be due in part to differences in the types of receptors located in the skin (for review, see Necker, 1983a). In contradistinction to mammals, birds apparently possess unique projections that ascend from major sensory input channels. These projections seem to bypass thalamic relay nuclei (see p. 45).

### A. SENSOMOTOR REFLEX SYSTEMS SUPPLYING BODY, WINGS AND LEGS

Afferent fibers that conduct information from receptors in the skin, the muscles, and the joints enter the spinal cord via the dorsal roots. Within the dorsal roots large-diameter fibers can be found in more medial positions, whereas small-diameter fibers occupy more lateral positions. Most notably the large-diameter fibers bifurcate in a T-shaped fashion upon entering the spinal cord. Their axon collaterals form ascending and descending fiber bundles that carry inputs to intersegmental and intrasegmental neuronal circuits (for further details, see Fig. 16). Major projections of large-diameter fibers become components of the posterior funiculi. They ascend ipsilaterally to terminate in the nuclei of the posterior funiculi located at the border between the spinal cord and the medulla oblongata (Fig. 17). These nuclei (nucleus gracilis and nucleus cuneatus) are composed of second-order sensory neurons the axons of which cross the midline and ascend as

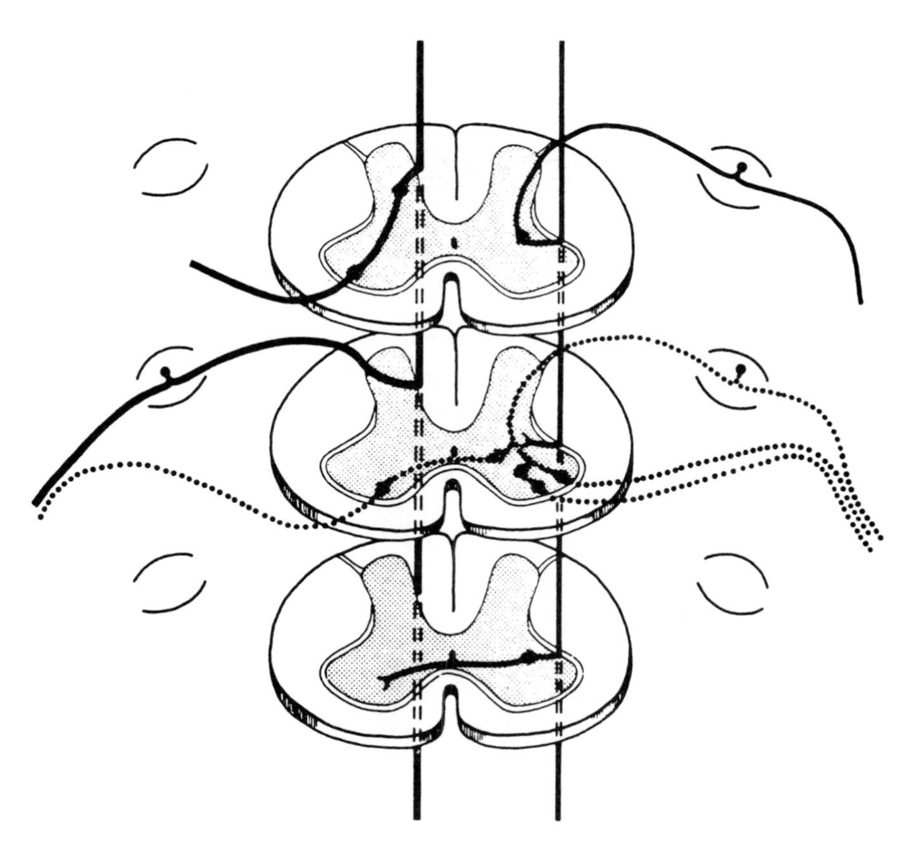

FIG. 16. Simplified diagram showing three segments of the spinal cord and selected pseudounipolar spinal ganglion cells. On the left the T-shaped bifurcation of a fiber ascending and descending in the dorsal funiculus. Note that axon collaterals may contact interneurons. On the right an afferent fiber drawn with a continuous line contacts an interneuron, thus establishing an intrinsic reflex circuit in the spinal cord. Neurons drawn with dotted lines represent the simplest reflex arcs controlling activities of ipsilateral and contralateral muscles.

bulbothalamic and bulbohypothalamic tracts (Karten and Hodos, 1967). The latter tracts correspond to the medial lemniscal system in mammals. In birds the majority of the second-order sensory axons seem to terminate in the reticular formation of the brain stem and the red nucleus of the mesencephalon; in mammals this is not the case. Afferent information may be relayed to the cerebellum from the tegmental nuclei of the brain stem. Moreover, fiber tracts descend as bulbospinal systems from the reticular formation and the bed nuclei to the spinal cord (for the locations of these fiber tracts in a cross section of the spinal cord, see Fig. 18). Second-order sensory neurons in birds also project to the prosencephalon, including the hypothalamus (Fig. 17).

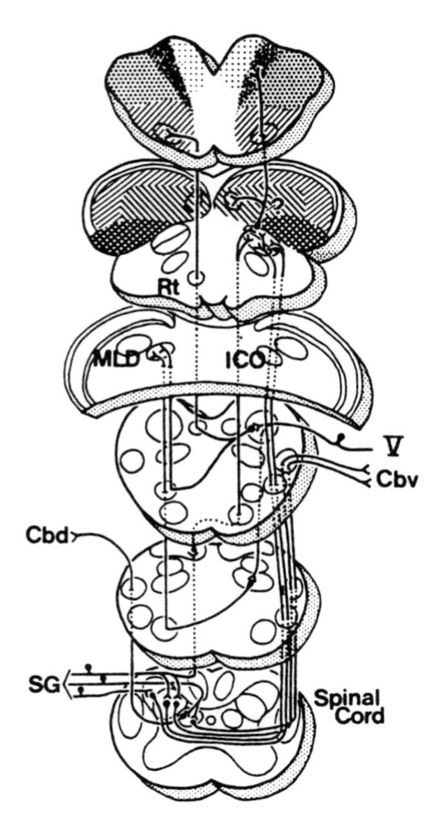


FIG. 17. Series of transverse sections arranged in a rostrocaudal sequence to show major ascending somatosensory pathways in a pigeon (after Necker, 1983a; for somatosensory projections of trigeminal system, see Fig. 19). See Fig. 14 for shading. Cbd and Cbv, dorsal and ventral spinocerebellar tract; ICO, intercollicular nucleus; MLD, nucleus mesencephalicus lateralis pars dorsalis; Rt, nucleus rotundus; SG, spinal ganglion; V, trigeminal nerve; ? indicates equivocal anatomical evidence (second section from top).

Small-diameter fibers that occupy more ventrolateral positions within the dorsal roots of the spinal cord terminate on neurons within laminae I–VI (Fig. 6). These are second-order neurons and have been shown to project ipsilaterally to the cerebellum. Their ascending fibers can be found within the dorsal part of the lateral funiculus (for review, see Pearson, 1972). The origins of other ascending projections in the lateral funiculi still remain to be determined. Following hemitransection of the spinal cord in pigeons, Karten (1963) observed ascending pathways in a spinothalamic tract or spinal lemniscus. These fibers projected to the nucleus dorsolateralis posterior and to the nucleus superficialis parvocellularis in the thalamus (Fig. 17). They also projected to areas of neuropil in the nucleus rotundus, which seems to be an important relay station for integrating somatosensory and visual information. For telencephalic projections of the nucleus rotundus,

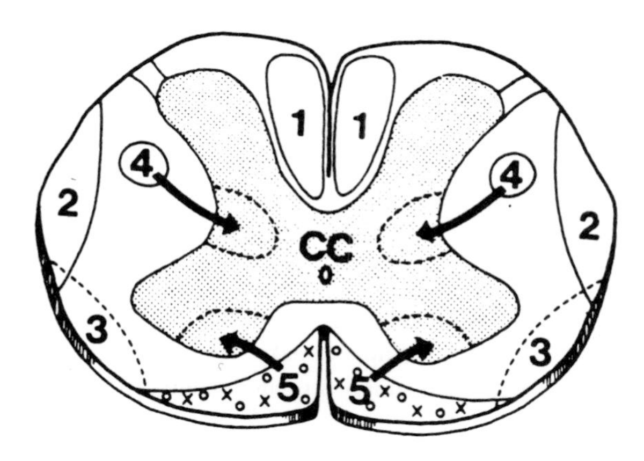


FIG. 18. Simplified diagram showing a transverse section of the spinal cord in the pigeon at the level of the cervical enlargement. Major ascending and descending fiber tracts are indicated (1–5). The boundaries of the tracts are in reality much more blurred so that the corresponding fields in the lateral and ventral funiculi might show substantial overlapping. 1, funiculus posterior, containing predominantly a system of ascending axons of primary spinal ganglion cells, the spino-bulbo-thalamic system crossing midline at bulbar levels; in addition, descending fibers are found in the posterior funiculi (Cabot *et al.*, 1982); 2, lateral funiculus containing ascending fiber systems of spinocerebellar system; axons of second-order neurons ascending primarily ipsilaterally and crossing at higher levels. In a similar location descending hypothalamo-spinal fibers arising primarily from nucleus periventricularis magnocellularis have been observed (Cabot *et al.*, 1982; Berk and Finkelstein, 1983); 3, spino-mesencephalic (tectal?) system of Friedländer (cf. Kuhlenbeck, 1975) comprised of axons of second-order neurons crossing midline approximately at the level of the second perikaryon; 4, rubro-spinal system crossed at bulbar levels; in a similar location descending dorsolateral ponto-spinal fiber systems have been reported (Cabot *et al.*, 1982); 5, vestibulo- and reticulo-(bulbo)-spinal fibers with variable crossing behavior (cf. Wold, 1978a for vestibulo-spinal fibers). Medialmost parts of the field labeled 5 contain descending medial pontospinal fibers (Cabot *et al.*, 1982). Adjacent to the dorsal parts of the ventral fissure, fibers descending from periaqueductal gray matter of the mesencephalon have been located (Cabot *et al.*, 1982); CC, central canal. Arrows indicate major projection sites of fibers leaving tracts 4 and 5.

see Karten and Hodos (1970). The nucleus dorsolateralis caudalis and the nucleus superficialis parvocellularis of the thalamus have been shown to receive tactile inputs from both spinothalamic and bulbothalamic tracts (electrophysiological investigations in pigeons by Delius and Bennetto, 1972). Somatosensory information is further transmitted from these thalamic nuclei to the hyperstriatum intercalatus and to the rostral and medial portions of the caudal neostriatum. The exact anatomical delineation and somatotopic arrangement of these projections, however, remain to be determined. In this context detailed tracing studies in the spinal cord and brain stem of the pigeon (Wild, 1985) deserve special attention. Wild (1987) critically reviewed the avian somatosensory system and described body representation in the forebrain of the pigeon that had been revealed in elegant studies combining electrophysiological investigations with the application of tracer substances. Unfortunately these most important studies were published after the figures accompanying the present chapter had already been completed.

## B. SENSOMOTOR REFLEX SYSTEMS OF THE HEAD REGION

General somatosensory information from the head is conducted primarily by trigeminal fibers. Afferent fiber bundles of the trigeminal nerve consist of three main branches:

(1) the ramus ophthalmicus, which arises from the ipsilateral orbita, the nasal area and the rostral part of the upper beak,
(2) the ramus maxillaris, which supplies predominantly distally-located regions of the upper beak, and
(3) the ramus mandibularis, which is composed mainly of afferents from the lower regions of the beak and the oral cavity.

Trigeminal fibers bifurcate in a T-shaped fashion upon entering the brain stem. The axons form ascending (mesencephalic) and descending (spinal) trigeminal tracts. These tracts are accompanied by relay neurons that form the terminal nuclei. Fibers of the ascending trigeminal tract terminate in part on the principal sensory neurons of the trigeminal nerve. Trigeminal nuclei have been shown to be organized according to somatotopic principles (Dubbeldam and Karten, 1978). They vary greatly in size among the taxonomic groups. The size of the principal sensory nucleus is apparently correlated with the size of the beak (Stingelin, 1961). Dubbeldam and Veenman (1978), Dubbeldam (1980), Dubbeldam *et al.* (1981), as well as Berkhoudt *et al.* (1981) and Kishida *et al.* (1985), examined in detail trigeminal projections in pigeons and domestic mallards.

The quintofrontal tract arises from the principal sensory nucleus of the trigeminal nerve. This tract projects, after partial decussation, to the nucleus basalis of the telencephalon. According to several investigators this (telencephalic) nucleus may represent a rostrally displaced thalamic relay station and not a nucleus of telencephalic origin (for details and review of older reports, see Cohen and Karten, 1974). Other investigators have suggested that the nucleus basalis is originally a telencephalic structure and that trigeminal afferents consequently bypass thalamic nuclei. In this context it is important to note that not only trigeminal afferent projections but also ascending fiber systems arising from all sensory input channels of the head and neck region have been shown to project to the nucleus basalis (for recent findings and review, see Schall *et al.*, 1986; for discussion, see p. 91). Fibers that arise from the neurons of the nucleus basalis project to the archistriatum via the fronto-archistriatal tract (Fig. 19). In addition to telencephalic projections, trigeminal afferent fiber systems possess terminal fields in the mesencephalon which overlap with terminal fields of perikarya located in the spinal cord (for review, see Necker, 1983a). Overlapping of trigeminal and spinal afferent systems has been observed predominantly in the mesencephalic region that corresponds to the torus semicircularis of poikilothermic vertebrates or to the inferior colliculus of mammals. Afferent and efferent projections from the nucleus of the descending trigeminal tract have been extensively studied in the duck by Arends *et al.* (1984) and Arends and Dubbeldam (1984).

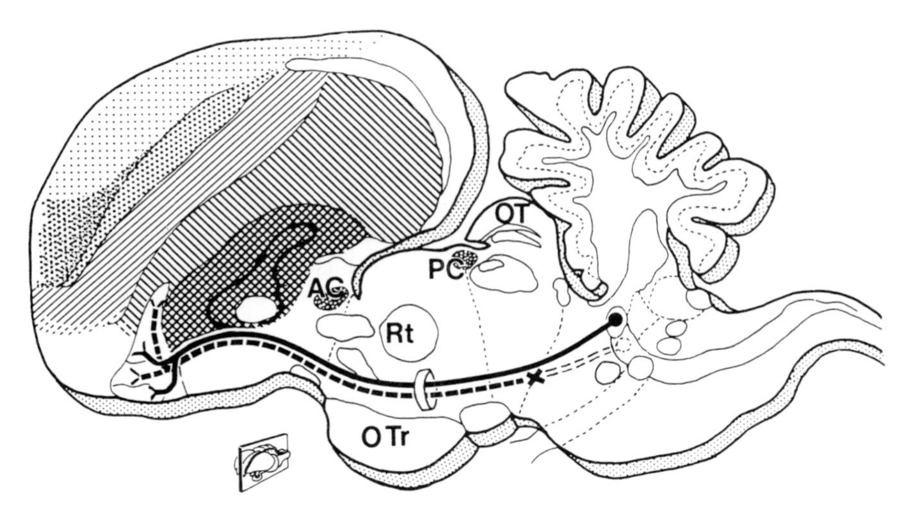


FIG. 19. Parasagittal section of a pigeon brain showing secondary trigeminal neurons ascending to the nucleus basalis located in the telecephalon. Continuous line represents ipsilateral, interrupted line contralateral component of the quintofrontal tract which is diagrammatically shown ventral to the nucleus rotundus (Rt). AC, anterior commissure; OT, optic tectum; OTr, optic tract; PC, posterior commissure; (after Zeigler and Karten, 1973). See Fig. 14 for shadings. X indicates fibers of the contralateral quintofrontal tract crossing the midline. Crossing fibers of the ipsilateral quintofrontal tract are not shown. Note that the nucleus basalis receives further afferent fibers from the brain stem that bypass the thalamus.

It is interesting to note the direct projections from certain multipolar cells of the mesencephalic nucleus of the trigeminal nerve to the cerebellar cortex via the rostral and caudal cerebellar peduncles (Bortolami *et al.*, 1982). Experimental evidence shows that neurons of the mesencephalic nucleus of the trigeminal nerve originate from neural crest cells but not from neural tube material (Rogers and Cowan, 1973; Narayanan and Narayanan, 1978). For a review of functions related to the mesencephalic nucleus of the trigeminal nerve and for its development under normal and experimental conditions in Domestic Mallards, see Narayanan and Narayanan (1987).

Motoneurons of the trigeminal nerve belong to the column of perikarya that innervates muscles originating from branchial arches. Their development and relationships to the trigeminal ganglion have been studied extensively in the chicken (Heaton and Moody, 1980; Moody and Heaton, 1983). Wild and Zeigler (1980) investigated the somatotopic organization of trigeminal and facial motor nuclei that innervate jaw muscles in the pigeon.

Other nuclei that innervate branchial muscles (e.g., the motor nuclei of the glossopharyngeal, vagus, and accessory nerves) have been less thoroughly studied. As in mammals, the motor nuclei of the glossopharyngeal and vagus nerves of birds may form a common column of perikarya called the nucleus ambiguus. Precise identification of projections from individual motor nuclei of the brain stem that supply branchial muscles have been reported in cockatoos (Wold, 1981), chicken (Youngren and Phillips, 1983) and in budgerigars (Manogue and Nottebohm, 1982). These investigations include studies of the somatomotor projections of the hypoglossal nerves. For details of afferent projections of the motoneurons of the hypoglossal nerve, see Wild and Arends (1987). Generally, the efferent portions of the glossopharyngeal nerves arise from four rhombencephalic nuclei and a ventral portion of the dorsal motor nucleus of the vagus nerve. A ventrolateral motor nucleus of the vagus contributes to the nucleus ambiguus. It is interesting to note that motor neurons devoted to the control of singing are able to concentrate androgen hormones (Arnold *et al.*, 1976; Nottebohm and Arnold, 1976; see also p. 59).

Afferent central projections of glossopharyngeal and vagus sensory ganglia have been shown to be directed predominantly toward central nuclei in the solitary complex. In Domestic Mallards the terminal fields of glossopharyngeal projections (probably somatosensory afferent fibers) overlap with a group of cells that belong to the principal sensory nucleus of the trigeminal nerve. The fact that projections from glossopharyngeal and trigeminal afferent fiber systems converge may reflect the functional coherence of information from mechanoreceptors in the bill and the tongue (for details, see Cohen *et al.*, 1970; Dubbeldam *et al.*, 1979; Wild, 1981). Convergent facial and glossopharyngeal afferent fibers (probably viscerosensory) have, however, been seen in connection with gustatory functions, for example afferents from taste buds located in the bill and the soft palate. For electrophysiological evidence that the chorda tympani carries gustatory information in the chicken, see Gentle (1975).

Whereas prosencephalic representations of general somatosensory functions have been thoroughly investigated, this is not so for higher gustatory centers. At least in Domestic Mallards, however, central processing of taste information seems to take place at hypothalamic levels. These connections are similar to those found in poikilothermic vertebrates (for details, see Gentle, 1975; cf. also Wenzel, 1983).

Parasympathetic preganglionic neurons of the dorsal motor nucleus of the vagus in pigeons have been grouped into several subnuclei which are probably related to the heterogeneity of the target organs controlled by these neurons (Katz and Karten, 1983). Other preganglionic parasympathetic neurons of cranial nerves located in the rhombencephalon have been less extensively studied.

Two other accumulations of neurons associated with the motor nuclei of rhombencephalic cranial nerves seem to be less clearly defined in structure and function. These are: (1) the nucleus intercalatus, which lies between the dorsal motor nuclei of the vagus and the nucleus originis of the hypoglossal nerve, and (2) the nucleus intermedius, which is located ventral to the dorsal motor nucleus of the vagus nerve and dorsolateral to the nucleus of the hypoglossal nerve (cf. Breazile, 1979).

The somatomotor nucleus of the hypoglossal nerve, which is essentially a spinal nerve, together with the nuclei origines of the abducens, the trochlear, and the oculomotor nerves, form the somatomotor column of neurons in the brain stem. In addition to possessing a nucleus motorius nervi abducentis principalis, which is located at the caudal level of the pons region near the floor of the fourth ventricle, numerous species of birds have a nucleus motorius nervi abducentis accessorius located lateral to the principal motor nucleus of the abducens nerve. This accessory nucleus is believed to innervate the nictitating membrane (for details, see Kuhlenbeck, 1975). However, Bravo and Inzunza (1985) presented evidence that it is the oculomotor nuclei and not the abducens nucleus that innervate muscles that advance the nictitating membrane.

Whereas the perikarya of the abducens nerve form an isolated nucleus, the somatomotor perikarya of the oculomotor and the trochlear nerves form a common column of neurons that is located dorsal and medial to the fasciculus longitudinalis medialis in the mesencephalon. The nucleus nervi oculomotorius can be divided into a pars accessoria, a pars dorsalis, a pars ventralis, and a pars dorsolateralis. The pars accessoria consists of preganglionic parasympathetic neurons. Some efferents of the somatomotor neurons decussate prior to leaving the mesencephalon. This decussation is a consequence of the ontogenetic migration of neuroblasts, which in chicks takes place between the 4th and 9th days of incubation (for a review and findings, see Puelles, 1978; for observations in Domestic Mallards, see Sohal, 1977). By using tracer techniques, Heaton and Wayne (1983) demonstrated a highly discrete and non-overlapping organization of the oculomotor subnuclei in the chicken. Perikarya of the trochlear nerve can be found immediately caudal to the oculomotor nucleus. Efferent fibers of these perikarya decussate within the velum medullare anterius of the cerebellum and leave the brain stem from its dorsal surface. The perikarya were located by means of tracer techniques in Domestic Mallards (Sohal and Holt, 1978).

Finally, it should be noted that in the pigeon proprioceptive neurons have been located by retrograde transport of horseradish peroxidase from extraocular muscles. These neurons were found in the ipsilateral descending nucleus of the trigeminal nerve. For details and a discussion of earlier ideas as well as a comparison of contradictory observations in mammals, see Eden *et al.*, 1982.

### C. COMPLEX OF VESTIBULAR NUCLEI

In birds as well as in other vertebrates, vestibular nuclei represent dominating elements in the somatosensory column of neuronal aggregates at the level of the rhombencephalon, including the "pons region". In birds the vestibular nuclei seem to be more differentiated than in mammals (for details on the neurogenesis of these nuclear complexes in the chicken, see Peusner and Morest, 1977a,b,c). Up to 11 different vestibular nuclei have been distinguished on the basis of cytoarchitectonic features (for details and comparisons, see Breazile, 1979). Within the extramedullary vestibular nerve leading to the brain are spindle-shaped neurons that form two distinct interstitial vestibular nuclei located either medially or laterally. Vestibular fibers bifurcate within the brain tissue in the classical T-shaped manner to form descending and ascending tracts. The ascending tract terminates within the main mass of perikarya that forms the complex of vestibular nuclei; the descending tract is accompanied by the nucleus vestibularis descendens. Sensitive tracing techniques have so far been applied only to a limited extent in studies on the avian vestibular system (for new data concerning a primary vestibular projection to the cerebellar cortex in domestic pigeons, see Schwarz and Schwarz, 1983). The somewhat controversial results obtained in studies on terminal degeneration following lesions of the vestibular nerves are comparable to those obtained in similar experiments on laboratory rodents. Boord and Karten (1974) distinguished four major territories of vestibular nuclei in pigeons: superior, lateral, medial, and descending. Only limited areas of the four territories contained degenerating profiles after the vestibular nerves had been experimentally lesioned. Thus, most of the perikarya in the vestibular nuclei are third-order vestibular neurons. Second-order vestibular neurons are found in: (1) the quadrangular region of the superior territory, (2) the dorsal and lateral areas of the descending territory, including the magnocellular division, (3) the dorsomedial area of the medial territory, and (4) the ventrolateral area of the lateral territory. Wold (1976) distinguished six major vestibular nuclei in chickens as well as two additional groups of nuclei close to the major vestibular nuclei (for further details obtained with tracer techniques, see Wold, 1978a; Eden and Correia, 1982).

Major descending fiber systems that arise from the complex of vestibular nuclei form the vestibulo-spinal tract (for details, see Fig. 18). These phylogenetically old fiber systems arise predominantly from lateral and descending vestibular nuclei and represent an important component of the descending supraspinal motor system (for a general overview, see Section V(E)). According to Wold (1978a), who mapped vestibulo-spinal projections in the chicken after having injected horseradish-peroxidase into circumscribed areas of the spinal cord, some of the vestibulo-spinal projections are somatotopically organized.

Another important fiber tract that contains vestibulo-efferent axons is the fasciculus longitudinalis medialis. This tract is also phylogenetically old. It contains ascending and descending vestibular projections and extends to the level of the spinal cord. The fibers are responsible for co-ordinating movements of the eyes with movements of the head and neck. In addition, they establish communication channels between the nuclei of the cranial nerves III–XII. In birds this important tract contains predominantly large-diameter fibers. Recently Correia *et al.* (1983) studied ascending and descending vestibular pathways in pigeons by using anterograde transneuronal transport of tracer substances (tritiated proline and fucose injected into the endolymphatic space of the membraneous labyrinth). They found the heaviest labeling in the contralateral vestibulo-ocular pathways. These pathways are composed of: (1) the fasciculus longitudinalis medialis, (2) the motor nuclei of the abducens and trochlear nerves, and (3) the dorsolateral and dorsomedial components of the nuclear areas that form the nucleus of the oculomotor nerve. Ipsilateral vestibulo-ocular pathways were less heavily labeled. Similar results were obtained by Wold (1978b) in chickens when the retrograde axonal transport of tracer injected into oculomotor and trochlear nuclei was investigated (for vestibular projections apparently bypassing thalamic nuclei and terminating in the nucleus basalis of the telencephalon, see p. 45).

Groups of neurons have been found in the thalamus of pigeons (nucleus posteroventralis and nucleus principalis praecommissuralis) that receive information from vestibular nuclei (Vollrath and Delius, 1976). Finally, vestibular nuclei have complex connections with the cerebellum (see following sections).

### D. CEREBELLUM

Numerous publications on the descriptive anatomy of the avian cerebellum have been critically summarized by Pearson (1972) and by Kuhlenbeck (1975). In the following, recent anatomical findings pertaining to the functional aspects of cerebellar circuitry will be briefly summarized (cf. also Section IV, C; for visual and auditory inputs to the cerebellum, see Sections VI and VII). The cerebellum of birds is a major cortical center in the metencephalon involved with the co-ordination of higher-order sensomotor reflex activities. Similarities in the fine structure of the cerebellum in birds and mammals become evident when the findings of ultrastructural morphometric investigations on the terminals of (afferent) mossy fibers are compared. For findings in pigeons, see Paula-Barbosa and Sobrinho-Simoes, 1976; for a detailed analysis of climbing and mossy fibers in the cerebellum of chickens, see Freedman *et al.*, 1977; Vielvoye and Voogd, 1977).

Efferent fibers in the cerebellar cortex (axons of Purkinje cells) are directed toward cerebellar nuclei (Fig. 20) including parts of vestibular nuclei (lateral territory or nucleus Deiters, a displaced cerebellar nucleus). Bird species differ with respect to the structure and number of neurons in the cerebellar nuclei

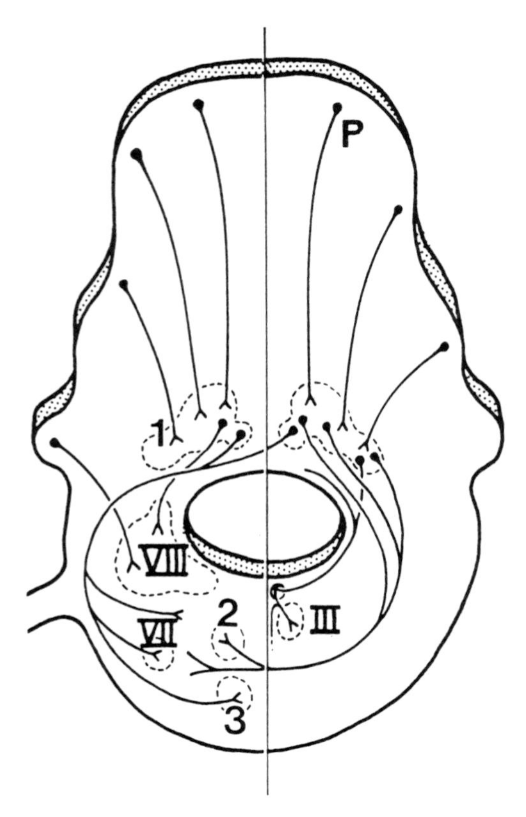


FIG. 20. Diagrammatic transverse section of the cerebellum and brain stem at a more caudal (left half) and a more rostral level (right half) showing classical efferent projections of the bird cerebellum (after Nieuwenhuys, 1967). P, Purkinje cells and their axons; 1, cerebellar nuclei; 2, nucleus ruber; 3, oliva inferior; III, nucleus originis nervi oculomotorii; VII, nucleus originis nervi facialis; VIII, complex of vestibular nuclei. Note cerebellar projection to the fasciculus longitudinalis medialis indicated dorsal to III.

(investigations considering 18 orders of birds; Renggli, 1967). Some cerebellar nuclei in species belonging to the Passeriformes, the Psittaciformes, the Piciformes, the Meropidae, the Trochilidae, and the Falconidae are folded in such a way that their profiles resemble the fissuration pattern of the cerebellar cortex (Renggli, 1967).

Major afferent and efferent fiber projections of the cerebellum in birds, summarized by Nieuwenhuys in 1967, are shown in Figs 20 and 21. Efferent projections of the cerebello-vestibular type in chickens have been studied more recently by Wold (1981), who discovered longitudinal zones of fibers in the ipsilateral cortex cere-

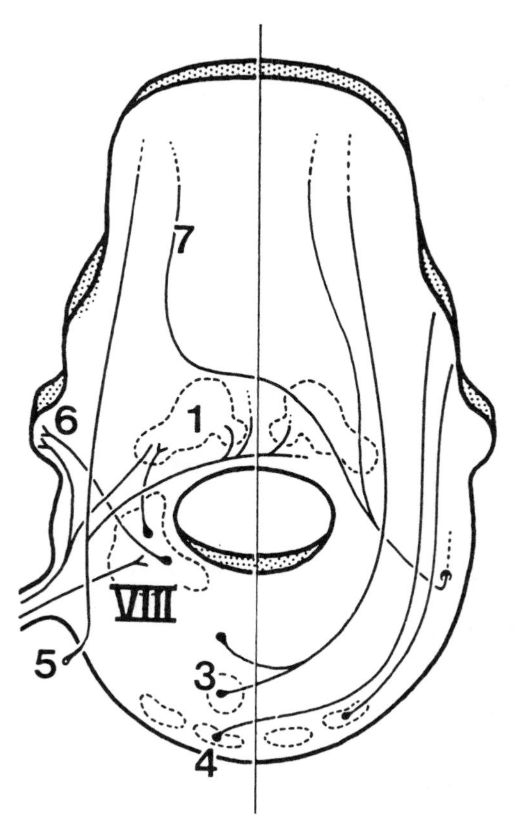


FIG. 21. Idealized transverse section of the cerebellum and brain stem at a more caudal (left half) and a more rostral level (right half) showing classical afferent projections (after Nieuwenhuys, 1967). 1, cerebellar nuclei; 3, oliva inferior with olivocerebellar and reticulocerebellar projections; 4, pontine nuclei with crossed and uncrossed pontocerebellar projections (not comparable to cortico-ponto-cerebellar system of mammals!); 5, spino-cerebellar projection; 6, flocculus receiving direct and indirect vestibular afferents; 7, tectocerebellar projection; VIII, complex of vestibular nuclei with corresponding projections.

belli that project to various cerebellar nuclei as well as a bilateral projection from ventral parts of medial and intermediate cerebellar nuclei to vestibular nuclei. The latter projection overlaps in part the cortical cerebellar projections. Thus as in mammals, well-defined regions of the cortex cerebelli and the cerebellar nuclei send fibers to partly separated and partly overlapping areas of vestibular nuclei. Other cerebellar projections have been discussed in Section IV, C.

### E. "BULBO-SPINAL" PROJECTIONS IN GENERAL

Descending projections arising from sensomotor reflex centers located in the diencephalon, the mesencephalon, the metencephalon and the rhombencephalon have been called "bulbo-spinal" projections in birds. Using retrograde transport of horseradish-peroxidase injected into circumscribed areas of the spinal cord, Cabot *et al.* (1982) mapped the origins of bulbo-spinal projections. They concluded that in general the fundamental organizational patterns of descending spinal pathways are quite comparable among birds, reptiles and mammals.

At the level of the *diencephalon* the following ipsilateral nuclear regions give rise to spinal projections:

(1) the paraventricular nucleus of the hypothalamus (neurophysin-containing neurons);
(2) the caudal part of the lateral hypothalamic area;
(3) the stratum cellulare internum;
(4) perikarya located in the postero-medial hypothalamic area in a circumscribed region dorsal to the supramammillary decussation, and
(5) the ventrolateral nucleus of the thalamus.

This general pattern has been further substantiated by tracing studies in hatchling chicks (Gross and Oppenheim, 1985). Projections from lateral hypothalamic areas and from the bed nucleus of the stria terminalis to the complex of dorsal vagal nuclei have been studied in the pigeon by anterograde tracing techniques (Berk, 1987). Finally, it should be noted that in the course of ontogeny some of the bulbospinal projections have been shown to be of transitory nature (for findings in chickens, see Okado and Oppenheim, 1985).

At the level of the *mesencephalon* one finds the following dominating fiber tracts that project to the spinal cord:

(1) crossed rubro-spinal projections that descend to lumbar levels of the spinal cord (rubro-spinal projections are not generally found down to lumbar levels in mammals);
(2) ipsilateral projections from the nucleus interstitialis of Cajal;
(3) crossed colliculo-spinal projections from the mesencephalon;

(4) a few poorly defined ipsilateral projections that arise from tegmental areas including the medial reticular formation, and
(5) descending catecholamine-containing pathways that originate predominantly in the ipsilateral locus coeruleus and locus subcoeruleus.

The *metencephalon* gives rise to ipsilateral reticulo-spinal and vestibulo-spinal projections (for the latter, see p. 49). Serotonin-containing fibers have also been found to project from different raphe nuclei to the spinal cord.

Finally, at the level of the *medulla oblongata* descending projections arise from:

(1) the nucleus of the solitary tract (ipsi- and contralateral projection);
(2) the region of nucleus gracilis, nucleus cuneatus, and nucleus cuneatus externus (ipsilateral projections located in the dorsal funiculus);
(3) perikarya located in a region ventral to the dorsal nucleus of the vagus nerve (contralateral projection; apparently corresponding to the nucleus intercalatus et intermedius of Karten and Hodos, 1967).

In the chicken the general pattern of bulbo-spinal projections seems to confirm results obtained in pigeons, although in chickens the perikarya of the nucleus tractus solitarii apparently lack spinal projections (for details and discussion, see Chikazawa *et al.*, 1983). Gross and Oppenheim (1985) detected two additional spinal projections in hatchling chicks that had not been seen until then. One of them originates in the accessory vestibular nucleus, the other in the nucleus ambiguus. Glover and Petursdotter (1988) established the specificity of reticulo-spinal and vestibulo-spinal projections in 11-day-old White Leghorn embryos by retrograde tracing experiments in an *in vitro* approach. Single- and double-labelling experiments were combined with selective lesioning of certain pathways.

## VI. Auditory System

The auditory system exhibits astounding specializations among the different taxonomic forms of birds. Certain species of owls, for example, are able to localize sounds precisely. (For observations and a review of the literature, see Knudsen and Konishi, 1979, and Knudsen, 1980.) Morphological and functional aspects of the various components of the auditory system have also been extensively reviewed by Schwartzkopff (1960, 1973), Cohen and Karten (1974), Sachs *et al.* (1980), and, on a broad comparative basis, by Il'icev (1972). More recently Necker (1983b) reviewed hearing mechanisms in pigeons. Thus, the following is but a brief outline and is summarized in Fig. 22. Attention is focused on findings obtained by applying modern neurobiological techniques (e.g., anterograde and retrograde

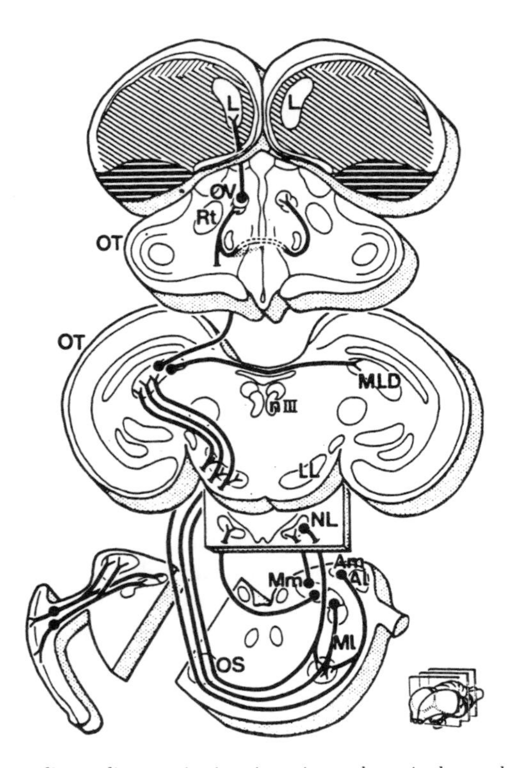


FIG. 22. Major ascending auditory projections in a pigeon shown in three subsequent representative transverse sections of brain stem and telencephalon. At the lowest level of the brain stem the primary auditory nuclei and their afferent projections have been separated on the left to provide a less complicated wiring diagram. On the right the primary auditory projection has not been indicated. Al, nucleus angularis pars lateralis; Am, nucleus angularis pars medialis; L, field of Rose; LL, lemniscus lateralis and nucleus of the lateral lemniscus; Ml, nucleus magnocellularis pars lateralis; MLD, nucleus mesencephalicus lateralis pars dorsalis; Mm, nucleus magnocellularis pars medialis; NL, nucleus laminaris; nIII, nucleus originis nervi oculomorii; OS, oliva superior with trapecoid fibers crossing the midline; OT, mesencephalic (optic) tectum; OV, nucleus ovoidalis; Rt, nucleus rotundus (modified after Boord, 1969). For graphic shadings, see Fig. 14.

axonal transport of tracer substances and mapping of metabolic activities by analyzing uptake of 2-deoxiglucose).

In contrast to past interpretations of histological and especially ultrastructural features of the avian cochlea and hair cells (sensory cells), more recent interpretations consider them to resemble in principle those of mammals (for a review of the older literature, see Schwartzkopff, 1973). Different types of hair cells have been distinguished on the basis of shape and size in pigeons and chickens (Takasaka and Smith, 1971; Hirokawa, 1978; Tanaka and Smith, 1978). All hair cells are polarized in such a manner that the stereocilia become shorter the further they stand from the single kinocilium. The kinocilium is always located at the pole of the hair cell that points to the loose end of the tectorial membrane. Hair cells have both afferent and efferent synaptic contacts. The sizes of the synaptic contacts correlate in part with the sizes and shapes of the hair cells (for recently detected mechanical functions of outer hair cells in the inner ear of mammals, see Zenner *et al.*, 1988). At the apical end of the cochlea there is a second sensory patch, the macula lagenae, the fine structure of which resembles that of the sensory apparatus of the vestibular system. The lagenar nerve apparently projects both to vestibular and to cochlear nuclei (Boord and Rasmussen, 1963). The function of this unique system is still not clearly understood.

Auditory pathways have been studied extensively by Boord and by Karten (see Karten, 1968; Boord, 1969; Cohen and Karten, 1974). The nuclei angularis and magnocellularis in the medulla oblongata are the primary terminal nuclei (seat of the second-order auditory neurons) of the cochlear nerve. Upon entering the medulla oblongata, the cochlear nerve bifurcates into medial and lateral branches (Fig. 22). Axons of the medial and lateral branches innervate in an orderly spatial pattern (tonotopic organization) the second-order auditory neurons in the primary auditory nuclei. The auditory nuclei can in turn be subdivided into three subunits. With the use of quantitative ultrastructural techniques axosomatic terminals have recently been examined in the various subunits of the nucleus magnocellularis in chickens. Furthermore, morphological findings have been correlated with results of electrophysiological recordings (Parks, 1981b). During the past few years, the physiology and the morphogenesis of perikarya in the nucleus magnocellularis have been studied in detail; the growth of dendrites in this nucleus most likely takes place when the animals are first exposed to airborne sound (for details and review, see Parks, 1979, 1981b; Conlee and Parks, 1981, 1983; Jhaveri and Morest, 1982a,b).

Projections of cochlear fibers to the nucleus laminaris have been found in investigations using specific silver impregnation of degenerating profiles after the cochlear nerve had been lesioned (see for example, Wold and Hall, 1975). Other investigations, however, in which modern tracing techniques were applied, did not confirm these observations (Parks and Rubel, 1978). Apparently the rapid transneural degeneration of deafferented second-order auditory neurons is responsible for the degeneration of terminals in the nucleus laminaris. These controversial results may have increased general interest in projections to the nucleus laminaris and in its ultrastructural characteristics (cf. Parks and Rubel, 1975; Benes *et al.*, 1977; Smith and Rubel, 1979; Parks, 1981a; Rubel *et al.*, 1981; Smith, 1981; Parks *et al.*, 1983; Smith *et al.*, 1983). It should be mentioned that the nucleus laminaris is the most distally located auditory nucleus that receives both ipsilateral and contralateral auditory information.

Apparently all the fibers that arise from primary auditory nuclei (medial part of the nucleus angularis and nucleus magnocellularis) and from third-order auditory neurons (nucleus laminaris) cross the midline (decussatio cochlearis dorsalis and corpus trapezoideum) and ascend contralaterally in the system of the lateral lemniscus (Fig. 22). Terminals of lateral lemniscal fibers or their collaterals have been observed in the nucleus olivaris superior. At the level of the midbrain, terminals of the lateral lemniscal fibers or their collaterals innervate neurons located in the ventral part of the nucleus lemnisci lateralis. Projections of the nucleus mesencephalicus lateralis pars dorsalis apparently arise only from perikarya located in the nucleus angularis or in the nucleus laminaris.

These briefly reviewed findings have been obtained primarily by applying specific silver impregnation techniques to locate degenerating fibers and terminals following stereotaxic lesions of various auditory nuclei. However, recent observations in pigeons indicate that ascending auditory pathways, at least up to the level of the mesencephalon, might be far more similar to those described in mammals than previously thought. For details of contralateral and newly observed ipsilateral pathways as revealed by autoradiographic tracing of transneuronal transport of tritiated substances injected into the endolymphatic space of the labyrinth, see Correia *et al.* (1982). The idea that the auditory fiber tracts in mammals and birds are quite similar at the level of the brain stem is further supported by the discovery of efferent perikarya that project to the cochlea (nucleus reticularis paragigantocellularis lateralis in pigeons; Schwarz *et al.*, 1981). Furthermore, Conlee and Parks (1986) applied anterograde and retrograde tracing techniques to detect the origin of ascending auditory projections to the nucleus mesencephalicus lateralis pars dorsalis in 3–12-week-old White Leghorn chickens. Their findings include a bilateral projection of the cochlear nuclei and demonstrate an important similarity between the brain stem auditory pathways of birds and other terrestrial vertebrates.

The nucleus mesencephalicus lateralis dorsalis may be compared with the colliculus inferior (caudalis) of mammals. Its size and neuronal composition apparently varies greatly among different taxonomic groups of birds. However, only limited numbers of species have been examined in detail (Cobb, 1964; Rylander, 1979). Only certain subdivisions of the nucleus mesencephalicus lateralis dorsalis receive ascending auditory fibers. In Barn Owls (*Tyto alba*) one subdivision of the nucleus mesencephalicus dorsalis lateralis is characterized by a tonotopic organization whereas in another subdivision the neurons are arranged according to the localization of their spatial receptive fields (Knudsen, 1983; for space-mapped auditory projections of the nucleus mesencephalicus dorsalis lateralis to the optic tectum, see Knudsen and Knudsen, 1983).

Ascending fibers arising from the nucleus mesencephalicus dorsalis lateralis form the tractus nuclei ovoidalis and terminate within the ipsilateral nucleus ovoidalis thalami and within the smaller nucleus semilunaris parovoidalis. Some efferent fibers cross the midline in the commissura supraoptica dorsalis (Fig. 22) to terminate in the contralateral nucleus ovoidalis. Other fibers descend to the ipsilateral trapezoid body (Karten, 1968). The nucleus ovoidalis thalami of birds seems to be comparable to the medial geniculate body of mammals. Afferent fibers from this thalamic relay nucleus of the auditory system join the medial part of the lateral forebrain bundle to enter the telencephalon rostral of the commissura rostralis in the tractus thalamofrontalis medialis. They terminate in the medial portion of the caudal neostriatum known as field L of Rose (Karten, 1968). In field L of Rose small groups of neurons form unit-like clusters (observations in European Starlings, *Sturnus vulgaris*; Saini and Leppelsack, 1977; for structural details on these neurons as seen in Golgi-impregnated material, see Saini and Leppelsack, 1981).

In other morphological studies dealing with the auditory system of birds, two different approaches were applied:

(1) the projections within the auditory system and related telencephalic structures were traced by means of anterograde transport of tritiated amino acids (for findings in domesticated canaries and in Guineafowl, see Kelley and Nottebohm, 1979; Bonke *et al.*, 1979), and
(2) metabolic activities were mapped by examining the uptake of 2-deoxiglucose following application of specific auditory stimuli (for a review of general phenomena of auditory perception in birds, see Dooling, 1982).

Efferent projections that arise from perikarya located in field L are shown in Fig. 23. Electrophysiological results and distributional patterns of 2-deoxiglucose uptake indicate that perikarya in field L are arranged tonotopically (for review and results in Guineafowl, see Bonke *et al.*, 1979). By acoustically stimulating monaurally deafened chickens, Scheich (1983) succeeded for the first time to demonstrate a two-columnar system in the auditory neostriatum resembling the columnar organization of the neocortex in mammals. Finally, Maier and Scheich (1983) showed in young Guineafowl that acoustic imprinting correlates with differential uptake of 2-deoxiglucose in the most rostral part of the forebrain (for recent findings in White Leghorn chickens and in Guineafowl chicks, see Heil and Scheich, 1986; Maier and Scheich, 1987; Wallhäusser and Scheich, 1987). Hitherto this part of the forebrain had not been considered to possess primarily auditory functions. (For a discussion of acoustically evoked electrical responses in the most rostral part of the forebrain of pigeons, see Necker, 1983b; for a description of possible auditory projections to the nucleus basalis, see p. 45.)

Neuronal circuitry between telencephalic auditory areas and telencephalic areas that control song remains to be elucidated in detail (for review, see Arnold, 1982). However, some quite interesting observations on the pertinent telencephalic nuclei that control song should be noted as follows:

(1) These nuclei (caudal nucleus of the hyperstriatum ventrale and nucleus robustus of the archistriatum) exhibit sexual dimorphism in several species of song birds (Nottebohm and Arnold, 1976; for review, see DeVoogd, 1984). In

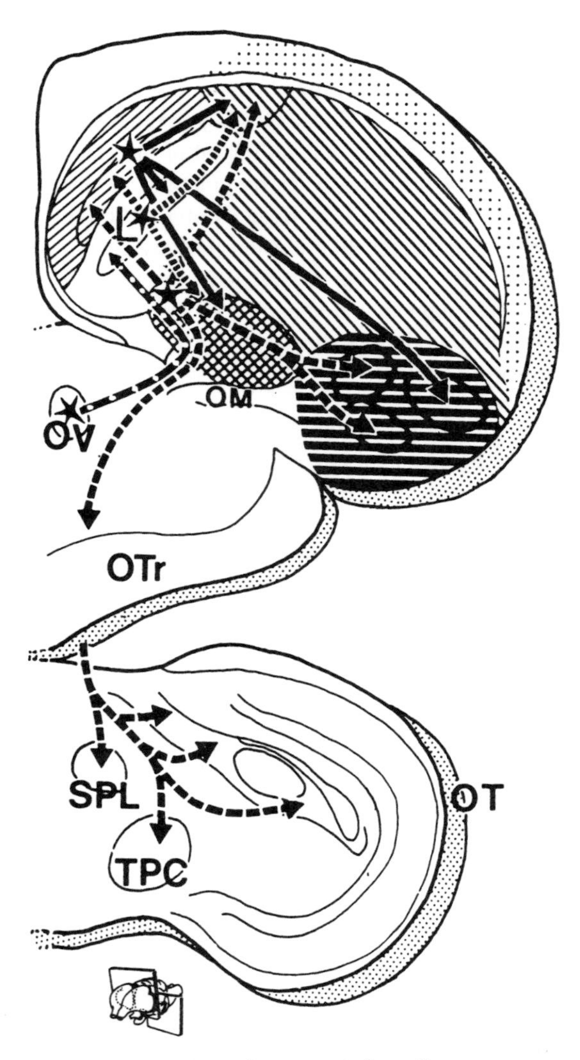


FIG. 23. Diagram showing connections of prosencephalic auditory centers as revealed by axonal transport of tritiated amino acids in Guineafowl (after Bonke *et al.*, 1979). Injection sites are indicated by asterisks. The directions taken by the axons differ depending upon which site in the hyperstriatum ventrale had been injected, field L (L), paleostriatum augmentatum, and nucleus ovoidalis (OV). OM, tractus occipitomesencephalicus; OT, mesencephalic (optic) tectum; OTr, optic tract; SPL, nucleus spiriformis lateralis; TPC, nucleus tegmenti pedunculopontinus. For graphic shadings, see Fig. 14.

European Starlings and in Song Sparrows (*Melospiza melodia*) telencephalic nuclei engaged in the control of song exhibit distinct patterns in the distribution of immunohistochemically detected vasoactive intestinal peptide, methionine-enkephalin, cholecystokinin, and substance P (Ball *et al.*, 1988).

(2) Some neurons that control song at different levels of the telencephalon, the mesencephalon, and the medulla oblongata are capable of concentrating sexual steroid hormones. This phenomenon is probably receptor mediated (Arnold *et al.*, 1976). In Zebra Finches (*Poephila guttata*) uptake is different in males and females (Arnold and Saltiel, 1979).
(3) Domesticated canaries have perikarya in the telencephalic nucleus robustus archistriatalis that exhibit sex-specific differences with respect to their dendritic morphology (DeVoogd and Nottebohm, 1981).
(4) The sexual dimorphism of telencephalic nuclei that control song shows seasonal changes in canaries but not in White-crowned Sparrows, *Zonotrichia leucophrys nuttalii* (for review and discussion, see Baker *et al.*, 1984).
(5) In the caudal nucleus of the hyperstriatum ventrale of canaries neurogenesis takes place continuously in a ventricular zone overlying the song-control nucleus and is followed by neuronal migration (Goldman and Nottebohm, 1983; cf. also Konishi and Akutagawa, 1985).

Burd and Nottebohm (1985) showed in adult male canaries that the synaptic terminals formed on newly generated neurons, which are known to be inserted into existing neuronal networks, belong to at least three morphological types. Learning of song control can be correlated with neurogenesis as well as with lateralization in several species of birds (Nottebohm *et al.*, 1981; Paton and Manogue, 1982). The phenomenon of lateralization has been extensively studied in a number of species. It can be observed at all levels of neuronal systems controlling song behavior including the final element of the common pathway—the motor neuron (for review, see Manogue and Nottebohm, 1982; Arnold, 1982; Konishi, 1985). Finally, in Zebra Finches the telencephalic song control areas exhibit delayed post-hatching development that is related to the development of song behavior (Herrmann and Bischof, 1986).

## VII. Visual System

Investigations on the retina and the central components of the visual system in birds have been undertaken by such well-known neurobiologists as Cajal, Pumphrey, Polyak, Cowan, Powell, and Karten, to name but a few. Their reports have been reviewed by Sillman (1973), Emmerton (1983a,b), and Henke (1983). In the following description only general aspects of the visual system will be considered.

### A. RETINA AND GENERAL ASPECTS OF CENTRAL VISUAL PATHWAYS

The eyes of a bird represent a considerable part of the total weight of the head (e.g.

in man less than 1% whereas in European Starlings, it is 15%). The general histological structure and ontogenetic development of the retina is quite similar in birds and in mammals. The fine structure of the retina, however, is remarkably inhomogeneous in numerous species of birds (for detailed review, see Polyak, 1957). These differences in structure can be correlated with differences in processing capabilities (cf. Sillman, 1973). Structural inhomogeneities have been best analyzed in pigeons in which the retina can be divided into two distinct areas, the red and the yellow fields. The colors of the fields are brought about by the uneven distribution of brightly colored oil droplets located between the inner and outer segments of cone receptor cells. (For review, see Emmerton, 1983a; for the role of oil droplets in color vision, see Bowmaker, 1977, 1980; for findings in passerine birds, see Mayr, 1972a,b; for observations in a nocturnal species, see Martin, 1978.) The colors, sizes and locations of oil droplets within the receptor layer can be used as criteria to differentiate cone receptors into morphologically distinct classes. It has been suggested that the orderly matrix-like arrangement of cone receptor cells may be the morphological basis for the sensitivity of several taxonomic forms of birds to the e-vector of linearly polarized light. Furthermore, yellow oil droplets have been reported to be fluorescent when exposed to UV light. In a number of species the lens is transparent to UV light. These droplets are therefore thought to mediate sensitivity to UV light which is known to be present in several taxonomic groups (for discussion and review, see Emmerton, 1983b; cf. also Goldsmith, 1980). However, Chen *et al.* (1984) demonstrated through electrophysiological recordings conducted on 15 species from 10 families of birds that they possess photoreceptors which contain a photopigment maximally sensitive at 370 nm; these species would not be dependent upon the fluorescence of the yellow oil droplets.

Retinal photoreceptor cells in birds, as in other vertebrates, continuously renew the photoreceptor membranes in a synchronized cycle (for review, see Young, 1978). The ontogenetic development of shedding of distal tips of outer segments depends on exposure to light (Teuchert amd Kretschel, 1985).

Structural inhomogeneities in the distributional patterns of rods and cones (for findings in chickens, see Morris, 1970, 1975, 1976; Meyer and May, 1973) are accompanied by regional variations in the sizes and numerical densities of the perikarya forming the ganglion of the optic nerve (for findings in chickens, see Ehrlich, 1981; in pigeons, see Binggeli and Paule, 1969; Hayes and Holden, 1980; for a review of older literature, see Sillman, 1973). Structural inhomogeneities are already present in the developing retina of the chick (DeMello and DeMello, 1985) and the pigeon (Bagnoli *et al.*, 1985). In this context it is interesting to note that in pigeons the optic pathways apparently develop before the receptor cells have begun functioning (Bagnoli *et al.*, 1985).

On average, the retina is thicker in birds than in mammals. Moreover, the avian retina is characterized by a more complex neuronal circuitry, first revealed by

Cajal, indicating that a great deal of visual processing already takes place there (for details and review, see Yazulla, 1974; Emmerton, 1983a,b; for comparative aspects, see Dubin, 1970).

Numerous taxonomic forms of birds possess large monocular, laterally directed visual fields combined with smaller, frontally directed binocular fields. Frequently there are two morphologically distinct depressions in each retina representing separate foveae for the binocular and the monocular fields. The depth of the fovea has been shown to be related to visual acuity. Visual acuity is greater in falconiform birds than in mammals (Snyder and Miller, 1978; for a general review, see Polyak, 1957; Pumphrey, 1961; for stereopsis in falconiform birds, see Fox *et al.*, 1977; for retinal binocular fields in pigeons, see Martin and Young, 1983). Where the optic nerve leaves the retina an elongated, highly vascularized (cf. Fischlschweiger and O'Rahilly, 1968) and frequently pleated, heavily pigmented (melanin) structure arises and penetrates into the vitreous body. This structure is called the pecten oculi. It has been shown to be responsible for nutritive and gas-exchange functions (Wingstrand and Munk, 1965) and supports the inner layers of the retina which, in contrast to those of mammals, lack a direct blood supply. In addition it may serve as a heat-exchanging device (cf. p. 15) and as a baffle for shading parts of the retina from intraocularly scattered light, e.g., light arising from a bright image of the sun; for details, see Emmerton (1983a).

In principle, the retina is composed of the following neurons: (1) various types of photoreceptor cells — single cones, double cones, and rods (primary sensory cells of the retina), (2) bipolar neurons, (3) cells of the ganglion nervi optici. Synaptic transmission between photoreceptor elements and bipolar cells is influenced by different types of horizontal cells (for comparative observations, see Gallego, 1976) whereas synaptic transmission from bipolar cells to ganglion cells is controlled by amacrine cells. It has been suggested that horizontal and amacrine cells arise from a common progenitor and then migrate actively from the site of origin (for morphological evidence in chick embryos, see Prada *et al.*, 1984). Retinal interneurons establish long-distance intra-retinal connections that span almost the entire extent of the retina in chicks and chick embryos (Catsicas *et al.*, 1987a). The way in which this complicated neuronal circuitry functions is not yet fully understood (for a review of findings in pigeons, see Emmerton, 1983a,b). Immunocytochemical observations have revealed the presence of subclasses of amacrine cells that contain enkephalin-immunoreactive material (Brecha *et al.*, 1979; Fukuda *et al.*, 1987; for light-dependent changes in this material observed in 5-day-old chicks, see Millar *et al.*, 1984). Amacrine cells containing monoamines have also been detected (indolamines or catecholamines; cf. Stoeckel *et al.*, 1976; Florén, 1979; Araki *et al.*, 1984; Kato *et al.*, 1984; Millar *et al.*, 1988). Monoamines and peptides in the avian retina exhibit light-dependent fluctuations in content, which can be demonstrated, for example in variations in immunoreactivity. Immunoreactivity of substance P is greatly reduced in newly hatched chicks that have been deprived of light (Millar and Chubb, 1984). Ehrlich *et al.* (1987) combined the immunocytochemical demonstration of substance P with the retrograde tracing of rhodamine-coupled latex beads that had been injected into the optic tectum. Control animals and animals with specific lesions within the optic tracts were investigated. The authors were able to demonstrate that certain retinal ganglion cells exhibiting a substance-P-like immunoreactivity project to individual layers of the optic tectum in the chicken. Katayama-Kumoi *et al.* (1985) observed the coexistence of immunoreactivity to antibodies raised against substance P and immunoreactivity to antibodies raised against pancreatic polypeptide in a subpopulation of amacrine cells in chickens of 50 g body weight.

Finally, the avian retina is characterized by the presence of so-called displaced ganglion cells, which were first detected nearly 100 years ago by Dogiel. They are located between the amacrine cells and give rise to the accessory optic system. (For findings, see Heaton *et al.*, 1979; Hayes and Holden, 1980; Fite *et al.*, 1981; for details on the accessory optic system, see p. 71). Displaced amacrine cells and so-called matching amacrine cells have been identified in the ganglion cell layer of the pigeon retina (Hayes, 1984).

The central portions of the avian visual system consist of three major components (the so-called autonomic projections are not included): (1) the tectofugal fiber tracts, (2) the only recently detected thalamofugal fiber systems, and (3) the accessory optic fiber systems. The principal features of these components are summarized in Figs 24–29 (for details on the neuronal substrates that mediate the pupillary light reflex in white Carneaux pigeons, see Gamlin *et al.*, 1984; Gamlin and Cohen, 1988a,b). Tectofugal and thalamofugal fiber systems project in a retino-topical order to different, precisely defined areas of the telencephalon, whereas accessory optic fibers project via the ectomammillary nucleus or nucleus of the basal optic root of the mesencephalon, to the vestibulo-cerebellum and to the complex of the oculomotor nuclei. In birds all of the fibers of the optic nerves seem to cross the midline. However, by using highly sensitive tracing techniques (e.g. anterograde axonal transport of wheat germ agglutinin-conjugated horseradish peroxidase), several groups of investigators have observed that approximately 1% of the fibers do not cross. There is a period during ontogenesis in which many more than 1% of the fibers do not cross the midline (for a review and findings in chickens, see O'Leary *et al.*, 1983; for observations in 13-day-old White Leghorn embryos, see Crossland, 1985). However, fibers that conduct visual information may recross the midline at various sites in the CNS (for a review, see Saleh and Ehrlich, 1984).

Since birds are at least equipotent to mammals in their visual abilities, it is interesting to compare the central nervous pathways that process visual information. The tectofugal and thalamofugal systems in birds seem to be comparable to the retino-thalamo-cortical and the retino-geniculo-(thalamo)-striatal pathways in mammals (for further details, see Henke, 1983). Comparisons of this kind will

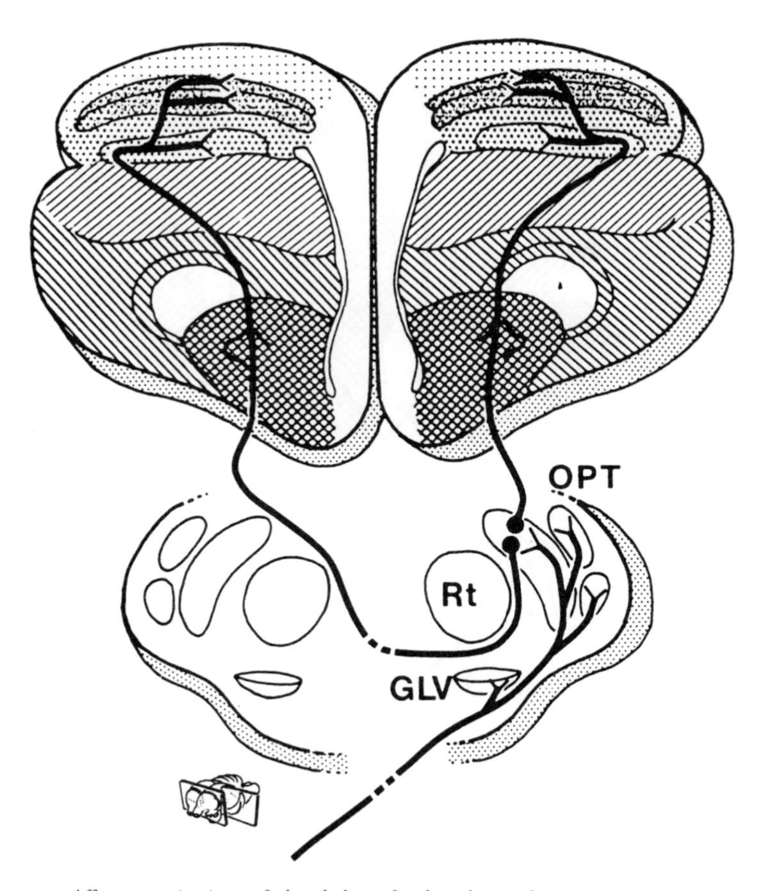


FIG. 24. Afferent projections of the thalamo-fugal pathway shown in representative transverse sections of the brain of a pigeon (after Karten *et al.*, 1973). GLV, nucleus geniculatus lateralis ventralis; OPT, nucleus opticus principalis thalami; Rt, nucleus rotundus. For graphic shadings, see Fig. 14.

remain speculative, however, until the structure and function of the avian telencephalon are better understood. (For a recent review of the functional anatomy of the visual system at the level of the telencephalon in birds and information on comparative anatomical aspects, see Watanabe *et al.*, 1985.) The evolution of the tectofugal and thalamofugal pathways in the visual systems of vertebrates has been most extensively studied by Ebbesson (1972; see also Ebbesson, 1980; for a comparison of diurnal and nocturnal birds, see Hirschberger, 1971). Unfortunately, the common nomenclature for visual fiber tracts and visual nuclei proposed by Ebbesson (1972) has not yet been accepted by the scientific community. For that reason the most widely accepted nomenclature based on Karten *et al.* (1973) has been adopted in the following description of pathways.

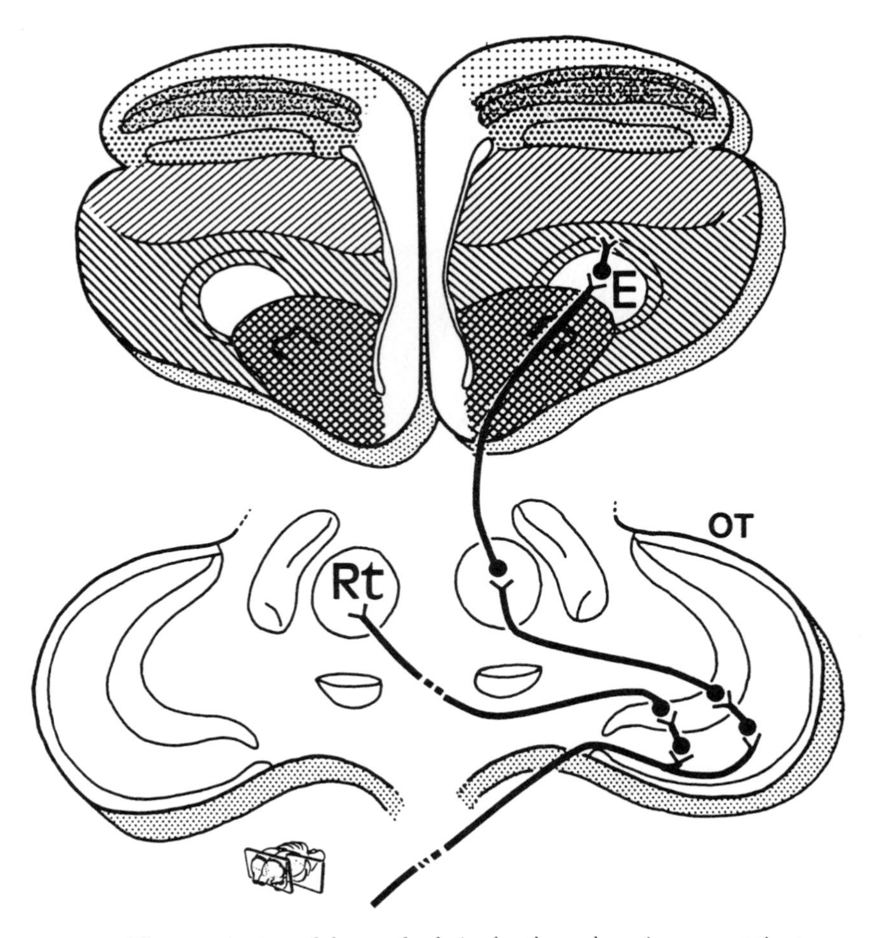


FIG. 25. Afferent projections of the tectofugal visual pathway shown in representative transverse sections of the brain of a pigeon (after Karten *et al.*, 1973). E, ectostriatum; OT, mesencephalic (optic) tectum; Rt, nucleus rotundus. For graphic shadings, see Fig. 14.

In contrast to mammals, birds possess a precisely defined system of retinotopically arranged efferent fibers (approximately 12,000 efferent fibers versus $2 \times 10^6$ afferent fibers per optic nerve in pigeons) that arise from the isthmo-optic nucleus. (For details, see p. 69; for quantitative data in pigeons, see Cowan, 1970; Duff and Scott, 1979; Duff *et al.*, 1981; in Domestic Mallards, see O'Flaherty, 1981; for a general discussion of the retinotopic arrangements of afferent and efferent visual fibers; see Bunt and Horder, 1983.) Weidner *et al.* (1987) investigated the nuclear origin of the centrifugal visual pathway in 11 species of birds of prey distributed among four families and belonging to two orders (Strigidae: *Otus choliba*, *Strix aluco*, *Athene noctua*, *Asio flammeus*; Falconiformes-Accipitridae: *Milvus migrans*,

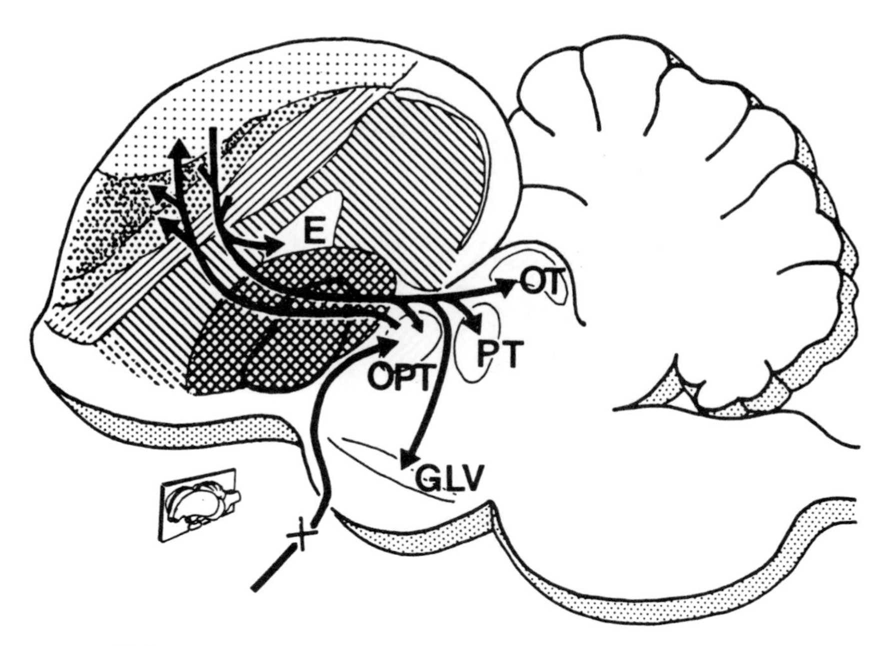


FIG. 26. Thalamofugal visual pathways and related efferent projections of the telencephalon of the pigeon shown in a schematic parasagittal section of the brain of a pigeon (after Emmerton, 1983a,b). Crossing of midline is indicated by +. E, ectostriatum; GLV, nucleus geniculatus lateralis ventralis; OPT, nucleus opticus principalis thalami; OT, mesencephalic (optic) tectum; PT, pretectal area or nuclei. For graphic shadings, see Fig. 14.

*Buteo buteo*, *Heterospizias meridionalis*, *Buteogallus* sp., *Aegypius monachus*; Cathartidae: *Coragyps atratus*; Falconidae: *Falco tinnunculus*). Cell counts of Nissl-stained perikarya in the isthmo-optic nucleus were reported. The authors confirmed the rather poor morphological development of the isthmo-optic nucleus in both diurnal and nocturnal raptors. They conclude that in these species of birds the centrifugal visual system plays a minor supplemental, rather than a central role in the visual process.

### B. TECTOFUGAL PATHWAY (FIG. 25)

Retinal ganglion cells belonging to different size classes give rise to retinotopically arranged, crossed axons. These penetrate in an oblique direction through the superficial layers of the optic tectum and arborize in Cajal's layers 2–7. Some fibers may penetrate into deeper layers. These loop back, however, before terminating in layers 2–7. (For details of distributional patterns of retinal fibers and terminals, see Repérant and Angaut, 1977; Acheson *et al.*, 1980; for observations

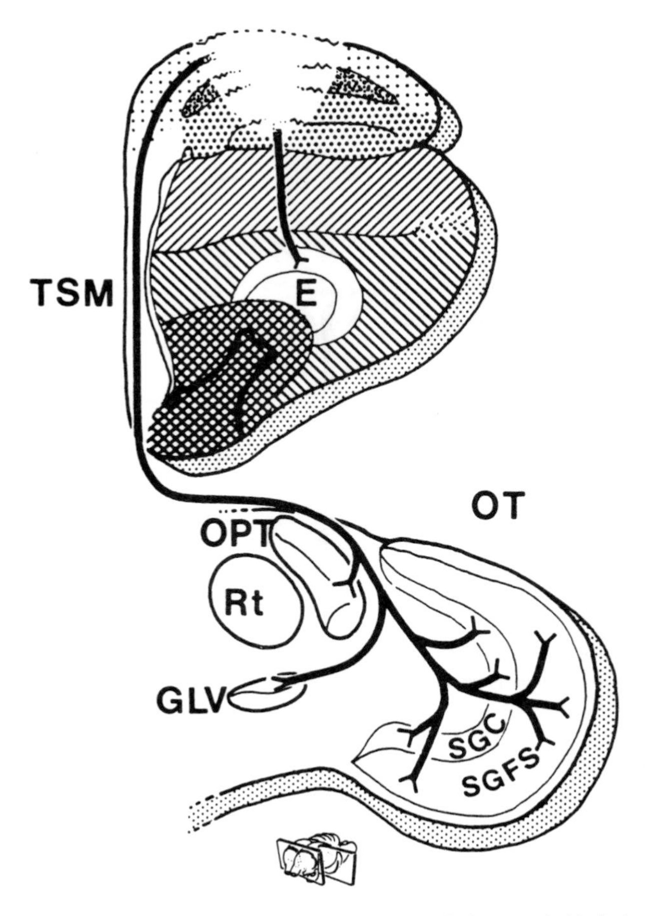


FIG. 27. Efferent projections of the visual Wulst in Burrowing Owl as revealed in lesion experiments, shown in two representative cross sections (after Karten *et al.*, 1973). E, ectostriatum; GLV, nucleus geniculatus lateralis ventralis; OPT, nucleus opticus principalis thalami; OT, mesencephalic (optic) tectum; Rt, nucleus rotundus; SGC, stratum griseum centrale; SGFS, stratum griseum et fibrosum superficiale; TSM, tractus septomesencephalicus; for graphic shadings, see Fig. 14.

on synaptogenesis in the optic tectum, see McGraw and McLaughlin, 1980a,b.) The development of retinotectal projections and the neurogenesis of the optic tectum under normal as well as under experimental conditions have, over the past years, been the subjects of intense neurobiological research (Thali, 1972; Goldberg, 1974; Domesick and Morest, 1977a,b; McGraw and McLaughlin, 1980a,b; McLoon and Lund, 1982; Alvarado-Mallart and Sotelo, 1984; Ehrlich and Mark, 1984; Fujisawa *et al.*, 1984; Thanos and Bonhoeffer, 1984). The experimental contributions and theoretical reflections about general ontogenetic principles of central nervous projections, especially those presented by Rager on retinotectal

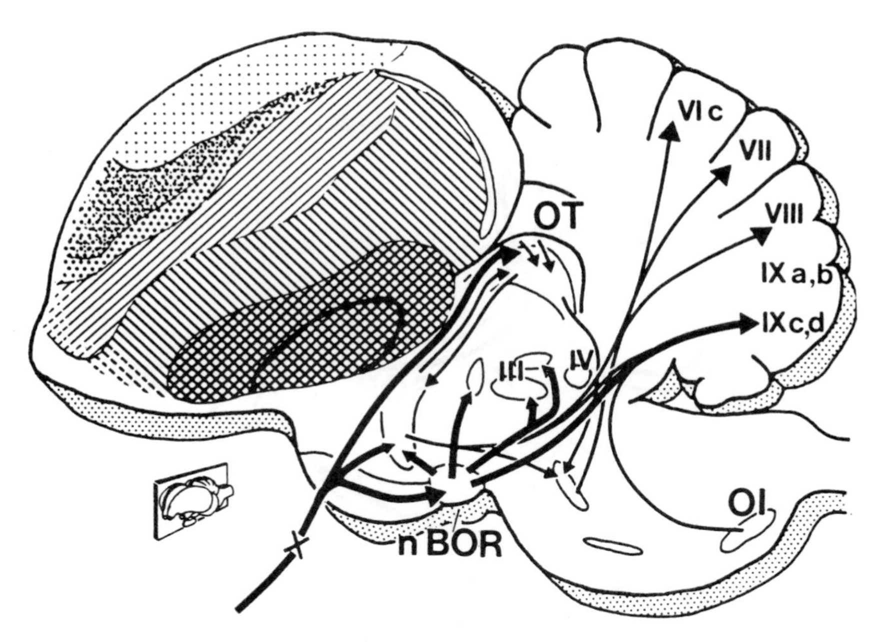


FIG. 28. Accessory optic pathways shown in a parasagittal section of a pigeon brain (after Emmerton, 1983a). Crossing of midline is indicated by an X. Thick lines and thin lines indicate major and minor projections as revealed by morphological methods. nBOR, nucleus of the basal optic root or nucleus ectomammillaris; OI, oliva inferior; OT, mesencephalic (optic) tectum; III and IV, motor nuclei of the oculomotor and trochlearis nerves; VI–IX, folia of the cerebellum. For graphic shadings, see Figure 14.

connections, are particularly interesting to neurobiologists (Rager, 1978; Rager and Rager, 1978; Rager, 1980a,b). Schwarz and Halfter (1984) and McLoon (1984) summarized the problems related to the development of retinotectal projections and interactions between retinal axons and their micro-environment in chick embryos. In a comparative study on developing chick, quail, and pigeon visual pathways Halfter *et al.* (1985) presented evidence that axon–axon interactions are *not* involved in initial axon orientation early in retinal morphogenesis.

In addition to the well-established input of retinal fibers, the superficial layers of the optic tectum receive a variety of fibers that arise from other sites in the CNS. These fibers probably release different types of transmitters (for a review, see Cuénod and Streit, 1979; for data on monoamines, see p. 83). In this context it should be noted that Mestres and Delius (1982) have demonstrated that septal areas project to the tectum mesencephali. While analyzing the possible functional significance of limbic projections to the tectum mesencephali, the presence of opiate receptors in the tectum deserves special attention (cf. Felix *et al.*, 1979).

Efferent fiber systems ascending to the prosencephalon arise from layers 10, 14, and 15 of the tectum. Perikarya in layer 10 send axons to the pars ventralis externa

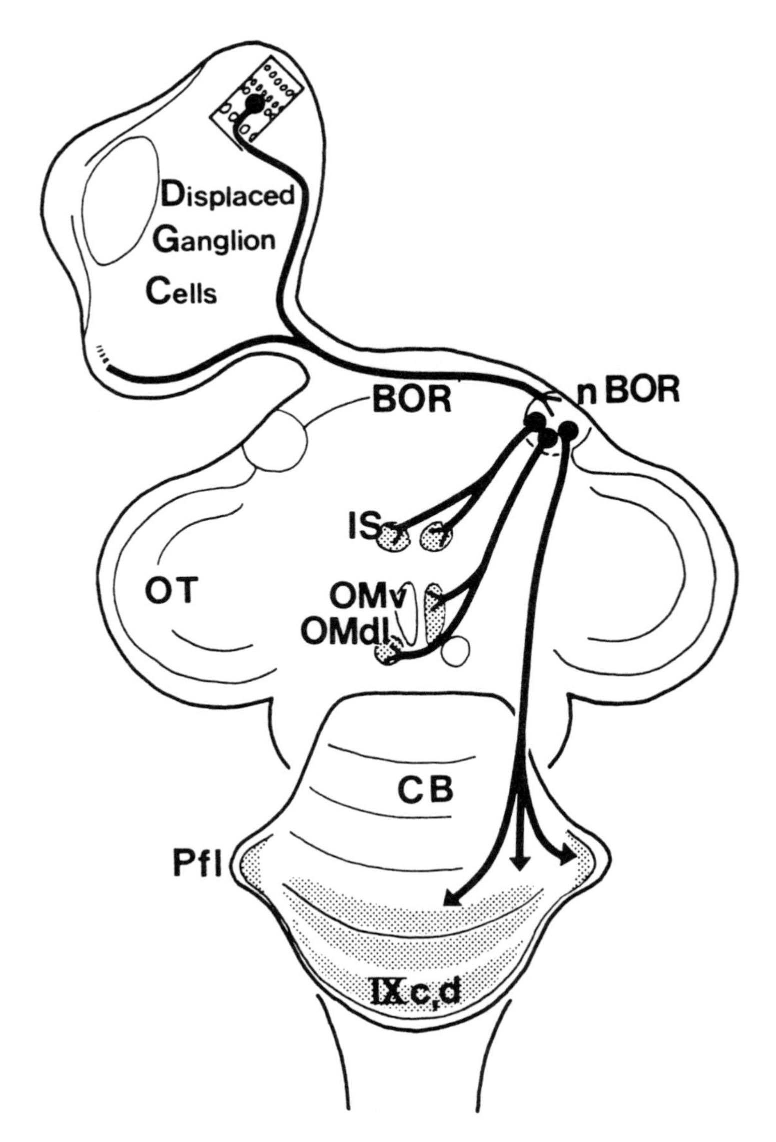


FIG. 29. Afferent projections of displaced retinal ganglion cells in an idealized bird brain (after Karten *et al.*, 1977; Brecha and Karten, 1979; Brecha *et al.*, 1980). BOR, basal optic root; nBOR, nucleus of the basal optic root or nucleus ectomammilaris; CB, cerebellum with folia (IXc, d) and paraflocculus (Pfl); IS, nucleus interstitialis (Cajal); OMv, nucleus originis nervi oculomotorii pars ventralis; OMdl, nucleus originis nervi oculomotorii pars dorsolateralis; OT, mesencephalic (optic) tectum.

of the ipsilateral geniculate nucleus in the thalamus. The majority of the ascending fibers terminate in an orderly arrangement at the level of the large nucleus rotundus thalami and the adjacent nucleus triangularis, which caps the former. Benowitz and Karten (1976) demonstrated the retinotopic arrangement within this ascending visual projection. Gamlin and Cohen (1986) reported about a second ascending visual pathway from the optic tectum to the nucleus dorsolateralis posterior of the thalamus in the pigeon. This thalamic nucleus projects to a discrete region of the ipsilateral neostriatum. Reiner and Karten (1982) showed that descending tectal projections arising from neurons located in layers 8–15 are arranged in a laminar pattern, and that only the perikarya located in layers 10, 14 and 15 give rise to ascending projections. Major ascending and descending tectofugal fiber systems in birds and mammals cannot be properly compared because in mammals the tectofugal ascending and descending fibers do not arise from separate but from the same layers.

The clearcut separation of tectal laminae that give rise to ascending or descending projections was not described in the earlier reports by Hunt and Künzle (1976) on the pigeon. In this context it is still an open question whether the lamination of the optic tectum in birds has led to a higher degree of differentiation or, within the framework of Ebbesson's theory (1980), to a higher degree of parcellation. Techniques that combine intracellular electrophysiological recordings with subsequent morphological analysis of cells filled with tracers after recording, will help to answer this question (for findings along this line in the pigeon, see Hardy *et al.*, 1985).

Perikarya located between layers 9 and 10 send retinotopically arranged axons to the isthmo-optic nucleus. This nucleus is located in the dorsal portion of the rostral mesencephalon. In cross sections the nucleus frequently appears convoluted. In pigeons approximately 12,000 perikarya of this nucleus send fibers to the contralateral retina to terminate on amacrine and displaced retinal ganglion cells (Cowan and Powell, 1963; for the appearance of isthmo-optic neurons in Golgi-impregnated material, see Güntürkün, 1987; for a review, see Emmerton, 1983b; Hayes and Holden, 1983; for findings in chickens, see Crossland and Hughes, 1978). Furthermore, it has been suggested that some perikarya in the isthmo-optic nucleus receive inputs from oculomotor and trochlear nuclei (Angaut and Repérant, 1978). In this context it might be interesting to note that isthmo-optic neurons in ketamin hydrochloride anesthetized Japanese Quail can be activated (extracellular recordings) with an average latency of 17 ms following electrical stimulation of the visual Wulst. Efferent perikarya of the isthmo-optic nucleus undergo spontaneous cell death during ontogenesis if they fail to reach target sites in the contralateral retina (Clarke, 1982a,b; McLoon and Lund, 1982; O'Leary and Cowan, 1982; Catsicas *et al.*, 1987b; Catsicas and Clarke, 1987) or if tectal afferents are prevented from reaching isthmo-optic perikarya (Clarke, 1985).

In addition to the ipsilaterally ascending major visual pathway from the tectum mesencephali that terminates mainly in the thalamic nucleus rotundus, a small contralateral thalamic projection to the nucleus rotundus has been described (Benowitz and Karten, 1976; Hunt and Künzle, 1976). Ascending fiber projections can be followed further to enter the ectostriatum and to pass from ectostriatal neurons to the peri-ectostriatal belt of the telencephalon. Thalamo-telencephalic visual projections make up a considerable portion of the lateral forebrain bundle. Projections from the nucleus rotundus to the ectostriatum are arranged in a retinotopic manner, whereas projections from the triangular nucleus to the peri-ectostriatal belt seem to be diffusely orientated. For a detailed review, see Henke (1983); for details on the complex afferent and efferent connections of the nucleus rotundus with other brain areas, see Benowitz and Karten (1976); Hunt and Künzle (1976).

## C. THALAMOFUGAL PATHWAY (FIGS 24 AND 26)

Retinal fibers, most probably from small-diameter perikarya of the ganglion nervi optici, cross the midline to terminate in a contralateral nuclear complex, the nucleus opticus principalis thalami, which may be subdivided into distinct subnuclei (for a detailed review, see Hirschberger, 1971; Henke, 1983). At least some of the retinal fibers have been shown to be arranged in retinotopic order. In addition, a small ipsilateral thalamofugal projection has been suggested to exist in pigeons (for evidence of uncrossed thalamofugal fibers in hatchling chicks, see O'Leary *et al.*, 1983; Crossland, 1985).

The thalamic relay nucleus (nucleus opticus principalis) mediates visual input to telencephalic regions—hyperstriatum accessorium, hyperstriatum dorsale, hyperstriatum intercalatum accessorius, and nucleus intercalatus hyperstriati accessorii, or, in simpler terminology, the visual Wulst (cf. p. 63). Recently Boxer and Stanford (1985) employed retrograde axonal transport of horseradish peroxidase in 8-day-old chicks and were able to describe a projection from thalamic nuclei to the posterior hyperstriatum which had hitherto not been regarded as having visual function. The size and development of cytoarchitectonic features in the visual Wulst vary considerably among taxonomic groups. There seems to be a positive correlation between the size of the visual Wulst and the size of the binocular visual field (see, in this context, Fig. 15). Thus, differences in the size of the visual Wulst probably reflect variations in the development of stereopsis. Finally, it is important to note that the projections from the thalamic relay station to the visual Wulst are thought to be bilaterally organized. Ehrlich and Stuchbery (1986), however, found no evidence for a contralateral telencephalic projection from the lateral anterior thalamic nucleus in 2–4-week-old Black Australorp chicks when they analyzed the retrograde transport of wheat germ agglutinin after having injected large amounts of it into the hyperstriatum.

Thalamofugal and tectofugal pathways in the avian visual system should not be considered as separate entities. They are morphologically interconnected at various sites in the CNS and their functional activities can be effectively mapped by using the $^{14}C$-2-deoxiglucose method (for findings in conscious adult pigeons, see Streit *et al.*, 1980). Interconnections between thalamofugal and tectofugal pathways, which have been discovered to be efferent projections from telencephalic visual centers, are summarized in Fig. 27. In this context the efferent projections of the visual Wulst to the ipsi- and to the contralateral optic tectum, as well as to the ipsilateral peri-ectostriatal belt deserve special attention (for further details and a review, see Henke, 1983). A recently detected third primary visual center in the caudolateral telencephalon of the pigeon (Güntürkün, 1984) will probably help to understand better further peculiarities in the visual system of birds. In this connection it should also be mentioned that the visual Wulst is rich in catecholamine-containing fibers and terminals (for details, see Bagnoli and Casini, 1985). In this respect it is similar to the striatal cortex of mammals.

### D. ACCESSORY OPTIC FIBER SYSTEM (FIGS 28 AND 29)

Remarkably thick axons within the accessory optic system were shown by Karten and co-workers (cf. Karten *et al.*, 1977; Brecha *et al.*, 1980) to arise from so-called displaced retinal ganglion cells and to project, via the basal optic root, to the contralateral nucleus of the basal optic root or nucleus ectomammillaris of the mesencephalon. Fibers that terminate as mossy fibers can be traced from these mesencephalic centers to the paraflocculus and folia IX c/d in the cerebellum. Other fibers that arise from different subunits of the ectomammillary nucleus terminate bilaterally in the nucleus interstitialis (Cajal) and in distinct compartments of the ipsi- and contralateral complex of the oculomotor nucleus. Britto *et al.* (1988) recently showed in adult pigeons that approximately 400 displaced retinal ganglion cells that project to the nucleus of the basal optic root (retrograde transport of rhodamine-labeled microspheres) react with antibodies raised against tyrosine-hydroxylase. This subpopulation of displaced retinal ganglion cells is thought to express a cathecholamine as neurotransmitter.

In addition to the above-mentioned visual projections, other retinal projections have been found which may in part serve to control autonomic functions (e.g., control of pupil diameter or control of light-dependent endocrine reflexes). Retinal fibers have been shown to terminate in the area praetectalis, the lateral geniculate nucleus, the lateral thalamic nucleus, the lentiform nucleus of the mesencephalon, and several other diencephalic and mesencephalic nuclei. (For a detailed review, see Henke, 1983; cf. in this context also Hartwig, 1980; Peduzzi and Crossland, 1983a,b; Gamlin and Cohen, 1988a,b; for the significance of retino-hypothalamic projections, see p. 34.)

## VIII. Olfactory System

Central nervous components that are involved with olfaction seem to be poorly developed in most species of birds (exception: *Apteryx*; see Fig. 13; for further details, see Starck, 1982). A functioning olfactory system has been described in different species of vulture (Bang, 1964). In addition, experimental findings in homing pigeons indicate that olfaction may be important for orientation (Papi *et al.*, 1972; cf. Wallraff, 1979; for a review, see Keeton, 1979). Up to a decade ago it was thought that only birds living primarily in aquatic environments possess well-developed olfactory abilities. Morphological observations on the olfactory apparatus of eight species of procellariform birds indicate that olfaction may be a vital sense in this group (Bang, 1966). In *Larus argentatus* Drenckhahn (1970) described well-developed olfactory epithelium and a well-developed olfactory nerve. Ultrastructural details of the olfactory epithelium in this species are similar to those in mammals.

In four different breeds of domestic pigeons including homing breeds, Müller *et al.* (1979) calculated the number of olfactory receptor cells to vary between 3 and $7.4 \times 10^6$. The number of olfactory receptor cells correlated well with body size and there were no differences in the numbers of olfactory receptors between the breeds independent of body size. As in mammals, olfactory cells are continuously renewed (for findings in pigeons, see Bedini *et al.*, 1976). Central components of the olfactory system in birds apparently have not yet been studied in detail by morphological techniques. Rebiere and Dainat (1981) investigated embryonic synaptogenesis in the olfactory bulb of Domestic Mallards. (For experimental evidence of sensitivity to odors in chick embryos, see Tolhurst and Vince, 1976.)

Ontogenetic synaptogenesis in the olfactory bulb seems to be completed in Domestic Mallards when the birds hatch. In principle, ultrastructural details of the olfactory bulb in Domestic Mallards seem to be comparable to those of mammals. It is to be anticipated that future investigations will reveal greater similarities between the olfactory systems of birds and mammals than were hitherto thought to exist. Such similarities may include the control of autonomic functions (e.g. reproduction; for the possible role of pheromones, see Balthazart and Schoffeniels, 1979) by central nervous centers closely related to the olfactory system. Rieke and Wenzel (1978) used electrophysiological and anatomical techniques (Fink–Heimer method) in adult pigeons to reveal ipsilateral olfactory projections from the olfactory lobe to the cortex piriformis (olfactory cortex), the hyperstriatum ventrale, and to the paraolfactory lobe. These studies also revealed a crossed projection that reaches the contralateral paleostriatum primitivum and the caudal part of the contralateral paraolfactory lobe via the anterior commissure.

## IX. Central Nervous Components of Autonomic Systems

### A. GENERAL ASPECTS

Birds successfully maintain homeostasis and reproduce under nearly all climatic conditions. Central as well as peripheral components of their autonomic nervous system have attained a degree of specialization, equivalent to that of mammals (cf. in this context Bennett, 1974). Patterned arrangements of different types of neurons, frequently those producing neuropeptides and monoamines, are the specific effectors of these autonomic components. In addition, the specific effectors serve as target sites for multiple nervous, humoral and neuroendocrine projections (Oksche and Hartwig, 1980; for a review on the peripheral autonomic nervous system, see Bennett, 1974).

In contrast to the manifold differences in somatosensory reflex systems, the components of the CNS that are devoted to the control of autonomic functions seem to be more uniform among different taxonomic forms of birds. Only a limited number of species, however, have been thoroughly investigated. Moreover, primarily hypothalamic centers of the autonomic system have received attention. Very little information is available on autonomic neuronal systems at the level of the brain stem or the spinal cord. Recently detected patterns of distribution of various neuropeptides and monoamines in the brain stem and the spinal cord indicate that the general morphological patterns of these "autonomic" structures may be quite similar in birds and mammals. In this context it should be noted that derivatives of alar plate material are especially rich in neuronal systems that produce monoamines and neuropeptides. For further details, consult Hartwig (1984); for a critical review of the sometimes astounding differences between the distributional patterns of monoamine-containing neuronal systems revealed by histofluorescence and neurotransmitter immunocytochemistry and the patterns revealed by tyrosine hydroxylase immunocytochemistry, see Kiss and Péczely (1987).

When considering the components of the CNS devoted to the control of autonomic functions, one must be aware of different patterns of interneuronal communication. Precisely defined point-to-point projections (somatotopic arrangement) make up the vast majority of neuronal circuits in somatosensory reflex systems. In the central nervous components devoted to the control of autonomic functions, however, two principally different patterns of interneuronal communication play crucial roles. These patterns are:

(1) Neurons projecting with ascending and descending axon collaterals to a variety of central nervous compartments; their axon terminals form swellings.

The latter frequently lack precisely defined postsynaptic sites. The transmitter is apparently released into the intercellular space in general and not into a clearly defined synaptic cleft. Aminergic perikarya in the brain stem frequently belong to this type of neuron. (For details and references, see Hartwig, 1981; Chikazawa *et al.*, 1983; Hartwig and Reinhold, 1983; for the locations of aminergic neurons, see p. 83.)

(2) Neurons releasing their transmitter substances into specialized transport channels or into the general circulatory system. This latter pattern of intercellular communication channels characterizes the type of neuroendocrine systems originally described by Scharrer and Bargmann (for references, see Leonhardt *et al.*, 1983; Hartwig and Reinhold, 1983; Scharrer, 1987; Scharrer *et al.*, 1987).

The structural principles of neuroendocrine systems have been studied extensively in a number of bird species (cf. Fig. 30; for detailed reviews, see Kobayashi and Wada, 1973; Oksche and Farner, 1974; Oksche, 1977). Finally, it should be mentioned that monoamine- and neuropeptide-containing neuronal systems appear to be phylogenetically ancient. Ontogenetically, monoamines can be found in primordial nervous tissue prior to the closure of the neural tube. Monoamines and peptides may regulate early embryonic development (for details in chickens, see Wallace, 1982). In the course of ontogeny monoamine and peptide transmitters seem to be of a transitory nature in a variety of central nervous structures (for a review and references, see Hartwig, 1984).

Numerous autonomic functions in birds exhibit rhythmic fluctuations of endogenous nature (cf. Gwinner, 1975; Farner, 1975, 1983, 1986; Farner *et al.*, 1977; for general aspects of circadian and, in some cases, apparently circannual rhythms, see Aschoff *et al.*, 1982). With as yet poorly understood receptor systems birds are able to detect subtle changes in their environment (e.g. changes in humidity or in the Earth's magnetic field) that influence their autonomic functions profoundly. In most species of birds so far investigated, seasonal changes in daylength play a crucial role as reliable entraining agents (*Zeitgeber*; Aschoff). A Zeitgeber induces autonomic functions or synchronizes endogenous rhythms. Changes in daylength also synchronize autonomic rhythms with periodic changes in the environment.

The information contained in changes of daylength is transduced into neuroendocrine signals (photoneuroendocrine systems; Scharrer, 1964). For details on the effects of light on reproduction in Domestic Mallards, initially detected by Benoit over 50 years ago, see Benoit, 1935, 1938, 1978; cf. also van Tienhoven and Planck, 1973; Glass and Lauber, 1981; for detailed reviews, see Farner 1975, 1983, 1986; Farner *et al.*, 1977. In contradistinction to mammals, but similar to other groups of vertebrates, birds generally possess enigmatic extraretinal photoreceptors that are most probably located in the tuberal hypothalamus (for detailed

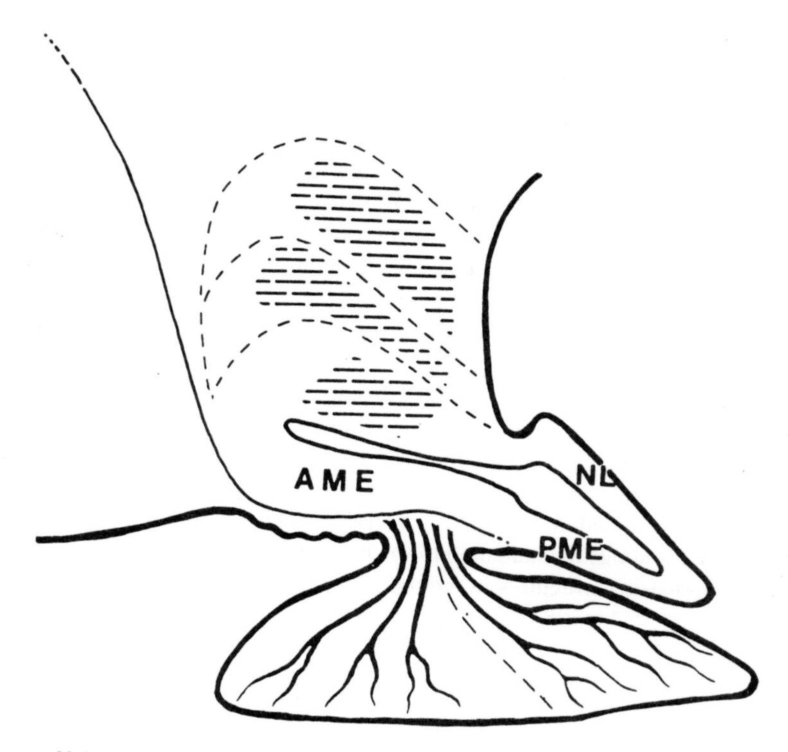


FIG. 30. Midsagittal section showing tuberal hypothalamus in a passerine bird with subunits of tuberal nuclei according to Oehmke (after Oksche and Farner, 1974). The mediobasal hypothalamus is the structural correlate of the final neuroendocrine pathway of the autonomic system. AME, anterior median eminence; NL, neural lobe of hypophysis; PME, posterior median eminence; note point-to-point projection of portal capillaries running from median eminence to adenohypophysis.

studies in the White-crowned Sparrow, *Zonotrichia leucophrys gambelii*, see Yokoyama *et al.*, 1978; for general comparisons and information about the penetration of light into the brain, see Hartwig and Oksche, 1982; Hartwig, 1982). Foster *et al.* (1985) recently showed in Japanese Quail that the sensitivity of this enigmatic receptor system resembles that of the photopigment rhodopsin with peak sensitivity in the range of approximately 492 nm. The level of overall sensitivity was comparable to that of vertebrate visual receptors. The enigmatic hypothalamic photoreceptors were unfortunately not revealed by immunocytochemical markers specifically labeling retinal and pineal photoreceptor systems in Japanese Quail (antibodies raised against opsin, retinal S-antigen, and alpha-transducin) (Foster *et al.*, 1987a). Furthermore, it has been shown that synthesis of indoleamines by cultured pineal organs from several species of birds can be directly influenced by light indicating that pinealocytes are photosensitive. For a review, see Binkley (1980); for the role of the pineal gland in circadian systems, see Menaker and Binkley

(1981). In this context it is interesting to note that Vigh *et al.* (1982) observed opsin-like immunoreactivity in the pineal glands of birds. For a review of photoreceptor differentiation and neuronal organization of the pineal organ from a comparative point of view, see Collin *et al.* (1987) and Korf and Ekström (1987). For information on light-dependent electrical responses in the pineal organ of homeothermic vertebrates, see Dodt (1987). There is experimental evidence indicating that dispersed cultured pinealocytes of 1–2-day-old White Leghorn chicks exhibit an endogenous component in the rhythmicity of melatonin production (for recent findings and review, see Zatz *et al.*, 1988). Light-dependent rhythms in melatonin content apparently do not disappear completely in eyeless (enucleated) and pinealectomized Japanese Quail (Underwood *et al.*, 1984; for the possible role of the suprachiasmatic nuclei in the circadian system of House Sparrows, see Takahashi and Menaker, 1982). On the other hand, Cogburn *et al.* (1987), who investigated young chickens, obtained results different from those of Underwood *et al.* (1984) in pinealectomized animals. It may be an important factor that the more recent study examined animals at 4-h intervals instead of only twice a day. These experimental findings indicate the presence of a multiplicity of photoreceptors and melatonin-releasing structures in the brains of birds.

## B. HYPOTHALAMUS AND PINEAL GLAND

The following section deals with two major components of the CNS that have been shown to play crucial roles in the control of rhythmic autonomic functions: the hypothalamus and the pineal gland. Both are parts of the diencephalon and are known to be rich in various peptides and biogenic monoamines. This section summarizes recent findings pertaining to these messenger molecules.

Although a great many morphological and experimental studies have dealt with the hypothalamus, the apparent superior integrative center of the autonomic nervous system, we are still far from understanding the distributional patterns of specifically identified neurons within the complex structure of this part of the brain. Knowledge about the complex "wiring" diagram of hypothalamic subunits has recently been greatly extended through the results of elegant experiments in which circumscribed injections of tracer substances were applied after electrophysiological recordings had been made in conscious Domestic Mallards (Korf *et al.*, 1982a). Numerous aspects about the functional interactions between hypothalamic neurons, however, still remain unclear (for a review, see Oksche, 1980a,b; cf. in this context also Panzica *et al.*, 1982; Korf *et al.*, 1983; Panzica and Viglietti-Panzica, 1983; Viglietti-Panzica and Contenti, 1983). Oksche convincingly described the phylogenetic development of cluster-like arrangements of hypothalamic neurons and the well-organized laminar patterns of their afferent and efferent projections (for details and a review, see Oksche, 1978a,b; 1980a,b). The afferent and efferent neuronal projections of the avian hypothalamus have been only poorly investigated when compared with those of mammals. However, similar to the hypothalamus in mammals, that of birds seems to be reciprocally interconnected with limbic structures or brain areas probably corresponding to components of the limbic system (for observations in pigeons and quail, see Bouillé *et al.*, 1977; Oliver *et al.*, 1977). By and large, the general patterns of efferent and afferent projections of the hypothalamus seem to be rather similar in birds and mammals. Nevertheless, specializations should not be neglected (e.g., somatosensory projections in birds; cf. p. 43). The nomenclature and location of hypothalamic nuclei (cf. Fig. 30) and associated circumventricular organs in birds have been dealt with from a primarily descriptive anatomical point of view by Kuenzel and van Tienhoven (1982). For Golgi studies and a review of different types of hypothalamic neurons in birds as revealed in silver-impregnated material, see Franzoni *et al.* (1984).

In birds the descending and ascending fiber projections from the hypothalamus have unfortunately been less intensively studied than the topography of the nuclei. (For findings in pigeons, see Cabot *et al.*, 1982; Berk and Finkelstein, 1983.) From a comparative anatomical point of view, the patterns of arrangement of the projections might elucidate functional relationships. Descending fibers arise primarily from phylogenetically ancient, periventricularly located magnocellular neurons. Similar to the apparent general morphological patterns of descending hypothalamic projections in vertebrates, these fibers terminate in autonomic regions of the brain stem (lateral border of the locus coeruleus, certain nuclei in the solitary tract and dorsal motor nucleus of the vagus nerve). They descend further within the lateral funiculus of the spinal cord but details about where they terminate remain to be elucidated (Berk and Finkelstein, 1983).

Finally, hypothalamic nuclei possess specific receptors for various messenger molecules of the nervous and endocrine systems (e.g., feedback loop mechanisms in the control of the gonads; for review and findings in the hen, see Sakurai *et al.*, 1986). Receptor systems engaged in the control of water metabolism have recently been extensively studied in pigeons (Thornton, 1986) and in Domestic Mallards (Gerstberger *et al.*, 1987; for a detailed analysis and a review, see Simon *et al.*, 1987).

In the following brief survey attention is focused on specifically identified autonomic neurons at the level of the hypothalamus which contain neuropeptides and monoamines. These neurons have been identified by immunocytochemical techniques. When considering immunocytochemical observations, one must be aware of certain technical problems which might lead to false negative or positive findings (e.g., unwanted cross reactions; for a critical recent review of control experiments, see Flucher, 1984). Furthermore, before immunocytochemical results can be properly interpreted, one must also consider by what techniques the antibodies were raised (e.g., monoclonal or polyclonal antibodies, species-specific or non-specific antigen). Despite these limitations, immunocytochemistry has increased tremendously our knowledge of the patterned arrangements of hypothalamic and extrahypothalamic neuronal systems that probably represent functional subunits of the CNS.

*Vasotocin and mesotocin.* Avian vasotocin and mesotocin, which cross-react with antibodies raised against mammalian vasopressin and oxytocin (Goossens *et al.*, 1977), can be detected in separate neurons of the supraoptic and paraventricular nuclei (Goossens *et al.*, 1977; Bons, 1980; Weindl *et al.*, 1980; Berk *et al.*, 1982; for the ontogeny of these nuclei in chickens, see Großmann and Ellendorf, 1987). Comparable to the morphological pattern known in mammals, vasotocin neurons occur predominantly in the rostral part of the supraoptic nucleus, whereas perikarya of the paraventricular nucleus are especially rich in mesotocin (Fig. 31).

The presence of mesotocin-like immunoreactivity in perikarya in the tuberal

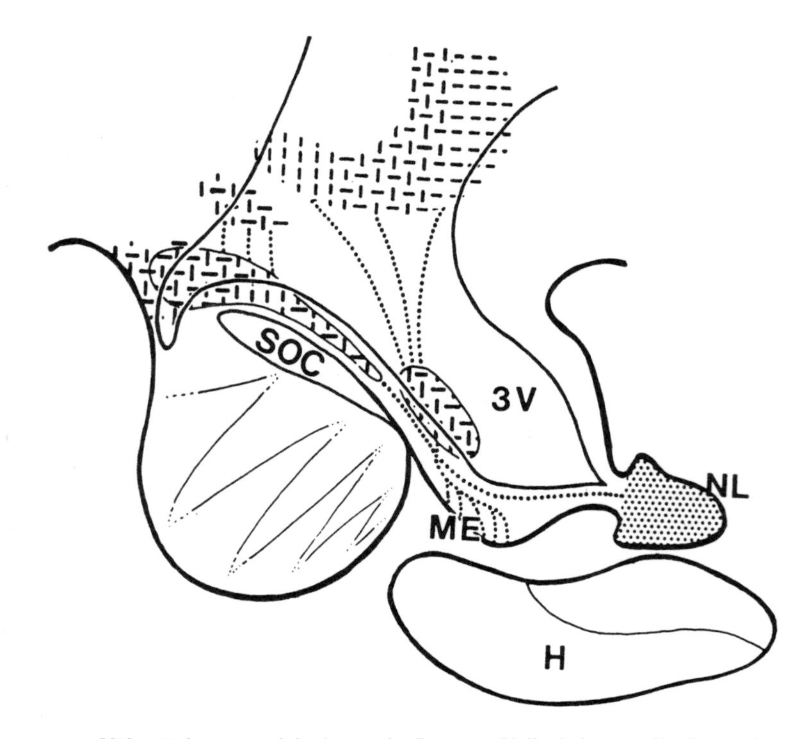


FIG. 31. Midsagittal section of the brain of a Domestic Mallard showing distributional patterns of vasotocin (perpendicular shading) and mesotocin (horizontal shading) immunoreactive perikarya and pathways (.......) to anterior median eminence (ME) and neural lobe (NL) of the hypophysis (after Bons, 1980). H, adenohypophysis; SOC, supraoptic commissure; 3V, third ventricle. A tuberomammillary area containing perikarya immunoreactive with antibodies raised against mesotocin, as observed by Bons (1980) has not been described in other species.

hypothalamus (Fig. 31) has been reported only by Bons (1980). Vasotocin- and mesotocin-containing perikarya project to the neural lobe of the hypophysis via axons arranged in an orderly fashion that course through the internal layer of the median eminence. In addition, the external layer of the rostral median eminence contains vasotocin-immunoreactive elements and to a much lesser degree mesotocin immunoreactivity. It is interesting to note that at the ultrastructural level fibers in the external layer of the median eminence are characterized by elementary granules of 130–180 nm in diameter whereas fibers in the internal layer contain dense-core vesicles with diameters between 200 and 250 nm. Furthermore, vasotocin-containing axons have been observed to be directed toward the telencephalon and toward certain layers of the optic tectum (for references and discussion, see Oksche and Hartwig, 1980). It remains to be elucidated whether in birds, as in mammals, vasotocin and mesotocin occur in association with different types of enkephalins. The amazing complexity of hypothalamic regulatory mechanisms has been revealed in the findings of Voorhuis *et al.* (1988) in adult canaries: vasotocin-immunoreactivity differs in males and females at the level of the dorsal diencephalon and in the lateral septum. Testosterone administration restored the substantial decrease in immunoreactivity brought about by castration. Finally it should be noted that the classical terminology for neurosecretory nuclei cannot be adopted for birds without certain modifications. The neuronal typology of vasotocin-immunoreactive systems can only be revealed in combined immunocytochemical and Golgi-impregnation studies (cf. Viglietti-Panzica and Panzica, 1987).

*Adrenocorticotropic Hormone (ACTH).* ACTH immunoreactivity has not yet been extensively studied in avian brains. Blähser (1988) raised an antibody against synthetic human ACTH 1–24 that did not cross-react with peptides of the proopiomelanocortin family and neurotensin. In male White-crowned Sparrows (*Zonotrichia leucophrys gambelii*) labeled perikarya were located primarily in the tuberal hypothalamic region and projected predominantly in dorsal directions into the striatum, into rostral diencephalic, septal, and thalamic areas, and into dorsal and ventral regions of the brain stem. Neurohemal zones were not supplied by ACTH-immunoreactive terminals.

*Corticotropin releasing factor (CRF).* Péczely and Antoni (1984) demonstrated the presence of perikarya and their processes that cross-react with antibodies raised against ovine CRF. The neurons were found in hypothalamic areas of the pigeon that are known to project especially to the external layer of the anterior median eminence. These neurophysiological findings indicate that CRF secretion influences anterior pituitary function. The investigation showed that the distributional patterns of CRF and neurophysin immunoreactivity are astonishingly similar.

*Luteinizing-hormone releasing hormone (LH-RH)*; Fig. 32. Perikarya exhibiting immunoreactivity to antibodies raised against synthetic or against mammalian LH-RH are found predominantly in the periventricular layer of the anterior or preoptic

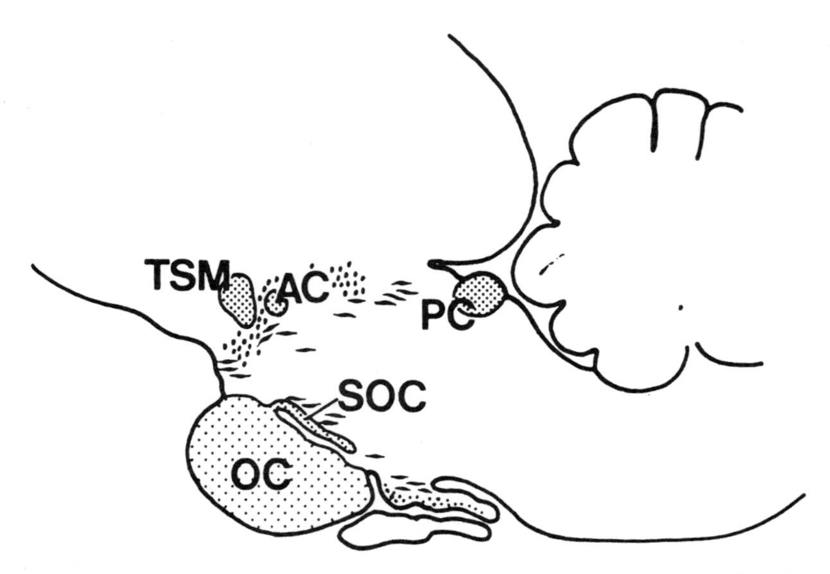


FIG. 32. Parasagittal section of diencephalon and adjacent brain regions of a domestic hen showing distributional patterns of LH-RH-immunoreactive perikarya (large dots), fibers (~~), and terminals (small dots) at hypothalamic neurohemal projection sites. Extrahypothalamic projection sites not indicated. AC, anterior commissure; PC, posterior commissure; OC, optic chiasma. SOC, supraoptic commissure; TSM, tractus septomesencephalicus. Simplified after Sterling and Sharp (1982).

area of the hypothalamus. In some taxonomic forms of birds, scattered immunoreactive perikarya have also been reported to occur in the paraolfactory lobe of the telencephalon (Bons *et al.*, 1978a,b; Józsa and Mess, 1982; Sterling and Sharp, 1982). Bons *et al.* (1978a) reported the existence of LH-RH-immunoreactive perikarya in the tuberal hypothalamus of Domestic Mallards. Lesioning experiments indicate the importance of both preoptic and tuberal hypothalamic areas in the control of gonadal function. The hierarchy and functional interaction between different gonadotropic centers remains to be elucidated (for review, see Yokoyama *et al.*,1978; Sterling and Sharp, 1982). The functional significance of immunoreactive fibers that project to the organum vasculosum of the lamina terminalis and of fibers that terminate in extrahypothalamic target sites also remains enigmatic. It should be noted that distributional patterns of LH-RH-immunoreactive perikarya and terminals indicate LH-RH has a bifunctional role as a neurohormone at neurohemal target sites and probably as a neurotransmitter at e.g., the olfactory bulb, in septal regions, and in the mesencephalic nucleus of the trigeminal nerve (for details, see Blähser, 1984). Foster *et al.* (1987b) reviewed data on the influences of photoperiods on LH-RH mechanisms. They found a striking reduction in the number of LH-RH-immunoreactive structures in male European Starlings that had been exposed to short photoperiods. Long photoperiods enhanced immunostaining. These authors state "The higher neural pathways regulating photorefractoriness induced by long days are unknown but clearly both production of LH-RH in the perikarya and release/storage of LH-RH in the terminals is being profoundly modified". In this context it should be noted that there are hormonal feedback loops between the hypothalamus and the gonads in both mammals and birds (for recent findings of nuclear progesterone receptors in the hypothalamus and forebrain of the domestic hen, see Sterling *et al.*, 1987). Some of the nuclei engaged in the control of gonadal functions exhibit clear cut morphological differences between the sexes (for a review and recent findings in Japanese Quails, see Panzica *et al.*, 1987). The complexity of hypothalamic regulatory mechanisms controlling gonadal functions (e.g., influence of serotonin and β-endorphin on the release of LH-RH) has been revealed by Sakurai *et al.* (1986) in their work on the hen.

*Somatostatin (SRIF).* SRIF seems to be biochemically similar in mammals and birds. Although SRIF-immunoreactive material was detected as early as 1974 in the external layer of the median eminence in chickens by Dubois and co-workers (cf. Dubois, 1976), the functional significance of this peptide is still a matter of controversy. Distributional patterns of immunoreactive perikarya at the level of the hypothalamus as well as at extrahypothalamic sites (telencephalon and lower brain stem) have been analyzed in Domestic Mallards by Blähser *et al.* (1978) and in parakeets by Takatsuki *et al.* (1981). Perikarya have been observed predominantly in phylogenetically ancient periventricular areas of the preoptic hypothalamus, whereas immunoreactive fibers and terminals are apparently widespread within the CNS of birds (for details and discussion, see Blähser, 1981; see also Shiosaka *et al.*, 1981).

*Met-enkephalin.* A similar widespread distribution has been reported for fibers and terminals immunoreactive to antibodies raised against met-enkephalin (Blähser and Dubois, 1980; Lanerolle *et al.*, 1981). Interestingly enough, enkephalin-like immunoreactivity can be found in close association with regions involved in vocal control in Zebra Finches of both sexes (Ryan *et al.*, 1981). It remains to be demonstrated whether met-enkephalin is associated with biogenic monoamines or with other neuropeptides.

*Thyrotropin-releasing hormone (TRH).* Using a commercially available TRH-antiserum and immunocytochemical methods, Thommes *et al.* (1985) investigated ontogenetic patterns of reactive neuronal elements in chick embryos. Between the 4th and the 5th day of incubation they observed immunoreactive structures in the *anlage* of the infundibulum. There was no abrupt change in immunoreactivity during the critical period of incubation (days 10.5–13.5), a period during which it had been established in earlier experiments that maturation of the pituitary–thyroid axis takes place. Józsa *et al.* (1988) investigated distributional patterns of TRH-immunoreactive structures in the brains of Domestic Mallards. They conclude that TRH may be involved in hypophysiotropic regulatory mechanisms and, in addition, may act as a neuromodulator or neurotransmitter in other areas of the brain lacking direct contact with the hemal milieu.

*Neurotensin.* Neurotensin, a neuropeptide of enigmatic functional significance, has been detected immunocytochemically in the hypothalamus of Japanese Quail (Yamada and Mikami, 1981). Immunoreactive perikarya occur predominantly in the medial preoptic nucleus; fiber tracts project primarily to the ventral septum and to the subfornical organ.

*Vasoactive intestinal polypeptide (VIP).* Immunoreactive neurons have been observed in the caudal portion of the hypothalamic infundibular (tuberal) nucleus and in the lateral mammillary nucleus of Japanese Quail (Yamada *et al.*, 1982). Only scattered immunoreactive perikarya occur in the preoptic area. Fibers apparently project to the external layer of the median eminence. Korf and Fahrenkrug (1984) recently detected a system of cerebrospinal fluid-contacting neurons that react with antibodies raised against VIP. They were located in the lateral ventricles of Domestic Mallards in an area adjacent to the nucleus accumbens and the basal pole of the lateral septum. In the spinal cords of embryonic and newly hatched chicks, VIP-immunoreactive perikarya have been detected in the dorsal horns (lamina I) and in the nucleus of the dorsolateral funiculus as well as in areas lateral to the central canal (thoracic levels only). Immunoreactive terminals were scattered primarily in the intermediate zone, especially around the central canal (Du *et al.*, 1988). In pigeons VIP- and TRH-immunoreactive structures are considerably altered after lactation or exposure to cold (Péczely and Kiss, 1988; for a review of control of prolactin secretion in birds, see Hall *et al.*, 1986).

*Substance P (SP).* In mammals this neuropeptide is known to be associated predominantly with peripheral neuronal systems that mediate pain. It has also been detected in extracts of pigeon brain (Reubi and Jessel, 1978). Immunoreactive fibers and terminals have been observed in the external layer of the median eminence as well as in circumscribed areas of neuropil in the reticular formation of the lower brain stem in Domestic Mallards and pigeons (Hartwig *et al.*, 1981). Detailed reports on the ontogeny of SP-like immunoreactivity in sensory ganglia of the lower brain stem and in the spinal cords of chickens and Japanese Quail were published by Fontaine-Perus *et al.* (1985). Lavalley and Ho (1983) observed the highest density of SP-immunoreactive fibers in laminae I and II in the spinal cords of chickens. These fibers were accompanied by structures containing SRIF and met-enkephalin immunoreactivity that were arranged in similar patterns of distribution. In lesion experiments conducted in domestic pigeons, Davis and Cabot (1984) collected evidence for major intraspinal networks of SP-immunoreactive fibers. At the level of the spinal cord, amine- and peptide-containing fibers can also be found in areas surrounding the central canal and, to a much lesser extent, in the ventral column. In the ventral column they apparently associate with the perikarya of motoneurons.

The transitory appearance of some of the above-mentioned neuropeptides during ontogeny may indicate that these substances play a role in the differentiation of the CNS (for studies in Domestic Mallards, chickens, and Japanese Quail, see Blähser and Heinrichs, 1982). This suggestion is supported by data on the expression of SRIF, met-enkephalin, and SP-like peptides in neurons that were isolated from cerebral hemispheres of 8-day-old chick embryos and maintained in serum-free culture (Louis *et al.*, 1983). TRH, LH-RH, and β-endorphin were not detected in that study.

Immunocytochemical techniques are providing more and more information on the distributional patterns of specific substances with as yet poorly understood functions. Here again possible cross-reactivities of peptides must be considered before accurate interpretations can be made. Furthermore, it remains to be determined whether certain peptides occur in association with other biologically active substances such as biogenic monoamines or endogenous opiates. Finally, the possibility that certain nervous structures react with antibodies as a consequence of the (specific?) uptake of immunoreactive material must be excluded (e.g., serotonin-immunoreactivity in sympathetic nerve fibers in the pineal gland).

The functional significance of immunoreactive fibers that project to neurohemal target sites in the median eminence and the neural lobe of the hypophysis is probably better understood than that of the fibers projecting to extrahypothalamic target sites. (See in this context a review on the structure and the functions of the pars tuberalis of the adenohypophysis by Fitzgerald, 1979.) Future investigations treating neuropeptides in birds must consider the distributional patterns of specific receptor sites as well as the distributional patterns and possible functional roles of (specific?) enzymes that degrade neuropeptides into smaller subunits (metabolites). Such metabolites may be biologically active but in a different way from that of the original substance.

*Biogenic monoamines.* This class of neurotransmitters has not yet been thoroughly examined in birds. The highly sensitive immunocytochemical techniques that have been developed for demonstrating monoamines and the enzymes that synthesize them have still to be applied in birds. For the immunocytochemical localization of serotonin in the paraventricular organ (PVO) of the chicken, see Sano *et al.* (1983); for recent detailed studies on the serotonin systems in the spinal cords and brains of embryos and post-hatchling chicks, see Sako *et al.* (1986a,b); Kojima *et al.* (1988). Thus the following brief description of distributional patterns of hypothalamic and extrahypothalamic monoamines is based on the less sensitive fluorescence-histochemical technique (for a more detailed review, see Oksche and Farner, 1974; Hartwig, 1981). General aspects of distributional patterns of hypothalamic and extrahypothalamic monoamine-containing perikarya and fiber tracts in the central nervous system of birds are summarized in Figs 33 and 34. When discussing the possible functional significance of distributional patterns of monoamines, receptor sites and turnover rates must be considered. For the distribution of dopamine receptors in pigeons, see Richfield *et al.* (1987); for the regulation of

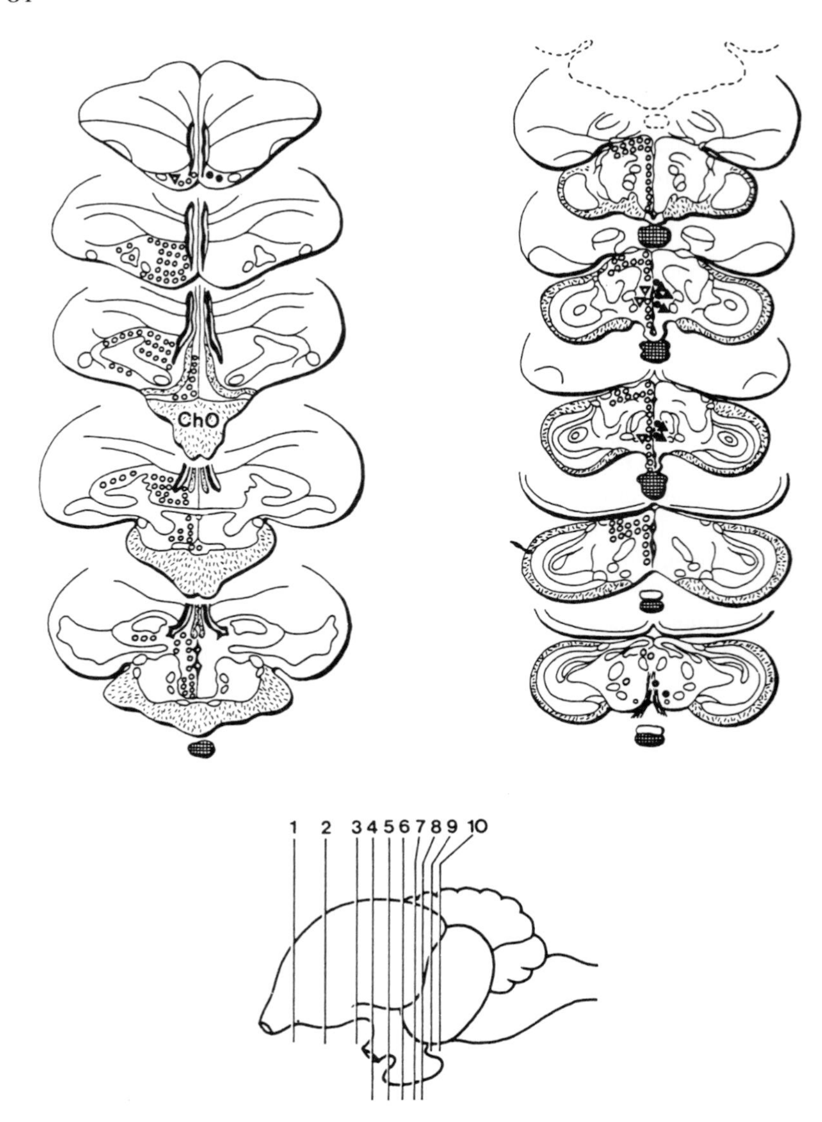


FIG. 33. Sequence of rostrocaudally arranged transverse sections through prosencephalon and mesencephalon of the chicken. Section levels indicated in sagittal view below, beginning with section 1 at top left. In the left halves of the sections the distributional patterns of fibers and terminals for catecholamine (○) and serotonin or 5-OH-tryptophan (▽) fluorescence are indicated. The right halves of the sections show distributional patterns of the perikarya: catecholamines (○), and serotonin or 5-OH-tryptophan (▲). Results were obtained following classical formaldehyde-induced monoamine fluorescence which is characterized by a limited sensitivity when compared to immunocytochemical techniques (after Muhibullah *et al.*, 1983).

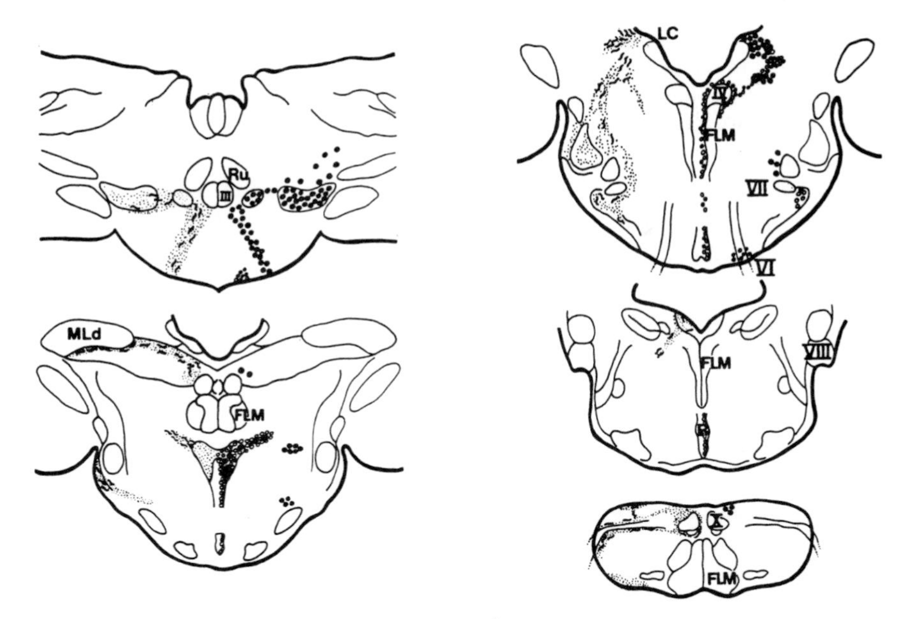


FIG. 34. Series of representative transverse sections through the brain stem of the chicken showing distributional patterns of monoamine-containing perikarya (right halves) as well as fibers and terminals (left halves). Sections are arranged in a rostrocaudal sequence. Roman numerals indicate cranial nerves or their nuclei. FLM, fasciculus longitudinalis medialis; LC, locus coeruleus; Ru, nucleus ruber; MLd, nucleus mesencephalicus dorsalis; filled circles indicate catecholamine-containing perikarya; open circles indicate serotonin or 5-OH-tryptophan containing neurons. The diagram is based on fluorescence-histochemical observations characterized by a low sensitivity to indolamines. Thus, only catecholamine fibers (~) and terminals (dots) appear in the left halves (after Dubé and Parent, 1981).

serotonin-binding sites by different drugs in chick embryos, see Soblosky *et al.* (1985). For the possible functional role of the CSF and extracellular space in circumscribed areas of the CNS, see Leonhardt *et al.* (1983), Hartwig and Reinhold (1983).

Last but not least, one should bear in mind that brain catecholamine content might differ considerably from one strain of a species to the next (for findings in pigeons, see Divac *et al.*, 1988). Recent findings obtained by applying antibodies raised against the enzyme tyrosine-hydroxylase (the first enzyme needed for catecholamine synthesis), have shown that there are considerable differences between the patterns of catecholamine-containing neurons revealed by histofluorescence or immunocytochemistry and the patterns exposing tyrosine-hydroxylase-immunoreactive neurons. It remains open to discussion whether neurons lacking tyrosine-hydroxylase are capable of taking up exogenous catecholamines (for a detailed discussion and findings in the pigeon, see Kiss and Péczely, 1987).

In birds most of the hypothalamic monoamine-containing perikarya seem to be concentrated in the paraventricular organ where they exhibit some of the structural features of CSF-contacting neurons (for a review on CSF-contacting neurons, see Vigh and Vigh–Teichmann, 1973). The sites to which these neurons project are still not precisely known. Most likely there is a projection to the external layer of the median eminence (cf. Calas *et al.*, 1974; Hartwig, 1981). In the European Common Quail (*Coturnix coturnix*) and in House Sparrows (*Passer domesticus*) dopamine- and serotonin-containing perikarya in the paraventricular organ (PVO) are present in a numerical ratio of approximately 5:3 (Hartwig, 1981). The dopamine neurons in the PVO of pigeons, however, do not react with antibodies raised against tyrosine-hydroxylase, the first enzyme involved in dopamine synthesis (Kiss and Péczely, 1987).

Monoamine-containing fibers and terminals are concentrated in phylogenetically ancient periventricular and basal areas of the hypothalamus including neurohemal target sites. Locally specialized patterns in the distribution of terminals suggest regional specializations in functions (for details, see Hartwig, 1981; Hartwig and Reinhold, 1983). At the level of the median eminence these patterns confirm the complex structural subdivisions revealed by classical neurohistological techniques (for an extensive review, see Oksche and Farner, 1974). Finally it should be noted that hypothalamic monoamine-containing perikarya may be transitory during ontogenesis (for findings in chickens, see Guglielmone and Panzica, 1984).

As mentioned earlier, numerous autonomic (hypothalamic) functions characterized by endogenous rhythmicity are known to be under the strong influence of the *pineal organ* (Chen *et al.*, 1980; Menaker and Binkley, 1981; for more recent observations in domestic hens, see Sharp *et al.*, 1984). As in mammals, the most thoroughly investigated messenger substance of the pineal organ in birds is the indoleamine melatonin. Although melatonin is an established messenger molecule that is released into the general circulation, it cannot be ruled out that it also possesses local functions within the pineal itself (for details see Hartwig, 1986, 1987). For observations on light-dependent changes in the activities of enzymes engaged in the control of melatonin synthesis in the retina and the pineal of 2–3-week-old chicks, see Hamm *et al.* (1983). In contradistinction to the situation in mammals, synthesis and release of melatonin in the pineal organ of several species of birds is apparently under the direct control of environmental light (cf. p. 76). For a review, see Binkley (1980), Deguchi (1979), for details on the metabolism of pineal indoleamines and speculations about biologically active pineal peptides, see Collin *et al.* (1982), Juillard *et al.* (1983), Collin *et al.* (1987). Pineal function may also be influenced by peripheral (sympathetic) and central innervation (for findings and a review, see Korf *et al.*, 1982b; Sato and Wake, 1983).

Pineal organs in different taxonomic forms of birds are remarkably diverse as to their absolute and relative sizes, their shapes and the fine structure of pinealocytes and pineal neurons. For details and an extensive review of earlier literature, see

Bargmann (1943), Menaker and Oksche (1974), and Vollrath (1980). In general, the avian pineal organ consists of (1) a broad distal part (the body) located between the telencephalic hemispheres and the cerebellum, and (2) a stalk-like proximal part. Frequently the body has the appearance of a gland. The slender pineal stalk connects the distal portion with the roof of the diencephalon in an area located between the habenular and posterior commissures (Fig. 2). Ueck (1981) suggested that, in general, the lumen of a pineal organ communicates with the circulating CSF in the third ventricle via the lumen of the stalk. In principle, the overall sizes of avian pineal organs do not seem to be correlated with body sizes. It should be noted that the pineal organs of most night-active species are relatively smaller than those of day-active species. For general information on the relative sizes of pineal organs in correlation with adaptation to different environments, see Ralph (1975). Central afferent and efferent projections of pineal neurons have recently been studied in House Sparrows by applying modern tracing techniques (Korf *et al.*, 1982b). The results indicate that pineal neurons project to the habenular ganglia and to a circumscribed periventricular area of the preoptic hypothalamus. Scattered perikarya in the medial habenular nucleus and the periventricular layer of the preoptic hypothalamus send axons to the pineal organ. New immunocytochemical techniques have helped reveal interesting details about the differentiation of pinealocytes with possible photoreceptive function and the neuronal organization of the pineal organ (for a review, see Korf and Ekström, 1987).

Electrophysiological investigations have so far failed to demonstrate direct photosensitivity in avian pineal organs that might be comparable to that recorded in pineal sense organs (for a detailed review, see Dodt, 1987). Nevertheless, it is well established that cultured pineal organs of galliform birds do synthesize and release indolamines in direct dependence upon light (for a review, see Binkley, 1980; Deguchi, 1982; Zatz *et al.*, 1988). In this context the immunocytochemical detection of opsin-like substances in pineal organs of birds deserves special mention (Vigh *et al.*, 1982). The close pineal–retinal relationships have been further confirmed in a comparative immunocytochemical study using antibodies raised against chick calbindin 27-kDa. This antibody reacted with pineal transducers and modified pineal photoreceptors in eight different species of birds. (For a definition of pineal transducers and modified pineal photoreceptors, see Collin *et al.* 1987.) It is interesting to note that this antibody did not react, with only a few exceptions, with photoreceptive structures in anamniotes. Finally, it should be noted that numerous taxonomic forms of birds apparently sense environmental photoperiods with enigmatic photoreceptors that are most probably located in the tuberal hypothalamus. For a review, see Benoit and Assenmacher (1953), Yokoyama *et al.* (1978), Glass and Lauber (1981), for comparative aspects of extraretinal and pineal photoreception, see Hartwig (1982), Hartwig and Oksche (1982), and Foster *et al.* (1987a).

## C. CIRCUMVENTRICULAR ORGANS AND CEREBROSPINAL FLUID-CONTACTING NEURONS

The avian CNS is rich in various types of CSF-contacting neurons and morphologically distinct circumventricular organs. In this respect the CNS of birds resembles that of poikilothermic vertebrates rather than that of mammals.

From the aspect of comparative anatomy, CSF-contacting neurons exhibit structural details that are characteristic of phylogenetically ancient nerve cells because they retain the positions originally occupied by developing neurons in the wall of the CNS. With a dendritic process bearing a cilium of the 9 x 2 + 0 type they contact the intraventricular CSF. At the level of the spinal cord their axonal processes may be in contact with the outer surface of the CNS. In contradistinction to the well-known intraventricular dendritic projections, the axonal projections of this ancient type of neurons have not yet been investigated in detail. Vigh, Vigh–Teichmann and coworkers have extensively studied CSF-contacting neurons in all classes of vertebrates (for reviews, see Vigh *et al.*, 1971, 1983). In chickens one finds conspicuous accumulations of these neurons in association with parvocellular areas in the preoptic recess, the neuronal part of the paraventricular organ, and in the central canal. It is interesting that retinal photoreceptor cells and pinealocytes have features in common with CSF-contacting neurons. The former possess sensory dendrites (outer segments) that project into a space originally filled with intraventricular CSF (for a detailed discussion, see Vigh *et al.*, 1975). Although numerous systems of accumulated or scattered CSF-contacting neurons and some of the transmitters they contain have been morphologically examined in several taxonomic forms of birds, the functions of these cells still remain to be elucidated in experiments, e.g., with electrophysiological techniques (cf. in this context Korf *et al.*, 1982a).

The term "circumventricular organ", as originally introduced by Bargmann and Hofer (for a detailed review, see Leonhardt, 1980), refers to specialized regions in the walls of the vertebrate CNS some of which represent sites of communication between the vascular bed and central nervous tissue. Several of these organs have already been discussed: (1) median eminence, (2) organum vasculosum of the lamina terminalis, (3) paraventricular organ, and (4) pineal gland. A blood–brain barrier as revealed by intravascular injections of different tracer substances appears to be lacking in most of the classical circumventricular organs that can be found in all vertebrate classes. For structural and histochemical features of the blood–brain barrier in birds, see Stewart and Wiley, 1981. The following brief survey deals with selected circumventricular organs that are apparently found in all taxonomic groups of birds (for detailed review, see Vigh, 1971; Leonhardt, 1980). Several circumventricular organs are characterized by nerve fibers that contain various neuropeptides (Weindl and Sofroniew, 1982) and monoamines (see below).

A number of ultrastructural details of the *organum vasculosum of the lamina terminalis (OVLT)* resemble those that characterize the neurohemal contact area and the inner layer of the median eminence (e.g., clearly recognizable stratified arrangement of structural components; Wenger and Törk, 1968; Calas *et al.*, 1975; for recent anatomical findings in Japanese Quail and domestic fowl, see Mikami, 1976; Viglietti-Panzica and Bessé, 1984). However, the OVLT does not seem to have specialized vessels for the supply and drainage of blood. Moreover, the median eminence is poor in neuronal perikarya whereas they can frequently be seen in association with the OVLT. In Domestic Mallards Bosler (1977) observed distinct distributional patterns of noradrenaline, serotonin and GABA in this functionally enigmatic organ.

The *subfornical organ (SFO)*, which is also located in the rostral wall of the third ventricle, contains even more neuronal perikarya with a variety of different ultrastructural features than the OVLT. A stratified arrangement of cellular components comparable to that of the median eminence, however, seems to be lacking (for ultrastructural observations in chickens and Japanese Quail, see Dellmann and Linner, 1979; Tsuneki *et al.*, 1978; Takei *et al.*, 1978). Monoamine- and neuropeptide-containing neuronal elements seem to be of crucial importance in one of the functional roles suggested for the SFO, i.e., the control of water metabolism (Takai, 1977a,b; Takei *et al.*, 1979).

The *subcommissural organ (SCO)* is an area of specialized ependyma (without neuronal perikarya) associated with the posterior commissure. Elongated ependymal cells secrete paraldehyde fuchsin-positive substances into the CSF. Within the ventricle these substances form threads that accumulate in distinct patterns as Reissner's fiber. Although structural details of the SCO have been extensively investigated in a variety of taxonomic forms of birds, the functional significance of this circumventricular organ and its secretory products remains completely enigmatic (for a review, see Ziegels, 1976). In contrast to the circumventricular organs mentioned above, the SCO does not have a neurohemal contact zone. Rodríguez and coworkers recently succeeded in developing specific antibodies raised against bovine Reissner's fiber (for a review, see Rodríguez *et al.*, 1987). These antibodies specifically label secretory products of the subcommissural organ in chickens and Domestic Mallards. Positive labeling can be detected in a dorsally located group of cells in the diencephalon as early as days 3–6 of incubation in these species (Schoebitz *et al.*, 1986).

In chickens and pigeons there is a regional specialization in the ependymal lining of the fourth ventricle that may be comparable to the *area postrema* of mammals (Pessacq, 1967; Böhme, 1970). The presence of a blood–brain barrier in this area is still a matter of controversy (see Böhme, 1972b). The scanning electron microscope has recently revealed other specialized areas in the surface morphology of ependymal cells lining the fourth ventricle in chickens (Hirunagi and Yasuda, 1979a).

The above mentioned circumventricular organs are found in all classes of vertebrates. A *paraventricular organ (PVO)*, however, has not been observed in mammals. The PVO is characterized by numerous CSF-contacting neurons that are arranged in morphologically distinct layers. This arrangement makes the PVO resemble to a certain degree the stratified patterns of the retina (Vigh, 1971). The ventricular surfaces of neuronal dendrites in the PVO are covered by a web-like material of unknown nature, possibly a precipitate that is formed when the tissue is processed for scanning electron microscopy (Hirunagi and Yasuda, 1979b). Biogenic monoamines have been detected in chick embryos from the 11th day of incubation onward (Gérard *et al.*, 1979; for further details concerning monoamines in the PVO, see p. 83).

The *choroid plexuses (CP)* of all four brain ventricles are also classified as circumventricular organs. The morphology of the ventricles and choroid plexuses has been less extensively studied by modern neurobiological techniques than other parts of the CNS. The general morphological pattern of the plexus epithelium seems to be similar in birds and mammals (for recent findings in chickens, see Doolin and Birge, 1969; el Gammal, 1981; for details on the ventricular system in chickens, Domestic Mallards, and pigeons, see Badawi, 1967; Böhme, 1969; for a review, see Pearson, 1972).

It should finally be noted that the morphological configuration of the CNS as a whole is strongly influenced by the shape of the skull (Hofer, 1952). Figure 35 presents an extreme example of the specialized arrangement of the bony skull and the brain.

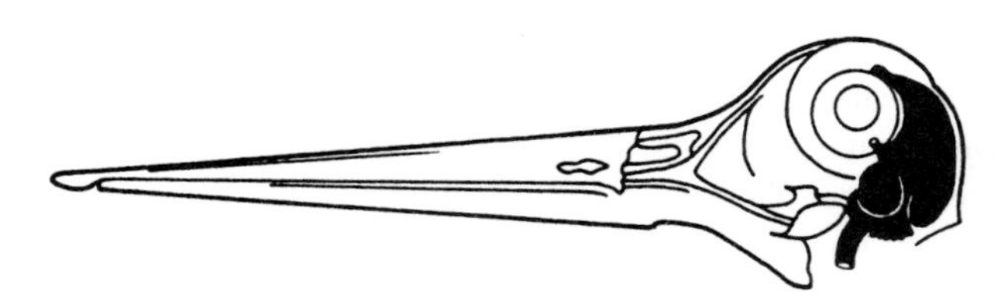

FIG. 35. European Woodcock, *Scolopax rusticola*, as an example of extreme position of brain with regard to the skull. Note dorsal (posterior) position of brain due to the development of large eyes (after Starck, 1982; for further details on configuration of skull and brain, see Hofer, 1952).

## X. Conclusions

Experimental findings obtained in numerous species of birds most of which are reviewed in this chapter have shown that most of the sensomotor reflex systems in birds exhibit structural characteristics more closely resembling those of mammals than had earlier been recognized (for details see Starck, 1962). Although birds evolved from an entirely different stock of reptiles than mammals did, and probably many million years after the first mammals had already emerged (Jerison, 1973), the functional morphology of the CNS of these two classes of vertebrates appears to be extraordinarily similar. This is especially true for recently detected, apparently phylogenetically ancient, neuronal systems that are characterized by the production of neuropeptides and monoamines. These neuronal systems frequently exhibit widespread and diffuse projections and differ from the precise long-distance point-to-point projections of sensomotor reflex systems. Although these more diffusely projecting neuronal systems seem to be integral components of the general morphological pattern of the CNS in vertebrates, their functions are only poorly understood.

The striking similarities between the general morphological patterns of the avian and mammalian spinal cord, the brain stem and the diencephalon are apparently not present at the level of the telencephalon. Since birds do not possess cortical areas directly comparable to those seen in mammals, it is rather difficult to compare telencephalic long-distance point-to-point projections.

Most likely our present knowledge about the general morphological pattern of the vertebrate telencephalon is not sufficient to clarify whether the avian telencephalon is indeed a unique and entirely new development or a specialization of components already established in the general morphological pattern of lower vertebrates. In addition to the apparent absence of cortical areas in the telencephalon of birds, another striking difference seems to characterize certain sensomotor reflex systems in birds and mammals: in birds somatosensory (especially trigeminal), auditory and vestibular afferent projections apparently bypass thalamic relay nuclei and terminate directly in the nucleus basalis telencephali which is located in the anterior region of the telencephalon. Furthermore, this nucleus most likely receives visual and olfactory inputs (Schall *et al.*, 1986). It should be noted that the nucleus basalis of birds does not correspond to the nucleus basalis of mammals. In birds this nucleus is an integral component of trigeminal sensomotor circuits essential for pecking, grasping, and feeding as determined by electrophysiological experiments and tracing studies (Wild and Arends, 1984; Wild *et al.*, 1985). At present it cannot be excluded that similar afferent pathways bypassing thalamic relay nuclei have been overlooked in mammals. The similarities that have been found in the structure and the functions of the avian and mammalian spinal cord, brain stem and diencephalon suggest that both mammalian and avian types of telencephalic projections fit well into the general morphological pattern of vertebrate telencephalic projections. These general patterns deserve further study in future research (cf. in this context the parcellation theory discussed by Ebbesson, 1980; for an excellent review of different ideas and concepts on the phylogenesis and ontogenesis of the vertebrate brain as well as for critical remarks on Ebbesson's presentation of the parcellation theory of brain development, see Rehkämper, 1984).

Although the telencephalon of birds does not exhibit a laminar pattern such as that characteristic of the cortex cerebri in mammals, the areal patterning found in birds can be compared with the laminar patterning in mammals (cf. Rehkämper *et al.*, 1985). In birds as well as in mammals telencephalic areas have been delineated that receive direct sensory inputs. These primary sensory areas are surrounded by secondary sensory areas that are probably devoted to associative functions (for details, see Rehkämper *et al.*, 1984, 1985). The analysis of areal patterns seen within the framework of functional specializations seems to be a most fruitful approach to finally elucidating the enigmatic structural and functional organization of the avian telencephalon (cf. in this context also Divac *et al.*, 1985).

In the analysis of the CNS of birds and other classes of vertebrates it is primarily the long-distance point-to-point projections that have been successfully mapped. However, the evolution of the CNS is of course not confined to these long-distance point-to-point projections in which neurons release their transmitters into a synaptic cleft. Continued analysis of the structure and the functions of the CNS must include studies on how neurons communicate with each other via messenger molecules they release into the extracellular spaces of the CNS itself or into the extracellular fluid of the body as a whole. Moreover, the evolution of the CNS is characterized by an increase in the number of glial cells that is considerably higher than the corresponding increase in the number of neurons. Unfortunately, with only a few exceptions, neither the functional aspects of neuronal systems that release messenger molecules into the extracellular fluid nor the functional meaning of the increase in glial cells are well understood. Both of these aspects of functional anatomy deserve attention in future research devoted to the CNS of birds and other vertebrates.

ACKNOWLEDGMENTS

Preparing this chapter was only possible with the help of many colleagues at the Departments of Anatomy of the Universities of Kiel and Düsseldorf. Several colleagues from other universities checked the text and provided useful comments. Among the numerous persons deserving my sincere thanks two are doubtlessly outstanding. Mr R. Clemens, Kiel, prepared the diagrams with unending patience and skill and Dr K. Rascher, Düsseldorf, succeeded in making the chapter readable. Last, but not least, I wish to express my deep gratitude to the late Professor Donald S. Farner who revised the first version and paternally supported my efforts in the field of science for two decades.

## References

Acheson, D. W., Kemplay, S. K., and Webster, K. E. (1980). Quantitative analysis of optic terminal profile distribution within the pigeon optic tectum. *Neuroscience* **5**, 1067–1084.

Agulleiro, B. (1975). Estructura del organo glucogenico en la medula lumbar de *Gallus gallus*. *Trab. Inst. Cajal Invest. Biol.* **67**, 53–74.

Alvarado-Mallart, R.-M., and Sotelo, C. (1984). Homotopic and heterotopic transplantations of quail tectal primordia in chick embryos: Organization of the retinotectal projections in the chimeric embryos. *Dev. Biol.* **103**, 378–398.

Angaut, P., and Repérant, J. (1978). A light and electron microscopic study of the nucleus isthmo-opticus in the pigeon. *Arch. d'Anat. micros.* **67**, 63–78.

Arad, Z., Mitgard, U., and Bernstein, M. H. (1987). Posthatching development of the rete ophthalmicum in relation to brain temperature of mallard ducks (*Anas platyrhynchos*). *Am. J. Anat.* **179**, 137–142.

Araki, M., Saito, T., Takeuchi, Y., and Kimura, H. (1984). Differentiation of monoamine accumulating neuron systems in cultured chick retina: An immunohistochemical and fluorescence histochemical study. *Dev. Brain Res.* **15**, 229–237.

Arends, J. J. and Dubbeldam, J. L. (1984). The subnuclei and primary afferents of the descending trigeminal system in the mallard (*Anas platyrhynchos* L.). *Neuroscience* **13**, 781–795.

Arends, J. J., Woelders-Blok, A. and Dubbeldam, J. L. (1984). The efferent connections of the nuclei of the descending trigeminal tract in the mallard (*Anas platyrhynchos* L.). *Neuroscience* **13**, 797–817.

Arnold, A. P. (1982). Neural control of passerine song. *In* "Acoustic Communication in Birds" (D. E. Kroodsma, and E. H. Miller, eds), Vol. 1, pp. 75–94. Academic Press, New York.

Arnold, A. P., and Saltiel, A. (1979). Sexual difference in pattern of hormone accumulation in the brain of a songbird. *Science* **205**, 702–705.

Arnold, A. P., Nottebohm, F., and Pfaff, D. W. (1976). Hormone concentrating cells in vocal control and other areas of the brain of the zebra finch (*Poephila guttata*). *J. Comp. Neurol.* **165**, 487–511.

Aschoff, J., Daan, S., and Groos, G. A. (1982). "Vertebrate Circadian Systems". Springer, Berlin.

Badawi, H. (1967). Das Ventrikelsystem des Gehirnes von Huhn (*Gallus domesticus*), Taube (*Columba livia*) und Ente (*Anas boschas domestica*), dargestellt mit Hilfe des Plastoid-Korrosionsverfahrens. *Zentralbl. Veterinärmed.* (A), **14**, 628–650.

Bagnoli, P., and Casini, G. (1985). Regional distribution of catecholaminergic terminals in the pigeon visual system. *Brain Res.* **337**, 277–286.

Bagnoli, P., Porciatti, V., Lanfranchi, A., and Bedini, C. (1985). Developing pigeon retina: Light-evoked responses and ultrastructure of outer segments and synapses. *J. Comp. Neurol.* **235**, 384–394.

Baker, M. C., Bottjer, S. W., and Arnold, A. P. (1984). Sexual dimorphism and lack of seasonal changes in vocal control regions of the white-crowned sparrow brain. *Brain Res.* **295**, 85–89.

Ball, G. F., Faris, P. L., Hartman, B. K., and Wingfield, J. C. (1988). Immunohistochemical localization of neuropeptides in the vocal control regions of two songbird species. *J.Comp. Neurol.* **268**, 171–180.

Balthazart, J., and Schoffeniels, E. (1979). Pheromones are involved in the control of sexual behaviour in birds. *Naturwissenschaften* **66**, 55–56.

Bang, B. G. (1964). The nasal organs of the Black and Turkey Vultures; a comparative study of the Cathartid species *Coragyps atratus atratus* and *Cathartes aura septentrionalis* (with notes on *Cathartes aura Falklandica*, *Pseudogyps bengalensis*, and *Neophron percnopterus*). *J. Morph.* **115**, 153–184.

Bang, B. G. (1966). The olfactory apparatus of tube-nosed birds (Procellariiformes). *Acta Anat.* **65**, 391–415.

Bargmann, W. (1943). Die Epiphysis cerebri. *In* "Handbuch der mikroskopischen Anatomie des Menschen" (W. von Möllendorff, ed.), Vol. 6, pp. 309–502. Springer, Berlin.

Bargmann, W., Oksche, A., Fix, J. D., Adams, R. D., and Haymaker, W. (1980). Meninges, choroid plexuses, ependyma, and their pathological reactions. *In* "Cellular Pathology of the Nervous System" (W. Haymaker *et al.*, eds.), pp. 560–641. Academic Press, San Diego.

Baumel, J. J. (1962). Asymmetry of encephalic arteries in the pigeon (*Columba livia*). *Anat. Anz.* **111**, 91–102.

Baumel, J. J. (1967). The characteristic asymmetrical distribution of the posterior cerebral artery of birds. *Acta Anat.* **67**, 523–549.

Baylé, J.-D., Ramade, F., and Oliver, J. (1974). Stereotaxic topography of the brain of the quail (*Coturnix coturnix japonica*). *J. Physiol.* (Paris) **68**, 219–241.

Bedini, C., Flaschi, V., and Lanfranchi, A. (1976). Degenerative and regenerative processes in the olfactory system of homing pigeons. *Arch. Ital. Biol.* **114**, 376–388.

Benes, F. M., Parks, T. N., and Rubel, E. W. (1977). Rapid dendritic atrophy following deafferentation: An EM morphometric analysis. *Brain Res.* **122**, 1–13.

Bennett, T. (1974). The peripheral and autonomic nervous system. *In* "Avian Biology" (D. S. Farner *et al.*, eds.), Vol. IV, pp. 1–79. Academic Press, San Diego.

Benoit, J. (1935). Stimulation par la lumière artificielle du développement testiculaire chez des canards avenglés par section du nerf optique. *C. R. Soc. Biol.* **120**, 133–136.

Benoit, J. (1938). Rôle des yeux et de la voie nerveuse oculo-hypophysaire dans la gonadostimulation par la lumière artificielle chez le canard domestique. *C. R. Soc. Biol.* **129**, 231–234.

Benoit, J. (1978). Chronobiologic study in the domestic duck. *Chronobiologia* **V**, 147–168.

Benoit, J., and Assenmacher, I. (1953). Action des facteurs externes et plus particulièrement du facteur lumineux sur l'activité sexuelle des oiseaux. *Rapport à la IIe réunion des endocrinologistes de langue française*, 33–80.

Benowitz, L. (1980). Functional organization of the avian telencephalon. *In* "Comparative Neurology of the Telencephalon" (S. O. E. Ebbesson, ed.), pp. 389–421. Plenum, New York.

Benowitz, L., and Karten, H. J. (1976). The tractus infundibuli and other afferents to the parahippocampal region of the pigeon. *Brain Res.* **102**, 174–180.

Benzo, C., De Gennaro, L. D., and Stearns, S. B. (1975). Glycogen metabolism in the developing chick glycogen body: Functional significance of the direct oxidative pathway. *J. Exp. Zool.* **193**, 161–166.

Bergquist, H. (1957). Comments on the architype of the vertebrate brain. *K. Fysiogr. Sällsk. Lund Förh.* **27**, 153–160.

Berk, M. L. (1987). Projections of the lateral hypothalamus and bed nucleus of the stria terminalis to the dorsal vagal complex in the pigeon. *J. Comp. Neurol.* **260**, 140–156.

Berk, M.-L., and Finkelstein, J. A. (1983). Long descending projections of the hypothalamus in the pigeon, (*Columba livia*). *J. Comp. Neurol.* **220**, 127–136.

Berk, M. L., Reaves, T. A., Hayward, J. N., and Finkelstein, J. A. (1982). The localization of vasotocin and neurophysin neurons in the diencephalon of the pigeon (*Columba livia*). *J. Comp. Neurol.* **204**, 392–406.

Berkhoudt, H., Dubbeldam, J. L., and Zeilstra, S. (1981). Studies on the somatotopy of the trigeminal system in the mallard (*Anas platyrhynchos* L.). IV. Tactile representation in the nucleus basalis. *J. Comp. Neurol.* **196**, 407–420.

Bernstein, M. H., Sandoval, I., Curtis, M. B., and Hudson, D. M. (1979). Brain temperature in pigeons: Effects of anterior respiratory bypass. *J. Comp. Physiol.* **129**, 115–118.

Bernstein, M. H., Duran, H. L., and Pinshaw, B. (1984). Extrapulmonary gas exchange enhances brain oxygen in pigeons. *Science* **226**, 564–566.

Bertler, A., Falck, B., Gottfries, C. G., Ljunggren, L., and Rosengren, E. (1964). Some observations on adrenergic connections between mesencephalon and cerebral hemispheres. *Acta Pharmacol. Toxicol.* **21**, 283–289.

Binggeli, R. L., and Paule, W. J. (1969). The pigeon retina: Quantitative aspects of the optic nerve and ganglion cell layer. *J. Comp. Neurol.* **137**, 1–18.

Binkley, S. (1980). Functions of the pineal gland. *In* "Avian Endocrinology" (A. Epple and M. H. Stetson, eds.), pp. 53–74. Academic Press, New York.

Blähser, S. (1981). Zur räumlichen Beziehung zwischen Somatostatin- und Enkephalin-immunoreaktiven Neurosystemen im Kaudalen Hirnstamm von *Gallus domesticus*. *Verh. Anat. Ges.* **75**, 793–795.

Blähser, S. (1984). Peptidergic pathways in the avian brain. *J. exp. Zool.* **232**, 397–403.

Blähser, S. (1988). The ACTH-immunoreactive system in the brain of the white-crowned sparrow, *Zonotrichia leucophrys gambelli* (Passeriformes, Emberizidae). *Histochemistry* **88**, 309–312.

Blähser, S., and Dubois, M. P. (1980). Immunocytochemical demonstration of met-enkephalin in the central nervous system of the domestic fowl. *Cell Tissue Res.* **213**, 53–68.

Blähser, S., and Heinrichs, M. (1982). Immunoreactive neuropeptide systems in avian embryos (domestic mallard, domestic fowl, Japanese quail). *Cell Tissue Res.* **223**, 287–303.

Blähser, S., Fellmann, D., and Bugnon, C. (1978). Immunocytochemical demonstration of somatostatin-containing neurons in the hypothalamus of the domestic mallard. *Cell Tissue Res.* **195**, 183–187.

Böhme, G. (1969). Vergleichende Untersuchungen am Gehirnventrikelsystem: Das Ventrikelsystem des Huhnes. *Acta Anat.* **73**, 116–126.

Böhme, G. (1970). Eine organartige Bildung im IV. Ventrikel beim Huhn. *Verh. Anat. Ges.* **64**, 245–250.

Böhme, G. (1972a). Granulationes leptomeningicae bei *Gallus domesticus*. *Experientia* **28**, 677–678.

Böhme, G. (1972b). Untersuchungen an der Area postrema von *Gallus domesticus*. II. Das morphologische Problem der Existenz einer Bluthirnschranke. *Acta Neuropathol.* **21**, 308–315.

Böhme, G. (1973). Lichtmikroskopische Untersuchungen über die Struktur der Leptomeninx encephali bei *Gallus domesticus*. *Z. Anat. Entwicklungsgesch.* **140**, 215–230.

Böhme, G. (1974). Untersuchungen an den Meningen des Huhnes (Granulationes leptomeningicae). Lichtmikroskopische Untersuchungen. *Anat. Histol. Embryol* (Berlin) **3**, 233–242.

Bonke, B. A., Bonke, D., and Scheich, H. (1979). Connectivity of the auditory forebrain nuclei in the guinea fowl (*Numida meleagris*). *Cell Tissue Res.* **200**, 101–121.

Bons, N. (1980). The topography of mesotocin and vasotocin systems in the brain of the domestic mallard and Japanese quail: Immunocytochemical identification. *Cell Tissue Res.* **213**, 37–51.

Bons, N., Kerdelhué, B., and Assenmacher, I. (1978a). Mise en évidence d'un deuxième système neurosécrétoire à LH-RH dans l'hypothalamus du canard. *C.R. Acad. Sc. Paris* **287**, 145–151.

Bons, N., Kerdelhué, B., and Assenmacher, I. (1978b). Immunocytochemical identification of the LHRH-producing system originating in the preoptic nucleus of the duck. *Cell Tissue Res.* **188**, 99–106.

Boord, R. L. (1969). The anatomy of the avian auditory system. *Ann. N.Y. Acad. Sci.* **31**, 165.

Boord, R. L., and Karten, H. J. (1974). The distribution of primary lagenar fibers within the vestibular nuclear complex of the pigeon. *Brain Behav. Evol.* **10**, 228–235.

Boord, R. L., and Rasmussen, G. L. (1963). Projection of the cochlear and lagenar nerves on the cochlear nuclei of the pigeon. *J. Comp. Neurol.* **120**, 463–475.

Bortolami, R., Callegari, E., and Lucchi, M. L. (1982). Anatomical relationship between mesencephalic trigeminal nucleus and cerebellum in the duck. *Brain Res.* **47**, 317–329.

Bosler, O., (1977). The organum vasculosum laminae terminalis. A cytophysiological study in the duck (*Anas platyrhynchos*). *Cell Tissue Res.* **182**, 383–399.

Bouillé, C., Raymond, J., and Baylé, J. D. (1977). Retrograde transport of horseradish peroxidase from the nucleus posterior medialis hypothalami to the hippocampus and the medial septum in the pigeon. *Neuroscience* **2**, 435–439.

Bowmaker, J. K. (1977). The visual pigments, oil droplets and spectral sensitivity of the pigeon. *Vision Res.* **17**, 1129–1138.

Bowmaker, J. K. (1980). Colour vision in birds and the role of oil droplets. *TINS 8/1980*, 196–199.

Boxer, M. I. and Stanford, D. (1985). Projections to the posterior visual hyperstriatal region in the chick: An HRP study. *Exp. Brain Res.* **57**, 494–498.

Brauth, S. E., Ferguson, J. L., and Kitt, C. A. (1978). Prosencephalic pathways related to the paleostriatum of the pigeon (*Columba livia*). *Brain Res.* **147**, 205–221.

Bravo, H. and Inzunza, O. (1985). The oculomotor nucleus, not the abducent, innervates the muscles which advance the nictitating membrane in birds. *Acta Anat.* **122**, 99–104.

Breazile, J. E. (1979). Systema Nervosum Centrale. *In* "Nomina Anatomica Avicum" (J. Baumel, A. S. King, J. Breazile, and H. E. Evans, eds), pp. 417–472. Academic Press, New York.

Brecha, N., and Karten, H. J. (1979). Accessory optic projections upon oculomotor nuclei and vestibulocerebellum. *Science* **203**, 913–916.

Brecha, N., Karten, H. J., and Laverack, C. (1979). Enkephalin-containing amacrine cells in the avian retina: Immunohistochemical localization. *Proc. Natl. Acad. Sci. USA* **76**, 3010–3014.

Brecha, N., Karten, H. J., and Hunt, S. P. (1980). Projections of the nucleus of the basal optic root in the pigeon: An autoradiographic and horseradish peroxidase study. *J. Comp. Neurol.* **189**, 615–670.

Brinkman, R., and Martin, A. H. (1973). A cytoarchitectonic study of the spinal cord of the domestic fowl *Gallus gallus domesticus*. I. Brachial region. *Brain Res.* **56**, 43–62.

Britto, L. R., Keyser, K. T., Hammasaki, D. E., and Karten, H. J. (1988). Catecholaminergic subpopulation of retinal displaced ganglion cells projects to the accessory optic nucleus in the pigeon (*Columba livia*). *J. Comp. Neurol.* **269**, 109–117.

Brodal, A. (1981). "Neurological Anatomy". Academic Press, San Diego.

Brodal, A., Kristiansen, K., and Jansen, J. (1950). Experimental demonstration of a pontine homologue in birds. *J. Comp. Neurol.* **92**, 23–69.

Bunt, S. M., and Horder, T. J. (1983). Evidence for an orderly arrangement of optic axons within the optic nerves of the major nonmammalian vertebrate classes. *J. Comp. Neurol.* **213**, 94–114.

Burd, G. D., and Nottebohm, F. (1985). Ultrastructural characterization of synaptic terminals formed on newly generated neurons in a song control nucleus of the adult canary forebrain. *J. Comp. Neurol.* **240**, 143–152.

Cabot, J. B., Reiner, A., and Bogan, N. (1982). Avian bulbuspinal pathways: Anterograde and retrograde studies of cells of origin, funicular trajectories and laminar terminations. *Progr. Brain Res.* **57**, 79–108.

Calas, A, Hartwig, H.-G., and Collin, J. P. (1974). Noradrenergic innervation of the median eminence. Microspectrofluorimetric and pharmacological study in the duck, *Anas platyrhynchos*. *Z. Zellforsch.* **147**, 491–504.

Calas, A., Bosler, O., and Assenmacher, I. (1975). Approche cytophysiologique de deux interfaces neuroendocrines de l'hypothalamus: L'éminence médiane et l'organe vasculaire de la lame terminale étude radioautographique chez le canard. *Arch. Anat. Histol. Embryol.* **58**, 107–120.

Cassone, V. M., and Moore, R. Y. (1987). Retinohypothalamic projection and suprachiasmatic nucleus of the house sparrow, *Passer domesticus*. *J. Comp. Neurol.* **266**, 171–182.

Catsicas, S. and Clarke, P. G. (1987). Abrupt loss of dependence of retinopetal neurons on their target cells, as shown by intraocular injections of kainate in chick embryos. *J. Comp. Neurol.* **262**, 523–534.

Catsicas, S., Catsicas, M., and Clarke, P. G. (1987a). Long-distance intraretinal connections in birds. *Nature* **326**, 186–187.

Catsicas, S., Thanos, S., and Clarke, P. G. (1987b). Major role for neuronal death during brain development: Refinement of topographical connections. *Proc. Natl. Acad. Sci. (USA)* **84**, 8168–8265.

Chen, H. J., Brainard, G. C., and Reiter, R. J. (1980). Melatonin given in the morning prevents the suppressive action on the reproductive system of melatonin in late afternoon. *Neuroendocrinology* **32**, 129–132.

Chen, D.-M., Collins, J. S., and Goldsmith, T. H. (1984). The ultraviolet receptor of bird retinas. *Science* **225**, 337–340.

Chikazawa, H., Fujioka, T., and Watanabe, T. (1983). Bulbar catecholaminergic neurons projecting to the thoracic spinal cord of the chicken. *Anat. Embryol.* **167**, 411–423.

Chu-Wang, I.-W., and Oppenheim, R. W. (1978a). Cell death of motoneurons in the chick embryo spinal cord. I. A light and electron microscopic study of naturally occurring and induced cell loss during development. *J. Comp. Neurol.* **177**, 33–58.

Chu-Wang, I.-W., and Oppenheim, R. W. (1978b). Cell death of motoneurons in the chick embryo spinal cord. II. A quantitative and qualitative analysis of degeneration in the ventral root, including evidence for axon outgrowth and limb innervation prior to cell death. *J. Comp. Neurol.* **177**, 59–86.

Clarke, P. G. (1977). Some visual and other connections to the cerebellum of the pigeon. *J. Comp. Neurol.* **174**, 535–552.

Clarke, P. G. (1982a). The generation and migration of the chick's isthmic complex. *J. Comp. Neurol.* **207**, 208–222.

Clarke, P. G. (1982b). The genuineness of isthmo-optic neuronal death in chick embryos. *Anat. Embryol.* **165**, 389–404.

Clarke, P.G. (1985). Neuronal death during development in the isthmo-optic nucleus of the chick: Sustaining role of afferents from the tectum. *J. Comp. Neurol.* **234**, 365–379.

Cobb, S. (1964). A comparison of the size of an auditory nucleus (N. mesencephalicus lateralis, pars dorsalis) with the size of the optic lobe in twenty-seven species of birds. *J. Comp. Neurol.* **122**, 271–279.

Cogburn, L. A., Wilson-Placentra, S., and Letcher, R. L. (1987). Influence of pinealectomy on plasma and extrapineal melatonin rhythms in young chickens (*Gallus domesticus*). *Gen. Comp. Endocrinol.* **68**, 343–356.

Coggeshall, R. E. (1980). Law of separation of function of the spinal roots. *Physiol. Reviews* **60**, 716–753.

Cohen, D. H., and Karten, H. J. (1974). The structural organization of avian brain. *In* "Birds, Brain and Behavior" (I. J. Goodman, and M. W. Schein, eds), pp. 29–73. Academic Press, New York.

Cohen, D. H., Schnall, A. M., MacDonald, R. L., and Pitts, L. W. (1970). Medullary cells of origin of vagal cardioinhibitory fibers in the pigeon. I. Anatomical studies of peripheral vagus nerve and the dorsal motor nucleus. *J. Comp. Neurol.* **140**, 299–320.

Collin, J. P., Balemans, M., Juillard, M.-T., Legerstee, W. C., van Benthem, J., and Voisin, P. (1982). Indole metabolism at the cellular level: An *in vivo* combined radiobiochemical and high-resolution autoradiographic study in the avian pineal, with special reference to melatonin/5-methoxytryptophol. *Biol. Cell* **44**, 25–34.

Collin, J. P., Voisin, P., Falcón, J., and Brisson, P. (1987). Evolution and environmental control of secretory processes in pineal transducers. *In* "Functional Morphology of Neuroendocrine Systems" (B. Scharrer, H. W. Korf, and H. G. Hartwig, eds), pp. 105–121. Springer, Berlin.

Conlee, J .W., and Parks, T. N. (1981). Age- and position-dependent effects of monaural acoustic deprivation in nucleus magnocellularis of the chicken. *J. Comp. Neurol.* **202**, 373–384.

Conlee, J. W. and Parks, T. N. (1983). Late appearance and deprivation-sensitive growth of permanent dendrites in the avian cochlear nucleus (Nuc. magnocellularis). *J. Comp. Neurol.* **217**, 216–226.

Conlee, J. W., and Parks, T. M. (1986). Origin of ascending auditory projections to the nucleus mesencephalicus lateralis pars dorsalis in the chicken. *Brain Res.* **367**, 96–113.

Correia, M. J., Eden, A. R., Westlund, K. N., and Coulter, J. D. (1982). Organization of ascending auditory pathways in the pigeon (*Columba livia*) as determined by autoradiographic methods. *Brain Res.* **234**, 205–212.

Correia, M. J., Eden, A. R., Westlund, K. N., and Coulter, J. D. (1983). A study of some of the ascending and descending vestibular pathways in the pigeon (*Columba livia*) using anterograde transneuronal autoradiography. *Brain Res.* **278**, 53–61.

Cowan, W. M. (1970). Centrifugal fibers to the avian retina. *Br. Med. Bull.* **26**, 112–118.

Cowan, W. M. (ed.) (1981). "Studies in Developmental Neurobiology. Essays in Honor of Viktor Hamburger". Oxford University Press, New York.

Cowan, W. M., and Powell, T. P. (1963). Centrifugal fibers in the avian visual system. *Proc. Roy. Soc. B* **158**, 232–252.

Crossland, W. J. (1985). Anterograde and retrograde axonal transport of native and derivatized wheat germ agglutinin in the visual system of the chicken. *Brain Res.* **347**, 11–27.

Crossland, W. J., and Hughes, C. P. (1978). Observations on the afferent and efferent connections of the avian isthmo-optic nucleus. *Brain Res.* **145**, 239–256.

Cuénod, M., and Streit, P. (1979). Amino acid transmitters and local circuitry in optic tectum. *In* "The Neurosciences", Fourth Study Program (Schmitt, F. O. and Worden, F. G., eds), pp. 989–1004. MIT Press, Massachusetts.

Cunningham, T. J. (1982). Naturally occurring neuron death and its regulation by developing neural pathways. *Int. Rev. Cytol.* **74**, 163–187.

Dahl, D., and Bignami, A. (1977). Preparation of antisera to neurofilament protein from chicken brain and human sciatic nerve. *J. Comp. Neurol.* **176**, 645–657.

Davis, B. M., and Cabot, J. B. (1984). Substance P-containing pathways to avian sympathetic preganglionic neurons: Evidence for major spinal–spinal circuitry. *J. Neuroscience* **4**, 2145–2159.

DeGennaro, L. D. (1982). The glycogen body. *In* "Avian Biology" (Farner, D. S. *et al.*, eds), Vol. VI, pp. 34–372. Academic Press, San Diego.

DeGennaro, L. D., and Benzo, C. A. (1976). Ultrastructural characterization of the accessory lobes of Lachi (Hofmann's nuclei) in the nerve cord of the chick. I. Axoglial synapses. *J. Exp. Zool.* **198**, 97–107.

DeGennaro, L. D., and Benzo, C. A. (1987). Development of the glycogen body of the Japanese quail, *Coturnix coturnix japonica*. I. Light microscopy and early development. *J. Morphol.* **194**, 209–217.

Deguchi, T. (1979). A circadian oscillator in cultured cells of chicken pineal gland. *Nature* **282**, 94–96.

Deguchi, T. (1982). Endogenous oscillator and photoreceptor for serotonin *N*-acetyltransferase rhythm in chicken pineal gland. *In* "Vertebrate Circadian Systems" (J. Aschoff, S. Daan, and G. Groos, eds), pp. 164–172. Springer, Berlin.

Delius, J. D., and Bennetto, K. (1972). Cutaneous sensory projections to the avian forebrain. *Brain Res.* **37**, 205–221.

Dellmann, H.-D., and Linner, J. G. (1979). Ultrastructure of the subfornical organ of the chicken (*Gallus domesticus*). *Cell Tissue Res.* **197**, 137–153.

Delorme, P., Grignon, G., and Gayet, J. (1970). Étude en microscopie d'électronique de l'histogenèse des capillaires du télencéphale chez l'embryon du poulet et le poussin et de leur perméabilité à la peroxydase. *C.R. Assoc. Anat.* **149**, 725–732.

DeMello, M. C., and DeMello, F. G. (1985). Topographical organization of the dopamine-dependent adenylate cyclase of the chick embryo retina. *Brain Res.* **328**, 59–63.

De Voogd, T. J. (1984). The avian song system: Relating sex differences in behavior to dimorphism in the central nervous system. *Prog. Brain Res.* **61**, 171–183.

De Voogd, T. J., and Nottebohm, F. (1981). Sex differences in dendritic morphology of a song control nucleus in the canary: A quantitative Golgi study. *J. Comp. Neurol.* **196**, 309–316.

Dingler, E. C. (1965). Einbau des Rückenmarks im Wirbelkanal bei Vögeln. *Verh. Anat. Ges.* **115**, 71–84.

Divac, I., Mogensen, J., and Björklund, A. (1985). The prefrontal "cortex" in the pigeon. Biochemical evidence. *Brain Res.* **332**, 365–368.

Divac, I., Mogensen, J., and Björklund, A. (1988). Strain differences in catecholamine content in pigeon brains. *Brain Res.* **444**, 371–373.

Dodt, E. (1987). Light sensitivity of the pineal organ in poikilothermic and homeothermic vertebrates.

*In* "Functional Morphology of Neuroendocrine Systems". (B. Scharrer, H. W. Korf, and H. G. Hartwig, eds), pp. 123–132. Springer, Berlin.

Domesick, V. B., and Morest, D. K. (1977a). Migration and differentiation of ganglion cells in the optic tectum of the chick embryo. *Neuroscience* **2**, 459–475.

Domesick, V. B., and Morest, D. K. (1977b). Migration and differentiation of shepherd's crook cells in the optic tectum of the chick embryo. *Neuroscience* **2**, 477–491.

Doolin, P. F., and Birge, W. J. (1969). Ultrastructural differentiation of the junctional complex of the avian choroidal epithelium. *J. Comp. Neurol.* **136**, 253–268.

Dooling, R. J. (1982). Auditory perception in birds. *In* "Acoustic Communication in Birds" (D. E. Kroodsma, and E. H. Miller, eds), Vol. 1, pp. 95–130. Academic Press, San Diego.

Douglas, J. H., and Ribchester, R. R. (1986). Dibutyryl cyclic guanine monophosphate inhibits natural but not limb amputation-induced motoneurone death in the chick embryo. *Dev. Brain Res.* **26**, 271–275.

Drenckhahn, D. (1970). Untersuchungen an Regio olfactoria und Nervus olfactorius der Silbermöve (*Larus argentatus*). *Zellforsch.* **106**, 119–142.

Dū, F., Chayvialle, J. A., and Dubois, P. (1988). Distribution and development of VIP immunoreactive neurons in the spinal cord of the embryonic and newly hatched chick. *J. Comp. Neurol.* **268**, 600–614.

Dubbeldam, J. L. (1968). On the shape and the structure of the brain stem in some species of birds. Thesis, Rijksuniversiteit Leiden, The Netherlands.

Dubbeldam, J. L. (1980). Studies on the somatotopy of the trigeminal system in the mallard (*Anas platyrhynchos* L.) II. The morphology of the principal sensory nucleus. *J. Comp. Neurol.* **191**, 557–571.

Dubbeldam, J. L., and Karten, H. J. (1978). The trigeminal system in the pigeon (*Columba livia*). I. Projections of the Gasserian ganglion. *J. Comp. Neurol.* **180**, 611–678.

Dubbeldam, J. L., and Veenman, C. L. (1978). Studies on the somatotopy of the trigeminal system in the mallard (*Anas platyrhynchos* L.). I. The ganglion trigeminale. *Neth. J. Zool.* **28**, 150–160.

Dubbeldam, J. L., Brus, E. R., Menken, S. B. J., and Zeilstra, S. (1979). The central projection of the glossopharyngeal and vagus ganglia in the mallard, *Anas platyrhynchos* L. *J. Comp. Neurol.* **183**, 149–168.

Dubbeldam, J. L., Brauch, C. S. M., and Don, A. (1981). Studies on the somatotopy of the trigeminal system in the mallard, *Anas platyrhynchos* L. III. Afferents and organization of the nucleus basalis. *J. Comp. Neurol.* **196**, 391–405.

Dubé, L., and Parent, A. (1981). The monoamine-containing neurons in avian brain: I. A study of the brain stem of the chicken (*Gallus domesticus*) by means of fluorescence and acetylcholinesterase histochemistry. *J. Comp. Neurol.* **196**, 695–708.

Dubin, M. W. (1970). The inner plexiform layer of the vertebrate retina: A quantitative and comparative electron microscopic analysis. *J. Comp. Neurol.* **140**, 479–506.

Dubois, M. P. (1976). Le système à somatostatine. *Ann. Endocr. (Paris)* **37**, 277–278.

Duff, T. A., and Scott, G. (1979). Electron microscopic evidence of a ventronasal to dorsotemporal variation in fiber size in pigeon optic nerve. *J. Comp. Neurol.* **183**, 679–688.

Duff, T. A., Scott, G., and Mai, R.(1981). Regional differences in pigeon optic tract, chiasm, and retino-receptive layers of optic tectum. *J. Comp. Neurol.* **198**, 231–247.

Ebbesson, S. O. E. (1972). A proposal for a common nomenclature for some optic nuclei in vertebrates and the evidence for the common origin of two such cell groups. *Brain Behav. Evol.* **6**, 75–91.

Ebbesson, S. O. E. (1980). The parcellation theory and its relation to interspecific variability in brain organization, evolutionary and ontogenetic development, and neuronal plasticity. *Cell Tissue Res.* **213**, 179–212.

Eden, A. R., and Correia, M. J. (1982). Identification of multiple groups of efferent vestibular neurons in the adult pigeon using horseradish peroxidase and DAPI. *Brain Res.* **248**, 201–208.

Eden, A. R., Correia, M. J., and Steinkuller, P. G. (1982). Medullary proprioceptive neurons from extraocular muscles in the pigeon identified with horseradish peroxidase. *Brain Res.* **237**, 15–21.

Ehrlich, D. (1981). Regional specialization of the chick retina as revealed by the size and density of neurons in the ganglion cell layer. *J. Comp. Neurol.* **195**, 643–657.

Ehrlich, D., and Mark, R. (1984). The course of axons of retinal ganglion cells within the optic nerve and tract of the chick (*Gallus gallus*). *J. Comp. Neurol.* **223**, 583–591.

Ehrlich, D., and Mills, D. (1985). Myelogenesis and estimation of the number of axons in the anterior commissure of the chick (*Gallus gallus*). *Cell Tissue Res.* **239**, 661–666.

Ehrlich, D. and Stuchberry, J. (1986). A note on the projection from the rostral thalamus to the visual hyperstriatum of the chicken (*Gallus gallus*). *Exp. Brain Res.* **62**, 207–211.

Ehrlich, D., Keyser, K. T., and Karten, H. J. (1987). Distribution of substance P-like immunoreactive retinal ganglion cells and their pattern of termination in the optic tectum of chick (*Gallus domesticus*). *J. Comp. Neurol.* **266**, 220–233.

Ehrlich, D., Zappia, J. V., and Saleh, C. N. (1988). Development of the supraoptic decussation in the chick (*Gallus gallus*). *Anat. Embryol.* **177**, 361–370.

Emmerton, J. (1983a). Functional morphology of the visual system. *In* "Physiology and Behavior of the Pigeon" (M. Abs, ed.), pp. 221–244. Academic Press, London.

Emmerton, J. (1983b). Vision. *In* "Physiology and Behavior of the Pigeon" (M. Abs, ed.), pp. 245–266. Academic Press, London.

Farner, D. S. (1975). Photoperiodic controls in the secretion of gonadotropins in birds. *Amer. Zool.* **15** (Suppl. 1), 117–135.

Farner, D. S. (1983). Some recent advances in avian physiology. *J. Yamashina Inst. Ornith.* **15**, 97–140.

Farner, D. S. (1986). Generation and regulation of annual cycles in migratory passerine birds. *Amer. Zool.* **26**, 493–501.

Farner, D. S., Donham, R. S., Lewis, R. A., Mattocks, P. W., Darden, T. R., and Smith, J. P. (1977). The circadian component in the photoperiodic mechanism of the house sparrow, *Passer domesticus*. *Physiol. Zool.* **50**, 247–268.

Feirabend, H. K. and Voogd, J. (1986). Myeloarchitecture of the cerebellum of the chicken (*Gallus domesticus*): An atlas of the compartmental subdivision of the cerebellar white matter. *J. Comp. Neurol.* **251**, 44–66.

Feirabend, H. K., Vielvoye, G. J., Freedman, S. L., and Voogd, J. (1976). Longitudinal organization of afferent and efferent connections of the cerebellar cortex of the white Leghorn (*Gallus domesticus*). *Exp. Brain Res.* (Suppl.) **1**, 72–78.

Felix, D., Henke, H., and Frangi, U. (1979). Opiate receptors in the pigeon optic tectum. *Brain Res.* **175**, 145–149.

Fischlschweiger, W., and O'Rahilly, R. (1968). The ultrastructure of the pecten oculi in the chick. II. Observations on the bridge and its relation to the vitreous body. *Z. Zellforsch.* **92**, 313–324.

Fite, K.V., Brecha, N., Karten, H. J., and Hunt, S. P. (1981). Displaced ganglion cells and the accessory optic system of pigeon. *J. Comp. Neurol.* **195**, 279–288.

Fitzgerald, K. T. (1979). The structure and function of the pars tuberalis of the vertebrate adenohypophysis. *Gen. and Comp. Endocrinol.* **37**, 383–399.

Florén, I. (1979). Indolamine accumulating neurons in the retina of chicken and pigeon. A comparison with the dopaminergic neurons. *Acta Ophthalmol.* **57**, 198–210.

Flucher, B. E. (1984). Detecting the cause of non-specific staining in immuno-cytochemistry (PAP-methode). *Mikroskopie* **41**, 219–226.

Fontaine-Perus, J., Chanconie, M., and le Deuarin, N. M. (1985). Embryonic origin of substance P containing neurons in cranial and spinal sensory ganglia of the avian embryo. *Dev. Biol.* **107**, 227–238.

Foster, R. G., Follett, B. K., and Lythgoe, J. N. (1985). Rhodopsin-like sensitivity of extra-retinal photoreceptors mediating the photoperiodic response in quail. *Nature* **313**, 50–52.

Foster, R. G., Korf, H.-W., and Schalken, J. J. (1987a). Immunocytochemical markers revealing retinal and pineal but not hypothalamic photoreceptor systems in the Japanese quail. *Cell Tissue Res.* **248**, 161–167.

Foster, R. G., Plowman, G., Goldsmith, A. R., and Follet, B. K. (1987b). Immunohistochemical demonstration of marked changes in the LHRH system of photosensitive and photorefractory European starlings (*Sturnus vulgaris*). *J. Endocr.* **115**, 211–220.

Fox, R., Lehmkuhle, S. W., and Bush, R. C. (1977). Stereopsis in the falcon. *Science* **197**, 79–81.

Franzoni, M. F., Viglietti-Panzica, C., Ramieri, G., and Panzica, G. C. (1984). A Golgi study on the neuronal morphology in the hypothalamus of the Japanese quail (*Coturnix coturnix japonica*). I. Tuberal and mammillary regions. *Cell Tissue Res.* **236**, 357–364.

Freedman, S. L., Feirabend, H. K., Vielvoye, G. J., and Voogd, J. (1975). Re-examination of the ponto-cerebellar projection in the adult white leghorn (*Gallus domesticus*). *Acta Morphol. Neerl.-Scand.* **13**, 236–238.

Freedman, S. L., Voogd, J., and Vielvoye, G. J. (1977). Experimental evidence for climbing fibers in the avian cerebellum. *J. Comp. Neurol.* **175**, 243–252.

Fujisawa, H., Thanos, S., and Schwarz, U. (1984). Mechanisms in the develop ment of retinotectal projections in the chick embryo studied by surgical deflection of the retinal pathway. *Dev. Biol.* **102**, 356–367.

Fukuda, M., Yeh, H. H., and Puro, D. G. (1987). Avian retinal cells express enkephalin-like immunoreactivity in culture. *Dev. Brain Res.* **31**, 147–150.

Gadow, A., and Selenka, G. (1891). Vögel. *In* "Klassen und Ordnungen des Thier-Reichs" Bd. **6**, 4. Teil (H. G. Bronn, ed.), pp.1–375. Winter'sche Verlagshandlung, Leipzig.

Gallego, A. (1976). The horizontal cells of the terrestrial vertebrate retina: I. Mammals and birds. *In* "The Structure of the Eye III" (E. Yamada, and S. Mishima, eds), pp. 273–280. *Jap. J. Ophthalmol.* University of Tokyo, Tokyo.

Gamlin, D. R., and Cohen, D. H. (1986). A second ascending visual pathway from the optic tectum to the telencephalon in the pigeon (*Columba livia*). *J. Comp. Neurol.* **250**, 296–310.

Gamlin, D. R., and Cohen, D. H. (1988a). Retinal projections to the pretectum in the pigeon (*Columba livia*). *J. Comp. Neurol.* **269**, 1–17.

Gamlin, D. R., and Cohen, D. H. (1988b). Projections of the retinorecipient pretectal nuclei in the pigeon (*Columba livia*). *J. Comp. Neurol.* **269**, 18–46.

Gamlin, P. D., Reiner, A., Erichsen, J. T., Karten, H. J., and Cohen, D. H. (1984). The neural substrate for the pupillary light reflex in the pigeon (*Columba livia*). *J. Comp. Neurol.* **226**, 523–543.

el-Gammal, S. (1981). The development of the diencephalic choroid plexus in the chick. *Cell Tissue Res.* **219**, 297–311.

Gentle, J. (1975). Gustatory behavior of the chicken and other birds. *In* "Neural and Endocrine Aspects of Behavior in Birds" (P. Wright, P. G. Caryl, and D. M. Vowles, eds), pp. 305–318. Elsevier, Amsterdam.

Gérard, A., Gérard, H., and Grignon, G. (1979). Innervation sérotoninergique de l'organe paraventriculaire de l'embryon de Poulet et du jeune Poussin; étude par histofluorescence et par autoradiographie en microscopie optique et électronique. *C. R. Acad. Sc.* (Paris) **289**, 355.

Gerstberger, R., Healy, D. P., Hammel, H. T., and Simon, E. (1987). Autoradiographic localization and characterization of circumventricular angiotensin II receptors in duck brain. *Brain Res.* **400**, 165–170.

Glass, J. D., and Lauber, J. K. (1981). Sites and action spectra for encephalic photoreception in the Japanese quail. *Am. J. Physiol.* **240**, R220–228.

Glover, J. C. and Petursdotter, G. (1988). Pathway specificity of reticulospinal and vestibulospinal projections in the 11-day chicken embryo. *J. Comp. Neurol.* **270**, 25–38.

Goldberg, S. (1974). Studies on the mechanics of development of the visual pathways in the chick embryo. *Dev. Biol.* **36**, 24–43.

Goldman, S. A., and Nottebohm, F. (1983). Neuronal production, migration, and differentiation in a vocal control nucleus of the adult female canary brain. *Proc. Natl. Acad. Sci. USA* **80**, 2390–2394.

Goldsmith, T. H. (1980). Hummingbirds see near ultraviolet light. *Science* **207**, 786–788.

Goller, H. (1972). Versorgungsgebiete und Zentren der Gehirnnerven vom Huhn (*Gallus domesticus*). *Berliner Münchner Tierärztl. Wochenschr.* **22**, 432–436.

Goodman, D. C., Horel, J. A., and Freemon, F. R. (1964). Functional localization in the cerebellum of the bird and its bearing on the evolution of cerebellar function. *J. Comp. Neurol.* **123**, 45–54.

Goossens, N., Blähser, S., Oksche, A., Vandesande, F., and Dierickx, K. (1977). Immunocytochemical investigation of the hypothalamo-neurohypo physial system in birds. *Cell Tissue Res.* **184**, 1–13.

Gross, G. H., and Oppenheim, R. W. (1985). Novel sources of descending input to the spinal cord of the hatchling chick. *J. Comp. Neurol.* **232**, 162–179.

Großmann, R., and Ellendorf, F. (1987). The chick embryo: A model to study the functional ontogeny of the magnocellular system. *In* "Functional Morphology of Neuroendocrine Systems" (B. Scharrer, H. W. Korf, and H. G. Hartwig, eds), p. 69. Springer, Berlin.

Guglielmone, R., and Panzica, G.C. (1984). Typology, distribution and development of the catecholamine-containing neurons in the chicken brain. *Cell Tissue Res.* **237**, 67–79.

Güntürkün, O. (1984). Evidence for a third primary visual area in the telencephalon of the pigeon. *Brain Res.* **294**, 247–254.

Güntürkün, O. (1987). A Golgi study of the isthmic nuclei in the pigeon (*Columba livia*). *Cell Tissue Res.* **248**, 439–448.

Gurusinghe C. J., and Ehrlich, D. (1985). Sex-dependent structural asymmetry of the medial habenular nucleus of the chicken brain. *Cell Tissue Res.* **240**, 149–152.

Gwinner, E. (1975). Circadian and circannual rhythms in birds. *In* "Avian Biology" (D. S. Farner *et al.*, eds), Vol. V, pp. 221–288. Academic Press, San Diego.

Haase, E., Otto, C., and Murbach, H. (1977). Brain weight in homing and "non-homing" pigeons. *Experientia* **33**, 606.

Halfter, W., Deiss, S., and Schwarz, U. (1985). The formation of the axonal pattern in the embryonic avian retina. *J. Comp. Neurol.* **232**, 466–480.

Hall, T. R., Harvey, S., and Chadwick, A. (1986). Control of prolactin secretion in birds: A review. *Gen. Comp. Endocrinol.* **62**, 171–184.

Haller, V. Graf v. Hallerstein (1934). Zerebrospinales Nervensystem. I. Äußere Gliederung des Zentralnervensystems. *In* "Handb. vergl. Anatomie der Wirbeltiere" (L. Bolk *et al.*, eds), Vol. 2, pp. 1–318 and pp. 817–832. Urban & Schwarzenberg, Berlin Wien.

Hamburger, V. (1975). Cell death in the development of the lateral motor column of the chick embryo. *J. Comp. Neurol.* **160**, 535–546.

Hamm, H. E., Takahashi, J. S., and Menaker, M. (1983). Light-induced decrease of serotonin *N*-acetyltransferase activity and melatonin in the chicken pineal gland and retina. *Brain Res.* **266**, 287–293.

Hardy, O., Leresche, N., and Jassik-Gerschenfeld, D. (1985). Morphology and laminar distribution of electrophysiologically identified cells in the pigeon's optic tectum: An intracellular study. *J. Comp. Neurol.* **233**, 390–404.

Hartwig, H.-G. (1980). Hypothalamic and extrahypothalamic brain centers involved in the control of circadian and circannual photoneuroendocrine mechanisms. *In* "Acta XVII Congressus Internationalis Ornithologici" (R. Nöhring, ed.), Vol. I, Berlin, 1978, pp. 417–424. Verlag der Deutschen Ornithologen-Gesellschaft, Berlin.

Hartwig, H.-G. (1981). Hypothalamic distribution of monoamine-containing neurons in homeothermic vertebrates. *In* "Neurosecretion—Molecules, Cells, Systems" (D. S. Farner, and K. Lederis, eds), pp. 93–103. Plenum Press, New York.

Hartwig, H.-G. (1982). Comparative aspects of retinal and extraretinal photosensory input channels entraining endogenous rhythms. *In* "Vertebrate Circadian Systems" (J. Aschoff, S. Daan, and G. Groos, eds), pp. 25–30. Springer, Berlin.

Hartwig, H.-G. (1984). General organization of the nervous system and its development: Introductory remarks. *In* "Fetal Neuroendocrinology" (F. Ellendorff, and P. D. Gluckman, eds), pp. 1–8. Perinatology Press, New York.

Hartwig, H.-G. (1986). Turnover of pineal photoreceptive membranes in the frog, *Rana esculenta* complex. *In* "Pineal and Retinal Relationships" (P. J. O'Brien, and D. C. Klien, eds), pp. 47–55. Academic Press, San Diego.

Hartwig, H.-G. (1987). Structure and function of retinal and extraretinal photoreceptive organs. A comparative approach. *In* "Comparative Physiology of Environmental Adaptations 3. Adaptations to Climatic Changes" (P. Pevét, ed.), pp. 45–55. Karger, Basel.

Hartwig, H.-G., and Oksche, A. (1982). Neurobiological aspects of extraretinal photoreceptive systems: Structure and function. *Experientia* **38**, 991–996.

Hartwig, H.-G., and Reinhold, C. (1983). Advanced histochemical analysis of locally specialized distribution patterns of hypothalamic monoamines. *In* "Structure and Function of Peptidergic and Mono-aminergic Neurons" (Sano *et al.*, eds), pp. 323–334. DSSP, Tokyo.

Hartwig, H.-G., and Wahren, W. (1982). Anatomy of the hypothalamus. *In* "Stereotaxy of the Human Brain" (G. Schaltenbrand, and E. Walker, eds), pp. 87–106. Thieme, Stuttgart.

Hartwig, H.-G., Pradelles, P., Christolomme, A., and Calas, A. (1981). Distribution des structures neuronales reconnues par des anticorps contre la met-enképhaline et la substance P dans le systéme nerveux central d'oiseaux. *J. Physiol.* **77**, 8B-9B.

Hayes, B.P. (1984). Cell populations of the ganglion cell layer: Displaced amacrine and matching amacrine cells in the pigeon retina. *Exp. Brain Res.* **56**, 565–573.

Hayes, B. P., and Holden, A. L. (1980). Size classes of ganglion cells in the central yellow field of the pigeon retina. *Exp. Brain Res.* **39**, 269–275.

Hayes, B. P., and Holden, A. L. (1983). The distribution of centrifugal terminals in the pigeon retina. *Exp. Brain Res.* **49**, 189–197.

Heaton, M. B. (1977). Retrograde axonal transport in lateral motor neurons of the chick embryo prior to limb bud innervation. *Dev. Biol.* **58**, 421–427.

Heaton, M. B., and Moody, S. A. (1980). Early development and migration of the trigeminal motor nucleus in the chick embryo. *J. Comp. Neurol.* **189**, 61–99.

Heaton, M. B., and Wayne, D. B. (1983). Patterns of extraocular innervation by the oculomotor complex in the chick. *J. Comp. Neurol.* **216**, 245–252.

Heaton, M. B., Alvarez, I. M., and Crandall, J. E. (1979). The displaced ganglion cell in the avian retina: Developmental and comparative considerations. *Anat. Embryol.* **155**, 161–178.

Heil, P., and Scheich, H. (1986). Effects of unilateral and bilateral cochlea removal on 2-deoxyglucose patterns in the chick auditory system. *J. Comp. Neurol.* **252**, 279–301.

Henke, H. (1983). The central part of the avian visual system. *In* "Progress in Nonmammalian Brain Research" (G. Nistic, and L. Bolis, eds), Vol. 1, pp. 113–158. CRC Press, Boca Raton, Florida.

Herrmann, K., and Bischof, H.-J. (1986). Delayed development of song control nuclei in the zebra finch is related to behavioral development. *J. Comp. Neurol.* **245**, 167–175.

Hirokawa, N. (1978). The ultrastructure of the basilar papilla of the chick. *J. Comp. Neurol.* **181**, 361–374.

Hirschberger, W. (1971). Vergleichend experimentell-histologische Untersuchungen zur retinalen Repräsentation in den primären visuellen Zentren einiger Vogelarten. Thesis, University of Frankfurt, FRG.

Hirunagi, K., and Yasuda, M. (1979a). Scanning electron microscopic analysis of the linings of the fourth ventricle in the domestic fowl. *Cell Tissue Res.* **197**, 169–173.

Hirunagi, K., and Yasuda, M. (1979b). Scanning electron microscopy of the ventricular surface of the paraventricular organ in the domestic fowl. *Cell Tissue Res.* **197**, 539–543.

Hofer, H. (1952). Der Gestaltwandel des Schädels der Säugetiere und der Vögel, mit besonderer Berücksichtigung der Knickungstypen und der Schädelbasis. *Anat. Anz.* **99**, 102–112.

Hollyday, M. (1980). Organization of motor pools in the chick lumbar lateral motor column. *J. Comp. Neurol.* **194**, 143–170.

Horstmann, E. (1954). Die Faserglia des Selachiergehirns. *Z. Zellforsch.* **39**, 588–617.

Hunt, S. P., and Künzle, H. (1976). Observations on the projections and intrinsic organization of the pigeon optic tectum: An autoradiographic study based on anterograde and retrograde, axonal and dendritic flow. *J. Comp. Neurol.* **170**, 153–172.

Ikeda, H., and Goto, J. (1974). Distribution of monoamine-containing terminals and fibers in the central nervous system of the chicken. *Jap. J. Pharmac.* **24**, 831–841.

Il'icev, V. (1972). "Bio-Acoustics of Birds". Moscow University Press, Moscow.

Inoue, Y. (1970). The glioarchitectonics of the chicken brain. I. The glial cells in the optic tract. *Okajimas Folia anat. jpn.* **47**, 229–265.

Inoue, Y. (1971). The glioarchitectonics of the chicken brain. II. Microglia. *Okajimas Folia anat. jpn.* **48**, 53–82.

Inoue, Y., Nakagawa, S., Sugihara, Y., and Shimai, K. (1971). The glioarchitectonics of the chicken brain. III. Astrocytes, oligodendroglia and other neuroglial cells. *Okajimas Folia anat. jpn.* **48**, 237–270.

Inoue, Y., Sugihara, Y., Nakagawa, S., and Kazuyo, S. (1973). Atypical glial cells in avian brains – Golgi study. *Okajimas Folia anat. jpn.* **50**, 295–306.

Jansen, J. (1972). Features of cerebellar morphology and organization. *Acta Neurol. Scand. Suppl.* **51**, 197–217.

Jerison, H. J. (1973). "Evolution of the Brain and Intelligence". Academic Press, New York.

Jhaveri, S., and Morest, D. K. (1982a). Neuronal architecture in nucleus magnocellularis of the chicken auditory system with observations on nucleus laminaris: A light and electron microscope study. *Neuroscience* **7**, 809–836.

Jhaveri, S., and Morest, D.K. (1982b). Sequential alterations of neuronal architecture in nucleus magnocellularis of the developing chicken: An electron microscope study. *Neuroscience* **7**, 855–870.

Józsa, R., and Mess, B. (1982). Immunohistochemical localization of the luteinizing hormone releasing hormone (LHRH)-containing structures in the central nervous system of the domestic fowl. *Cell Tissue Res.* **227**, 451–458.

Józsa, R., Korf, H.-W., Csernus, V., and Mess, B. (1988). Thyrotropin-releasing hormone (TRH)-immunoreactive structures in the brain of the domestic mallard. *Cell Tissue Res.* **251**, 441–449.

Juillard, M.-T., Balemans, M., Collin, J.-P., Legerstee, C., and VanBenthem, J. (1983). An *in vitro* combined radiobiochemical and high-resolution autoradiographic study predicting evidence for indolergic cells in the avian pineal organ. *Biol. Cell* **47**, 365–378.

Källén, B. (1962). Embryogenesis of brain nuclei in the chick telencephalon. *Ergebn. Anat. Entwicklungsgesch.* **36**, 62–82.

Karten, H. J. (1963). Ascending pathways from the spinal cord in the pigeon (*Columba livia*). *Proc. Int. Cong. Zool.* **2**, 23.

Karten, H. J. (1968). The ascending auditory pathway in the pigeon (*Columba livia*). II. Telencephalic projections of the nucleus ovoidalis thalami. *Brain Res.* **11**, 134–153.

Karten, H. J. (1969). The organization of the avian telencephalon and some speculations on the phylogeny of the amniote telencephalon. *Ann. N.Y. Acad. Sci.* **167**, 164–179.

Karten, H. J., and Dubbeldam, J. L. (1973). The organization and projections of the paleostriatal complex in the pigeon (*Columba livia*). *J. Comp. Neurol.* **148**, 61–90.

Karten, H. J., and Hodos, W. (1967). "A Stereotaxic Atlas of the Brain of the Pigeon". Johns Hopkins Press, Baltimore.

Karten, H. J., and Hodos, W. (1970). Telencephalic projections of the nucleus rotundus in the pigeon (*Columba livia*). *J. Comp. Neurol.* **140**, 35–52.

Karten, H. J., Hodos, W., Nauta, W. J. H., and Revzin, A. M. (1973). Neural connections of the "visual wulst" of the avian telencephalon. Experimental studies in the pigeon (*Columba livia*) and owl (*Speotyto cunicularia*). *J. Comp. Neurol.* **150**, 253–278.

Karten, H. J., Fite, K. V., and Brecha, N. (1977). Specific projection of displaced retinal ganglion cells upon the accessory optic system in the pigeon (*Columba livia*). *Proc. Natl. Acad. Sci. USA* **74**, 1753–1756.

Katayama-Kumoi, Y., Kiyama, H., Emson, P. C., Kimmel, J. R., and Tohyama, M. (1985). Coexistence of pancreatic polypeptide and substance P in the chicken brain. *Brain Res.* **361**, 25–35.

Kato, S., Negishi, K., and Teranishi, T. (1984). Embryonic development of monoaminergic neurons in the chick retina. *J. Comp. Neurol.* **224**, 437–444.

Katz, D. M., and Karten, H. J. (1983). Subnuclear organization of the dorsal motor nucleus of the vagus nerve in the pigeon, *Columba livia*. *J. Comp. Neurol.* **217**, 31–46.

Keeton, W. T. (1979). Avian orientation and navigation. *Ann. Rev. Physiol.* **41**, 353–366.

Kelley, D. B., and Nottebohm, F. (1979). Projections of a telencephalic auditory nucleus—field L—in the canary. *J. Comp. Neurol.* **183**, 455–470.

Keynes, R. J., and Stern, C. D. (1984). Segmentation in the vertebrate nervous system. *Nature* **310**, 786–789.

Kilgore, D. L., Boggs, D. F., and Birchard, G. F. (1979). Role of the rete mirabile ophthalmicum in maintaining the body-to-brain temperature difference in pigeons. *J. Comp. Physiol.* **129**, 119–122.

King, J. S. (1966). A comparative investigation of neuroglia in representative vertebrates: A silver carbonate study. *J. Morphol.* **119**, 435–466.

Kishida, R., Dubbeldam, J. L., and Goris, R. C. (1985). Primary sensory ganglion cells projecting to the principal trigeminal nucleus in the mallard, *Anas platyrhynchos*. *J. Comp. Neurol.* **240**, 171–179.

Kiss, J. Z., and Péczely, P. (1987). Distribution of tyrosine-hydroxylase (TH)-immunoreactive neurons in the diencephalon of the pigeon (*Columba livia*). *J. Comp. Neurol.* **257**, 333–346.

Kitt, C.A., and Brauth, S.E. (1982). A paleostriatal–thalamic–telencephalic path in pigeons. *Neuroscience* **7**, 2735–2751.

Kitt, C. and Brauth, S. E. (1986). Telencephalic projections from midbrain and isthmal cell groups in the pigeon. I. Locus coeruleus and subcoeruleus. *J. Comp. Neurol.* **247**, 69–91.

Kitt, C., and Brauth, S. E. (1986). Telencephalic projections from midbrain and isthmal cell groups in the pigeon. II. The nigral complex. *J. Comp. Neurol.* **247**, 92–110.

Klintworth, G. K. (1968). The comparative anatomy and phylogeny of the tentorium cerebelli. *Anat. Rec.* **160**, 635–642.

Knudsen, E. J. (1980). Sound localization in birds. *In* "Comparative Studies of Hearing in Vertebrates" (A. N. Popper, and R. R. Fay, eds), pp. 289–322. Springer, Berlin.

Knudsen, E. J. (1983). Subdivisions of the inferior colliculus in the barn owl (*Tyto alba*). *J. Comp. Neurol.* **218**, 174–186.

Knudsen, E. J., and Knudsen, P. F. (1983). Space-mapped auditory projections from the inferior colliculus to the optic tectum in the barn owl (*Tyto alba*). *J. Comp. Neurol.* **218**, 187–196.

Knudsen, E. J., and Konishi, M. (1979). Mechanisms of sound localization in the barn owl (*Tyto alba*). *J. Comp. Physiol A.* **133**, 13–21.

Kobayashi, H., and Wada, M. (1973). Neuroendocrinology in birds. *In* "Avian Biology" (D. S. Farner *et al.*, eds), Vol. III, pp. 287–348. Academic Press, San Diego.

Kojima, T., Homma, S., Sako, H., Shimizu, I., Okado, A., and Okado, N. (1988). Developmental changes in density and distribution of serotoninergic fibers in the chick spinal cord. *J. Comp. Neurol.* **267**, 580–589.

Konishi, M. (1985). Birdsong: From behavior to neuron. *Ann. Rev. Neuosci.* **8**, 125–170.

Konishi, M., and Akutagawa, E. (1985). Neuronal growth, atrophy and death in a sexually dimorphic song nucleus in the zebra finch brain. *Nature* **315**, 145–147.

Korf, H.-W., and Ekström, P. (1987). Photoreceptor differentiation and neuronal organization of the pineal organ. *In* "IV Colloquium of the European Pineal Study Group, Modena, Italy" (G. P. Trentini *et al.*, eds), pp. 35–47. Raven Press, New York.

Korf, H.-W., and Fahrenkrug, J. (1984). Ependymal and neuronal specializations in the lateral ventricle of the Pekin duck, *Anas platyrhynchos*. *Cell Tissue Res.* **236**, 217–227.

Korf, H.-W., Simon-Oppermann, C., and Simon, E. (1982a). Afferent connections of physiologically identified neuronal complexes in the paraventricular nucleus of conscious Pekin ducks involved in regulation of salt- and water-balance. *Cell Tissue Res.* **226**, 275– 300.

Korf, H.-W., Zimmermann, N.H., and Oksche, A. (1982b). Intrinsic neurons and neural connections of the pineal organ of the house sparrow, *Passer domesticus*, as revealed by anterograde and retro grade transport of horseradish peroxidase. *Cell Tissue Res.* **222**, 243–260.

Korf, H.-W., Viglietti-Panzica, C., and Panzica, G.C. (1983). A Golgi study on the cerebrospinal fluid (CSF)-contacting neurons in the paraventricular nucleus of the Pekin duck. *Cell Tissue Res.* **228**, 149–163.

Kuenzel, W. J., and van Tienhoven, A. (1982). Nomenclature and location of avian hypothalamic nuclei and associated circumventricular organs. *J. Comp. Neurol.* **206**, 293–313.

Kuhlenbeck, H. (1968). "The Central Nervous System of Vertebrates. A General Survey of its Comparative Anatomy with an Introduction to the Pertinent Fundamental and Logical Concepts". Vol. 1: Propaedeutics to comparative neurology. Karger, Basel.

Kuhlenbeck, H. (1970), "The Central Nervous System of Vertebrates. A General Survey of its Comparative Anatomy with an Introduction to the Pertinent Fundamental and Logical Concepts". Vol. 3/I: Structural elements: Biology of nervous tissue. Karger, Basel.

Kuhlenbeck, H. (1973). "The Central Nervous System of Vertebrates. A General Survey of its Comparative Anatomy with an Introduction to the Pertinent Fundamental and Logical Concepts". Vol. 3/II: Overall morphologic pattern. Karger, Basel.

Kuhlenbeck, H. (1975). "The Central Nervous System of Vertebrates. A General Survey of its Comparative Anatomy with an Introduction to the Pertinent Fundamental and Logical Concepts". Vol. 4: Spinal cord and deuterencephalon. Karger, Basel.

Kuhlenbeck, H. (1977), "The Central Nervous System of Vertebrates. A General Survey of its Comparative Anatomy with an Introduction to the Pertinent Fundamental and Logical Concepts". Vol. 5/I: Derivatives of the prosencephalon, diencephalon and telencephalon. Karger, Basel.

Kuhlenbeck, H. (1978), "The Central Nervous System of Vertebrates. A General Survey of its Comparative Anatomy with an Introduction to the Pertinent Fundamental and Logical Concepts". Vol. 5/II: Mammalian telencephalon: Surface morphology and cerebral cortex. The vertebrate neuroaxis as a whole. Karger, Basel.

Lanerolle de, N. C., Elde, R. P., Sparber, S. B., and Frick, M. (1981). Distribution of methionine-enkephalin immunoreactivity in the chick brain: An immunohistochemical study. *J. Comp. Neurol.* **199**, 513–533.

Lanser, M. E., and Fallon, J. F. (1987). Development of the brachial lateral motor column in the wingless mutant chick embryo: Motoneuron survival under varying degrees of peripheral load. *J. Comp. Neurol.* **261**, 423–434.

Lavalley, A. L., and Ho, R. H. (1983). Substance P, somatostatin and methionine enkephalin immunoreactive elements in the spinal cord of the domestic fowl, *Gallus domesticus*. *J. Comp. Neurol.* **213**, 406–413.

Lemmon, V. (1985) Monoclonal antibodies specific for glia in the chick nervous system. *Dev. Brain Res.* **23**, 111–120.

Leonard, R. B., and Cohen, D. H. (1975a). A cytoarchitectonic analysis of the spinal cord of the pigeon (*Columba livia*). *J. Comp. Neurol.* **163**, 159–180.

Leonard, R. B., and Cohen, D. H. (1975b). Spinal terminal fields of dorsal root fibers in the pigeon (*Columba livia*). *J. Comp. Neurol.* **163**, 181–192.

Leonhardt, H. (1980). Ependym und circumventriculäre Organe. *In* "Handbuch der mikroskopischen Anatomie des Menschen" (A. Oksche, and L. Vollrath, eds), Vol. 4, part 10, Neuroglia I, pp. 177–666. Springer, Berlin.

Leonhardt, H., Krisch, B., and Hartwig, H.-G. (1983). Circumventricular organs as targets and release sites for peptide hormones and monoamines. *In* "Structure and Function of Peptidergic and Aminergic Neurons" (J. Sano *et al.*, eds), pp. 51–71. Jap. Sci. Soc. Press, Tokyo.

Leonhardt, H., Krisch, B., and Erhardt, H. (1987). Organization of the neuroglia in the midsagittal plane of the central nervous system: A speculative report. *In* "Functional Morphology of Neuroendocrine Systems". (B. Scharrer, H.-W. Korf, and H.-G. Hartwig, eds), pp. 175–187. Springer, Berlin.

Louis, J. C., Rougeot, C., Bepoldin, O., Vulliez, B., Mandel, P., and Dray, F. (1983). Presence of somatostatin, enkephalins, and substance P-like peptides in cultured neurons from embryonic chick cerebral hemispheres. *J. Neurochem.* **41**, 930–938.

Lyser, K. M. (1972). The fine structure of glial cells in the chicken. *J. Comp. Neurol.* **146**, 83–94.

Lyser, K. M. (1973). The fine structure of the glycogen body of the chicken. *Acta Anat.* **85**, 533–549.

MacDonald, R. L., and Cohen, D. H. (1970). Cells of origin of sympathetic pre- and postganglionic cardioacceleratory fibers in the pigeon. *J. Comp. Neurol.* **40**, 343–358.

Maier, V., and Scheich, H. (1983). Acoustic imprinting leads to differential 2-deoxy-D-glucose uptake in the chick forebrain. *Proc. Natl. Acad. Sci. USA* **80**, 3860–3864.

Maier, V., and Scheich, H. (1987). Acoustic imprinting in guinea fowl chicks: age dependence of 2-deoxyglucose uptake in relevant forebrain areas. *Dev. Brain Res.* **31**, 15–27.

Manogue, K. R., and Nottebohm, F. (1982). Relation of medullary motor nuclei to nerves supplying the vocal tract of the budgerigar (*Melopsittacus undulatus*). *J. Comp. Neurol.* **204**, 384–391.

Martin, G. R. (1978). Spectral sensitivity of the red and yellow oil droplet fields of the pigeon (*Columba livia*). *Nature* **274**, 620– 621.

Martin, G. R., and Young, S. R. (1983). The retinal binocular field of the pigeon (*Columba livia*: English racing homer). *Vision Res.* **23**, 911–916.

Matsushita, M. (1968). Zur Zytoarchitektonik des Hühnerrückenmarks nach Silberimprägnation. *Acta Anat.* **70**, 238–259.

Mayr, I. (1972a). Verteilung, Lokalisation und Absorption der Zapfenölkugeln bei Vögeln (Ploceidae). *Vision Res.* **12**, 1477–1484.

Mayr, I. (1972b). Das Farbunterscheidungsvermögen einiger Ploceidae (Passeriformes, Aves) im kurzwelligen Bereich des Spektrums. *Vision Res.* **12**, 509–517.

McGraw, C. F., and McLaughlin, B. J. (1980a). Fine structural studies of synaptogenesis in the superficial layers of the chick optic tectum. *J. Neurocytol.* **9**, 79–93.

McGraw, C. F., and McLaughlin, B. J. (1980b). A freeze-fracture study of synaptic junction development in the superficial layers of the chick optic tectum. *J. Neurocytol.* **9**, 95–106.

McLoon, S. C. (1984). Development of the retinotectal projection in chicks. *In* "Organizing Principles of Neural Development" (S. C. Sharma, ed.), pp. 325–342. NATO ASJ Series, Series A: Life Sciences **78**.

McLoon, S. C., and Lund, R. D. (1982). Transient retinofugal pathways in the developing chick. *Exp. Brain Res.* **45**, 277–284.

McMillan-Carr, V., and Simpson, S. B. (1978a). Proliferative and degenerative events in the early development of chick dorsal root ganglia. I. Normal development. *J. Comp. Neurol.* **182**, 727–740.

McMillan-Carr, V., and Simpson, S. B. (1978b). Proliferative and degenerative events in the early development of chick dorsal root ganglia. II. Responses to altered peripheral fields. *J. Comp. Neurol.* **182**, 741–756.

Melian, A. P., Fonolla, J. P., and Loyzaga, P. G. (1986). The ontogeny of the cerebellar fissures in the chick embryo. *Anat. Embryol.* **175**, 119–128.

Menaker, M., and Binkley, S. (1981). Neural and endocrine control of circadian rhythms in the vertebrates. *In* "Handbook of Behavioral Neurobiology" (J. Aschoff, ed.), Vol. 4, Biological rhythms, pp. 243–255. Plenum Press, New York.

Menaker, M., and Oksche, A. (1974). The avian pineal organ. *In* "Avian Biology" (D. S. Farner *et al.*, eds), Vol. 4, pp. 79–118. Academic Press, San Diego.

Mestres, P., and Delius, J. D. (1982). A contribution to the study of the afferents to the pigeon optic tectum. *Anat. Embryol.* **165**, 415–423.

Mestres, P. and Rascher, K. (1987). The ependyma in the brain of the adult pigeon. A SEM study. *In* "Functional Morphology of Neuroendocrine Systems" (B. Scharrer, H.-W. Korf, and H.-G. Hartwig, eds), p. 173. Springer, Berlin.

Meyer, D. B., and May, H.C. (1973). The topographical distribution of rods and cones in the adult chicken retina. *Exp. Eye Res.* **17**, 347–355.

Midtgard, U. (1984). The blood vascular system in the head of the herring gull (*Larus argentatus*). *J. Morphol.* **179**, 135–152.

Mikami, S. (1976). Ultrastructure of the organum vasculosum of the lamina terminalis of the Japanese quail, *Coturnix coturnix japonica*. *Cell Tissue Res.* **172**, 227–243.

Millar, T. J., and Chubb, I. W. (1984). Substance P in the chick retina: Effects of light and dark. *Brain Res.* **307**, 303–309.

Millar, T. J., Salipan, N., Oliver, J. O., Morgan, I. G., and Chubb, I. W. (1984). The concentration of enkephalin-like material in the chick retina is light dependent. *Neuroscience* **13**, 221–226.

Millar, T. J., Winder, C., Ishimoto, I., and Morgan, I. G. (1988). Putative serotonergic bipolar and amacrine cells in the chicken retina. *Brain Res.* **439**, 77–87.

Möller, W. (1978). Circumventriculäre Organe in der Gewebekultur. *Adv. Anat. Embryol. Cell Biol.* **54**, 3–95.

Möller, W. (1989). Immuncytochemische Zelltypisierung des Glykogenkörpers der Vögel. *Verh. Anat. Ges.* **82**. (Anat. Anz. Suppl. 164), 979–980.

Moody, S. A., and Heaton, M. B. (1983). Ultrastructural observations of the migration and early development of trigeminal motoneurons in chick embryos. *J. Comp. Neurol.* **216**, 20–35.

Morris, V. B. (1970). Symmetry in a receptor mosaic demonstrated in the chick from the frequencies, spacing and arrangement of the types of retinal receptor. *J. Comp. Neur.* **140**, 359–398.

Morris, V. B. (1975). Non-randomness in the sequential formation of principal cones in small areas of the developing chick retina. *J. Comp. Neurol.* **164**, 95–104.

Morris, V. B. (1976). Pattern analysis of developing photoreceptors in the chick retina. *In* "Structure of the Eye III" (E. Yamada, and S. Mishima, eds), pp. 239–246. *Jap. J. Ophthalmology*. University of Tokyo, Tokyo.

Morris, D. G. (1978). Development of functional motor innervation in supernumerary hindlimbs of the chick embryo. *J. Neurophysiol.* **41**, 1450–1465.

Müller, H., Drenckhahn, D., and Haase E. (1979). Vergleichend quantitative und ultrastrukturelle Untersuchungen am Geruchsorgan von vier Haustaubenrassen. *Z. mikrosk.-anat. Forsch.* **93**, 888–900.

Mugnaini, E., and Dahl, A. L. (1975). Mode of distribution of aminergic fibers in the cerebellar cortex of the chicken. *J. Comp. Neurol.* **162**, 417–432.

Muhibullah, M., Gargiulo, G., Nisticò, G., and Stephenson, J.D. (1983). Distribution of monoamine-containing neurons in the fowl brain (*Gallus domesticus*). *In* "Progress in Nonmammalian Brain Research" (G. Nisticò, and L. Bolis, eds), Vol I, pp. 81–112. CRC Press, Boca Raton, Florida.

Narayanan, C. H., and Narayanan, Y. (1978). Determination of the embryonic origin of the mesencephalic nucleus of the trigeminal nerve in birds. *J. Embryol. Exp. Morphol.* **43**, 85–105.

Narayanan, Y., and Narayanan, C.H. (1987). Neuronal development in the trigeminal mesencephalic nucleus of the duck under normal and hypothyroid states: I. A light microscopic morphometric analysis. *Anat. Rec.* **217**, 79–89.

Navascués, J., Rodríguez-Gallardo, L., Martin-Partido, G., and Alvarez, I. S. (1985). Proliferation of glial precursors during the early development of the chick optic nerve. *Anat. Embryol.* **172**, 365–373.

Necker, R. (1983a). Somatosensory system. *In* "Physiology and Behaviour of the Pigeon" (M. Abs, ed.), pp. 169–192. Academic Press, San Diego.

Necker, R. (1983b). Hearing. *In* "Physiology and Behaviour of the Pigeon" (M. Abs, ed.), pp. 193–219. Academic Press, San Diego.

Niessing, K., Scharrer, E., Scharrer, B., and Oksche, A. (1980). Die Neuroglia: Historischer Überblick. *In* "Handbuch der mikroskopischen Anatomie des Menschen" (A. Oksche, and C. Vollrath, eds), Vol. IV/10, Neuroglia, pp. 1–156. Springer, Berlin.

Nieuwenhuys, R. (1967). Comparative anatomy of the cerebellum. *Prog. Brain Res.* **25**, 1–93.

Nottebohm, F., and Arnold, A. P. (1976). Sexual dimorphism in vocal areas of the songbird brain. *Science* **194**, 211–213.

Nottebohm, F., Kasparian, S., and Pandazis, C. (1981). Brain space for a learned task. *Brain Res.* **213**, 99–109.

O'Flaherty, J. J. (1981). The optic nerve of the mallard duck: Fiber-diameter frequency distribution and physiological properties. *J. Comp. Neurol.* **143**, 17–24.

Okado, N., and Oppenheim, R. W. (1985). The onset and development of descending pathways to the spinal cord in the chick embryo. *J. Comp. Neurol.* **232**, 143–161.

Oksche, A. (1977). The neuroanatomical basis of avian neuroendocrine mechanisms. *In* "Proc. 1st Internat. Symp. Avian Endocrinology", Calcutta, Jan. 1977 (Conference Publication). (B.K. Follett, ed.), pp. 17–19.

Oksche, A. (1978a). Pattern of neuroendocrine cell complexes, (subunits), in hypothalamic nuclei: Neurobiological and phylogenetic concepts. *In* "Neurosecretion and Neuroendocrine Activity, Evolution, Structure, and Function" (W. Bargmann *et al.*, eds), pp. 64–71. Springer, Berlin.

Oksche, A. (1978b). Evolution, differentiation and organization of hypothalamic systems controlling reproduction: Neurobiological concepts. *In* "Neural Hormones and Reproduction. Brain Endocrine Interaction III" (D. E. Scott *et al.*, eds), pp. 1–15. Karger, Basel.

Oksche, A. (1980a). Structural organization of avian neuroendocrine systems. *In* "Biological Rhythms in Birds. Neural and Endocrine Aspects" (Y. Tanabe *et al.*, eds), pp. 3–15. Jpn. Sci. Soc. Press, Tokyo.

Oksche, A. (1980b). Structural principles of central neuroendocrine systems. *In* "Acta VII Congressus Internationalis Ornithologici" (R. Nöhring, ed.), pp. 217–222. *Verh. Deutsche Ornithol. Ges.*, Berlin.

Oksche, A., and Farner, D. S. (1974). Neurohistological studies of the hypothalamo-hypophysial system of *Zonotrichia leucophrys gambelli. Adv. Anat. Embryol. Cell Biol.* **48**, 1–136.

Oksche, A., and Hartwig, H.-G. (1980). Structural principles of central neuroendocrine systems. *In* "Avian Endocrinology" (A. Epple, and M. H. Stetson, eds), pp. 75–84. Academic Press, San Diego.

O'Leary, D. D., and Cowan, W. M. (1982). Further studies on the development of the isthmo-optic nucleus with special reference to the occurrence and fate of ectopic and ipsilaterally projecting neurons. *J. Comp. Neurol.* **212**, 399–416.

O'Leary, D. D., Gerfen, C. R., and Cowan, W. M. (1983). The development and restriction of the ipsilateral retinofugal projection in the chick. *Dev. Brain Res.* **10**, 93–109.

Oliver, J., Bouillé, C., Herbuté, S., and Baylé, J. D. (1977). Horseradish peroxidase study of intact or deafferented infundibular complex in *Coturnix* quail. *Neuroscience* **2**, 989–996.

Oppenheim, R. W. (1975). The role of supraspinal input in embryonic motility: A reexamination in the chick. *J. Comp. Neurol.* **160**, 37–50.

Oppenheim, R. W. (1984). Cell death of motoneurons in the chick embryo spinal cord. VIII. Motoneurons prevented from dying in the embryo persist after hatching. *Dev. Biol.* **101**, 35–39.

Oppenheim, R. W., and Majors-Willard, C. (1978). Neuronal cell death in the brachial spinal cord of the chick is unrelated to the loss of polyneuronal innervation in wing muscle. *Brain Res.* **154**, 148–152.

Oppenheim, R. W., Chu-Wang, I., and Maderdrut, J. L. (1978). Cell death of motoneurons in the chick embryo spinal cord. III. The differentiation of motoneurons prior to their induced degeneration following limb bud removal. *J. Comp. Neurol.* **177**, 87–112.

Ossenkopp, K.-P., and Barbeito, R. (1978). Bird orientation and the geomagnetic field: A review. *Neurosci. Biobehav. Rev.* **2**, 255–270.

Pannese, E. (1978). The response of the satellite and other non-neuronal cells to the degeneration of neuroblasts in chick embryo spinal ganglia. *Cell Tissue Res.* **190**, 1–14.

Panzica, G. C., and Viglietti-Panzica, C. (1983). A Golgi study of the parvocellular neurons in the paraventricular nucleus of the domestic fowl. *Cell Tissue Res.* **231**, 603–613.

Panzica, G. C., Viglietti-Panzica, C., and Contenti, E. (1982). Synaptology of neurosecretory cells in the nucleus paraventricularis of the domestic fowl. *Cell Tissue Res.* **227**, 79–92.

Panzica, G. C., Viglietti-Panzica, C., Calacagni, M. Anselmetti, G. C., and Balthazart, J. (1987). Sexual differentiation and hormonal control of the sexually dimorphic medial preoptic nucleus in the quail. *Brain Res.* **416**, 59–68.

Papi, F., Fiore, L., Fiasche, V., and Benvenuti, S. (1972). Olfaction and homing in pigeons. *Monit. Zool. Ital.* **6**, 85–95.

Parks, T. N. (1979). Afferent influences on the development of the brain stem auditory nuclei of the chicken: Otocyst ablation. *J. Comp. Neurol.* **183**, 665–678.

Parks, T. N. (1981a). Changes in the length and organization of nucleus laminaris dendrites after unilateral otocyst ablation in chick embryos. *J. Comp. Neurol.* **202**, 47–57.

Parks, T. N. (1981b). Morphology of axosomatic endings in an avian cochlear nucleus: Nucleus magnocellularis of the chicken. *J. Comp. Neurol.* **203**, 425–440.

Parks, T. N., and Rubel, E. W. (1975). Organization and development of brain stem auditory nuclei of the chicken: Organization of projections from *N. magnocellularis* to *N. laminaris*. *J. Comp. Neurol.* **164**, 435–448.

Parks, T. N., and Rubel, E. W. (1978). Organization and development of the brain stem auditory nuclei of the chicken: Primary afferent projections. *J. Comp. Neurol.* **180**, 439–448.

Parks, T. N., Collins, P., and Conlee, J. W. (1983). Morphology and origin of axonal endings in nucleus laminaris of the chicken. *J. Comp. Neurol.* **214**, 32–42.

Paton, J. A., and Manogue, K. R. (1982). Bilateral interactions within the vocal control pathway of birds: Two evolutionary alternatives. *J. Comp. Neurol.* **212**, 329–335.

Paul, E. (1971). Neurohistologische und fluoreszenzmikroskopische Untersuchungen über die Innervation des Glycogenkörpers der Vögel. *Z. Zellforsch.* **112**, 516–525.

Paul, E. (1973). Histologische und quantitative Studien am lumbalen Glykogenkörper der Vögel. *Z. Zellforsch.* **145**, 89–101.

Paula-Barbosa, M. M., and Sobrinho-Simoes, M. A. (1976). An ultrastructural morphometric study of mossy fiber endings in pigeon, rat and man. *J. Comp. Neurol.* **170**, 365–380.

Pearson, R. (1972), "The Avian Brain". Academic Press, San Diego.

Péczely, P. and Antoni, F. A. (1984). Comparative localization of neurons containing ovine corticotropin releasing factor (CRF)-like and neurophysin-like immunoreactivity in the diencephalon of the pigeon (*Columba livia domestica*). *J. Comp. Neurol.* **228**, 69–80.

Péczely, P., and Kiss, J. Z. (1988). Immunoreactivity to vasoactive intestinal polypeptide (VIP) and thyrotropin-releasing hormone (TRH) in hypothalamic neurons of the domesticated pigeon (*Columba livia*). Alterations following lactation and exposure to cold. *Cell Tissue Res.* **251**, 485–494.

Peduzzi, J. D., and Crossland, W. J. (1983a). Anterograde transneuronal degeneration in the ectomamillary nucleus and ventral lateral geniculate nucleus of the chick. *J. Comp. Neurol.* **213**, 287–300.

Peduzzi, J. D., and Crossland, W. J. (1983b). Morphology of normal and deafferented neurons in the chick ectomamillary nucleus. *J. Comp. Neurol.* **213**, 301–309.

Pessacq, T. P. (1967). Un organe paraventriculaire situé dans l'angle inférieur du quatrième ventricule des oiseaux. *C.R. Soc. Biol. (Paris)* **161**, 229–230.

Petrovicky, P. (1966). Reticular formation of the pigeon. *Folia Morphol.* **14**, 334–346.

Peusner, K. D., and Morest, D. K. (1977a). The neuronal architecture and topography of the nucleus vestibularis tangentialis in the late chick embryo. *Neuroscience* **2**, 189–207.

Peusner, K. D., and Morest, D. K. (1977b). A morphological study of neurogenesis in the nucleus vestibularis tangentialis of the chick embryo. *Neuroscience* **2**, 209–227.

Peusner, K. D., and Morest, D. K. (1977c). Neurogenesis in the nucleus vestibularis tangentialis of the chick embryo in the absence of the primary afferent fibers. *Neuroscience* **2**, 253–270.

Pfaff, D. W. (1976). The neuroanatomy of sex hormone receptors in the vertebrate brain. *In* "Neuroendocrine Regulation of Fertility". (T. C. Anand Kumar, ed.), pp. 30–45. Karger, Basel.

Polyak, S. (1957), "The Vertebrate Visual System". (H. Klüver, ed.). Chicago University Press, Chicago.

Popper, A. N., and Fay, R. R. (eds), (1980). "Comparative Studies of Hearing in Vertebrates". Springer, Berlin.

Portmann, A. (1950). Systèmes nerveux. *In* "Traité de Zoologie" Tome 15 "Les Oiseaux". (J. Benoit *et al.*, eds), pp. 184–203. Masson, Paris.

Prada, F. A., Armengol, D. A., and Genis-Gálvez, J. M. (1984). Displaced horizontal cells in the chick retina. *J. Morphol.* **182**, 221–225.

Puelles, L. (1978). A Golgi-study of oculomotor neuroblasts migrating across the midline in chick embryos. *Anat. Embryol.* **152**, 205–215.

Pumphrey, R. J. (1961). Sensory organs, vision. *In* "Biology and Comparative Physiology of Birds" (A. J. Marshall, ed.), pp. 55–68. Academic Press, San Diego.

Rager, G. (1978). System-matching by degeneration. II. Interpretation of the generation and degeneration of retinal ganglion cells in the chicken by a mathematical model. *Exp. Brain Res.* **33**, 79–90.

Rager, G. H. (1980a). Development of the retinotectal projection in the chicken. *Adv. Anat. Embryol. Cell Biol.* **63**, 1–92.

Rager, G. H. (1980b). Die Ontogenese der retinotopen Projektion. Beobachtung und Reflexion. *Naturwissenschaften* **67**, 280–287.

Rager, G., and Rager, U. (1978). System-matching by degeneration. I. A quantitative electron microscopic study of the generation and degeneration of retinal ganglion cells in the chicken. *Exp. Brain Res.* **33**, 65–78.

Ralph, C. L. (1975). The pineal gland and geographical distribution of animals. *Int. J. Biometeorol.* **19**, 289–303.

Rebiere, A., and Dainat, J. (1981). Quantitative study of synapse formation in the duck olfactory bulb. *J. Comp. Neurol.* **203**, 103–120.

Rehkämper, G. (1984). Remarks upon Ebbesson's presentation of a parcellation theory of brain development. *Z. zool. System. Evolutionsf.* **22**, 321–327.

Rehkämper, G., Zilles, K., and Schleicher, A. (1984). A quantitative approach to cytoarchitectonics. IX. The areal pattern of the hyperstriatum ventrale in the domestic pigeon, *Columba livia* f.d. *Anat. Embryol.* **169**, 319–327.

Rehkämper, G., Zilles, K., and Schleicher, A. (1985). A quantitative approach to cytoarchitectonics. X. The areal pattern of the neostriatum in the domestic pigeon, *Columba livia* f.d. A cyto- and myeloarchitectonical study. *Anat. Embryol.* **171**, 345–355.

Reid, F. A., Gasc, J.-M., Stumpf, W. E., and Sar, M. (1981). Androgen target cells in spinal cord, spinal ganglia, and glycogen body of chick embryos. Autoradiographic localization. *Exp. Brain Res.* **44**, 243–248.

Reiner, A., and Karten, H. J. (1982). Laminar distribution of the cells of origin of the descending tectofugal pathways in the pigeon (*Columba livia*). *J. Comp. Neurol.* **204**, 165–187.

Renggli, F. (1967). Vergleichend anatomische Untersuchungen über die Kleinhirn- und Vestibulariskerne der Vögel. *Rev. Suisse Zool.* **74**, 701–778.

Repérant, J., and Angaut, P. (1977). The retinotectal projections in the pigeon. An experimental optical and electron microscope study. *Neuroscience* **2**, 119–140.

Reubi, J. C., and Jessel, T. M. (1978). Distribution of substance P in the pigeon brain. *J. Neurochem.* **31**, 359–361.

Richfield, E. K., Young, A. B., and Penney, J. B. (1987). Comparative distribution of dopamine D-1 and D-2 receptors in the basal ganglia of turtles, pigeons, rats, cats, and monkeys. *J. Comp. Neurol.* **262**, 446–463.

Rieke, G. U., and Wenzel, B. M. (1978). Forebrain projections of the pigeon olfactory bulb. *J. Morphol.* **158**, 41–56.

Rodríguez, E. M., Hein, S., Rodríguez, S., Herrera, H., Peruzzo, B., Nualart, F., and Oksche, A. (1987). Analysis of the secretory products of the subcommissural organ. *In* "Functional Morphology of Neuroendocrine Systems" (B. Scharrer, H.-W. Korf, and H.-G. Hartwig, eds), pp. 189–202. Springer, Berlin.

Rogers, L. A., and Cowan, W. M. (1973). The development of the mesencephalic nucleus of the trigeminal nerve in the chick. *J. Comp. Neurol.* **147**, 291–320.

Rubel, E. W., Smith, Z. D., and Stewart, O. (1981). Sprouting in the avian brain stem auditory pathway: Dependence on dendritic integrity. *J. Comp. Neurol.* **202**, 397–414.

Ryan, S. M., Arnold, A. P., and Elde, R. P. (1981). Enkephalin-like immunoreactivity in vocal control regions of the zebra finch brain. *Brain Res.* **229**, 236–240.

Rylander, M. K. (1979). The nucleus mesencephalicus lateralis, pars dorsalis in neotropical passerines. *Brain Behav. Evol.* **16**, 315–318.

Sachs, M. B., Woolf, N. K., and Sinnott, J. M. (1980). Response properties of neurons in the avian auditory system: Comparisons with mammalian homologues and consideration of the neural encoding of complex stimuli. *In* "Comparative Studies of Hearing in Vertebrates" (A. N. Popper, and R. R. Fay, eds), pp. 323–353. Springer, Berlin.

Saini, K. D., and Leppelsack, H. J. (1977). Neuronal arrangement in the auditory field L of the neostriatum of the starling. *Cell Tissue Res.* **176**, 309–316.

Saini, K. D., and Leppelsack, H. J. (1981). Cell types of the auditory caudomedial neostriatum of the starling (*Sturnus vulgaris*). *J. Comp. Neurol.* **198**, 209–229.

Sako, H., Kojima, T., and Okado, N. (1986a). Immunohistochemical study on the development of serotoninergic neurons in the chick: I. Distribution of cell bodies and fibers in the brain. *J. Comp. Neurol.* **253**, 61–78.

Sako, H., Kojima, T., and Okado, N. (1986b). Immunohistochemical study on the development of serotoninergic neurons in the chick: II. Distribution of cell bodies and fibers in the spinal cord. *J. Comp. Neurol.* **253**, 79–91.

Sakurai, H. Kawashima, M., Kamiyoshi, M., and Tanaka, K. (1986). Effect of serotonin and β-endorphin on the release of luteinizing hormone in the hen (*Gallus domesticus*). *Gen. Comp. Endocrinol.* **63**, 24–30.

Saleh, C. N., and Ehrlich, D. (1984). Composition of the supraoptic decussation of the chick (*Gallus gallus*). A possible factor limiting interhemispheric transfer of visual information. *Cell Tissue Res.* **236**, 601–609.

Sano, Y., Ueda, S., Yamada, H., Takeuchi, Y., Goto, M., and Kawata, M. (1983). Immunohistochemical demonstration of serotonin-containing CSF-contacting neurons in the submammalian paraventricular organ. *Histochemistry* **77**, 423–430.

Sansone, F. M. (1977). The craniocaudal extent of the glycogen body in the domestic chicken. *J. Morphol.* **153**, 87–105.

Sansone, F. M., and Lebeda, F. J. (1976). A brachial glycogen body in the spinal cord of the domestic chicken. *J. Morphol.* **148**, 23–31.

Sarnat, H. B., and Netzky, M. G. (eds) (1981). "Evolution of the Nervous System". Oxford University Press, London.

Sarnat, H. B., Campa, J. F., and Lloyd, J. M. (1975). Inverse prominence of ependyma and capillaries in the spinal cord of vertebrates: A comparative histochemical study. *Am. J. Anat.* **143**, 439–450.

Sato, T., and Wake, K. (1983). Innervation of the avian pineal organ. A comparative study. *Cell Tissue Res.* **233**, 237–264.

Schall, U., Güntürkün, O., and Delius, J. D. (1986). Sensory projections to the nucleus basalis prosencephali of the pigeon. *Cell Tissue Res.* **245**, 539–546.

Scharrer, E. (1962). "Brain Function and the Evolution of Cerebral Vascularization", pp. 1–32. *Am. Mus. Nat. Hist.*, New York.

Scharrer, E. (1964). Photo-neuro-endocrine systems: General concepts. *Ann. N.Y. Acad. Sci.* **117**, 13–22.

Scharrer, B. (1987). Evolution of intercellular communication channels. *In* "Functional Morphology of Neuroendocrine Systems" (B. Scharrer, H.-W. Korf, and H.-G. Hartwig, eds), pp. 1–8. Springer, Berlin.

Scharrer, B., Korf, H.-W., and Hartwig, H.-G. (1987). Round table discussion. *In* "Functional Morphology of Neuroendocrine Systems" (B. Scharrer, H.-W. Korf, and H.-G. Hartwig, eds), pp. 223–225. Springer, Berlin.

Scheich, H. (1983). Two columnar systems in the auditory neostriatum of the chick: Evidence from 2-deoxyglucose. *Exp. Brain Res.* **51**, 199–205.

Scheich, H., Bonke, B. A., Bonke, D., and Langner, G. (1979). Functional organization of some auditory nuclei in the guinea fowl demonstrated by the 2-deoxyglucose technique. *Cell Tissue Res.* **204**, 17–27.

Schoebitz, K., Garrido, O., Heinrichs, M., Speer, L., and Rodríguez, E. M. (1986). Ontogenetical development of the chick and duck subcommissural organ. An immunocytochemical study. *Histochemistry* **84**, 31–40.

Schroeder, D. M., and Murray, R. G. (1987). Specializations within the lumbosacral spinal cord of the pigeon. *J. Morphol.* **194**, 41–53.

Schueler, R., and Lierse, W. (1970). Die Entwicklung der Kapillardichte nach der Geburt im Gehirn zweier Nestflüchter; eines Vogels (Haushuhn) und eines Säugers (Meerschweinchen). *Acta Anat.* **75**, 453–465.

Schwartzkopff, J. (1960). Physiologie der höheren Sinne bei Säugern und Vögeln. *J. Ornithol.* **101**, 61–91.

Schwartzkopff, J. (1973). Mechanoreception. *In* "Avian Biology" (D. S. Farner *et al.*, eds), Vol. 3, pp. 417–477. Academic Press, San Diego.

Schwarz, I. E., and Schwarz, D. W. (1983). The primary vestibular projection to the cerebellar cortex in the pigeon (*Columba livia*). *J. Comp. Neurol.* **216**, 438–444.

Schwarz, U., and Halfter, W. (1984). Interactions of axons with their environment: The chick retinotectal system as a model. *In* "Organizing Principles of Neural Development" (S. C. Sharma, ed.), pp. 343–359. NATO ASJ Series, Series A: Life Sciences, **78**.

Schwarz, I. E., Schwarz, D. W., Fredrickson, J. M., and Landolt, J. P. (1981). Efferent vestibular neurons: A study employing retrograde tracer methods in the pigeon (*Columba livia*). *J. Comp. Neurol.* **196**, 1–12.

Sedláček, J. (1977). Development of spontaneous motility in chick embryos. Sensitivity to aminergic transmitters. *Physiol. Bohemoslov.* **26**, 425–433.

Sedláček, J. (1979). Development of spontaneous motility in chick embryos. Significance of basic brain regions. *Physiol. Bohemoslov.* **28**, 193–200.

Sedláček, J., and Doskocil, M. (1978). Development of spontaneous motility in chick embryos. Supraspinal control. *Physiol. Bohemoslov.* **27**, 7–14.

Sharp, P. J., Klandorf, H., and Lea, R. W. (1984). Influence of lighting cycles on daily rhythms in concentrations of plasma tri-iodothyronine and thyroxine in intact and pinealectomized immature broiler hens (*Gallus gallus*). *J. Endocr.* **103**, 337–345.

Shiosaka, S., Takatsuki, K., Inagaki, S., Sakanaka, M., Takagi, H., Senba, E., Matsuzaki, T., and Tohyama, M. (1981). Topographic atlas of somatostatin-containing neuron systems in the avian brain in relation to catecholamine-containing neuron system. II. Mesencephalon, rhombencephalon, and spinal cord. *J. Comp. Neurol.* **202**, 115–124.

Sillman, A. J. (1973). Avian vision. *In* "Avian Biology" (D. S. Farner *et al.*, eds), Vol. 3, pp. 349–388. Academic Press, San Diego

Simon, E., Eriksson, S., Gerstberger, R., Gray, D. A., and Simon-Oppermann, C. (1987). Comparative aspects of osmoregulation. *In* "Functional Morphology of Neuroendocrine Systems" (B. Scharrer, H.-W. Korf, and H.-G. Hartwig, eds), pp. 37–49. Springer, Berlin.

Smith, Z. D. (1981). Organization and development of brain stem auditory nuclei of the chicken: Dendritic development in *N. laminaris*. *J. Comp. Neurol.* **203**, 309–333.

Smith, Z. D, and Rubel, E. W. (1979). Organization and development of brain stem auditory nuclei of the chicken: Dendritic gradients in nucleus laminaris. *J. Comp. Neurol.* **186**, 213–240.

Smith, Z. D., Gray, L., and Rubel, E. W. (1983). Afferent influences on brainstem auditory nuclei of the chicken: *N. laminaris* dendritic length following monaural conductive hearing loss. *J. Comp. Neurol.* **220**, 199–205.

Snyder, A. W., and Miller, W. H. (1978). Telephoto lens system of falconiform eyes. *Nature* **275**, 127–129.

Soblosky, J. S., DuMontier, G., and Jeng, I. (1985). Down-regulation of ($^3$H)5-hydroxytryptamine binding sites in chick embryo brain by monoamine oxidase inhibitors or fenfluramine and potentiation by d,l-5-hydroxytryptophan. *J. Neurochem.* **45**, 1923–1931.

Sohal, G. S. (1976). Effects of deafferentation on the development of the isthmo-optic nucleus in the duck (*Anas platyrhynchos*). *Exp. Neurol.* **50**, 161–173.

Sohal, G. S. (1977). Development of the oculomotor nucleus, with special reference to the time of cell origin and cell death. *Brain Res.* **138**, 217–228.

Sohal, G. S., and Holt, R. K. (1978). Identification of the trochlear motoneurons by retrograde transport of horseradish peroxidase. *Exp. Neurol.* **59**, 509–514.

Starck, D. (1962). Die Evolution des Säugetier-Gehirns. Sitzungsber. wiss. Gesellsch. J.W. Goethe University, Frankfurt/FRG, Vol. 1, pp. 23–60. Steiner Verlag, Wiesbaden.

Starck, D. (1982). "Vergleichende Anatomie der Wirbeltiere auf evolutionsbiologischer Grundlage". Vol. 3: Organe des aktiven Bewegungsapparates, der Koordination der Umweltbeziehungen, des Stoffwechsels und der Fortpflanzung. Springer, Berlin.

Stensaas, L. J., and Stensaas, S. S. (1968). Light microscopy of glial cells in turtles and birds. *Z. Zellforsch.* **91**, 315–340.

Sterling, R. J., and Sharp, P. J. (1982). The localisation of LH-RH neurones in the diencephalon of the domestic hen. *Cell Tissue Res.* **222**, 283–298.

Sterling, R. J., Gase, J. M., Sharp, P. J., Renoir, J. M., Tuohimaa, P., and Baulieu, E. E. (1987). The distribution of nuclear progesterone receptor in the hypothalamus and forebrain of the domestic hen. *Cell Tissue Res.* **248**, 201–205.

Stewart, P. A., and Wiley, M. J. (1981). Structural and histochemical features of the avian blood–brain barrier. *J. Comp. Neurol.* **202**, 157–167.

Stingelin, W. (1958). "Vergleichend morphologische Untersuchungen am Vorderhirn der Vögel auf cytologischer und cytoarchitektonischer Grundlage". Helbing & Lichtenhahn, Basel.

Stingelin, W. (1961). Größenunterschiede des sensiblen Trigeminuskerns bei verschiedenen Vögeln. *Rev. Suisse de Zool.* **68**, 247–251.

Stingelin, W. (1965). "Qualitative und quantitative Untersuchungen an Kerngebieten der Medulla oblongata bei Vögeln" (E. A.Boyden *et al.*, eds), Bibl. Anat. Vol. 6, pp. 1–116. Karger, Basel.

Stingelin, W., and Senn, D. G. (1969). Morphological studies on the brain of Sauropsida. *Ann. New York Acad. Sci.* **167**, 156–163.

Stirling, R. V., and Summerbell, D. (1977). The development of functional innervation in the chick wing-bud following truncations and deletions of the proximal-distal axis. *J. Embryol. exp. Morphol.* **41**, 189–207.

Stoeckel, M. E., Roussel, G., Zwiller, J., Madarasz, B., and Porte A. (1976). Concentration of dopaminergic fibres in the marginal zone of the bird retina. A non-visual photoreceptor system? *Cell Tissue Res.* **173**, 335–341.

Straznicky, Ch., and Tay, D. (1983). The localization of motoneuron pools innervating wing muscles in the chick. *Anat. Embryol.* **166**, 209–218.

Streit, P., Burkhalter, A., Stella, M., and Cuénod, M. (1980). Patterns of activity in pigeon brain's visual relays as revealed by the ($^{14}$C)2-deoxyglucose method. *Neuroscience* **5**, 1053–1066.

Tagawa, T., Ando, K., and Wasano, T. (1979). A histochemical study of the innervation of the cerebral blood vessels in the domestic fowl. *Cell Tissue Res.* **198**, 43–51.

Takahashi, J. S., and Menaker, M. (1982). Role of the suprachiasmatic nuclei in the circadian system of the house sparrow, *Passer domesticus. J. Neurosci.* **2**, 815–828.

Takasaka, T., and Smith, C. A. (1971). The structure and innervation of the pigeon's basilar papilla. *J. Ultrastruct. Res.* **35**, 20–65.

Takatsuki, K., Shiosaka, S., Inagaki, S., Sakanaka, M., Takagi, H., Senba, E., Matsuzaki, T., and Tohyama, M. (1981). Topographic atlas of somatostatin-containing neuron system in the avian brain in relation to catecholamine-containing neuron system. I. Telencephalon and diencephalon. *J. Comp. Neurol.* **202**, 103–113.

Takei, Y. (1977a). The role of the subfornical organ in drinking induced by angiotensin in the Japanese quail, *Coturnix coturnix japonica. Cell. Tissue Res.* **185**, 175–181.

Takei, Y. (1977b). Angiotensin and water intake in the Japanese quail (*Coturnix coturnix japonica*). *Gen. Comp. Endocrinol.* **31**, 364–372.

Takei, Y., Tsuneki, K., and Kobayashi, H. (1978). Surface fine structure of the subfornical organ in the Japanese quail (*Coturnix coturnix japonica*). *Cell Tissue Res.* **191**, 389–404.

Takei, Y., Kobayashi, H., Yanagisawa, M., and Bando, T. (1979). Involvement of catecholaminergic nerve fibers in angiotensin II-induced drinking in the Japanese quail, *Coturnix coturnix japonica. Brain Res.* **174**, 229–244.

Tanaka, K., and Smith, C. A. (1978). Structure of the chicken's inner ear: SEM and TEM study. *Amer. J. Anat.* **153**, 251–272.

Teuchert, G. and Kretschel, A. (1985). Imprinting of the Peking duck (*Anas platyrhynchos*) and dependence on exposure to light during ontogenesis. *Experientia* **41**, 400–402.

Thali, L. (1972). "Vergleichend morphologische und embryologische Untersuchungen am optischen Zentrum im Gehirn der Vögel". Thesis, University Basel, Switzerland.

Thanos, S., and Bonhoeffer, F. (1984). Development of the transient ipsilateral retinotectal projection in the chick embryo: A numerical fluorescence-microscopic analysis. *J. Comp. Neurol.* **224**, 407–414.

Thommes, R. C., Caliendo, J., and Woods, J. E. (1985). Hypothalamo-adenohypophyseal-thyroid interrelationships in the developing chick embryo. VII. Immunocytochemical demonstration of thyrotrophin-releasing hormone. *Gen. Comp. Endocrinol.* **57**, 1–9.

Thornton, S. N. (1986). Osmoreceptor localization in the brain of the pigeon (*Columba livia*). *Brain Res.* **377**, 96–104.

Tohyama, M., Maeda, T., Hashimoto, J., Shrestha, G. R., and Tamura, O. (1974). Comparative anatomy of the locus coeruleus. I. Organization and ascending projections of the catecholamine-containing neurons in the pontine region of the bird, *Melopsittacus undulatus*. *J. Hirnforsch.* **15**, 319–330.

Tolhurst, B. E., and Vince, M. A. (1976). Sensitivity to odours in the embryo of the domestic fowl. *Anim. Behav.* **24**, 772–779.

Tsai, H. M., Garber, B. B., and Larramendi, L. M. (1981a). 3H-thymidine autoradiographic analysis of telencephalic histogenesis in the chick embryo: I. Neuronal birthdates of telencephalic compartments in situ. *J. Comp. Neurol.* **198**, 275–292.

Tsai, H. M., Garber, B. B., and Larramendi, L. M. (1981b). $^3$H-thymidine autoradiographic analysis of telencephalic histogenesis in the chick embryo: II. Dynamics of neuronal migration, displacement, and aggregation. *J. Comp. Neurol.* **198**, 293–306.

Tsuneki, K., Takei, Y., and Kobayashi, H. (1978). Parenchymal fine structure of the subfornical organ in the Japanese quail, *Coturnix coturnix japonica*. *Cell Tissue Res.* **191**, 405–419.

Ueck, M. (1981). Variation in structure and function of the pineal systems. *In* "The Pineal Organ: Photobiology–Biochronometry–Endocrinology" (A. Oksche, and P. Pévet, eds), pp. 151–168. Elsevier, Amsterdam.

Uehara, M., and Ueshima, T. (1982). Development of the glycogen body through the whole length of the chick spinal cord. *Anat. Rec.* **202**, 511–519.

Underwood, H., Binkley, S., Siopes, T., and Mosher, K. (1984). Melatonin rhythms in the eyes, pineal bodies, and blood of Japanese quail (*Coturnix coturnix japonica*). *Gen. Comp. Endocrinol.* **56**, 70–81.

Van Rybroek, J. J., and Low, F. N. (1982). Intercellular junctions in the developing arachnoid membrane in the chick. *J. Comp. Neurol.* **204**, 32–43.

van Tienhoven, A., and Juhasz, L. P. (1962). The chicken telencephalon, diencephalon and mesencephalon in stereotaxic coordinates. *J. Comp. Neurol.* **118**, 185–198.

van Tienhoven, A., and Planck, R. J. (1973). The effect of light on avian reproductive activity. *In* "Handbook of Physiology—Endocrinology II, Part I" (R.O. Greep, ed.), pp. 79–107. Am. Physiol. Soc., Washington.

Vielvoye, G. J., and Voogd, J. (1977). Time dependence of terminal degeneration in spino-cerebellar mossy fiber rosettes in the chicken and the application of terminal degeneration in successive degeneration experiments. *J. Comp. Neurol.* **175**, 233–242.

Vigh, B. (1971), "Das Paraventrikularorgan und das zirkumventrikuläre System". *Stud. biol. Acad. Sci. hung.* Vol. X, Akadémiai Kiadó, Budapest.

Vigh, B., and Vigh-Teichmann, I. (1973). Comparative ultrastructure of the cerebrospinal fluid-contacting neurons. *Int. Rev. Cytol.* **35**, 189–251.

Vigh, B., Vigh-Teichmann, I., Koritsánszky, S., and Aros, B. (1971). Ultrastructure of the spinal CSF contacting neuronal system in the white leghorn chicken. *Acta morphol. Acad. Sci. hung.* **19**, 9–24.

Vigh, B., Vigh-Teichmann, I., and Aros, B. (1975). Comparative ultrastructure of cerebrospinal fluid-contacting neurons and pinealocytes. *Cell Tissue Res.* **158**, 409–424.

Vigh, B., Vigh-Teichmann, I., Röhlich, P., and Aros, B. (1982). Immunoreactive opsin in the pineal organ of reptiles and birds. *Z. mikroskop.-anat. Forsch.* **96**, 113–129.

Vigh, B., Vigh-Teichmann, I., Manzano e Silva, M. J., and van den Pol, A. N. (1983). Cerebrospinal fluid-contacting neurons of the central canal and terminal ventricle in various vertebrates. *Cell Tissue Res.* **231**, 615–621.

Viglietti-Panzica, C., and Bessé, M. C. (1984). The organum vasculosum laminae terminalis of the domestic fowl: A Golgi and Ultrastructural study. *Anat. Anz.* **155**, 341–353.

Viglietti-Panzica, C., and Contenti, E. (1983). Cytodifferentiation of the paraventricular nucleus in the chick embryo. A Golgi and electron-microscopic study. *Cell Tissue Res.* **229**, 281–297.

Viglietti-Panzica, C. and Panzica, G. C. (1987). Neuronal typology of the vasotocin-immunoreactive system. Golgi and immunocytochemical studies of the diencephalon in the quail. *In* "Functional

Morphology of Neuroendocrine Systems" (B. Scharrer, H.-W. Korf, and H.-G. Hartwig, eds), p. 68. Springer, Berlin.

Vitums, A., Michami, S. J., and Farner, D. S. (1965). Arterial blood supply to the brain of the white-crowned sparrow. *Anat. Anz.* **116**, 309–326.

Vollrath, L. (1980). The pineal organ. *In* "Handbuch der mikroskopischen Anatomie des Menschen" (A. Oksche and L. Vollrath, eds), Vol. 6/7, pp. 1–589. Springer, Berlin.

Vollrath, F. W., and Delius, J. D. (1976). Vestibular projections to the thalamus of the pigeon. *Brain Behav. Evol.* **13**, 56–68.

Voorhuis, T. A., Kiss, J. Z., deKloet, E. R., and deWied, D. (1988). Testosterone-sensitive vasotocin-immunoreactive cells and fibers in the canary brain. *Brain Res.* **442**, 139–146.

Vowels, D. M., Beazley, L., and Harwood, D. H. (1975). A stereotaxic atlas of the brain of the barbary dove (*Streptopelia risoria*). *In* "Neural and Endocrine Aspects of Behaviour in Birds" (P. Wright *et al.*, eds), pp. 351–394. Elsevier, Amsterdam.

Wallace, J. A. (1982). Monoamines in the early chick embryo: Demonstration of serotonin synthesis and the regional distribution of serotonin-concentrating cells during morphogenesis. *Am. J. Anat.* **165**, 262–276.

Wallhäusser, E., and Scheich, H. (1987). Auditory imprinting leads to differential 2-deoxyglucose uptake and dendritic spine loss in the chick rostral forebrain. *Dev. Brain Res.* **31**, 29–44.

Wallraff, H. G. (1979). Olfaction and homing in pigeons. A problem of navigation or of motivation? *Naturwissenschaften* **66**, 269.

Watanabe, M., Ito, H., and Ikushima, M. (1985). Cytoarchitecture and ultrastructure of the avian ectostriatum: Afferent terminals from the dorsal telencephalon and some nuclei in the thalamus. *J. Comp. Neurol.* **236**, 241–257.

Weidner, C., Repérant, J., Desroches, A.-M., Miceli, D., and Vesselkin, N. P. (1987). Nuclear origin of the centrifugal visual pathway in birds of prey. *Brain Res.* **436**, 153–160.

Weindl, A., and Sofroniew, M. V. (1982). Peptide neurohormones and circumventricular organs in the pigeon. *Front. Horm. Res.* **9**, 88–104.

Weindl, A., Sofroniew, M. V., Mestres, P., and Wetzstein, R. (1980). Immunhistochemische Lokalisation von neurohypophysären Peptiden im Gehirn der Taube (*Columba livia*). *Verh. Anat. Ges.* **74**, 769–774.

Welsch, U., and Wächtler, K. (1969). Zum Feinbau des Glykogenkörpers im Rückenmark der Taube. *Z. Zellforsch.* **97**, 160–169.

Wenger, T., and Törk, I. (1968). Studies on the organon vasculosum laminae terminalis. II. Comparative morphology of the organon vasculosum laminae terminalis of fishes, amphibia, reptilia, birds and mammals. *Acta biol. Acad. Sci. hung.* **19**, 83–96.

Wenzel, B. M. (1983). Chemical senses. *In* "Physiology and Behaviour of the Pigeon" (M. Abs, ed.) pp. 149–167. Academic Press, San Diego.

Wild, J. M. (1981). Identification and localization of the motor nuclei and sensory projections of the glossopharyngeal, vagus, and hypoglossal nerves of the cockatoo, *Cacatua roseicapilla*, Cacatuidae. *J. Comp. Neurol.* **203**, 351–377.

Wild, J. M. (1985). The avian somatosensory system. I. Primary spinal afferent input to the spinal cord and brainstem in the pigeon (*Columba livia*). *J. Comp. Neurol.* **240**, 377–395.

Wild, J. M. (1987). The avian somatosensory system: connections of regions of body representation in the forebrain of the pigeon. *Brain Res.* **412**, 205–223.

Wild, J. M., and Arends, J. J. (1984). A trigeminal sensorimotor circuit for pecking, grasping and feeding in the pigeon (*Columba livia*). *Brain Res.* **300**, 146–151.

Wild, J. M., and Arends, J. J. (1987). A respiratory–vocal pathway in the brainstem of the pigeon. *Brain Res.* **407**, 191–194.

Wild, J. M., and Zeigler, H. P. (1980). Central representation and somatotopic organization of the jaw muscles within the facial and trigeminal nuclei of the pigeon (*Columba livia*). *J. Comp. Neurol.* **192**, 175–201.

Wild, J. M., Arends, J. J., and Ziegler, H. P. (1985). Telencephalic connections of the trigeminal system in the pigeon (*Columba livia*): a trigeminal sensorimotor circuit. *J. Comp. Neurol.* **234**, 441–464.

Wingstrand, K. G. (ed.) (1951). "The Structure and Development of the Avian Pituitary". Gleerups, Lund.

Wingstrand, K. G., and Munk, O. (1965). The pecten oculi of the pigeon with particular regard to its function. *Biol. Skr. Dan. Vid. Selsk.* **14**, 1–64.

Wold, J. E. (1976). The vestibular nuclei in the domestic hen (*Gallus domesticus*). I. Normal anatomy. *Anat. Embryol.* **149**, 29–46.

Wold, J. E. (1978a). The vestibular nuclei in the domestic hen (*Gallus domesticus*). IV. The projection to the spinal cord. *Brain Behav. Evol.* **15**, 41–62.

Wold, J. E. (1978b). The vestibular nuclei in the domestic hen (*Gallus domesticus*). III. Ascending projections to the mesencephalic eye motor nuclei. *J. Comp. Neurol.* **179**, 393–405.

Wold, J. E. (1981). The vestibular nuclei in the domestic hen (*Gallus domesticus*). VI. Afferents from the cerebellum. *J. Comp. Neurol.* **201**, 319–341.

Wold, J. E., and Hall, J. G. (1975). The distribution of primary afferents to the cochlear nuclei in the domestic hen (*Gallus domesticus*). *Anat. Embryol.* **147**, 75–89.

Wright, L. L. (1981). Time of cell origin and cell death in the avian dorsal motor nucleus of the vagus. *J. Comp. Neurol.* **199**, 125–132.

Yamada, S., and Mikami, S. (1981). Immunocytochemical localization of neurotensin-containing neurons in the hypothalamus of the Japanese quail, *Coturnix coturnix japonica*. *Cell. Tissue Res.* **218**, 29–39.

Yamada, S., Mikami, S., and Yanaihara, N. (1982). Immunohistochemical localization of vasoactive intestinal polypeptide (VIP)-containing neurons in the hypothalamus of the Japanese quail, *Coturnix coturnix japonica*. *Cell. Tissue Res.* **226**, 13–26.

Yamamoto, K., Tohyama, M., and Shimizu, N. (1977). Comparative anatomy of the topography of catecholamine containing neuron system in the brain stem from birds to teleosts. *J. Hirnforsch.* **18**, 229–240.

Yazulla, S. (1974). Intraretinal differentiation in the synaptic organization of the inner plexiform layer of the pigeon retina. *J. Comp. Neurol.* **153**, 309–324.

Yokoyama, K., Oksche, A., Darden, T. R., and Farner, D. S. (1978). The sites of photoreception in photoperiodic induction of the growth of the testes in the white-crowned sparrow, *Zonotrichia leucophrys gambelii*. *Cell. Tissue Res.* **189**, 441–467.

Young, R. W. (1978). The daily rhythm of shedding and degradation of rod and cone outer segment membranes in the chick retina. *Invest. Ophthalmol. Vis. Sci.* **17**, 105–116.

Youngren, O. M., and Phillips, R. E. (1978). A stereotaxic atlas of the brain of the three-day-old domestic chick. *J. Comp. Neurol.* **181**, 567–599.

Youngren, O. M., and Phillips, R. E. (1983). Location and distribution of tracheosyringeal motorneuron somata in the fowl. *J. Comp. Neurol.* **213**, 86–93.

Zatz, M., Mullen, D. A., and Moskal, J.R. (1988). Photoendocrine transduction in cultured chick pineal cells: effects of light, dark, and potassium on the melatonin rhythm. *Brain Res.* **438**, 199–215.

Zeevalk, G. D. and Hyndman, A. G. (1987). Transferrin in chick retina: distribution and location during development. *Dev. Brain Res.* **37**, 231–241.

Zeigler, H. P., and Karten, H. J. (1973). Brain mechanisms and feeding behavior in the pigeon (*Columba livia*). I. Quinto-frontal structures. *J. Comp. Neurol.* **152**, 59–82.

Zeier, H., and Karten, H. (1971). The archistriatum of the pigeon: Organization of afferent and efferent connections. *Brain Res.* **31**, 313–326.

Zenner, H.-P., Arnold, W., and Gitter, A. H. (1988). Outer hair cells as fast and slow cochlear amplifiers with a bidirectional transduction cycle. *Acta Otolaryngol.* (*Stockh*) **105 (5–6)**, 457–462.

Ziegels, J. (1976). The vertebrate subcommissural organ: A structural and functional review. *Arch. Biol.* **87**, 429–476.

Zilles, K., Becker, C.-M., and Schleicher, A. (1981). Transmission blockade during neuronal development. Observations on the trochlear nucleus with quantitative histological methods and with ultrastructural and axonal transport studies in the chick embryo. *Anat. Embryol.* **163**, 87–123.

# Chapter 2

# THE ORIGIN OF FEATHERS: A NOVEL APPROACH

*Alan H. Brush*

*Department of Physiology and Neurobiology*
*University of Connecticut*
*Storrs, Connecticut 06268*



## I. Introduction

Feathers occur on all birds (at least as adults) but not on any other animals. Like many other structures feathers appeared abruptly in the fossil record and represent a morphological innovation. The form and biological role of their immediate structural precursors are unknown, but fossil feathers are found in essentially contemporaneous deposits on several continents, providing evidence for a rapid radiation of birds early in their history. The earliest bird fossils, *Archaeopteryx*, possessed feathers with differentiated shapes, presumably related to different functions (Feduccia and Tordoff, 1979). In addition to feathers, claws, which also have an epidermal origin, were present (Wellnhofer, 1974). Paradoxically, no scale fossils have been found, which may be a result of either their absence or their failure to fossilize. Nor is there evidence for a horny cover of the snout (ramphotheca). This might be important in elucidating the evolutionary history of feathers, a subject that remains controversial and has recently come under intensive investigation.



Avian Biology, Vol. IX

ISBN 0-12-249409-1

There is a long-held assumption that feathers arose directly from scales (Maderson, 1972a,b). In fact, the nature and function of structures intermediate between scales and feathers are speculative. Actually, there might have been multiple shapes and several roles for the intermediate structures. A few of the most obvious have been described (Regal, 1975; Dyck, 1985). Function at any given time might have varied and even been less than optimal. One important aspect of these structures was that rapid phenotypic change was possible in the early phases of evolution. The earliest fossilized structures were completely modern in structure (see figures in Hecht *et al.*, 1985). Further, no contemporary feather structure is identifiable unequivocally as "primitive". Thus there is no fossil evidence for the sequence of events or structures that led to feathers.

Feathers are, however, an exceptional model for the emergence of evolutionary novelty. Recent work on the biophysics, biochemistry and molecular biology of feather, scale, and claw has provided insights into their complex molecular nature and revealed some of the processes involved in their construction and evolution. As a result of these and other advances feathers provide an unexcelled opportunity for the investigation of the evolution of molecular structure and biophysical processing and illustrate how morphological and functional diversity can be derived from limited information and relatively simple structural units.

There are no genes for feathers. The information for their production resides in the timing for production of a set of unique protein molecules. This occurs only in follicles during well-defined periods of molt. The conversion of the protein into feathers involves a production-line-like series of events. The proteins and the processes have an evolutionary history that may be reconstructed, in part, from their contemporary manifestations. Some of the same processes may account for the origin of feathers and explain their morphological diversity. One approach is to understand the nature of the structural elements and the limits to their interactions. Their evolutionary history adds another dimension. We presume that their interactions follow a set of rules, perhaps determined by their shape. It is also clear that in any series of such interactions it is not possible to predetermine either the nature of the product or its subsequent behavior. Emergent properties typify many biological systems and ultimately provide insight into our understanding.

Unlike many biological systems, feathers consist of molecular assemblies of increasing complexity (Hinegardner and Engelberg, 1983; Wicken, 1984). Each level comprises only one or a few types of molecules. The processes by which components assemble and subsequently generate morphological features are still generally unknown (Cohen, 1979). Possibly much of the mechanism resides within the components themselves. The interactions among components are dictated by the nature of the constituent molecules and constrained by the information inherent in their structure. Protein folding, filament formation and the interactions of cytoskeletal elements (Sakai, 1980; Horwitz *et al.*, 1981; Jacobson, 1983) are general examples of this process. Organization at one level may be guided by structural characteristics or behavior at the antecedent level. Thus hierarchical organization emerges from the interactions of smaller and simpler elements. Levels are recognized and defined by suites of characteristics from the preceding level. A complex interaction exists among emergent characters, fabrication processes, molecular and supramolecular structure that must be carefully untangled (Løvtrup, 1984; Authur, 1984).

Although it is axiomatic that hierarchical organization exists in organisms, the nature of any particular level is rarely predictable totally from information contained in the antecedent level (Alberch, 1980). Because evolutionary processes involve a hierarchical organization (Salthe, 1975; Gould, 1980, 1982b; Webster and Goodwin, 1981; Klee, 1984), the transformations across level boundaries becomes an important consideration. Yet, a conceptual scheme based on a hierarchical organization of structure and complexity is still under development (Ho and Saunders, 1979; Arnold and Fristrup, 1982; Campbell, 1982). Each level in a hierarchy contains specific information, unique structures and is controlled by processes appropriate to the scale of that level. Only a portion of the information in the genome is "hard-wired" (Stent, 1981); an undetermined quantity is epigenetic in origin (Rachootin and Thompson, 1981). Epigenetic mechanisms modify the idea of simple linearly arranged genomes as they require fewer genes to produce complex structures or control physiological processes (Løvtrup, 1983). Hierarchies also afford a high degree of plasticity and amplification of the primary effects of genes. The models used to study these relationships are critically important to understanding theory (Gould, 1982a; Charlesworth *et al.*, 1982). For example, it is not yet possible to map any gross phenotypic feature directly onto the genotype, as is possible for single gene products. Nor is there convincing evidence for a central control program that specifies all the steps in the construction and final form of any morphological feature. Biologists have begun to model differentiation and development but no model has been tested adequately (Cheverud, 1984; Stubblefield, 1986; Bailey, 1986). Further, although information on the amino acid sequence of structural proteins occurs specifically in the genome, there is presently no reliable method for predicting subsequent supramolecular interactions and the behavior of molecules. Thus, many properties of structure and pattern emerge as a consequence of the constructional material and ontogenetic processes. Minor changes in processing might have extensive structural implications.

It is instructive to consider feathers as a clustered hierarchy of structures. As each structural level may have different rules for fabrication, a hierarchical approach can encompass differences in aspects of genetic organization and control, constructional materials, and fabricational processes and thereby attempt to account for emergent properties. In this chapter I will consider aspects of the genomic organization of feather production, the nature of the feather ($\phi-$)proteins, and the processes through which supramolecular structures are produced. I will use these features to attempt to understand the processes by which the molecular substructure of feathers might have evolved and influenced the present contemporary diversity. Potential relationships between processes and products will be explored (Fig. 1). I hope to provide a molecular perspective that will both complement and supplement the information currently available from morphological and developmental analysis. The analysis can be extended to include molecular processes in reconstructing the evolutionary events surrounding the appearance of feathers, a dimension not previously addressed. It is natural to consider feathers as a novelty whose emergence was the combined product of the emergence of a new set of proteins, changes in ontogeny and a restructuring of physical processes.

## II. Origin of Feathers

The orthodox explanation of the origin of feathers considers them to be descended directly from reptilian scale (Spearman, 1966; Bereiter-Hahn *et al.*, 1986). Various scenarios for the evolutionary pathways and possible selective forces have been constructed (Maderson, 1972a,b; Regal, 1975). The arguments for the traditional viewpoint were developed before the vast differences among reptilian structures and between reptilian and avian structures were demonstrated at the molecular level (Wyld and Brush, 1979, 1983). Although all reptiles possess scale-like structures, the variation within and between species remains unappreciated at both the molecular and morphological levels. Furthermore, there have been no convincing demonstrations of mechanisms for the conversion of a scale to a feather (Fig. 2). Such a model would have to account for both the molecular and morphological changes and contain pathways by which the observed variation could be generated and maintained. Ewart (1921) was probably the first to dispute the traditional scale–feather relationship. He argued from an ontogenetic viewpoint and attempted to marshall evidence for a separate origin of feathers. It is now realized that much of the geometry and processing involved in such a change in the production of epidermal structures can be the result of purely physical forces (Oster *et al.*, 1980; Oster and Alberch, 1982; Ettenshon, 1985). The control mechanisms are not necessarily derived from genetic similarity, but reflect the organization of the cells, constraints in fabrication inherent in fibrous molecules, and the nature of the physical or mechanical interactions of cell and molecules at several possible hierarchical levels. It may therefore be appropriate to reconsider models of the emergence of feathers and the relationships among epidermal appendages.

Explanations of the role of early structures (or protofeathers) often depend on behavioral interpretations of fossil evidence. The traditional hypotheses (Parkes,

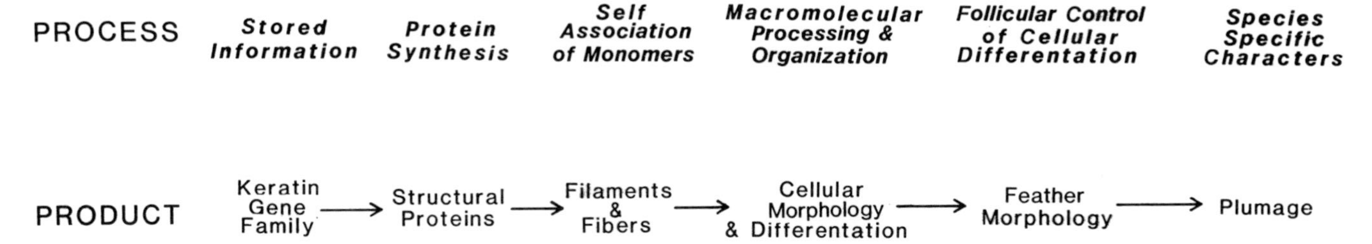


FIG. 1. Nature of the hierarchical organization of feathers. The chemical and physical characteristics in each step are unique, depend on the previous events, and constrain subsequent processing. Both processes and products are indicated. Some transitions (the horizontal arrows) may occur as physical processes (e.g., protein folding, filament formation, tissue organization) and do not require genetic information.

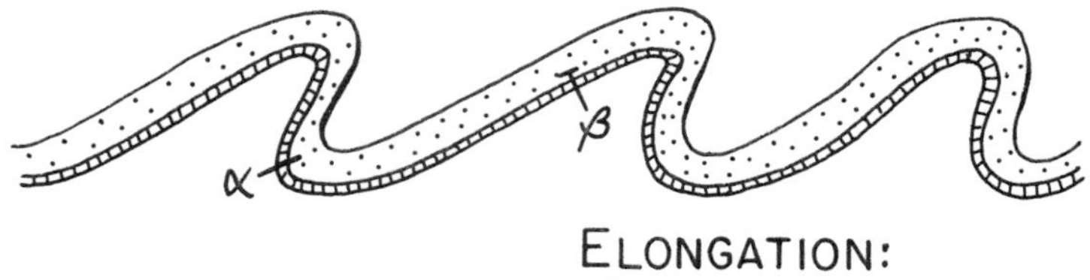


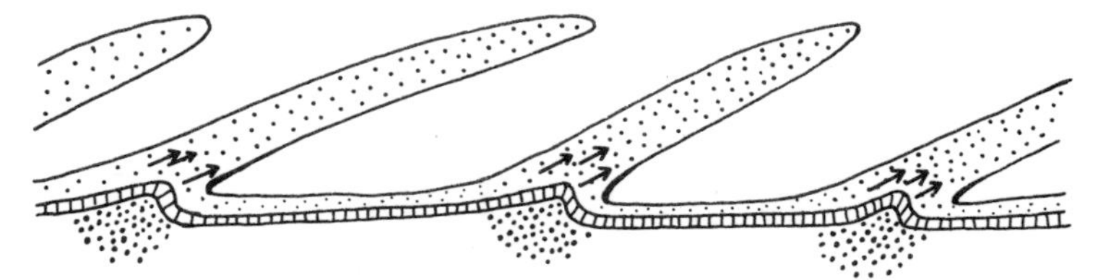


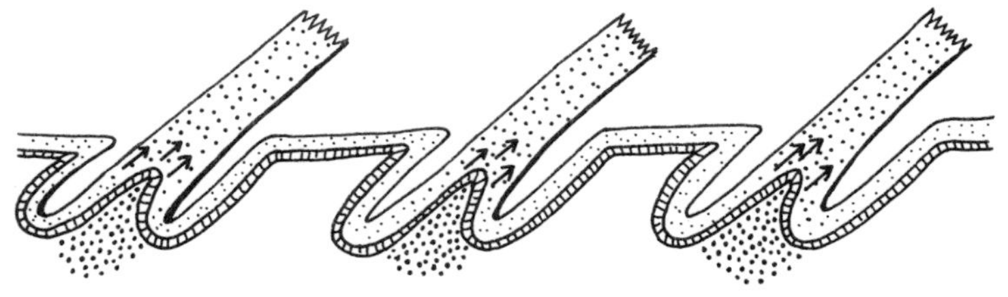


FIG. 2. Traditional scheme for the evolution of a feather from a reptilian scale (modified from Maderson, 1972b). Alternation of β-(heavy dots) and α-keratin indicate spacial and temporal separation of gene activity. The model does not account for changes in protein structure that would have accompanied morphological change. Nor are mechanisms suggested to account for the necessary ontogenetic conversions. An alternative hypothesis is that feathers arose independently, rather than serially, from the reptilian-type scale.

1966; Maderson, 1972a,b; Regal, 1975; Dyck, 1985) deal with potential morphological intermediates and the selective forces that drove their evolution. In each case there is a serious problem of cause and effect, which is complicated by the paucity of fossil evidence. Indeed, no fossil of a structure intermediate between a scale and a feather has been uncovered. Almost certainly different selective forces were involved at various stages in the evolution of the structure. The habitat and behavior of the early avian ancestor, and the attendant climatic conditions suggest several different functions for protofeathers. Feathers are thought to have served as insulation, flight surfaces, protection from solar radiation, and mechanical protection, or functioned in trapping insects, display, waterproofing, and other less obvious roles (reviewed in Hecht *et al.*, 1985). Still, no coherent framework exists to correlate the processes.

It is not my intention to attempt to choose among the possible hypotheses of feather evolution or to suggest an alternative "cause". Rather, I intend to deal with available knowledge of the genes involved in the production of constructional materials, the characteristics of the structural proteins, and aspects of their behavior. At the gene level it is possible to discuss the probable origin, the nature and extent of their diversity, and their organization in the genome. At the polypeptide level the relationships between sequence and configuration are introduced and provide for the structural characteristics at the gross morphological level. These features will determine the potential pathways of development and channel possible morphological solutions. The system appears to be designed to take advantage of fibrous molecular structures produced by cells that are derived from a simple, multilayered tissue. Because feathers and scales are produced from the same embryonic tissues and share some geometric features in development, it has been traditionally assumed that reptilian scales were the immediate precursors of feathers. However, neither the selective pressures involved in this change, nor the early roles of the modified ancestral structures, are known. As a result, the causes of evolution of feathers are unknown and are perhaps unknowable. This continues to be a point of serious debate (Hecht *et al.*, 1985). Nevertheless, the design, materials, and mechanisms of contemporary structures can be investigated and used to reconstruct some pathways and eliminate others to explain *how* feathers evolved. It is already obvious that any phenotypic transition from a reptilian scale to a feather could not have been simple. Indeed, it is not clear if feathers arose directly from a reptilian scale, secondarily through an avian scale, or if they shared a common ancestor. There is already evidence that only certain avian scales share biochemical and ontological features with reptilian scales, while others are as diverse as feathers, claws, and beaks.

A molecular model for the evolution of avian epidermal structures would complement the traditional morphological approach. A morphological model for the evolution of feathers must accommodate the related structures that share homologous proteins, similar organizational features, fabricational processes, and constructional constraints. Similarly, a model at the molecular level must account for the origin and organization of any unique genes, accommodate protein heterogeneity, account for the shape of the polypeptides and describe their function at each level of interaction. Each level of organization will be included within a series of integrated ontogenetic processes, each of which in turn is subject to evolutionary change. Elucidation of the molecular nature of feathers, the initial steps in the fabricational processes, and the genomic organization of the structural genes present an extraordinary opportunity to attempt to reconstruct an important evolutionary event. The notion of feathers as an evolutionary novelty on the molecular level provides a perspective of relative rates of potential evolutionary change at the genetic and phenotypic levels (Schopf, 1981). Because changes within multigene families have extensive regulatory consequences as well (Dover and Flavell, 1982), the molecular origin of morphological novelties might involve processes analogous to those of speciation (Rose and Doolittle, 1983; Ginzburg *et al.*, 1984). Such effects are magnified by any additional influence that higher-order genomic organization might have on gene function and expression. These simple changes may precipitate rapid phenotypic changes and, when coupled with the availability of new genes, produce novel structures (Hall, 1983, 1984; Erwin and Valentine, 1984).

If the best morphological models of the origin of feathers from scales are unsupported, it is appropriate to investigate other systems. Evidence from other systems might suggest alternative hypotheses. One logical approach would be to consider possible relationships to the structural proteins of skin and epidermal appendages. The α-keratins of vertebrates have been compared on only a superficial level, and adult chicken epidermis produces a set of four major proteins that resemble those of human and mouse skin (Brush, 1985). The available evidence suggests that the genes for the α-keratins of vertebrate skin are homologous (Fraser and MacRae, 1980a). Wool protein genes are derived from the skin α-keratins (Weber and Geisler, 1982).

Wool and feather proteins share no common amino acid sequences, and have remarkably different secondary structures. Feather, down, scute, claw, and beak share a common set of proteins found in no other vertebrate. These proteins are, for all intents and purposes, an innovative character of birds. The proteins of the epidermal appendages in reptiles are generally larger and more heterogeneous in size than in birds (Wyld and Brush, 1979, 1983). The avian epidermal appendage protein most like that in reptiles is in the reticulate scales of the plantar surface of the feet (Brush and Wyld, 1982; Brush, 1985). In reptiles, α-keratins provide the structural proteins of the skin and the flexible portions of scales (Alexander, 1970; Baden *et al.*, 1974). A protein with an X-ray diffraction pattern similar to avian ϕ-keratins (Baden and Maderson, 1970) is present in the harder structures, but the structural polypeptide differs from the avian ϕ-keratin in size, amino acid composition, and chemical behavior. The structure may be convergent simply as a result of functional demands. Indeed, the distribution of these two protein types in the reptilian epidermis can be correlated with function (Wyld and Brush, 1983). These data imply that the α-keratins are certainly phylogenetically older than the avian ϕ-keratin and that reptilian and avian ϕ-keratin type proteins are only distantly related, if at all. No strong evidence supports the hypothesis that avian ϕ-keratins are descended directly from a reptilian epidermal protein.

An alternative hypothesis to a direct relationship is that the divergence of birds from their reptilian ancestor was accompanied by the appearance of a completely new set of genes which produce the structural proteins that constitute the epidermal appendages. Simultaneously, a novel structure, or set of structures, appeared. These are different aspects of the same evolutionary event. The uniqueness of feathers is widely appreciated at the morphological level. I submit that an equally important series of events occurred at the molecular level. The uniqueness of feathers was afforded by a series of events and changes at the genomic level (Schopf, 1981; Hewett-Emmet *et al.*, 1982). It is now possible to explore the molecular basis of this event. If there was not a direct progression from reptilian scale through avian scale to feathers, what are the options and what are the implications? I present here biochemical and biophysical aspects of this process as reconstructed from contemporary structures, review the nature of the avian ϕ-keratin genes and proteins, and attempt to develop a theoretical framework to understand some of the processes involved.

## III. The ϕ-Keratin Genes

Characterization of the soluble carboxymethylated (SCM-) proteins of feathers (Harrap and Wood, 1964, 1967; O'Donnell, 1973) has revealed differences in the proteins among the feathers and scales and a chemical heterogeneity among the polypeptides within species. These proteins represent the products of a family of related genes (Brush, 1975). The precise differences among the proteins were documented when isolated monomers were subjected to peptide mapping and amino-acid composition analysis (Busch and Brush, 1979; see also Section IV). Almost simultaneously ϕ-keratin messenger RNA (mRNA) was isolated from chick feather (Kemp *et al.* 1974a,b) and subsequently scale tissue (Wilton *et al.*, 1985). This established the existence of mRNAs coding for about 20 proteins and corroborated the observed protein heterogeneity. Hybridization experiments with DNA (cDNA) made from this message indicated that the genes for ϕ-keratins existed as moderate repeats, perhaps 150 times through the genome (Kemp, 1975). Rogers (1978) reviewed the genomic organization of the keratin genes and speculated on the mechanism of expression. Since the avian genes function in a co-ordinated manner and represent a family of moderate repeats, interest developed in the nature of the possible control of their action. It also became increasingly apparent that the proteins, and thus the structural genes of the epidermal appendages, were quite unlike the genes and proteins of the skin. Access to the gene products and information on gene organization within the genome, coupled with the mechanisms of expression and the processes by which the products interact, have made it possible to study their organization, function, and evolution (Fig. 3). Thus, the system can be explored at both the level of genic and genomic evolution (Wright, 1982).

The isolation and cloning of avian ϕ-keratin genes has been partly completed (Molloy *et al.*, 1982). Parts of five of the approximately 20 expressed genes have been completely or partly sequenced (Gregg *et al.*, 1983, 1984). The pattern of distribution within the genome of the chick ϕ-keratin protein genes is in a tandemly repeating cluster with a center to center spacing of approximately 3.3 kb (Molloy *et al.*, 1982). Gene-sequencing data provide evidence of coding for a protein composed of 97 amino acids. This is in excellent agreement with the length of soluble protein monomers as determined by amino-acid sequence analysis (O'Donnell, 1974; O'Donnell and Inglis, 1974; reviewed by Brush, 1980b). There are no introns within the coding region, and little sequence similarity among the spacer regions. Introns are non-coding nucleotide regions that are copied in the RNA and then excised. A single intron occurs in the 5′ non-coding region. These attributes, especially the absence of intron in the coding region, suggested to Rogers that the ϕ-keratin genes were not constructed from separate functional domains brought together to form a single protein product. The alternative explanation would depend on molecular mechanisms that splice divergent structural domains to produce new structures (Shapiro and Cordell, 1982; Craik *et al.*, 1983; Janin and Wodak, 1983). Splicing mechanisms would allow wide latitude in molecular evolution. The possibility of molecular assembly fits well into a hierarchical model as the stability of any element is derived from the stability of its constituent parts and the strengths of the association among the parts. This would also provide a pathway for rapid evolutionary change. Exons (the expressed portion of the gene) could be shuffled among genes to produce entirely new products. The absence of introns between coding regions is no longer considered a compelling argument against this mechanism.

Rogers placed great significance on the position of the single intron as it probably provides information on the timing of gene expression during development. This would be an important factor in the process of differentiation (Sengel, 1976). Timing in the expression of avian epidermal structural proteins is critical. However, split genes are common in systems where protein diversity is generated (Janin and Wodak, 1983). Thus, introns may provide regulatory information regarding structural gene activity, indicate splicing sites in the coding region, or

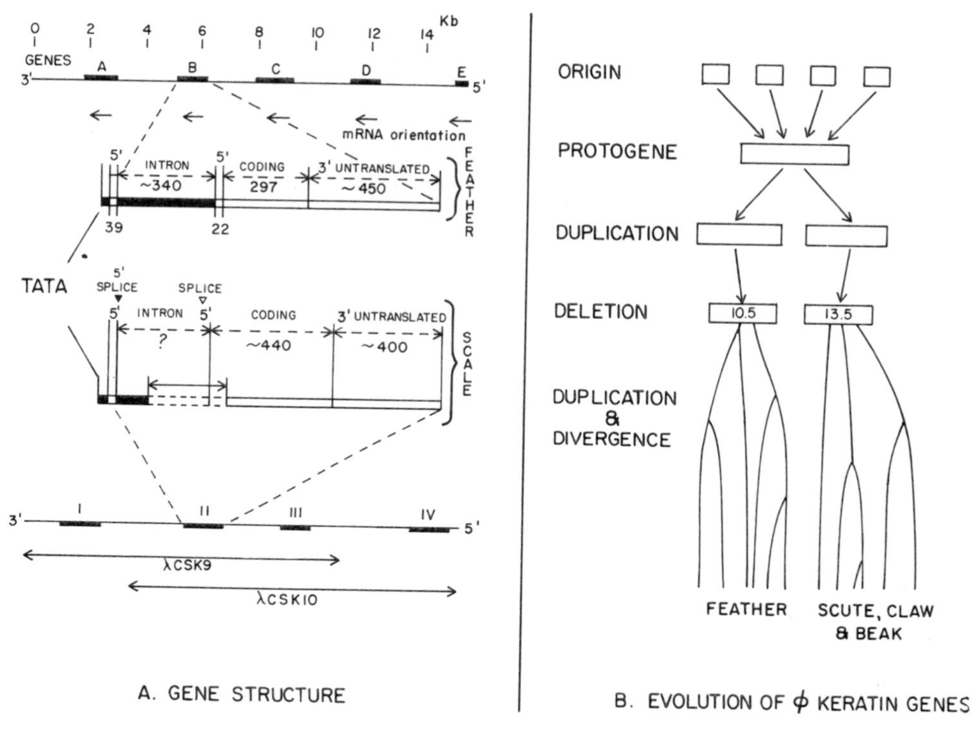


FIG. 3. A, Organization of the avian ϕ-keratin genes (after Gregg *et al.*, 1983, 1984). The genes for feather (top) and scale (bottom) exist as tandem repeats on the chromosome. The size of the coding region match the size of the polypeptide and account for observed tissue differences. Coding differences in the genes produce sequence heterogeneity. TATA signifies initiation sequence, Kb are numbers of bases counted from initiation site; B, Proposed model of the evolution of avian ϕ-keratin genes. The original organization of the protogene was followed by a simple duplication. After a deletion in one gene, each line undergoes multiple duplications and minor point mutations. This accounts for the tissue-specific sizes and the extensive molecular heterogeneity.

provide some yet unknown function (Sharp, 1985). One overriding characteristic of the epidermal tissue is that a series of morphologically divergent structures is produced sequentially from similar, but not identical, sets of protein. Thus an individual follicle might first produce natal down, then a definitive feather that, in subsequent molts, might vary in the basic and alternate plumages. Further, different structures, such as scales, claws, and beaks, contain proteins that share all the important molecular, functional, and conformational characteristics of feather proteins but have different ontogenetic pathways and shapes. This most obvious differences among tissue categories is in polypeptide size and distribution.

The origin of the ϕ-keratin gene family is unknown. The molecular processes were related intimately to the concurrent morphological events. The first step in the evolution of the ϕ-keratins was their modification from an ancestral gene. A simple assumption, because of their inclusion in the general category of filamentous proteins (Lazarides, 1980, 1982; Fuchs *et al.*, 1981), is that they shared a common ancestor with the α-keratins. The α-keratins are considerably more complex in both chemical and size heterogeneity (Cohen, 1966; Crewether *et al.*, 1982) and certainly phylogenetically older (Fraser *et al.*, 1972). The two categories share chemical properties that relate to the general functions concomitant with their role in the epidermis. Among these are general insolubility and cysteine residues involved in disulfide bonds. Proof of homology would require extensive comparisons of either the amino acids of the polypeptides or the nucleotide sequences of the genes. There is no direct evidence from amino acid sequence that ϕ- and α-keratins diverged from a common ancestral gene. Nor is there evidence that the ϕ-keratins were derived from existing α-keratin genes.

One alternative explanation is that the ϕ-keratins were derived from a cytokeratin protein. Possible candidates include intermediate filament (10 nm) proteins of the epidermis, mesenchymally derived vimentin (Wu *et al.*, 1982; Mahrle *et al.*, 1983) or some other intracellular structural proteins produced under modified or unusual conditions (reviewed by Moll *et al.*, 1982). Cytokeratins are a type of intermediate filament of epithelial cells produced by a family of related genes. The polypeptides are independently translated products which exist in several size categories, all larger than avian ϕ-proteins (Fuchs and Green, 1979, 1980). Terminal differentiation in these tissues is characterized by changes in specific mRNAs for proteins of different size. The cytokeratin proteins are sensitive to environmental (Lavker and Sun, 1983) and translational (Magia *et al.*, 1983) modification. The proteins of the mammalian α-keratins exist in two distinctive size classes, both of which are necessary for filament formation (Steinert *et al.*, 1976; Steven *et al.*, 1983; Sun *et al.*, 1984). Although the evolutionary relationship between cytoskeletal and microfibrillar keratins (Hanukoglu and Fuchs, 1982), and between an intermediate filament protein and wool keratins (Weber and Geisler, 1982) have been explored in mammals, no such data exist for birds. Fuchs *et al.* (1981) concluded that avian ϕ-genes are "very different from (mammalian) epidermal keratins . . . and their genes would not be expected to contribute to the cross-hybridization seen between human keratin probes and chicken DNA". The cross-reactivity, about 20% between human keratin probes hybridized with chicken DNA, is presumably for the α-keratins only. Thus, the relationship between mammalian and avian epidermal α-keratins is distant, and between avian α- and ϕ-keratins is unknown.

Because the α-keratin proteins are larger than the ϕ-keratin, the ϕ-keratin genes might have originated from either a single α-gene or as the result of the stepwise fusion of several functional domains (Gilbert, 1978; Crick, 1979). The first possibility is unlikely due to the difficulty incumbent in generating the degree of change necessary for the conversion. Rogers argued against this second type of "mosaic" evolution mainly on the grounds of the absence of introns in the coding sequence of the ϕ-genes. Yet, the decrease in size of avian ϕ-genes could have easily resulted from intron exclusion (Hewett-Emmet *et al.*, 1982). This option, a gene produced from exon shuffling, is attractive. The unique configuration and structuro-functional relationship within the ϕ-keratin polypeptide suggests a domain-like organization. The domains are unlike those in α-keratins (Steinert *et al.*, 1985) in size, shape, and redundancy. The details of the avian ϕ-keratins define proteins unique in sequence, shape, and size. I contend it is only remotely possible that the gene arose gradually from an ancestral sequence in either the cytoskeletal or microfibrillar α-keratin family to produce a protein with an entirely new configuration and vaguely similar chemical properties. Rather, a new epidermal gene appeared in association with a new structure—feathers. Compositional features such as high tyrosine, cystine, glycine, and low histidine, methionine, and lysine are likely to be shared as they define basic chemical properties and are associated with the process of chemical cross-linkage.

The proto-ϕ-keratins featured a predominant β-pleated sheet structure, unlike the α-helical configuration of all the other epidermal proteins. Despite the changes in primary structure and configuration, the proteins retained the ability to self-associate and produce filaments. Filament diameter was different from all other keratins and intermediate filaments, features of the proteins that were conserved across the changes in the gene and primary protein structure. It is possible that the similarities in chemical composition are due simply to convergence. Selection may have favored amino-acid compositional features related to the internal structure such as proline in the pleated sheet or glycine and alanine as space fillers. Other residues might be included to increase the hydrophobic nature of the proteins, facilitate polypeptide interactions in filament formation or influence the solubility and mechanical properties of the surface structures. It is noteworthy that ϕ-keratins are high in glycine content which is correlated negatively with amino acid substation rates (Graur, 1985) and is another feature of an evolutionarily conservative molecule.

Zuckerkandl (1975) argued for evolutionary conservatism in the generation of protein families at the molecular level. This is a traditional position. Zuckerkandl concluded that new structural genes were almost always derived from pre-existing structural genes. He contended further that new functions evolved "only on the basis of old proteins". Zuckerkandl's arguments predated the recent appreciation of the mechanisms that allow for rapid gene evolution. These include the extent of repeated DNA sequences, their mobility in evolutionary time (King, 1985) and the enhanced rates of recombination, transposition, and other genomic changes now recognized as common.

Gene rearrangement now appears to be an important mechanism for the generation of molecular diversity. Its rapid incorporation into the genome is a consequence of molecular drive such that the tendency towards homogenization should be synchronized throughout populations (Dover, 1982a,b; Dover and Flavell, 1982). The behavior of the genes is independent of population size or local selective regimes. Thus when new and different gene combinations arise within a population a relatively large number should emerge at approximately the same time. This could produce the genetic information necessary for a radical phenotypic change. The implication is that evolutionary change may occur quite rapidly at the levels of gene structure and genomic organization. The possibility exists that new genes and gene families may arise quickly, producing new types of proteins and, through modified ontogenetic steps, support significant morphological change. Within families, exons, as discrete domains of structure or function, can also be rearranged to increase diversity further (Crick, 1979; Craik *et al.*, 1983) and accelerates rates of change. I submit that the avian φ-keratins arose by a type of molecular mosaic evolution. The two globular, cystine-rich portions and the antiparallel β-pleated sheets were from pre-existing molecular structures spliced together. This was accomplished by mechanisms of self-splicing of elements and was subject to Darwinian evolution. The introns were subsequently lost prior to any gene duplication. The absence of introns at domain junctions also occurs in the 50-kDa human α-keratin (Marchuk *et al.*, 1984) and are not a necessary condition for this process to occur.

Based on the current analysis of avian feather proteins (Section IV) it is likely that the original gene product had a molecular mass of about 13,500 Da, a β-pleated sheet structure and a moderate number of cystine residues capable of forming intermolecular disulfide bonds. One of several products of the differentiation of epidermal tissues, it tended to be hydrophobic and to engage in self-associative behavior that would lead to fibrous macromolecular structures. The primitive φ-protein gene had no introns or lost them soon after its appearance. Shortly thereafter, or very early in the line leading to birds, a series of duplications occurred. As a result the genes assumed their present-day tandem arrangement. Thus the φ-keratin genes probably resembled those now associated with the tissues that produce scutes, claw, and beak (Fig. 3).

The φ-keratins now exist in two distinct sizes, 13,500 and 10,500 Da, each of which is typically associated with one morphological lineage. Hence, a second important event in the early evolution of these genes was a deletion of a portion in at least one of the early duplicates. The segment was lost from the globular portion on the carboxy-terminal side of the β-sheet. The divergence–deletion event presumably occurred prior to the completion of the morphological (phenotypic) divergence. The scale–claw–beak group retained the larger peptide (Walker and Bridgen, 1976). It has a simple, repetitive sequence, rich in glycine and other hydrophobic residues (Walker and Rogers, 1976; Powell and Rogers, 1979). The peptide, which is not found in the proteins of either adult feather or natal down, was presumably deleted (Brush and Wyld, 1982). The apparent consistency of this feature suggests its presence in the ancestral gene and deletion in the derived state. If it were the result of a later insertion event in a primitive smaller gene, then its distribution would be less consistent, its composition more variable, and its presence more disruptive of structure.

These events produced two basic lines within the avian ϕ-keratins: the larger polypeptides typical of scutes, claw, and beak and the smaller one typical of feathers and down. Each line, in turn, underwent multiple duplications and subsequent diversification. Sequence replacements were limited to sites outside of the pleated sheet, reinforcing the importance of this feature in the function of the monomer. The minimal chemical differences among monomers is reflected in their distribution within and between tissues. However, many differences are small and presumably are functionally equivalent. The evidence for this evolutionary sequence is found in the structure of the proteins.

## IV. ϕ-Keratin Structure: Proteins and Filaments

The electrophoretic heterogeneity of the soluble feather proteins was related to structure by peptide mapping (O'Donnell, 1973) and, ultimately, by sequence studies of one of the major components from the Emu (*Dromaius novaehollandiae*) (O'Donnell, 1973) and Silver Gull (*Larus novaehollandiae*) (O'Donnell and Inglis, 1974). Additional sequences of proteins from chicken (Arai *et al.*, 1983), domestic duck, and pigeon (Arai *et al.*, 1986) show an 85% similarity. The internal β-pleated sheet and cysteine sites are highly conserved. Chromatographic fractionation (Brush, 1974; Akahane *et al.*, 1977) has confirmed the basic structural similarity of the feather proteins, verified their molecular weight, and provided information on their chemistry. The nature and degree of intraspecific heterogeneity was resolved when the essential chemical nature of the soluble feather polypeptides was determined (Busch and Brush, 1979). The results also verified the estimates of the number of active genes based on the mRNA data. The distribution of polypeptides among morphologically distinctive parts and interspecific differences was reviewed by Brush (1978a).

The nature of the ϕ-keratins in other tissues such as scale (Walker and Bridgen, 1976), natal down (Walker and Rogers, 1976) and beaks (Frenkel and Gillespie, 1976) has been characterized to varying degrees. Adequate information on amino acid sequence is available to establish the overall relatedness of the genes in these diverse structures. In addition to sequence similarity the ϕ-keratins all have identical three-dimensional configurations. The heterogeneity within tissues is generated primarily by sequence differences among the monomers, but some minor bands may be the result of deamination or other post-translational chemical modifications. A variety of techniques has been used to detect potential functional differences in the proteins of various tissues (Brush, 1980a); furthermore, the divergence among the tissues has been investigated by comparisons of amino acid similarity indices (Brush, 1980b). The similarity among known sequences, commonly 75% or more, is adequate to conclude that all the avian φ-keratins shared a common ancestor (Fig. 4).

The demonstration of a high degree of sequence heterogeneity in feather keratin monomers raises questions about possible functional differences. Since all monomers within tissues have an essentially equal mass, size variation is not a factor. In α-keratins I and II, the size categories are typified by differences in chemistry, conformation, and function (Moll *et al.*, 1982; Sun *et al.*, 1984). Several explanations for the relatively high redundancy of φ-keratins are possible. First, the demands of protein synthesis, where large quantities of protein are produced over a short time, may have favored gene amplification. This would sustain multiple gene duplications. The quantities of feather keratin are large; feathers constitute on average approximately 6–8% of the total body mass (Turçek,

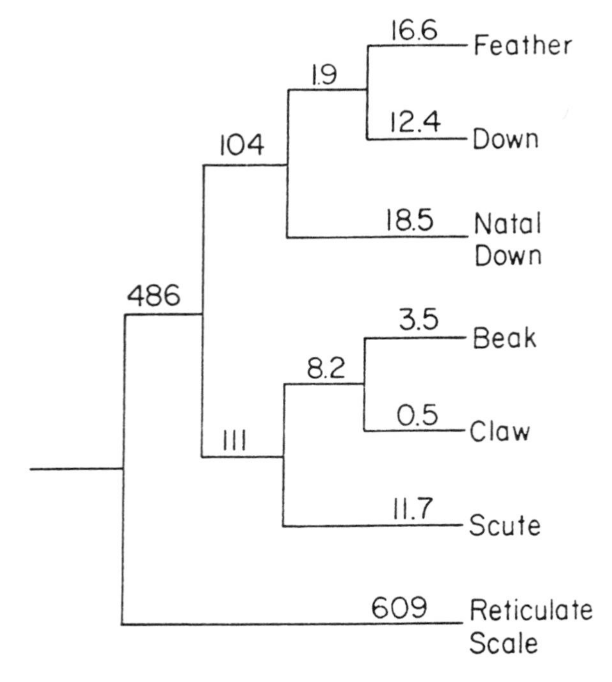


FIG. 4. Dendrogram of the φ-keratin epidermal structures of birds. Distances derived from amino acid compositions (Brush, 1980b).

1966). The demands of a repetitive series of molts must be met by synthesis of new protein. Although protein synthesis has been studied most thoroughly in embryos, it is repeated with each bout of feather formation through the life of the bird. One solution to the demands of rapid production of large quantities of a specialized protein would be to have multiple structural genes. Minor sequence divergence over time would produce electrophoretic heterogeneity. If they were simple point mutations, size would be conserved.

A second possible explanation for the existence of structural heterogeneity of feather keratins relates to design and microstructural requirements. Each of the major morphological units of feathers has a characteristic electrophoretic pattern. This implies differences in gene activation and presumably reflects constructional demands. Although electrophoretic patterns of soluble proteins are characteristic of the feather parts, they might (Brush, 1972) or might not (Cheng and Brush, 1984) be subject to changes in mutations that affect feather structure. Specific molecular requirements might be a design feature of fine structure. Yet scales and claws that superficially appear less complex morphologically than feathers have equivalent genetic heterogeneity. At a higher level in feather fine structure where electrophoretic patterns were identical the three-dimensional configuration as judged by X-ray diffraction can differ and is related to gross structure (Brush, 1978b). Although there could be functional differences among the monomers, this possibility is difficult to test. In one functional attribute, the ability to self-associate, the monomers within tissues appear to be functionally equivalent (Brush, 1983, and below). By contrast, mammalian α-keratin requires both acidic and basic units which differ in size to filament (Sun *et al.*, 1984). In birds filament formation, a function essential to feather formation, occurs among equi-size monomers. It may be possible that in the case of both tissue microheterogeneity and the differences in morphological parts the sequence variation in polypeptides is so slight that selection does not discriminate among them. That is, most of the sequence differences are inconsequential to the chemistry, physical properties, or shape of the polypeptide.

Interspecific sequence differences among monomers suggests that all sites on the polypeptide are not equally likely to undergo substitution. Constraints on the evolution of some proteins exist due to molecular interactions (Zuckerkandl, 1975, 1978) as well as amino acid sequence. Selection to maintain certain structural domains in proteins must be quite strong. None of the sequence replacements among the feather keratins occurs in the β-pleated sheet. Indeed, the X-ray diffraction evidence indicates that this portion is critical to the recognition process among the molecules and their subsequent self-association. Sequence changes presumably would disrupt the geometry of the sheet and adversely affect the function. Here, one structural aspect, the pleated sheet which is important in design and function, is conserved. The globular arms which have fewer constraints are the site of the sequence substitutions.

Evolution of intracellular proteins generally may be slow because of their role in the formation of "stable, functional, multimolecular complexes" (McConkey, 1982). The fibrous structures characteristic of feathers is one such complex. On the other hand, the globular, wing-like portion of feather keratins seems to be the site of point mutations and subsequent amino acid replacements. These portions are considered to be the matrix-forming parts of the molecule. Single amino acid replacements in this portion of the gene are less likely to be disruptive of molecular morphology and function than in the pleated sheet portion. The functional requirements are apparently limited to maintaining a certain length or volume and a minimal number of cystine residues. Indeed, the cystine residues are invariant. This allows for the formation of a matrix space between filaments and the chemical cross-linkage necessary for keratinization.

The existence of genetic redundancy and protein heterogeneity is compatible at the molecular level with the idea of "momentary excessive construction" (Gans, 1979). Feathers may differ from Gans's general case in that redundancy was not related to performance for survival under extreme conditions but in the development of a programmed process. The excess construction involved large amounts of protein, with considerable structural heterogeneity, and the genetic redundancy and multiple gene products. Gans envisioned the excessive construction as permissive of "degrees of freedom in (an animal's) behavior and physiology; the excess allows novel behavior and, with this, adaptive shifts". Although Gans focused on the morphological aspects of the problem, the same reasoning can be applied to the molecular level and the genotype. The presence of numerous genes provides an enriched mechanism for morphological differentiation and specialization. The duplicated genes enhance the capacity for protein production and multiple genes enhance protein heterogeneity. The presence of potential plasticity at the genomic level would afford phenotypic divergence, whether it was between tissues or among the structural elements within tissues. For example, the morphological differentiation among feather parts such as pennaceous and plumulose feathers, or portions of flight vs. contour feathers, might be the result of the assortive expression of genes that produce slight, but significant, differences in function and distribution of proteins.

One characteristic of paramount importance to feather keratins is their ability to form filaments. Fraser (1977) reviewed the conformation of polypetide chains, microfibrils, and fibrous macromolecular assemblages. The three-dimensional configuration at each level was derived primarily from physical measurements (Fraser and MacRae, 1973, 1976), and validated by chemical evidence. The models developed by Fraser and MacRae point out the importance of the β-pleated sheets in self-assembly and the secondary role of the globular portions in spacing the filaments. Filament structures similar to those in feathers are found in scale (Stewart, 1977), beak, and claw. The formation of the filament is associated with the packing characteristics of the pleated sheet. The α-keratins of skin, hair, and wool and the ϕ-keratins of feathers and scale have converged in the ability to form filaments, in the use of disulfide bonding for strength, solubility characteristics, and other physio-chemical properties associated mostly with their function. This has occurred in spite of monomer size differences, divergent helical designs, and different amino acid compositions.

β-pleated sheets are widespread in proteins (Richardson, 1977) and their structure (Lifson and Sander, 1980), conformation, and geometrical properties (Saleme and Weatherford, 1981) have been determined. The packing of β-sheets in globular proteins has been studied by Chothia and Janin (1981, 1982) but similar studies of equal detail have not been undertaken in fibrous proteins. Although Chothia *et al.* (1977) elucidated the general nature of packing of helices and pleated structures, Fraser and MacRae (1976) identified the portion of the primary sequence that produces the pleated sheets in ϕ-keratins (Fig. 5). The pleated sheets are stacked to produce a filament and the globular portions project out into the surrounding space and provide the matrix structure.

Recently, Brush (1983) succeeded in obtaining filaments of soluble monomers of feathers and claws. Filaments were produced from unfractionated samples, isolated individual monomers, and various mixtures of monomers (Fig. 6). The degree of filamentation, as per cent filament produced, varied, but averaged about 70% over all trials. All tests were done with monomers in the thiol (SH-) form that most closely approximates native conditions. The technique has been extended to numerous species and other tissues such as adult and natal downs. Results of these experiments have several implications. First, no enzyme, template, phosphate compound, or other cofactor is necessary for *in vitro* filamentation of the ϕ-keratin proteins. Second, all monomers are functionally equivalent in regard to filament formation. This is unlike the situation in mammalian α-keratins in which only size heteropolymers will form filaments (Steinert *et al.*, 1976). Third, the mechanism is universal among avian epidermal tissues. Despite the differences in polypeptide size, the filaments produced and the kinetics of formation are structurally identical. It is possible that differences in size of monomers confer specific properties on the tissue matrix, but without influence on the form of the filaments.

The mechanical properties of epidermal structures are determined by both form and intrinsic properties of the constructional materials. The main physical and mechanical properties of a structure of filaments embedded in a matrix have been analyzed for keratin structures (Fraser and McRae, 1980b). Among tissues one difference is the global organization of the filaments. Feathers, which are linear, branching structures, are flexible (Purslow and Vincent, 1978; MacLeod, 1980) but tend to abrade easily. The problem of feather wear is solved by the annual replacement through molt. The flexibility is confirmed by the general alignment of the filaments on the long axis of the feather structure. In scales and claw the fibers are organized loosely into interwoven layers that lie at different angles (Stewart, 1977). Thus, a "plywood"-like infrastructure, analogous to scales in other forms

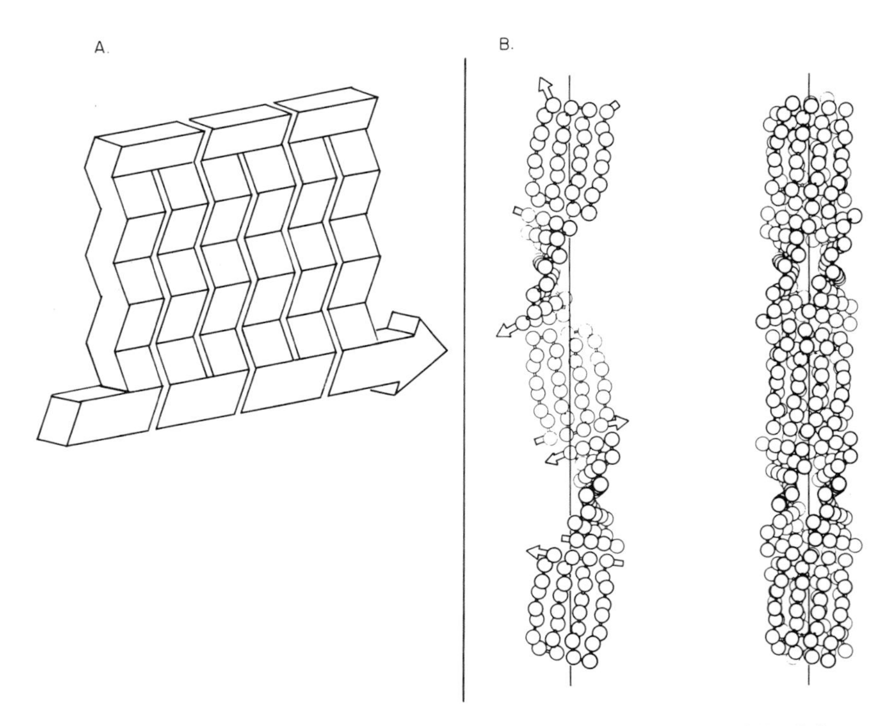


FIG. 5. A, Stylized model of β-pleated sheet. Individual boxes represents the peptide bonds between amino acids and there are eight residues/leg in the ϕ-keratins. Amino acid side groups extend into the plane of the pages. Chains run in an antiparallel direction and are cross-linked by weak hydrogen bonds. Surface of the actual sheets are more twisted (see B). The entire polypeptide includes 100–100 residues that run from left to right, as indicated by arrow. These form the globular portions; B, Organization of ϕ-keratin monomers into filaments. Only the β-pleated sheets of polypeptides are shown. Polypeptides are organized about a helical axis with four units/twist and a pitch of 9.5 MM. The diagram at the left illustrates the internal twist of the individual molecule and their pitch about the axis. On the right the four-fold structure of the actual microfilament is illustrated. It incorporates four of the basic strands and is the unit observed *in vivo*.

(Giraud *et al.*, 1978) is generated. This increases the resistance to abrasion, but reduces the flexibility. The wear associated with function is compensated for by continual growth. In this way, avian claws, scales, and beaks are analogous to human finger nails. In general, filament organization correlates with the mechanical demands of the tissue, and relates macromolecular organization to function.

The molecular mechanisms of protein production and filament formation are but two processes in the morphogenesis of the epidermal appendages. Additional processes occur; some are genetically determined but information may be derived from other tissues as chemical messages (Sengel, 1976; Haake and Sawyer, 1982; Sawyer *et al.*, 1986) or various physical forces (Oster *et al.*, 1980; Oster and

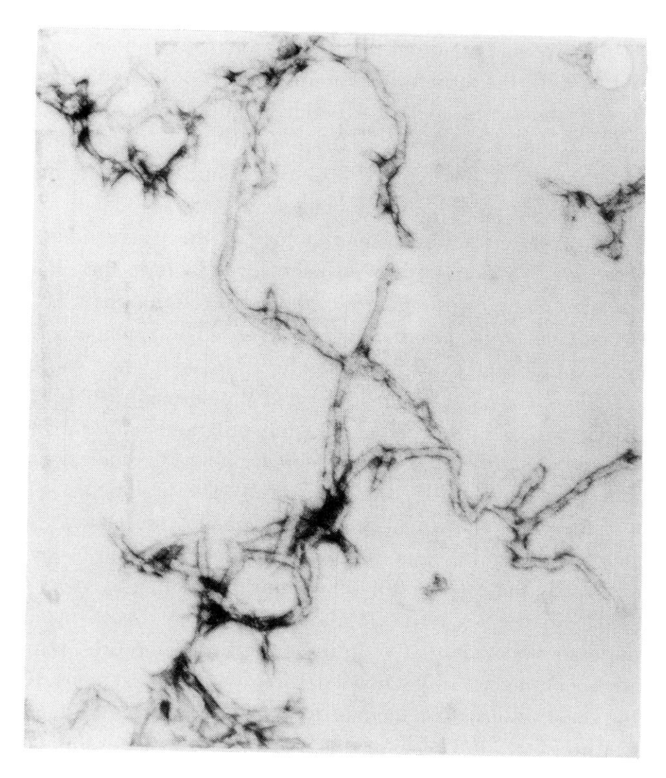

FIG. 6. Filaments of ϕ-keratin proteins reconstructed *in vitro* from SH-polypeptides. Primary structure is identified to that of native filaments formed *in vivo*. Electron micrograph approximately 115,000 ×.

Alberch, 1982). Self-assembly of filaments without a template implies that the inherent constraints of the precursor elements (the polypeptides) alone are sufficient to specify the final structure (Katz and Chow, 1985). Further, follicular geometry may influence the processing of both information and material. The origin of the feather follicle and its organization are understood in only the most general terms. Follicular design is especially significant because it affects the spatial relationships among cells and groups of cells that are otherwise similar. The potential influence of the location of cells and the forces imposed on them by physical changes in the follicle are unknown.

The feather follicle is a classic example of problems of co-ordinating cell shape, location, and mechanical behavior (Lillie, 1942). The role of physical forces acting over short distances in initiating and directing global process (O'Dell *et al.*, 1980; Oster *et al.*, 1980; Oster and Alberch, 1982; Ettenshon, 1985) must be matched with the role of the follicle in providing the environment for biochemical processing and control of the size and shape of the final product. Morphological changes are accommodated by both the structural proteins and co-ordinated processing of the various products (Schopf *et al.*, 1975; Alberch, 1980). The description and assessment of form is complex and, minimally, involves functional, historical, and fabricational parameters (Hickman, 1980). Biochemical differentiation is under separate control from the morphological changes at the tissue level; thus emphasis is placed on the emergent nature of morphology as the consequence of complex cellular interactions. The constraints on each level include the physical forces, genetically encoded changes, the nature of the constitutional material, and factors at both the ecological and "programmatic" levels. The broader view considers mosaic changes, differential rates, and the effect of canalization within each level of the system (Alberch *et al.*, 1979; Clowes and Wasserman, 1984).

The traditional paradigm that developmental programmes are a linear, causal chain progressing inevitably along a hierarchical pathway is undergoing challenge (Rachootin and Thompson, 1981; Bonner, 1982). The arguments are essentially that numerous steps in biochemical processing, phenotypic change, and phyletic transformations are the consequence of physical processes that occur in development, not necessarily the sole product of a central directing agency (Webster and Goodwin, 1981). In Stent's terms (1981) they are not "hard-wired". Cellular aggregates undergo behavior that requires co-ordination representing processes and information not necessarily encoded directly in the genome (Chandebois, 1976, 1980). These alternative models for such processing (Robertson, 1979; Nardi, 1981; Mittenthal, 1981) share a basic premise—processing though physical forces. Just as the soluble polypeptides self-associate into filaments, directed changes at the tissue level occur that require only physical or chemical information. As an example, Oster's model of the epithelium accounts for the generation of structural primordia as the result of viscoelastic properties of the cell cytoskeleton and mechanical interactions between tissues. Bifurcation in developmental sequences leading to morphologically quite diverse endpoints is not programmed genetically but is considered the result of physical forces that act upon the placode and related early structures. Relatively small forces will produce perturbations and cause the cell sheets to invaginate or evaginate. These directional changes ultimately produce massive morphological differences. The models imply that small genetic differences within cells in the type or timing of protein production and behavior can be subsequently amplified. The result may be a radically modified endproduct or, more importantly, increased diversity within an existing system.

The driving forces envisioned in this processing are generated intracellularly and transmitted across cells. In sheet-like tissues such as the epithelium, these forces may act over distances of several cell lengths. The primary distortion of cell shape is produced by cytoskeletal changes. These forces are of a different magnitude than the electrostatic forces that act on molecules and are limited to molecular distances. They differ also from the phase transitions within cells that may affect chemical organization such as filament formation, or the subunit interactions of polymeric molecules. At the cellular level the forces are magnified by numerous cells acting in concert. Feedback and lateral interactions may be involved. Direction, or polarity, may be provided by differential growth or forces related to stress or strain within tissues. These are emergent properties and not predictable from the properties of a single cell. The diverse nature of the processes, their capacity to act over variable distances, with different time scales, and their diverse modes of generation give physical forces an important role in the determination of phenotype.

It is widely accepted that avian epidermal appendages were derived in a gradualistic fashion from the archetypical reptilian scale (Maderson, 1972a,b). If this were true then a high degree of similarity among the constituent proteins would follow. Many of the developmental events of avian and reptilian scales, e.g., placode formation, invagination, or evagination and surface pattern formation, are shared. Except for some similarities in the crystallinity of the microfibrils, some general chemical properties such as relative enrichment in certain amino acids and deficiencies of others, and the ability to form filaments, the reptilian and avian non-α-keratins have little in common. I submit that the molecular similarities between classes are the result of convergence on common functional demands rather than homology.

Consider once again the morphological diversity of the structures on the body of a bird. All structures are formed from a single category of protein, the ϕ-keratins. The morphological differences are presumed to be the result of divergence from a common structural ancestor. The ontogenetic evidence to indicate the nature of the earliest structure is scanty; derived from phylogenetic reconstructions it is not adequate even to tell the shape of the ancestral feather (Lucas and Stettenheim, 1972). Because the divergence among the ϕ-keratin-containing structures must have been rapid, some type of identifiable common ancestor is hypothesized to exist. The alternative is that very early in development a simple, basic change occurred whose wide distribution, e.g., throughout the entire epidermis, produced a series of changes that, when modified by the local conditions, produced several different structures. This implies that different structures emerged from similar proteins, the morphological differences were the result of the conditions and rules under which they are produced and organized (Webster and Goodwin, 1981). An interactive relationship exists between structures and their constitutive elements. Structures within each system are governed by laws derived from the nature of the proteins and the laws governing their interaction. In the case of avian epidermal structures a new set of genes (ϕ-keratins) appeared. The size, shape, sequence, and abundance of the structural proteins had long-range influences on the derived structures.

The evolution of feathers must have involved modification of available processes and the production of new materials (Fig. 1). The processes of molecular synthesis, protein folding, and filament formation are shared widely in living systems. The occurrence of the φ-keratins, proteins unique to feathers, presumably followed a series of mutations, reorganizations, and duplication of genetic material (Fig. 2). Gene change (nucleotide sequences) led unavoidably to new gene products. The polypeptides folded according to universal rules but produced a set of gene products typified by a β-pleated sheet, a single molecular weight, and a characteristic amino acid composition. These molecules automatically assumed a certain shape. This is a derived property that emerges from the primary structure. The polypeptides spontaneously form filaments which, in turn, determine the structural and functional aspects of the cells and tissues. Neither final molecular structure, filament formation, nor subsequent cellular behavior were directly determined genetically. Like the plumage pattern derived from the individual feathers, they are emergent properties and unpredictable. Just as one cannot predict the shape of a feather from its component parts, the structure of a protein does not follow from knowing the sequence of nucleotides that encode it in the genome.

## V. Emergence of Evolutionary Novelty

It is informative to consider the hierarchical organization of feathers to understand their evolution, since organization represents an important component of evolutionary processes (Gould, 1977; Davidson, 1982). The problem is to account for the existence of diverse morphologies and to derive the structures at each level from the nature of the preceding level (Webster and Goodwin, 1981; Klee, 1984). Each level or unit controls its subunits and is subservient to a larger unit (Layzer, 1980). Single measures of evolutionary change are limited to one level and thus restrict the recognition of hierarchical structure, epigenetic processes and interactions across levels (Arnold and Fristrup, 1982). The organization of the system, its history, and its composition determine the potential effects of external selection and interactions with the environment. Thus the interactions, existing complexity, and internal constraints become important determining and limiting factors in potential evolutionary change and reorganization (Ho and Saunders, 1979; Saunders and Ho, 1981; Dwyer, 1984). One potential attribute of the hierarchical approach is the inclusion of emergent features.

The φ-keratins of birds are a family of genes. One significant advantage of the organization of gene families is that large sets of information are placed under a common control system (Hunkapiller *et al.*, 1982). As a result, the activity of these genes can be controlled by relatively simple steps. Small changes in timing sequences may invoke large morphological effects. Important as this is in ontogeny (Gould, 1977, 1982b; McNamara, 1982; Hall, 1984), it is also becoming widely recognized as a cellular control mechanism (Goldman *et al.*, 1984). Timing differences in molecular processes in the cell cycle are related to the control and expression of tissue-specific genes. Consequently, major phylogenetic changes can be produced from relatively few genetic changes. These need not be point mutations exclusively but can include chromosomal rearrangements and intragene (exon-) shuffling (Südhof *et al.*, 1985; van den Heuvel *et al.*, 1985). Rapid morphological change can occur under favorable selection but may be the product of minor genomic change (Alberch *et al.*, 1979). It is possible that neutral changes and drift contribute to such events as substrate for the direction of change.

We assume that to survive organisms must function well. We assume further that adaptation as a process is responsible for morphological diversity (Maynard Smith, 1978). However, selection does not necessarily act to maximize efficiency or to design each feature independently (Lewontin, 1979; Gould and Lewontin, 1979). Because organisms are constrained by their history, modes of development, and the properties of their building material, adaptations are probably never maximized for all environmental conditions. Rather, animals should be considered optimizing machines. Without question evolutionary design and the mechanisms that generate it have complex interactions, yet each stage of change need not be perfect, i.e., maximally adapted. Partridge (1982) points out that evolution is not engineering and need be only "good enough" for the co-ordination of structure and function at any given stage. I would add "at any given level" as well. This may be particularly relevant in the early stages of the appearance of novel structures (Lewin, 1985). Feathers are a prime example. Each structure contains combinations of preexisting information and limited material resources. The ways in which they are modified, combined, integrated, and expressed provides, and limits, the potential for change. Understanding adaptations usually focuses on the morphological phenotype, but requires analysis at other levels; it involves epigenetic processes, and explanations that may not be intuitively obvious (Charlesworth *et al.*, 1982; Meyr, 1983; Cheverud, 1984).

Genomic changes, no matter how small or incidental, must be compatible with the survival and success of the organism. The mechanism by which such genetic integration is accomplished is the subject of debate (Bock, 1979; Wright, 1982). Because of the difference in the scope of change at the genomic and phenotypic levels, rates of evolution at the two levels can differ (Prager and Wilson, 1975; Eldredge, 1979; Wyles *et al.*, 1983; Avise and Ball, 1991). Such changes occur at all levels and have important consequences in the evolution of higher categories (Riedl, 1977), in speciation (Mayr, 1982) and within species as in the emergence of novelties. Modern synthetic theory holds that in the course of this change there are neither periods of maladaptiveness nor times when selection does not operate. This implies that each change must be incorporated within the appropriate system with minimal amount of disruption and maximum gain. This may not always be the case. Perhaps the least disruptive changes lead simply to polymorphic features or modifications of existing structures. At least two events are represented in these changes. One involves "co-ordinate reprogramming" of complex developmental pathways (Britten and Davidson, 1971; Davidson, 1982; Dover, 1982a,b) and another involves timing (McNamara, 1982; Satoff, 1982; Raff and Kaufman, 1983; Soll, 1983). Either may provide events not subject immediately to selection or equally functional alternative conditions. In addition, Lindenmayer (1982) pointed out that "the control factors involved in multicellular growth and morphogenesis are not necessarily the same as the heredity factors". All of the above imply the importance of regulatory and epigenetic changes in the processes of evolutionary change.

Because complete information on the history of many morphological features and most biochemical processes is generally unavailable, it is not always possible to identify cause and effect or to predict the consequences of changes in a series of events. Nor is a simple reductionist explanation sufficient to relate the two (Eldredge and Cracraft, 1980). Larson *et al.* (1981) discuss the rapid appearance of morphological features based on the concept of a "key innovation" (Miller, 1949; Lauder, 1981). Another approach is Gans's theme (1979) in which mechanical design is directed toward prevention of failure under extreme conditions. Both ideas, in an evolutionary sense, provide opportunity for change. Major morphological change may occur on a short time scale compared to the life of the taxon, followed by slower, gradual adjustment. The same type of event may occur at the molecular level. In a sense a punctuated event occurs. Goodman *et al.* (1982) have argued for something like this in the evolution of molecules, especially hemoglobins. However, there is no paradigm to predict how such events at different levels will interact. Many of the genomic processes, such as gene duplication, could occur without immediately affecting protein synthesis or gross morphology. Changes at one level may or may not accompany rapid phenotypic change in other systems that represent radical ecological or life-style changes.

The reconstruction of the evolution of feathers must consider several levels. This brings into focus the nature of micro- and macroevolutionary changes and the processes that produce them, at least as invoked at the molecular and cellular levels (Bock, 1979; Charlesworth *et al.*, 1982; Rose and Doolittle, 1983). Rates of change within each level were probably different; the processes occurred asynchronously. The magnitude of each step and the immediate result were also small in order to be accommodated within the existing function. Interactions between levels (e.g., across organizational boundaries) and the effect of emerging properties were more plastic and perhaps more influential in the sense that larger changes in function, structure, and direction were possible. Regardless of the mechanisms involved in major new adaptations, the novelties of evolution represent a mutually adapted, complex set of features. Each step, in an engineering sense, need only be good enough, not maximally adapted. The functional advantage outweighs energetic design or structural disadvantage. Solutions tend towards operational optima rather than some "ideal state". Several phenotypes may satisfy the criteria of the problem, i.e., they are equally satisfactory solutions. Optimization of this sort is an important factor in explaining the diversity of life (Maynard Smith, 1978).

### A. ESTABLISHING GENETIC DIVERSITY

Genetic diversity of φ-keratins appears to be the result of at least two mechanisms. The original φ-keratin gene probably coded for a relatively small polypeptide with a central β-pleated sheet, a moderate number of cystines and an amino acid composition rich in glycine, proline, alanine, and tyrosine. It might have been derived directly from a pre-existing keratin gene, or modified extensively from some other source (Ilyin and Georgiev, 1982; Shapiro and Cordell, 1982). A *de novo* construction of a protein assembled from existing domains already present in the genome is clearly possible (Schopf, 1981; Burnett, 1982; Janin and Wodak, 1983). It is the putative mechanism that I favor. The degree of shuffling within gene families is unclear but obviously extensive (Blake, 1985; Südhof *et al.*, 1985; van den Heuvel *et al.*, 1985).

Regardless of the nature of its source and the details of its origin, the earliest φ-keratin gene was expressed preferentially in epithelial tissues. Models for the appearance of specialized proteins that involve structural heterogeneity and complex supramolecular organization are available for systems as diverse as insect vitellin (Harnish and White, 1982), chorionic proteins (Rodakis and Kafatos, 1982) and tubulin (Alexandraki and Ruderman, 1981; Little *et al.*, 1982). The classic, most complex and best-known case is that of the immunoglobulins. In the case of the φ-keratins, if they were derived from α-keratins then domains or sequence strings were lost, as the feather keratins are significantly smaller than proteins of skin or mammalian epidermal appendages. Among φ-keratins the minimal functional requirements include the ability to recognize selectively other φ-keratins, to produce filaments with restricted dimensions and to form intermolecular disulfide bonds. Minor sequence differences were presumably accommodated easily once these criteria were met. It is likely that a high percentage of the variation was at sites of splicing, between domains, or where overall shape was more important than chemical behavior.

One likely example of a "shuffled" exon in φ-keratins is the insoluble tryptic peptide of the scale–claw–beak genes. This peptide constitutes the size difference between these genes and those expressed in feathers. It always occurs in the same location on the gene and produces the same amino acid sequence in the protein product. Exclusion was non-disruptive to the greater structure as it survived selection. This peptide alone accounts for the size difference between these and the feather genes.

The second factor in the development of variability was duplications of the φgene. The presence of two size categories that are tissue specific implies that size difference was a primitive trait, and may have accompanied the origin of the genes. Multiple duplications subsequently occurred within the genes each of the two size lines. The duplicated genes appear in tandem arrays on the chromosome with no introns in the coding region. Multiple copies may not all be expressed (McKnight and Kingsbury, 1982), and the genes within the family that are not frequently translated may be replicated without transcription (MacLean, 1973). Hence, the potential for accumulation of genes that yield functionally competent products while conserving normal function is high. These arguments support a conservative mode of gene evolution and especially preservation of function. Given the increased genetic heterogeneity, new functional processes seem likely. The role of initial random sorting of gene products in the origin of new structures or functions has yet to be determined. Integration of parts is conserved even though changes occur in the structure and organization of the constituent elements.

### B. MAINTENANCE OF PROTEIN HETEROGENEITY

We have no way of estimating the time course of the duplication process of the tissue-specific ϕ-keratin genes. It is likely that the process of substitution began soon after the appearance of the first gene products. Sequence data show that substitutions in the protein were not random within the molecule, but occur preferentially in the globular portions. The original appearance of multiple genes may have had the effect of amplification of protein production. Differences in sequence and composition as they now exist may also be concerned with functional and design characters of the structures. It is possible that most or all of the existing substitutions were the result of nucleotide substitutions that had no significant effect on structure or function of polypeptides. Substitutions of this type are accepted as long as neither size, conformation, nor solubility of the protein are affected. This type of 'soft' selection is tolerated and would produce the type of heterogeneity characteristic of epidermal proteins. Unlike the other vertebrate epidermal and insect chorionic proteins, the avian ϕ-keratin system does not contain size categories within tissue types. Gene diversity and protein heterogeneity provide a potential for emergence of new structures.

### C. TRANSLATION OF GENETIC INFORMATION

The organization of the monomers into multimolecular functional units involves supramolecular processes. The first is the folding of the monomer into the appropriate conformation. The second is tonofibril (40 nm) production from the monomers.

The same units interact simultaneously to form fibers and filaments. The third process is the formation of disulphide bonds among the filaments during keratinization. The spacing of the globular portions of the filamented monomers supports interfilament linkage and determines fibrous packing dimensions. Similarly, changes in the organization of epithelial sheets, based on the contractions of intercellular proteins and conditioned by the physical nature of the sheets, produce the first supracellular morphological changes (Oster and Alberch, 1982). Little of this activity need be controlled through specific genetic information. Each step can be reproduced *in vitro* or explained by simple physical processing.

The forces that produced and modified the feather follicle are not known in detail. The earliest ontogenetic steps, presumably related to the evolution of the system, were reviewed by Wessells (1982). Much of the early descriptive morphology has been worked out in chickens (Lillie, 1942). However, we lack even rudimentary knowledge of the physical forces involved. Certainly a differential growth rate of cells in the generative collar and their inevitable elongation have a primary effect on size and shape. Other forces that act on the surface of the epithelium, and pressure from the dermis, must also be involved. Considerable force can be produced from filament reorganization in normal cell activity that can be transmitted across cell boundaries (Horwitz *et al.*, 1981; Jacobson, 1983). Thus at molecular and cellular levels extensive non-genetic activity occurs among the ϕ-keratins. It is becoming increasingly evident that non-genetic (epigenetic) mechanisms and simple changes in timing occur at this, and higher levels (Katz, 1983; Klee, 1984).

The processes in follicles include the events and interactions within tissues and the emergent properties of each organizational level. Tissue interactions are often cited for their role in pattern formation which implies some type of specific message. Early histochemical changes (Thompson, 1964) and the chemical influence of the dermis (Haake and Sawyer, 1982) have been studied but fail to identify the existence of tissue-specific chemical morphogens. The nature of tissue interactions in scale morphogenesis provides basic data and some insights into the general processes in development. We lack similar studies on feathers. The evidence indicates that differentiation involves reciprocal interactions between the dermis and epidermis (Sawyer, 1983). In fact, much of scale morphology is regulated by messages from the dermis. But the chemicals that act as messages are not tissue specific. Furthermore the system is plastic enough that the primary interactions can be modified to produce a wide range of appendages.

## D. MORPHOGENESIS

Numerous elements determine the morphological phenotype: molecular regulation, ontogenetic mechanisms, and timing. Changes in each are mutually interactive and act both within and across hierarchical levels. Molecular processes encompass the nature of the polypeptides and their physicochemical interactions (Klug, 1968; Kirschner and Mitchison, 1986). These include protein folding, macromolecular structures, and processes such as keratinization. Chemical processing and macromolecular geometry at the level of tissues are responsible for an impressive array of biological phenomena (Harrison, 1981), especially pattern formation (Harrison, 1982), and the generation of structure (Meinhardt, 1982). Other molecules, such as vitamin A (Dhouailly *et al.*, 1980; Fuchs and Green, 1981) and fibronectin (Haake and Sawyer, 1982) influence epidermal differentiation, but their precise role in individual morphogenetic processes is unclear. The role of mechanical forces, especially in the epithelium, has been investigated primarily at the tissue level (Oster *et al.*, 1980; Oster and Alberch, 1982), but interactions at higher levels also exist (Maderson, 1975, 1983; Cohen, 1979). External forces may also influence intracellular organization. One element, timing, may be the easiest to change phylogenetically and the most influential phenotypically. This occurs because the mechanism of heterochronism is simple in the sense that only periods within segments are changed, as opposed to the rearrangement of segments or the appearance of new steps or mechanisms. The generation of feather primordia patterns across the surface of the embryo appears to develop as the consequence of temporal co-ordination and local spacing mechanisms (Davidson, 1983). The emergence of the pattern over the epidermis demands no chemical information (e.g., a morphogen) and acts across several levels of hierarchical organization. Similar processes, acting in three dimensions, presumably influence the morphology of the follicle and subsequently the morphology of the feather.

Beyond the earliest molecular stages, such as the translation of keratin genes and production of mRNA for keratin synthesis, only rudimentary information is available regarding chemical differentiation of avian epidermal appendages. Epidermal thickening, placode formation and the histological organization of the epidermis for growth are essentially universal processes. At a higher level, the epithelio-mesenchymal interactions provide both limits and controls to processes in the epidermis (Sawyer, 1983; Maderson, 1983). The level of structure and types of processes define the universe within which morphogenesis occurs (Maderson, 1975; Davidson, 1983; Konig and Sawyer, 1985). Another significant evolutionary feature is the importance of the maintenance of the physical parameters of microfibrillar structures while the population of constituent polypeptides varies. Generally then, shape or organization at one level is conserved despite internal compositional change. This can be confusing because the process of convergence can lead to a similar situation where a common feature is the product of similar selective pressures but is built from a dissimilar infrastructure.

Thus, the nature and control of specific ontogenetic events must be set in an appropriate environment. The influence of higher levels and the constraints of lower levels, which effect the patterns of ontogeny, must be considered (Cooke,

1982; Maynard Smith *et al.*, 1985). It is also necessary to distinguish changes in the genome from changes in ontogeny. As Salthe (1975) contended, "structural transformations at the organismic level are merely *post hoc* results of meaningful changes occurring at other levels of organization". While this position may seem extreme, certainly there are reciprocal relationships among levels, and levels may evolve simultaneously, albeit at different rates. Because of the hierarchical effects, a small genomic change when translated can have broad morphological consequences. This may give the appearance of very rapid morphological change. A simple example of this in the avian epidermis are the two types of scales. Although they are similar morphologically, the reticulate and scutellate scales may have different phylogenetic histories as judged by their protein composition and ontogeny (Brush and Wyld, 1982; Brush, 1985). Some common features occur at the level of fibrillar production, keratinization and early tissue level processing presumably as the result of canalized pathways that influence morphology. But the protein evidence indicates large historical differences in the structural genes. The situation became more complex as the reticulate morphology has evolved several times, but includes the scute-like protein (Brush, 1985).

A more complex example in which morphology and molecular levels of organization indicate evolutionary differences is that of the down-like plumages of birds. Riddle (1908) concluded that the differences between pennaceous and plumulose structures in the definitive plumage are due essentially to differential growth rates. Clearly the relationship is not this simple. Using electrophoretic patterns of the soluble proteins and scanning electron microscopy to study "frizzle" and other mutants, I demonstrated that, in addition to the morphological differences, these feather parts contain different populations of proteins as well (Brush, 1972). Adult downs, as compared with natal downs, also have identifiably different electrophoretic patterns, but are obviously more closely related to definitive feathers (Brush, 1980a,b; Brush and Wyld, 1982). It appears that the function of the downy structure is conserved while two sets of different, but related, proteins are involved in their fabrication. The proteins involved are as much a function of the feather generation produced by the follicle as they are of the structure produced. Finally, definitive feathers represent extreme morphological diversity based on a single set of proteins and identical supramolecular organization (filaments). The processes that produce the diversity must reside in the follicular organization and processing. This involves interactions across cells, a dependence on geometry and the emergence of shapes not predictable from the properties of any of the constituent elements.

It is important to appreciate that the genome alone is inadequate to explain completely the organism, its ontogeny, and the pathways of evolution that yield the final product. Each step in development is executed in the light of a phylogenetic history and material constraints, depends on materials available and the capacity of each level to use or modify molecules, and is the result of the continuous interaction between existing and new patterns, timing and products (Gould, 1977; Katz, 1983; Maynard Smith *et al.*, 1985). Each pathway is itself potentially modulated by extra-cellular information capable of eliciting differential responses and inducing changes in the role or products of the cells. This may be especially true in the epidermis (Lavker and Sun, 1983). The timing of these processes has assumed new significance in evolution generally and in the emergence of novel structures in particular (McNamara,1982; Hall, 1984).

In the emergence of feathers, the earliest processes—cellular organization of the epidermis, placode formation and invagination—are primitive characters. They are shared widely among vertebrates. They are critical to the initiation and subsequent execution of the developmental processes. The genes that direct synthesis of the avian ϕ-keratins represent a significant divergence from those of their reptilian ancestor. The exact nature of their origin is still unknown but they are a derived character of birds. Thus, in the earliest stages of feather production, phylogenetically older histological patterns accommodated new protein structures and behavior. At another level, the processing of the keratin proteins during terminal differentiation, e.g., the formation of filaments, is also a primitive feature as it is shared by all fibrous protein systems (Cohen, 1966). Filament formation and organization are physical processes and not directed genetically. The structure of the polypeptides in bird ϕ-keratins is radically different from wool proteins or cytokeratins, yet the fibrillar organization persists. Here the structure is derived, but the process is primitive. At yet another level, the organization and mechanics of the follicle are uniquely avian, without a close homolog among reptiles or mammals. The origin of the geometry and growth patterns in follicles are unexplained but, logically, are derived from the general processes or organization associated with epidermal appendages. Regulation of morphological details of feathers is, to a great degree, the result of follicular timing and differential growth rates. The production of the morphological differences is the result of similar, if not identical, molecular processes. Thus a plasticity of shape emerged based on control of timing, while the constructional materials and processing remained relatively unchanged. The role of the fabrication process itself is influenced both genetically and by the cellular environment.

When considered as a hierarchical system, it becomes apparent that feather evolution was interactive among the levels and mosaic-like in its progression. Both the object and its history are complex (Hinegardner and Engelberg, 1983). Certain features, perhaps those considered most primitive, were assimilated from general epithelial processing. Other features, the ϕ-keratin genes and follicular geometry, were novel. The processing of the proteins in filament formation and the chemical changes in keratinization were shared with other systems, but modified to accommodate both functional and constituent molecules. Permissive developmental changes, and modification of the timing of events, combined with changes in patterns of synthetic activity, produced the heritable elements subjected to continued selection. Together these events result in a transfer of function (Mayr, 1960), increased developmental complexity and integration, while selection aspects of the organization were conserved.

## ACKNOWLEDGMENTS

Numerous people have been involved in this work in my laboratory. Ms Erika Kares and Dr Jean Wyld provided technical support and encouragement for at least a decade. A series of generous grants from the National Science Foundation and the University Research Foundation provided financial support. I appreciate the patience of my colleagues who listened, criticized, made suggestions, and have shared my amazement at feathers.

## ADDENDUM

A model for the evolution of complex morphological features has been developed by Atchley & Hall (1991). The entire matter of evolutionary novelties was reviewed by Müller & Wagner (1991) who discuss many of the conceptual issues. They derive innovative pathways to define and analyze novelties.

Atchley, W. R., and Hall, B. K. (1991) A model for development and evolution of complex morphological structures. *Biol. Rev.* **66**, 101–157.

Müller, G. B. and Wagner, G. P. (1991) Novelty in evolution: restructuring the concept. *Annu. Rev. Ecol. Syst.* **22**, 229–256.

## REFERENCES

Akahane, K., Murozono, S., and Murayama, K. (1977). Soluble proteins from fowl feather keratins. I & II. *J. Biochem.* **81**, 11–18, 19–24.

Alberch, P. (1980). Ontogenesis and morphological diversification. *Amer. Zool.* **20**, 653–667.

Alberch, P., Gould, S. J., Oster, G. F., and Wake, D. B. (1979). Size and shape in ontogeny and phylogeny. *Paleobiology* **5**, 296–317.

Alexander, N. J. (1970). Comparison of the α- and β-keratin in reptiles. *Z. Zellforsch.* **110**, 153–165.

Alexandraki, D., and Ruderman, J. V. (1981). Sequence heterogeneity, multiplicity, and genomic organization of α- and β-tubulin genes in sea urchins. *Molec. Cellul. Biol.* **1**, 1125–1137.

Arai, K. M., Takahashi R., Yolote, Y., and Akahane, K. (1983). Amino-acid sequence of feather keratin from fowl. *Eur. J. Biochem.* **132**, 501–507.

Arai, K. M., Takahashi, R., Yokote, Y., and Akahane, K. (1986). The primary structure of feather keratins from duck (*Anas platyrhynchos*) and pigeon (*Columba livia*). *Biochim. Biophys. Acta* **873**, 6–12.

Arnold, A. J., and Fristrup, K. (1982). The theory of evolution by natural selection: a hierarchical expansion. *Paleobiology* **8**, 113–129.

Authur, W. (1984). "Mechanisms of Morphological Evolution". Wiley, New York.

Avise, J. C., and Ball, Jr., R. M. (1991) Mitochondrial DNA and avian microevolution. *Acta xx Congressus Internationalis Ornithologici*, 514–524.

Baden, H. P., and Maderson, P. F. A. (1970). Morphological identification of fibrous proteins in the amniote epidermis. *J. Exptl. Zool.* **174**, 249–262.

Baden, H. P., Svioka, S., and Roth, I. (1974). The structural protein of reptilian scales. *J. Exptl. Zool.* **187**, 287–294.

Bailey, D. W. (1986). Genetic programming of development: A model. *Differentiation* **33**, 89–100.

Bereiter-Hahn, J., Matoltsy, A. G., and Richards. K. S. (1986). "Biology of The Integument. 2. Vertebrates". Springer-Verlag, Berlin.

Blake, C. C. F. (1985). Exons and the evolution of proteins. *Internat. Rev. Cytol.* **93**, 149–185.

Bock, W. J. (1979). The synthetic explanation of microevolutionary change—A reductionistic approach. *In* "Models and Methodologies in Evolutionary Theory" (J. H. Schwartz and H. B. Rollins, eds), pp. 20–69. *Bull. Carnegie Mus. Nat. Hist. No. 13.*, Pittsburgh.

Bonner, J. T. (1982). "Evolution and Development". Dahlem Worshop Report 22. Springer-Verlag, Berlin.

Britten, R. J., and Davidson, E. H. (1971). Repetitive and nonrepetitive DNA sequence and a speculation on the origins of evolutionary novelty. *Quart. Rev. Biol.* **46**, 111–146.

Brush, A. H. (1972). Correlation of protein electrophoresis pattern with morphology of normal and mutant feathers. *Biochem. Genet.* **7**, 87–93.

Brush, A. H. (1974). Feather keratin—Analysis of subunit heterogeneity. *Comp. Biochem. Physiol.* **48**, 661–670.

Brush, A. H. (1975). Molecular heterogeneity of the structure of feathers. *Isozymes* **4**, 901–914.

Brush, A. H. (1978a). "Feather Keratins". *Chemical Zoology* **X**. 117–139. Academic Press. New York.

Brush, A. H. (1978b). Structural aspects of the duck speculum. *Ibis* **120**, 523–527.

Brush, A. H. (1980a). Chemical heterogeneity in keratin proteins of avian epidermal structures: Possible relationships to structure and function. *In* "The Skin of Vertebrates" (R.I.C. Spearman and P. A. Riley, eds), pp. 87–109. *Linnean Soc. Lond. Symp.* **9**. London.

Brush, A. H. (1980b). Patterns in the amino-acid composition of avian epidermal proteins. *Auk* **97**, 742–753.

Brush, A. H. (1983). Self-assembly of avian $\phi$-keratins. *J. Prot. Chem.* **2**, 63–75.

Brush, A. H. (1985). Convergent evolution of reticulate scales. *J. Exptl. Zool.* **36**, 303–308.

Brush, A. H., and Wyld, J. A. (1982). Molecular organization of avian epidermal structures. *Comp. Biochem. Physiol.* **73**, 313–325.

Burnett, L. (1982). A model for the mechanism and control of eukaryote gene splicing. *J. Theor. Biol.* **97**, 351–366.

Busch, N. E., and Brush, A. H. (1979). Avian feather keratins: molecular aspects of structural heterogeneity. *J. Exptl. Zool.* **210**, 39–47.

Campbell, J. H. (1982). Autonomy in evolution. *In* "Perspectives in Evolution" (R. Milkman, ed.), pp. 190–200. Sinauer Associates, Sunderland, Ma.

Chandebois, R. (1976). Cell sociology: a way of reconsidering the current concepts of morphogenesis. *Acta Biotheor.* **25**, 71–102.

Chandebois, R. (1980). Cell sociology and the problem of automation in the development of pluricellular animals. *Act Biotheor.* **29**, 1–35.

Charlesworth, B., Lande, R., and Slatkin, M. (1982). A neo-Darwinian commentary on microevolution. *Evolution* **36**, 474–498.

Cheng, K. M., and Brush, A. H. (1984). Feather morphology of four different mutants in the Japanese Quail. *Poult. Sci.* **63**, 391–400.

Cheverud, J. M. (1984). Quantitative genetics and developmental constraints on evolution by selection. *J. Theor. Biol.* **110**, 115–171.

Chothia, C., and Janin, J. (1981). Relative orientation of close-packed $\beta$-pleated sheets in proteins. *Proc. Nat. Acad. Sci.* **78**, 4146–4150.

Chothia, C., and Janin, J. (1982). Orthogonal packing of β-pleated sheets in proteins. *Biochemistry* **21**, 3955–3965.

Chothia, C., Levitt, M., and Richardson, D. (1977). Structure of proteins: packing of α-helices and pleated sheets. *Proc. Nat. Acad. Sci.* **74**, 4130–4134.

Clowes, J. S., and Wasserman, G. D. (1984). Genetic control theory of developmental events. *Bull. Math. Biol.* **46**, 785–825.

Cohen, C. (1966). Design and function of fibrous proteins. *Ciba Symp.* pp. 101–135. CIBA, Basle.

Cohen, J. (1979). Maternal constraints on development. *In* Maternal Effects in Development" (D. H. Newth, and M. Bulls, eds), pp. 1–28. British Soc. Dev. Biol. Symp. No. 4.

Cooke, J. (1982). The relation between scale and the completeness of pattern in vertebrate embryogenesis: models and experiments. *Amer. Zool.* **22**, 91–104.

Craik, C. S., Rutter, W. J., and Fletterick. R. (1983). Splice junctions: Association with variation in protein structure. *Science* **220**, 1125–1129.

Crewether, W. G., Dowling, L. M., Gough, K. H., Marshall, R. C., and Sparrow, L. G. (1982). The microfibrillar proteins of α-keratin. *In* "Fibrous proteins: Scientific, Industrial and Medical Aspects" (D. A. D. Perry, and L. K. Creamer, eds), Vol. 2, pp. 151–159. Academic Press, London.

Crick, F. (1979). Split genes and RNA splicing. *Science* **204**, 264–271.

Davidson, D. (1983). Mechanism of feather pattern development in the chick. I & II. *J. Embryol. exp. Morph.* **74**, 245–259, 261–273.

Davidson, E. H. (1982). Evolutionary change in genomic regulatory organization: Speculations on the origins of novel biological structure. *In* "Evolution and Development" (J. T. Bonner, ed.), pp. 65–84. Springer-Verlag, Berlin.

Dhouailly, D., Hardy, M. H., and Sengel, P. (1980). Formation of feathers on chick foot scales: a state-dependent morphogenetic response to retionic acid. *J. Embryol. exp. Morph.* **58**, 63–78.

Dover, G. (1982a). Molecular drive: a cohesive mode of species evolution. *Nature* **299**, 111–117.

Dover, G. (1982b). A molecular drive through evolution. *Bioscience* **32**, 526–533.

Dover, G., and Flavell, R. B. (1982). "Genomic Evolution". Syst. Assoc. Special Vol. 20. Academic Press, London.

Dwyer, P. D. (1984). Functionalism and Structuralism: two programs for Evolutionary Biologists. *Amer. Nat.* **124**, 745–750.

Dyck, J. (1985). The evolution of feathers. *Zoologica Scripta* **14**, 137–154.

Eldredge, N. (1979). Alternative approaches to evolutionary theory. In "Models and Methodologies in Evolutionary Theory" (J. H. Schwartz, and H. B. Rollins, eds), pp. 7–19. Bull. Carnegie Mus. Nat. Hist. No. 13., Pittsburgh.

Eldredge, N., and Cracraft, J. (1980). "Phylogenetic Patterns and the Evolutionary Process". Columbia University Press, New York.

Erwin, D. H., and Valentine, J. W. (1984). "Hopeful monsters", transponds and metazoan radiation. *Proc. Nat. Acad. Sci.* **81**, 5482–5483.

Ettenshon, C. A. (1985). Mechanisms of epithelial invagination. *Quart. Rev. Biol.* **60**. 289–307.

Ewart, J. C. (1921). The nestling feathers of the Mallard, with observations on the composition, origin, and history of feathers. *Proc. Zool. Soc. London*, 609–642.

Feduccia, A., and Tordoff, H. B. (1979). Feathers of *Archaeopteryx*: Asymmetric vanes indicate aerodynamic function. *Science* **203**, 1021–1022.

Fraser, R. D. B. (1977). The structure of fibrous proteins. In "First Cleveland Symp. Macromolecules" (A. C. Wilton, ed.), pp. 1–21. Elsevier, Amsterdam.

Fraser. R. D. B., and MacRae, T. P. (1973). Feather keratin. In "Conformation of Proteins" (R. D. B. Fraser, and T. P. MacRae, eds), pp. 525–540. Academic Press, New York.

Fraser, R. D. B., and MacRae, T. P. (1976). The molecular structure of feather keratin. In "Proc. 16th Internat. Ornith. Congress" (H. J. Frith, and J. H. Calaby, eds), pp. 443–452. Canberra, 1974.

Fraser, R. D. B., and MacRae, T. P. (1980a). Current views on the keratin complex. *In* "The Skin of Vertebrates" (R. I. C. Spearman, and P. A. Riley, eds), pp. 67–86. Linnean Soc. Sym, No. 9, London.

Fraser, R. D. B., and MacRae, T. P. (1980b). Molecular structure and mechanical properties of biological materials. *Symp. Soc. Exptl. Biol.* **34**, 211–246.

Fraser, R. D. B., MacRae, T. P., and Rogers, G. E. (1972). "Keratins. Their Composition, Structure and Biosynthesis". C. C. Thomas, Springfield, Ill.

Frenkel, M. I., and Gillespie, J. M. (1976). The proteins of the keratin component of birds beaks. *Aust. J. Biol. Sci.* **29**, 467–479.

Fuchs, E., and Green, H. (1979). Multiple keratins of cultured human epidermal cells are translated from different mRNA molecules. *Cell* **17**, 573–582.

Fuchs, E., and Green, H. (1980). Changes in keratin gene expression during terminal differentiation of the keratinocyte. *Cell* **19**, 1033–1042.

Fuchs, E., and Green, H. (1981). Regulation of terminal differentiation of cultured human keratinocytes by Vitamin A. *Cell* **25**, 617–625.

Fuchs, E. V., Coppock, S. M., Green, H., and Cleveland, D. W. (1981). Two distinct classes of keratin genes and their evolutionary significance. *Cell* **27**, 75–84.

Gans, C. (1979). Momentarily excessive construction as the basis for protoadaptation. *Evolution* **33**, 227–233.

Gilbert, W. (1978). Why genes in pieces? *Nature* **221**, 501.

Ginzburg, L. P., Bingham, P. M., and Yoo, S. (1984). On the theory of speciation induced by transposable elements. *Genetics* **107**, 331–341.

Giraud, M. M., Castenet, J., Meunier, F. J., and Bouligand, Y. (1978). The fibrous structure of Coelacanth scales: a twisted "plywood". *Tissue & Cell* **10**, 671–686.

Goldman, M. A., Holmquist, G. P., Gray, M. C., Caston, L. A., and Nag, A. (1984). Replication timing of genes and middle repetitive sequences. *Science* **224**, 686–692.

Goodman, M., Weiss, M. L., and Czelusniak, J. (1982). Molecular evolution above the species level: Branching pattern, rates and mechanism. *Syst. Zool.* **31**, 376–399.

Gould, S. J. (1977). "Ontogeny and Phylogeny". Belknap Press, Harvard University, Cambridge, Ma.

Gould, S. J. (1980). Is a new and general theory of evolution emerging?*Paleobiology* **6**, 119–130.

Gould, S. J. (1982a). Darwinism and the expansion of evolutionary theory. *Science* **216**, 380–387.

Gould, S. J. (1982b). Change in developmental timing as a mechanism of macroevolution. *In* "Evolution and Development" (J. T. Bonner, ed.), pp. 333–346, Springer-Verlag, Berlin.

Gould, S. J., and Lewontin, R. C. (1979). The spandrels of San Marco and the panglossian paradigm: a critique of the adaptationist program. *Proc. Royal Soc. Lond. B* **205**, 581–598.

Graur, D. (1985). Amino acid composition and the evolutionary rates of protein-coding genes. *J. Mol. Biol.* **22**, 53–62.

Gregg, K., Wilson, S. D., Rogers, G. E., and Molloy, P. I. (1983). Avian keratin genes: organization and evolutionary interrelationships. *In* "Manipulation and Expression of Genes in Eukaryotes" (P. Nagley, A. W. Linname, W. J. Peacock, and J. A. Pateman, eds), pp. 65–72. Academic Press, Sydney.

Gregg, K., Wilton, S. D., Perry, D. A. D., and Rogers, G. E. (1984). A comparison of genomic coding sequences for feather and scale keratin: Structural and evolutionary implications. *EMBO J.* **301**, 175–179.

Haake, A. R., and Sawyer, R. H. (1982). Avian feather morphogenesis: fibronectin-containing anchor filaments. *J. Exptl. Zool.* **221**, 119–124.

Hall, B. K. (1983). Epigenetic control in development and evolution. *In* "Development and Evolution" (B. C. Goodwin, N. Holder, and C. C. Wyle, eds), pp. 352–379. Sym. Brit. Soc. Dev. Biol., Cambridge University Press.

Hall, B. K. (1984). Developmental processes underlying heterochrony as an evolutionary mechanism. *Can. J. Zool.* **62**, 1–7.

Hanukoglu, I., and Fuchs, E. (1982). The cDNA sequence of a human epidermal keratin: Divergence of sequence but conservation of structure among intermediate filament proteins. *Cell.* **31**, 243–252.

Harrap, B. S., and Wood, E. F. (1964). Soluble derivatives of feather keratin. I & II. *Biochem. J.* **8**, 8–18, 19–26.

Harrap, B. S., and Wood, E. F. (1967). Species differences in the proteins of feathers. *Comp. Biochem. Physiol.* **20**, 449–460.

Harrison, L. G. (1981). Physical chemistry of biological morphogenesis. *Chem. Soc. Rev.* **10**, 491–528.

Harrison, L. G. (1982). An overview of kinetic theory in developmental modeling. *In* "Developmental Order: its Origin and Regulation" (S. Subtelny and P. B. Green, eds), pp. 3–33. A. R. Liss, New York.

Harnish, D. G., and White, B. N. (1982). An evolutionary model for the insect vitellins. *J. Mol. Evol.* **18**, 405–413.

Hecht, M. K., Ostrom, J. H., Viohl, G., and Wellnhofer, P. (1985). "The Beginnings of Birds". *Proc. Internal. Archaeopteryx Conf.*, Friends of the Jura Museum, Eichstatt, 1984.

Hewett-Emmet, D., Venta, P. J., and Tashian, R. E. (1982). Features of gene structure, organization and expression that are providing unique insights into molecular evolution and systematics. *In* "Macromolecular Sequences in Systematics and Evolutionary Biology" (M. Goodman, ed.), pp. 357–405. Plenum, New York.

Hickman, C. S. (1980). Gastropod radulae and the assessment of form in evolution. *Paleobiology* **6**, 276–294.

Hinegardner, R., and Engelbert, J. (1983). Biological complexity. *J. Theor. Biol.* **104**, 7–20.

Ho, M. W., and Saunders, P. T. (1979). Beyond neo-Darwinism—An epigenetic approach to evolution. *J. Theor. Biol.* **78**, 573–591.

Horwitz, B., Kupfer, H., Eshhar, Z., and Geiger, B. (1981). Reorganization of arrays of prekeratin filaments during mitosis. *Exptl. Cell Res.* **134**, 181–191.

Hunkapiller, T., Hung, H., Hood, L., and Campbell, J. H. (1982). The impact of modern genetics on evolutionary theory. *In* "Perspectives in Evolution" (R. Milkman, ed.), pp. 164–189. Sinauer, Sunderland, Ma.

Ilyin, Y. V., and Georgiev, G. P. (1982). The main types of organization of genetic materials in eukaryotes. *CRC Critical Reviews in Biochemistry*, pp. 237–287.

Jacobson, B. S. (1983). Interaction of the plasma membrane with the cyto-skeleton: an overview. *Tissue & Cell* **15**, 829–852.

Janin, J., and Wodak, J. (1983). Structural domains in proteins and their roles in the dynamics of protein function. *Progr. Biophs. Molec. Bio.* **42**, 21–78.

Katz, M. J. (1983). Ontophyletics: Studying evolution beyond the genome. *Persp. Biol. & Med.* **26**, 323–333.

Katz, M., and Chow, Y-S. (1985). Templating and self-assembly. *J. Theor. Biol.* **113**, 1–13.

Kemp, D. J. (1975). Unique and repetitive sequence in multiple genes for feather keratins. *Nature* **254**, 573–577.

Kemp, D. J., Partington, G. A., and Rogers, G. E. (1974a). Isolation and molecular weight of pure feather keratin mRNA. *B.B.R.C.* **60**, 1006–1014.

Kemp, D. J., Schwinghammer, M. W., and Rogers, G. E. (1974b). Translation of pure feather keratin mRNA in a wheat embryo cell-free system. *Mol. Biol. Reports* **1**, 441–447.

King, C. C. (1985). A model for transposon-based eucaryote regulatory evolution. *J. Theor. Biol.* **114**, 462–477.

Kirschner, M., and Mitchison, T. (1986). Beyond self-assembly: from microtubles to morphogenesis. *Cell* **45**, 329–342.

Klee, R. L. (1984). Micro-determinism and concepts of emergence. *Philos. Science* **51**, 44–63.

Klug, A. (1968). The design of self-assembling systems of equal units. *Symp. Int. Soc. Cell Biol.* **6**, 1–18.

Konig, G., and Sawyer, R. (1985). Analysis of morphogenesis and keratinization in transfilter recombinants of feather forming skin. *Devel. Biol.* **109**, 381–392.

Larson, A., Wake, D. B., and Highton, R. (1981). A molecular perspective on the origins of novelties in the salamanders of the tribe Plethodontini (Amphibia, Plethodontidae). *Evolution* **35**, 405–422.

Lauder, G. V. (1981). Forms and function: Structural analysis in evolutionary morphology. *Paleobiology* **7**, 430–442.

Lavker, R. M., and Sun, T.-T. (1983). Rapid modulation of keratinocyte differentiation by the external environment. *J. Invest. Dermat.* **80**, 228–237.

Layzer, D. (1980). Genetic variation and progressive evolution. *Amer. Nat.* **115**, 809–826.

Lazarides, E. (1980). Intermediate filaments as mechanical integrators of cellular space. *Nature* **283**, 249–256.

Lazarides, E. (1982). Intermediate filaments: A chemically heterogeneous, developmentally regulated class of proteins. *Ann. Rev. Biochem.* **51**, 219–250.

Lewin, R. (1985). How does half a bird fly? *Science* **230**, 530–531.

Lewontin, R. C. (1979). Sociobiology as an adaptationist program. *Behav. Sci.* **24**, 1–10.

Lifson, S., and Sander, C. (1980). Specific recognition in the tertiary structure of β-sheet of proteins. *J. Mol. Biol.* **139**, 627–640.

Lillie, F. R. (1942). On the development of feathers. *Biol. Rev.* **17**, 247–266.

Lindenmayer, A. (1982). Developmental algorithms: Lineage versus interactive mechanisms. *In* "Developmental Order: Its Origin and Regulation" (S. Subtelny and P. B. Green, eds), pp. 219–245. A. R. Liss, New York.

Little, M., Luduena, R. F., Kennan, R., and Asnes, C. F. (1982). Tubulin evolution: Two major types of α-tubulin. *J. Mol. Evol.* **19**, 80–86.

Løvtrup, S. (1983) Reduction and emergence. *Riv. Biologia* **76**, 437–461.

Løvtrup, S. (1984). Ontogeny and phylogeny. *In* "Beyond NeoDarwinism" (M.-W. Ho, and P. T. Saunders, eds), pp. 159–190. Academic Press, New York.

Lucas, A. M., and Stettenheim, P. R. (1972). "Avian Anatomy". Agriculture Handbook No. 362. US Government Printing Office, Washington, DC.

MacLean, N. (1973). Suggested mechanism for increase in size of the genome. *Nature, New Biology* **246**, 205–206.

MacLeod, G. D. (1980). Mechanical properties of contours feathers. *J. Exptl. Biol.* **87**, 65–71.

Maderson, P. F. A. (1972a) When? Why? and How?: Some speculations on the evolution of the vertebrate integument. *Amer. Zool.* **12**, 159–171.

Maderson, P. F. A. (1972b). On how an archosaurian scale might have given rise to an avian feather. *Amer. Nat.* **106**, 424–428.

Maderson, P. F. A. (1975). Embryonic tissue interactions as the basis for morphological change in evolution. *Amer. Zool.* **15**, 315–327.

Maderson, P. F. A. (1983). An evolutionary view of epithelial–mesenchymal interactions. *In* "Epithelial–Mesenchymal Interactions in Development" (R. H. Sawyer, and J. F. Fallon, eds), pp. 215–242. Prager Scientific, New York.

Magia, T. M., Jorcano, J. L., and Franke, W. W. (1983). Translational products of mRNAs coding for non-epidermal cytokeratin. *EMBO J.* **2**, 1387–1392.

Mahrle, B., Bolling, R., Osborn, M., and Weber, K. (1983). Intermediate filaments of the Vimentin and prekeratin type in human epidermis. *J. Invest. Dermat.* **81**, 46–48.

Marchuk, D., McCrohon, S., and Fuchs, E. (1984). Remarkable conservation of structure among intermediate filament genes. *Cell* **39**, 491–498.

Maynard Smith, J. (1978). Optimization theory in evolution. *Ann. Rev. Ecol. & Syst.* **9**, 31–56.

Maynard Smith, J., Burain, R., Kauffman, S., Alberch, P., Campbell, J., Goodwin, B., Lande, R., Raup, D., and Wolpert, L. (1985). Developmental constraints and evolution. *Quart. Rev. Biol.* **60**, 265–287.

Mayr, E. (1960). The emergence of evolutionary novelties. *In* "Evolution after Darwin" (S. Tax, ed.), pp. 349–380. University Chicago Press, Chicago.

Mayr, E. (1982). "The Growth of Biological Thought", Belknap Press, Harvard University, Cambridge, Ma.

Mayr, E. (1983). How to carry out the adaptationist program. *Amer. Nat.* **121**, 324–334.

McConkey, E. H. (1982). Molecular evolution, intracellular organization, and the quinary structure of proteins. *Proc. Nat. Acad. Sci.* **79**, 3236–3240.

McKnight, S. L., and Kingsbury, R. (1982). Transcriptional control signals of a eukaryotic protein-coding gene. *Science* **217**, 316–324.

McNamara, K. J. (1982). Heterochrony and phylogenetic trends. *Paleobiology* **8**, 130–142.

Meinhardt, H. (1982). Generation of structure in a developing organism. *In* "Developmental Order: Its Origin and Regulation" (S. Subtelny and P. B. Green, eds), pp. 439–461. A. R. Liss, New York.

Miller, A. H. (1949). Some ecological and morphological considerations in the evolution of high taxonomic categories. *In* "Ornithologie als Biologische Wissenschaft" (E. Mayr, ed.), pp. 84–88. Carl Winter Universitaetsverlag, Heidelberg.

Mittenthal, J. E. (1981). The role of normal neighbors: A hypothesis for morphogenetic pattern regulation. *Devel. Biol.* **88**, 15–26.

Moll, R. W., Franke, W., Schiller, D. L., Geiger, B., and Kropler, R. (1982). The catalogue of human cytokeratins: Patterns of expression in normal epithela, tumors and cultured cells. *Cell* **31**, 11–24.

Molloy, P. L., Powell, B. D., Gregg, K., Barone, E. D., and Rogers, G. E. (1982). Organization of feather keratin genes in the chick genome. *Nucleic Acid Res.* **10**, 6007–6021.

Nardi, J. B. (1981). Epithelial invagination: adhesive properties of cells can govern position and directionality of epithelial folding. *Differen.* **20**, 97–102.

O'Dell, G. M., Oster, G., Alberch, P., and Burnside, B. (1980). The mechanical basis of morphogenesis. *Devel. Biol.* **85**, 446–462.

O'Donnell, I. J. (1973). A search for a simple keratin—Fractionation and peptide mapping of proteins from feather keratin. *Aust. J. Biol. Sci.* **26**, 416–437.

O'Donnell, I. J. (1974). The complete amino acid sequence of a feather keratin from Emu (*Dromaius novae-hollandiae*). *Aust. J. Biol. Sci.* **26**, 369–382.

O'Donnell, I. J., and Inglis, A. S. (1974). Amino acid sequence of a feather keratin from Silver Gull (*Larus novae-hollandiae*) and comparison with one from Emu (*Dromaius novae-hollandiae*). *Aust. J. Biol. Sci.* **27**, 369–382.

Oster, G., and Alberch, P. (1982). Evolution and bifurcation of developmental programs. *Evolution* **36**, 444–459.

Oster, G., O'Dell, G., and Alberch, P. (1980). Mechanics, morphogenesis and evolution. *Lect. Mathematics in the Life Sciences* **13**, 165–255.

Parkes, K. C. (1966). Speculations on the origin of feathers. *Living Bird* **5**, 77–86.

Partridge, L. D. (1982). The good enough calculi of evolving control systems: evolution is not good engineering. *Amer. J. Physiol.* **242**, R173–R255.

Powell, B. C., and Rogers, G. E. (1979). Isolation of messenger RNA coding for the "fast" protein of embryonic chick feather. *Nucleic Acid Res.* **7**, 2165–2176.

Prager, E. M., and Wilson, A. C. (1975). Slow evolutionary loss of the potential for interspecific hybridization in birds: a manifestation of slow regulatory evolution. *Proc. Nat. Acad. Sci.* **72**, 200–204.

Purslow, P. P., and Vincent, J. F. V. (1978). Mechanical properties of primary feathers from the pigeon. *J. Exptl. Biol.* **72**, 251–260.

Rachootin, S. P., and Thompson, K. S. (1981). Epigenetics, paleontology and evolution. *In* "Evolution Today. Proc. Second. Internat. Cong. Syst. & Evol. Biol." (G. G. Scudder, and E. Reveal, eds), pp. 181–193. Hunt Institute, Pittsburgh.

Raff, R. A., and Kaufman, T. C. (1983). "Embryos, Genes, and Evolution". Macmillan, New York.

Regal, P. J. (1975). The evolutionary origin of feathers. *Quart. Rev. Biol.* **50**, 35–66.

Richardson, J. S. (1977). Beta-pleated topology and relatedness of proteins. *Nature* **268**, 495–500.

Riddle, O. (1980). The cause of the production of "down" and other downlike structures in the plumage of birds. *Biol. Bull.* **14**, 163–176.

Riedl, R. (1977). A systems analytical approach to macroevolutionary phenomena. *Quart. Rev. Biol.* **52**, 351–370.

Robertson, A. (1979). Waves propagated during vertebrate development: observations and comments. *J. Embryol. Exp. Morph.* **50**, 155–167.

Rodakis, G. C., and Kafatos, F. C. (1982). Origin of evolutionary novelty in proteins: How a high-cysteine chorion protein has evolved. *Proc. Nat. Acad. Sci.* **79**, 3551–3555.

Rogers, G. E. (1978). Keratins viewed at the nucleic acid level. *TIBS.* **3**, 131–133.

Rose, M. R., and Doolittle, W. F. (1983). Molecular biological mechanisms of speciation. *Science* **220**, 157–162.

Sakai, H. (1980). Regulation of microtubule assembly *in vitro*. *Biomed. Res.* **1**, 359–375.

Saleme, F. R., and Weatherford, D. W. (1981). Conformational and geometrical properties of beta-sheets in proteins. I, II & III. *J. Mol. Biol.* **146**, 101–117, 119–141, 143–156.

Salthe, S. N. (1975). Problems of macroevolution (molecular evolution, phenotype definition and canalization) as seen from a hierarchical viewpoint. *Amer. Zool.* **15**, 295–314.

Satoff, N. (1982). Timing mechanisms in early development. *Differ.* **22**, 165–173.

Saunders, P. T., and Ho, M. W. (1981). On the increase in complexity in evolution. II. The relativity of complexity and the principle of minimum increase. *J. Theor. Biol.* **90**, 515–530.

Sawyer, R. H. (1983). The role of epithelial–mesenchymal interactions in regulating gene expression during avian morphogenesis. *In* "Epithelial–Mesenchymal Interactions in Development" (R. H. Sawyer, and J. F. Fallon, eds), pp. 115–146. Praeger Scientific, New York.

Sawyer, R. H., Knapp, L. W., and O'Guin, W. M. (1986). Epidermis, Dermis and appendages. *In* "Biology of the Integument. 2. Vertebrates, V. The Skin of Birds" (J. Bereiter-Hahn, A. G. Matoltsy, and K. S. Richards, eds), pp. 194–238. Springer-Verlag, Berlin.

Schopf, T. J. M. (1981). Evidence from findings of molecular biology with regard to the rapidity of genomic change: Implications for species duration. *In* "Paleobotany, Paleoecology and Evolution" (K. J. Niklas, ed.), pp. 135–192. Prager, New York.

Schopf, T. J. M., Raup, D. M., Gould, S. J., and Simberloff, D. S. (1975). Genomic versus morphological rates of evolution influence of morphological complexity. *Paleobiology* **1**, 63–70.

Sengel, P. (1976). "Morphogenesis of Skin". Develop. & Cell Biol. Vol. 3. Cambridge University Press, Cambridge.

Shapiro, J. A., and Cordell, B. (1982). Eukaryotic mobile and repeated genetic elements. *Bio. Cell* **43**, 31–54.

Sharp, P. A. (1985). On the origin of RNA splicing and introns. *Cell* **42**, 397–400.

Soll, D. R. (1983). A new method for examining the complexity and relationships of "timers" in developing systems. *Devel. Biol.* **95**, 73–91.

Spearman, R. I. C. (1966). The keratinization of epidermal scales, feathers and hair. *Biol. Rev.* **41**, 53–96.

Steinert, P. M., Idler, W. W., and Zimmerman, S. B. (1976). Self-assembly of bovine epidermal keratin filaments *in vitro*. *J. Mol. Biol.* **108**, 547–567.

Steinert, P. M., Steven, A. C., and Roop, D. R. (1985). The molecular biology of intermediate filaments. *Cell* **42**, 411–419.

Stent, G. S. (1981). Strength and weakness of the genetic approach to the development of the nervous system. *Ann. Rev. Neurosci.* **4**, 163–194.

Steven, A. C., Hainfeld, J. F., Trus, B. L., Wall, J. S., and Steinert, P. M. (1983). Epidermal keratins assembled *in vitro* have masses-per-unit-length that scale according to average subunit mass: structural basis for homologous packing of subunits in intermediate filaments. *J. Cell Biol.* **97**, 1939–1944.

Stewart, M. (1977). The structure of chicken scale keratin. *J. Ultrastruct. Res.* **60**, 27–33.

Stubblefield, E. (1986). A theory for developmental control by a program encoded in the Genome. *J. Theor. Biol.* **118**, 124–143.

Südhof, T. C., Goldsmith, J. L., Brown, M. S., and Russell, D. W. (1985). The LDL receptor gene: A mosaic of exons shared with different proteins. *Science* **228**, 815–822.

Sun, T-T, Eichner, R., Schermer, A., Cooper, D., Nelson, W. G., and Weiss, P. A. (1984). Classification, expression and possible mechanisms of evolution of mammalian epithelial keratins: A unifying model. *Cancer Cells* **1**, 169–1176. Cold Spring Harbor Laboratory.

Thompson, J. L. (1964). Morphogenesis and histochemistry of scales in the chicken. *J. Morph.* **115**, 207–223.

Turçek, F. J. (1966). On plumage quantity in birds. *Ecol. Polska. Ser. A.* **14**, 1–18.

van den Heuvel, R., Hendricks, W., Quax, W., and Bioemendal, H. (1985). Complete structure of the hamster αA crystallin gene. Reflection of an evolutionary history by means of exon shuffling. *J. Mol. Biol.* **185**, 273–284.

Walker, I. A., and Bridgen, J. (1976). Differentiation in avian keratinocytes. *Eur. J. Biochem*, **67**, 329–339.

Walker, I. A., and Rogers, G. E. (1976). The structural basis for the heterogeneity of chick down feather. *Eur. J. Biochem.* **69**, 341–350.

Weber, K., and Geisler, J. (1982). The structural relation between intermediate filament proteins in living cells and the α-keratins of sheep wool. *EMBO J.* **1**, 1155–1160.

Webster, G., and Goodwin, B. (1981). History and structure in biology. *Perspec. in Biol. & Med.* **27**, 39–62.

Wellnhofer, P. (1974). Das fünfte Skelettexamplar von *Archaeopteryx*. *Palaeontographica* **147**, 169–216.

Wessells, N. K. (1982). A catalogue of processes responsible for metazoan morphogenesis. *In* "Evolution and Development" (J. T. Bonner, ed.), pp. 115–154. Springer-Verlag, Heidelberg.

Wicken, J. S. (1984). On the increase of complexity in evolution. *In* "Beyond Neo-Darwinism" (M.-W. Ho, and P. Saunders, eds). pp. 89–112. Academic Press, New York.

Wilton, S. D., Crocker, L. A., and Rogers, G. E. (1985). Isolation and characterization of keratin nRNA from the scale epidermis of the embryonic chick. *Biochem. Biophys. Acta* **824** (3), 201–208.

Wright, S. (1982). The shifting balance theory and macroevolution. *Ann Rev. Genetics* **16**, 1–20.

Wu, Y-J, Parker, L. M., Binder, N. E., Beckett, M. A., Sinard, J. H., Griffiths, C. T., and Rheinwold, J. G. (1982). The mesothelial keratins. A new family of cytoskeletal proteins identified in cultured mesothelial cells and nonkeratinizing epithelia. *Cell* **31**, 693–703.

Wyld, J. A., and Brush, A. H. (1979). The molecular heterogeneity and diversity of reptilian keratins. *J. Mol. Evol.* **12**, 331–347.

Wyld, J. A., and Brush, A. H. (1983). Keratin diversity in the reptilian epidermis. *J. Exptl. Zool.* **225**, 387–396.

Wyles, J. S., Kunkel, J. G., and Wilson, A. C. (1983). Birds, behavior and anatomical evolution. *Proc. Nat. Acad. Sci.* **80**, 4394–4398.

Zuckerkandl, E. (1975). The appearance of new structures and functions in proteins during evolution. *J. Molec. Evol.* **7**, 1–57.

Zuckerkandl, E. (1978). Multilocus enzymes, gene regulation, and genetic stability. *J. Molec. Evol.* **12**, 57–89.

# Chapter 3

# ENDOCRINOLOGY OF REPRODUCTION IN WILD SPECIES

*John C. Wingfield and Donald S. Farner**
*Department of Zoology*
*University of Washington*
*Seattle, Washington 98195*





Avian Biology, Vol. IX

ISBN 0-12-249409-1

* Professor Donald S. Farner passed away on 18 May, 1988 in Marysville, Washington, before this manuscript was completed. It was Professor Farner's wish that the manuscript provide a broad perspective of the reproductive endocrinology of wild avian species for lay readers, graduate students and postdoctoral associates, as well as a reference for established researchers in this field. It is a tragedy that Professor Farner died before the manuscript was completed because, in many ways, it is a monument to his epochal contributions to avian biology in general. It should be stated at the outset that any omissions or shortcomings of this chapter are the responsibility of JCW and should in no way reflect upon Donald Farner's contribution. Nevertheless, it is hoped that most, if not all, of the original goals have been achieved and, as Professor Farner wished, many readers (especially younger generations) will be stimulated to pursue investigations in this rich and fascinating field.

## I. Introduction

Our knowledge of natural history, ecology, ethology, population biology, and breeding systems of birds is more extensive than for any other class of vertebrates. This provides significant advantages for investigations of wild species in the field of both an empirical and experimental nature, including adaptations to changes in environmental conditions. Most avian species are diurnal and conspicuous making field observations relatively easy. They are hardy and easily withstand more invasive investigations using capture, marking, or even surgery (e.g., Oring *et al.*, 1988a). Many thrive in captivity allowing experimental procedures under rigorously controlled conditions. Thus it is not surprising that investigations of birds, including wild species, have made major contributions to biology in general (see Konishi *et al.*, 1989 for discussion).

The avian neuroendocrine and endocrine systems are, by and large, very similar to those of mammals. Thus it has been possible to use techniques developed for mammals and clinical purposes, on avian species. As a result there is a broad and extensive literature on avian endocrinology that goes back to the very beginnings of this branch of biology. Nevertheless, most textbooks of endocrinology focus on mammalian studies and when referring to birds do so by citing the literature on domesticated species almost exclusively. It is our aim here to provide a broad, historical base and current account of the endocrinology (emphasis on reproductive) of wild species, referring to the classical studies of domesticated species only to provide background and context.

### A. HISTORICAL PERSPECTIVES

The relationship between gonads and secondary sex characteristics deduced from castration is not only ancient, as indicated by the writings of Aristotle, but also widely known among the lay public. Without the replacement of gonads into castrated animals, however, observations of the effects of gonadectomy remained only suggestive with respect to the possibility that these organs release into circulating blood an active principle that causes development and maintenance of secondary sex characteristics. A very careful and thoughtful review by Barker Jørgensen (1971) of the experiments of John Hunter published by himself (e.g., 1771, 1792), and as described by others (e.g., Palmer, 1837, 1841), indicates that he performed successful transplants of testes of cockerels, *Gallus gallus*, into the body cavities of hens, and also auto- and heterotransplants of these organs into the body cavity of males. Hunter, although asserting the success of such transplants, communicated relatively little concerning their effects and offered no precise explanations thereof. The history of endocrinology, and indeed that of endocrinology in general, is most frequently traced to classical experiments reported succinctly by Arnold A. Berthold of Göttingen (1849a,b). He removed the testes of cockerels and showed a reduction of male-like behavior, and reduced size of some secondary sex characteristics. Transplantation of testes into these castrates restored male reproductive behavior and secondary sex characters.

Perhaps the most significant effect of the experiments of Hunter was the subsequent stimulation of the above-noted experiment by Berthold with six castrated cockerels. Two of these received an autotransplanted testis in the body cavity; two received single heterotransplanted testes; the remaining two received none. The last two became typical capons; whereas the four with implanted testes developed normal male secondary sex characteristics. When the testis from one heterotransplanted bird was removed its behavior became that of a capon and it failed to regenerate a comb after it was surgically excised. Berthold (1849a,b), consistent with the philosophy of physiology of the times, explained these results as an effect of testes on blood, independent of the nervous system, that then affected the body in general. To us Berthold was, therefore, the first to express the principle of endocrine control.

Berthold published nothing further on this experiment despite the indication that his communications were preliminary. Several possible reasons therefore have been considered by Barker-Jørgensen (1971). To us the most probable is the failure of Berthold's esteemed colleague, Wagner (1851), to obtain viable transplants in an extensive series of experiments that included a repetition of those of Berthold. One cannot preclude the possibility that subsequent experiments by Berthold himself gave inconsistent results. It should be emphasized that, subsequently, several cases of successful transplantation of avian testes have been reported, but it also should be noted that partial successes and failures have been frequent. Thus among wild passerine species Novikov (1955) and Roudneva (1970) have successfully performed heterotransplantations of testes in House Sparrows, *Passer domesticus*, whereas in our laboratory (J. E. Erickson and D. S. Farner, unpublished results) such transplants in White-crowned Sparrows, *Zonotrichia leucophrys gambelii*, failed in a manner similar to those described by Wagner (1851).

For whatever reason, the experiments of Berthold faded rapidly into oblivion until their "rediscovery" a half-century later (e.g., Nussbaum, 1905) after the principle of endocrine control had been well established. In summary we agree with Barker-Jørgensen (1971) that Hunter, for his establishment of the feasibility of testicular transplantation, should be regarded as the precursor of endocrinology, and that Berthold, despite the possible fortuity of his experiments, is deservedly regarded as the founder of avian endocrinology, and indeed of endocrinology as a whole. Although not strictly parallel, the position of Berthold in the history of endocrinology is nevertheless very similar to that of Gregor Mendel in the history of genetics.

Through the ensuing century the development of avian endocrinology, although on a much more modest scale, proceeded in a pattern similar to that of mammalian endocrinology. For obvious reasons, however, an overwhelmingly large fraction of the research effort was effected on domesticated stocks of the Jungle Fowl (= chicken), *Gallus gallus*; Turkey, *Meleagris galloparvo*; Japanese Quail, *Coturnix coturnix*; Mallard, *Anas platyrhynchos*; Greylag Goose, *Anser anser*; and Rock Dove, *Columba livia*.

The very modest effort on truly wild species was confined largely to histological investigations of endocrine glands; experimental studies, such as those on photoperiodic control of gonadal function, thyroidectomy, and effects of exogenous hormones; and were performed on caged subjects. Thus, for over a century following the pioneer experiment of Berthold, endocrinology of wild species under natural conditions was confined to inferences from histological studies in laboratory experiments on caged birds, largely without evaluation of the effects of captivity on endocrine functions; and extrapolation from results of investigations of domesticated birds.

### B. OBJECTIVES

The function of this chapter is to review the development and current status of endocrinology of wild species with special attention to recent results of field investigations. Emphasis is placed on the endocrinology of reproduction and associated annual events such as migration, molt, and responses to stress; and endocrine aspects of behavior. This chapter is therefore both supplementary and complementary to earlier chapters in *Avian Biology* by Assenmacher (1973), Kobayashi and Wada (1973), Lofts and Murton (1973), Tixier-Vidal and Follett (1973) and Balthazart (1983), and also to the important symposium volumes edited by Epple and Stetson (1980), Mikami *et al.* (1983), Johnson *et al.* (1984), Follett *et al.* (1985), and Wada *et al.* (1990a,b), and to Murton and Westwood (1977).

Although they overlap in time, and frequently within single investigations, several phases in the development of endocrinology of wild species can nevertheless be identified. The most recent phases, which this chapter emphasizes, are those of descriptive and experimental field endocrinology. First we include a brief and selective review, primarily for non-endocrinologists, on the development of endocrinology in wild species as background to the more recent and exciting phases of avian endocrinology. We then go on to discuss the integration of field and laboratory investigations with primary attention given to endocrine elements involved in annual reproductive cycles and other cyclic functions associated therewith. The literature is truly voluminous and in the limited space here we can only report pertinent examples. Wherever possible we also cite appropriate reviews that will provide more detailed reading on each subtopic.

## II. Phases in the Development of Endocrinology in Wild Species

### A. HISTOLOGICAL AND CYTOLOGICAL INVESTIGATIONS

In this phase we include a broad spectrum of morphological studies that extends from those based on traditional tinctorial techniques through transmission and scanning electron microscopy. We emphasize functional implications through correlations with phases of annual cycles. However, because of the extent of the literature, which spans more than a century, we cite only illustrative examples.

#### *1. Endocrine Gonads*

Although seasonal changes in size of testes of birds were recorded by Aristotle, and Hunter (see Barker-Jørgensen, 1971), the first systematic investigation of the seasonal development of the testes of a reproductively periodic species from resting to fully functional state was apparently that of Etzold (1891) on the House Sparrow. Although his notes on the literature suggest that he was familiar with Leydig's description of interstitial cells, his communication is devoid of any description of intertubular components. In general, the early literature on secretory interstitial cells is difficult to use comparatively because of differences in techniques and interpretation.

An illustration of the confusion that evolved was the conclusion of Stieve (1919) that despite the enormous difference in size between the resting and breeding testes of the Jackdaw, *Corvus monedula*, the amount of non-vascular intertubular tissues remains relatively constant. This and the results of subsequent investigations on domestic geese led Stieve (e.g., 1920, 1921, 1926), contrary to the earlier conclusion of Firket (1920), to assert that the interstitial cells are not secretory. This conclusion was appropriately challenged by Benoit (1923) on the basis of investigations on the "Combasou" (presumably a West African species of widow bird *Hypochera* or *Vidua*), and by Orban (1929) using material taken from October to May from the Rook, *Corvus frugilegus*, and Jackdaw. Subsequently the careful investigations of Threadgold (1956a,b) resolved the issue for Jackdaws by demonstration of a real increase in size and number of the cells of Leydig in April and October, the latter being coincident with autumnal sexual activity without an increase in testicular volume. A similar bimodal curve in cells of Leydig has been described for the Mallard by Höhn (1947) and in the House Sparrow by Threadgold (1960). Reports on the absence of Leydig cells in resting testes, e.g., in White-crowned Sparrow (Blanchard and Erickson, 1949); Red-winged Blackbird, *Agelaius phoeniceus* (Wright and Wright, 1944); and California Gull, *Larus californicus* (Johnston, 1956a) probably reflect inadequacies of techniques used or differences in interpretation of light microscopic material. Lam and Farner (1976) have demonstrated by light and electron microscopy the occurrence of Leydig cells in the winter phase of the testes of *Z. l. gambelii.* Although identifying cyclic fluctuations in activity and numbers, the investigations of Röhss and Silverin (1983) on the Great Tit, *Parus major*, clearly demonstrate the presence of Leydig cells throughout the year, as had the light microscopic investigations of Marshall (1949a) on the Fulmar, *Fulmarus glacialis*, Marshall and Coombs (1957) on the Rook, Sarkar and Ghosh (1964) on the House Sparrow, and Payne (1969) on red-winged and Tricolored Blackbirds, *Agelaius tricolor*.

The investigations summarized above, as well as others, indicate that cells of Leydig are present in all phases of the annual cycle of periodically breeding species, and that their numbers and activity correlate reasonably well with sexual behavior. This conclusion is consistent, with few exceptions, with measured plasma levels of testosterone, e.g., in *Z. l. gambelii* (Wingfield and Farner, 1978a,b), European Starling, *Sturnus vulgaris* (Temple, 1974), Rook (Lincoln *et al.*, 1980; Péczely and Pethes, 1982), and Mallard (Paulke and Haase, 1978; Donham, 1979).

Although the morphological identity of the endocrine ovary of the Domestic Fowl has been the subject of extensive investigation, its nature is still not definitely resolved (for reviews see Dahl, 1970; Gilbert, 1971; van Tienhoven, 1983; Johnson and Tilly, 1990). Because it is highly probable that medullary and cortical interstitial cells of the theca are structurally identical, separate designations for them may be useful only because they may be functional at different times. Evidence suggests that interstitial cells of the theca interna and/or stratum granulosa of the follicle secrete estrogens, progesterone, and testosterone. Indeed,

Huang *et al.* (1979) have proposed a model that attributes an interaction between these two types of cells in the secretion of these hormones. Because of this uncertainty about the function, and for other reasons, it is not surprising that there have been few meaningful investigations thereon in wild species. Erpino (1969) from investigations of the Black-billed Magpie, *Pica pica*, has reported thecal "glands" in larger ovarian follicles and "stromal (interstitial) glands" throughout the year, both being most abundant during the nest-building period. In his detailed study of ovarian follicles of *Z. l. gambelii*, Kern (1970, 1972) described "gland" cells in the theca from March through July, and in the stroma apparently throughout the year. However, because of technique limitations, Kern and his associates were unable to detect changes in cellular activity associated with hormone secretion, or to find "gland" cells in the thecae of postovulatory follicles, although atretic follicles did apparently have cells with signs of secretory activity. Unlike investigations of interstitial cells of males, the much smaller number undertaken on these and other putative steroidogenic cells in females have contributed only modestly to our knowledge of the endocrine ovary of wild species.

### 2. *Thyroid Gland*

The thyroid gland is the only avian alveolar endocrine organ. Its epithelial cells transfer iodine from blood, after which it is combined with tyrosine of alveolar thyroglobulin either in the alveolar membrane of the cell or at its alveolar surface to form thyroid hormones. The epithelial cells also transfer iodinated thyroglobulin into blood capillaries. Since both of these processes are controllable, both the amount of colloid in the alveolus and the height and activity of the epithelial cells, by cytologic criteria, can vary (e.g., Gorbman *et al.*, 1983). Thus histologic examinations of thyroid gland have been used to infer its functional state. Height of epithelial cells, or the ratio of their total volume to that of the alveoli (follicles), indicates secretory activity. Diameter of alveoli and weight of the gland can be misleading since they reflect storage of colloid, rather than release of thyroid hormones.

Based on results of histological investigations of material from wild species taken under natural conditions, as well as from studies on domesticated species, several functions were attributed to hormones of the avian thyroid gland. These included adjustments in rate of metabolism in response to change in ambient temperature, induction of molt, control of physical structure and color of plumage, essential support for normal function of gonads in some species, and induction of migratory behavior and metabolism. It should also be noted that because release of thyroid hormones into circulating blood has been associated with differing histologic criteria, comparisons among results are fraught with problems, many of which have not yet been resolved by use of more sophisticated methods.

Interpretation of histological results from wild species under natural conditions has also been confounded by the lack of information, in most investigations, on functional state, other endocrine glands, and by the multiple functions of thyroid hormones.

*Adjustment of metabolic rates in response to changes in ambient temperature.* The literature on correlation of histologic changes with environmental adjustments presents conflicting results, very likely because of differences and changes in heat conductance of body surface and plumage, temporary nocturnal hypothermia, behavioral thermoregulation, and differences in the ratio of body mass to body temperature. For House Sparrows in climates with relatively cold winters, various investigators found correlations between low temperature and increased activity of the secretory cells and discharge of colloid from thyroid follicles (e.g., Watzka, 1934; Woitkevitsch and Novikov, 1936; Miller, 1939; Kendeigh and Wallin, 1966). Consistent with this assumed thermoregulatory role is the observation of only slightly variable secretory activity of thyroid epithelial cells of *P. domesticus* in the mild climate of southern California (Davis and Davis, 1954). However, Küchler (1935) reported relatively low cell height in House Sparrows taken in winter in Germany. Küchler also reported a mild increase in activity and discharge of colloid in the cogeneric European Tree Sparrow, *Passer montanus*, which appears to be consistent with the observations of Nakamura (1957, 1958) in Japan. Moreover, the response of this species to the severe Russian winter appears to be more conspicuous (Woitkewitsch and Novikov, 1936).

As assessed by histologic criteria, the relationship of the thyroid gland of White-crowned Sparrows to environmental temperature appears generally consistent with that of the House Sparrow. Wilson and Farner (1960) found an increase of thyroid activity in *Z. l. gambelii* wintering in eastern Washington, a relationship that was experimentally confirmed. This contrasts with slight or no increase in thyroid activity of wintering populations in the milder climate of the area around San Francisco Bay region; however, a more conspicuous increase in activity in winter was reported in the resident race of White-crowned Sparrow, *Z. l. nuttalli* (Oakeson and Lilly, 1960).

Histological evidence of a hibernal increase in thyroid activity in other passerine species has been reported for the Bullfinch, *Pyrrhula pyrrhula* (Gal, 1940), Red Crossbill, *Loxia curvirostra* (Putzig, 1937), Mistle Thrush, *Turdus viscivorus* (Gal, 1940), European Robin, *Erithacus rubecula* and Yellowhammer, *Emberiza citrinella* (Küchler, 1935). Although Gal reported increased activity of thyroids in Black-billed Magpies during winter in Hungary, a more extensive investigation of this species in Wyoming by Erpino (1968) found no significant changes in thyroid histology through the course of a year. No significant hibernal increase in thyroid activity, as indicated by histologic changes, were found in the European Blackbird, *Turdus merula* (Bigalke, 1956), Greenfinch, *Carduelis chloris*, and Bohemian Waxwing, *Bombycilla garrulus* (Putzig, 1937), Rook, (Gal, 1940), Carrion Crow,

*Corvus corone* (Haecker, 1926), and Tricolored Blackbird (Payne and Landolt, 1970).

Among wild galliform species, information is sparse. Elevated hibernal secretory activity of the thyroid has been reported for the Gray Partridge, *Perdix perdix* (Gal, 1940) and Gambel's Quail, *Lophortyx gambelii* (Raitt, 1968), whereas in the White-tailed Ptarmigan, *Lagopus leucurus*, the increase in thyroid activity was modest (Höhn and Braun, 1977). In other non-passerine species, Thybusch (1965) reported a slightly increased height of thyroid secretory cells in the Mew Gull, *Larus canus*, in winter.

Bearing in mind other thermoregulatory mechanisms cited above, one can conclude that histological studies, despite difficulties in interpretation and comparisons, established a role of the thyroid gland in thermoregulation in at least some wild species.

*Induction of molt and nature of plumage.* Attention has been directed most frequently to prebasic (= postnuptial) molts. Correlations with deduced functional condition have been variable, not only because of histological interpretation, but also because control of the process is clearly not unifactorial and apparently differs extensively among taxa (e.g., Assenmacher, 1958, 1973; Stresemann and Stresemann, 1966; Payne, 1972; Jallageas and Assenmacher, 1979, 1985; Farner, 1983), despite arguments for paramount thyroid control (e.g., Salomonsen, 1939; Woitkewitsch, 1940a; Gavrilov and Dolnik, 1974).

Among wild passerine species correlations between histologically active thyroid glands and molt have been reported for the House Sparrow (Küchler, 1935; Woitkewitsch and Novikov, 1936; Gal, 1940; Davis and Davis, 1954; Bigalke, 1956). Similar observations were made of the European Tree Sparrow by Küchler (1935) and Woitkewitsch and Novikov (1936), but not confirmed by Nakamura (1957, 1958). This relationship has also been described for the Yellowhammer (Küchler, 1935) and in the Ortolan Bunting, *Emberiza hortulana* by Woitkewitsch and Novikov (1936), who noted significantly that both molt and increased thyroid activity were delayed in renesting birds. For *Z. l. gambelii*, Wilson and Farner (1960) found no correlation between thyroid activity and prebasic molt whereas Oakeson and Lilley (1960) noted an increase in activity at the time of gonadal regression which persisted through the molt. For the Silvereye, *Zosterops lateralis*, Keast (1953) found increased thyroid activity during both spring and summer molts, and that the latter overlapped with spermatogenesis. In the Tricolored Blackbird there appears to be a slight elevation of thyroid activity before prebasic molt (Payne and Landolt, 1970). Erpino (1968) found no significant correlation in Black-billed Magpies, as has also been reported from investigations on several other passerine species. Wilson and Farner (1960) have further summarized the conflicting results of investigations on the temporal relationships between histologically assessed thyroid activity and molt in passerine species.

Among wild galliform species correlations between prebasic molt and thyroid activity have been described for Japanese quail, Gray Partridge (Gal, 1940), and Gambel's Quail (Raitt, 1968). For the complex multiple-molt cycles of ptarmigan species (e.g., Salomonsen, 1939; Stresemann and Stresemann, 1966), the situation is more complex. Histological examinations of thyroid glands in White-tailed Ptarmigan by Höhn and Braun (1977) suggest increased activity preceding and during the breeding season and summer molt, but not necessarily during the remaining two molts.

The pitfalls of judging the role of thyroid gland in molt through correlation of the histologically assessed activity of the gland therewith were well illustrated by Höhn (1950) in an investigation on the Mallard in which a period of histologically assessed activity preceded each molt. However, thyroidectomy of young Mallards 2 months before the molts from juvenile plumage to breeding plumage had little or no effect on either the molt nor the ensuing nuptial plumage.

*Thyro-gonadal relationships.* In retrospect, several of the earlier histological investigations cited above (e.g., Watzka, 1934; Küchler, 1935; Woitkewitsch and Novikov, 1936) contain observations on several species that suggest a reciprocal relationship between gonadal and thyroid function, despite a lack of information on interstitial endocrine cells in the former. However, it remained for Woitkewitsch (1940b) to hypothesize formally such a relationship for the European Starling. A reciprocal relationship was strongly emphasized by Vaugien (1948) for the Chaffinch, *Fringilla coelebs*, and Great Tit. Furthermore, morphological evidence of a depression in thyroid activity during the reproductive season has been reported for *Z. l. nuttalli* (Oakeson and Lilley, 1960), *Foudia madagascariensis* (Legendre and Rakotondrainy, 1963), Lal Munia, *Estrilda amandava* and Spotted Munia, *Lonchura punctulata* (Thapliyal, 1969). For several species, including the Mallard (Höhn, 1950), European Blackbird (Fromme-Bouman, 1962), European Starling (Burger, 1938), it has been reported that thyroid activity increases during seasonal regression of the gonads. Observations of Ljunggren (1968), with similar techniques, on the Woodpigeon, *Columba palumbus*, are consistent therewith. On the other hand, in Silvereyes, Keast (1953) found histologically active thyroid glands during spermatogenesis, whereas observations on the Black-billed Magpie (Erpino, 1968, 1969) showed no obvious correlations. Collectively these morphological observations demonstrated that among many species there is some sort of reciprocal relationship between the activities of thyroid glands and gonads. They were the precursors of subsequent investigations with more sophisticated methods.

Jallageas and Assenmacher (1985) now recognize two patterns of thyro-gonadal relationship. In the most common, the endocrine activities are reciprocal, i.e., thyroid activity is low during the reproductive season. In the less common pattern, secretion of thyroid hormones remains high during reproduction.

### *3. Adrenal Cortex (= Interrenal Tissue)*

Contributions of histological investigations of the adrenal cortex of wild species to knowledge of functions under natural conditions have been modest. In part this is because of the small size of the entire adrenal gland—0.01–0.04% of body mass. Also responsible for the limited contributions of such studies is the organization of the gland which generally consists of cylindrical strands of cortical (interrenal) steroidogenic cells among which are interspersed islets, or networks of islets, of medullary (chromaffin) cells which secrete epinephrin and norepinephrin. In most, if not all species, cortical tissue is zonated on an approximately surface-to-interior radius (e.g., Knouff and Hartman, 1951; for brief discussions and references see Ghosh, 1962; Haack *et al.*, 1972; Pearce *et al.*, 1979; Mikami *et al.*, 1980; Holmes and Cronshaw, 1980). Interpretations of histological changes in cortical tissue are difficult because it secretes at least three or four steroid hormones in response to changes in internal conditions and indirectly to a variety of changes in external environment, especially those of stressful nature (e.g., Holmes and Phillips, 1976; Harvey *et al.*, 1984; Wingfield, 1988b).

Since microanatomy of the adrenal gland changes with age in the domestic fowl and Rock Dove (Müller, 1929) and also in the Mew Gull (Thybusch, 1965), it can be suspected that such is general in avian species.

From an investigation of adrenal glands from adults of more than 400 species of several orders, Hartman and Albertin (1951) and Hartman *et al.* (1947) reported extensive interspecific differences among the relative amounts and spatial arrangements of cortical and medullary tissue. Such differences were also noted in an investigation of eight Indian species by Vyas and Jacob (1976).

Within our knowledge the first significant year-round histological investigation of seasonal changes in cortical tissue in a wild species was that of Fromme-Bouman (1962) on at least largely non-migratory individuals of the European Blackbird. On the basis of karyometry of cortical cells, she concluded that these cells were most active and occupied the greatest fraction of the gland in June–July, i.e., at the termination of seasonal gonadal activity and the onset of prebasic molt. She also presented evidence that low environmental temperature increases activity of cortical cells, which appears consistent with the observation that the annual increase in activity begins in the colder part of winter. In a rather similar investigation on the migratory *Z. l. gambelii*, Lorenzen and Farner (1964) also demonstrated a conspicuous annual cycle in histologically assessed cortical activity with a maximum level in winter and a minimum in June, i.e., early in the breeding season, thus well in advance of prebasic molt.

In adrenal glands of adult male Mew Gulls taken throughout the year, Thybusch (1965) from histological examination reported that cortical tissue was most active at the time of maximum testicular development, but much lower during the ensuing prebasic molt and enhanced thyroid activity. Similarly Hall (1968) reported a close correlation between fractional cortical volume of the adrenal gland and fractional volume of seminiferous tubules in seasonally breeding Eastern Rosellas, *Platycercus eximius*. In the Common Eider, *Somateria mollissima*, inter-renal activity increased to a maximum during incubation—a period of high energy utilization and mobilization of nutrient reserves. Note also that increasing activity of the interrenal cells in spring was accompanied by heavy fat deposition in April possibly by inducing hyperphagia which in turn prevents the catabolic effects of very high glucocorticoid production (Gorman and Milne, 1971). Contrarily, Ljunggren (1969a,b) from his investigations of endocrine correlates of the annual cycle concluded that "... the proportion of cortical tissue diminishes in spring and summer—the breeding time of woodpigeons ...". The changes were more conspicuous in males than in females. In House Sparrows, Moens and Coesseus (1970) found that nuclear size in adrenocortical cells increased from November to May and then declined through the rest of the breeding season.

### 4. *Adenohypophysis (= Anterior Lobe of the Pituitary Gland)*

The avian adenohypophysis, which consists of a relatively large pars distalis and a smaller pars tuberalis, is unique among vertebrates in that the former is divided into cytologically distinctive rostral and caudal lobes, and lacks a pars intermedia. The rostral and caudal lobes receive blood via separate sets of portal vessels from the anterior and posterior divisions of the median eminence of the hypothalamus, respectively. These two divisions are morphologically and functionally different. For treatises on these distinctive characteristics of the avian pars distalis see Rahn and Painter (1941), Wingstrand (1951), Tixier-Vidal (1963), Tixier-Vidal and Follett (1973), Oksche and Farner (1974), and Mikami (1986).

Although the gross anatomy of the avian adenohypophysis was described more than two centuries ago (Malacarne, 1782), and the differentiation of its pars distalis onto distinct cephalic and caudal lobes has been known since the 19th century (Hannover, 1824), progress on its functional cytology during the ensuing half century was markedly slower than that of peripheral endocrine organs such as thyroid gland, testis, adrenal cortex, and adrenal medulla. Among the conspicuous reasons therefore were:

(1) The multiple functions of the pars distalis.
(2) Lack of sufficiently specific tinctorial and cytochemical techniques for identification of types of cells with respect to hormones secreted. This was largely responsible for excessively confusing nomenclature (cf. Rahn and Painter, 1941; Wingstrand, 1951, 1954; Tixier-Vidal, 1963; Mikami, 1986; Tixier-Vidal and Follett, 1973). Some of the problems from these techniques were

subsequently ameliorated by electron-microscopic examination of secretory granules and other ultrastructural features of putative cell types.

(3) Assumption that each cell type secretes a single hormone. It now appears that this is only partially true since immunocytochemical studies suggest very strongly that the two gonadotropins are produced by a single type of cell that occurs in both the cephalic and caudal lobes (cf. Mikami, 1986). Similar problems emerge with the putative corticotropin cells.

(4) Insufficient studies on wild species with conspicuous annual cycles in endocrine function. Historically, histological and cytological investigations of the avian adenohypophysis have focused on five domesticated species. This concentration of effort involves not only the potential bias of species from only three orders of birds, but also variability that has developed with domestication (Obussier, 1948; Wingstrand, 1951). The the best of our knowledge the first study on seasonal changes in a wild species was that of Schildmacher (1937) on the European Blackbird.

A major advance in histology and cytology of the pars distalis came with the diversified investigation of Rahn and Painter (1941) on 18 species representing 12 families. The results of these observations permitted the designation of the "avian pituitary pattern" as outlined in the first paragraph of this section. Although four types of cells were described their functional significance remained obscure.

The following decade produced the classical monograph of Wingstrand (1951) based on histological investigations of the adenohypophysis, neurohypophysis, hypothalamus and hypothalamic portal system using material from 69 species representing 34 families in 18 orders. Not only did this extensive investigation provide a greatly expanded perception of the "avian pituitary pattern" of Rahn and Painter, it provided an enormously enhanced comprehension of its morphological relationships with the hypothalamus, and important rationalizations of cell types of the pars distalis described by other authors using various cytological and cytochemical techniques on different species. Wingstrand, as a cautious morphologist, made no suggestions concerning putative functions of cell types. The material available to him did not permit assessment of seasonal differences.

Even before the appearance of the classical work of Wingstrand, but more intensively thereafter, investigations began to assign putative functions to the cell types of the rostral and caudal lobes of the pars distalis of domesticated species by means of a variety of cytologic and cytochemical techniques enhanced by correlation with reproductive state and effects of ablation of glands (for reviews see Mikami, 1958, 1986; Tixier-Vidal, 1963; Tixier-Vidal and Follett, 1973). Some of these investigations were augmented by morphology of secretory granules and other electron-microscopic observations.

On the basis of the approaches outlined above, Andrée Tixier-Vidal and colleagues (for review see Tixier-Vidal and Follett, 1973), developed over the course of a decade a scheme of putative functions of seven cell types assuming that there are two types of gonadotropes in the avian pars distalis and that each cell produced a single unique hormone. Although based largely on the domesticated Rock Dove, Pekin Duck, and Japanese Quail, evidence was offered for the general applicability of the scheme to the domesticated Guineafowl, Ring Dove and domestic fowl; and to two wild species, the House Sparrow and Red Bishop, *Euplectes orix*. Briefly, the scheme of Tixier-Vidal contains the following putative cell types:

(1) *Thyrotropes*. These are basophilic, glycoprotein-containing cells that are activated by thyroidectomy and treated with antithyroid agents. They occur predominantly, if not exclusively, in the rostral lobe.

(2,3) *Gonadotropes*. Like the thyrotropes these cells contain glycoproteins. The Tixier-Vidal scheme recognizes two putative gonadotropin-secreting cells, a FSH type in cells of the rostral lobe and a LH type in the caudal lobe. This distinction rests primarily on correlations with testicular growth and circulating levels of testosterone, respectively, and to a lesser extent on responses of the two cell types to castration.

(4) *Lactotropes (= prolactin cells)*. Since prolactin has no easily ablatable target organ, the identification of lactotropes was effected primarily by correlation of tinctorial and electron-microscopic observations with phase of the annual cycle reproductive cycles, i.e., incubation and care of young, secretion of crop milk by doves, and prolactin content of the anterior pituitary. These erythrosinophilic cells occur predominantly, if not exclusively, in the rostral lobe.

(5) *Corticotropes*. These cells, which are conspicuously activated by adrenalectomy, have a slight affinity for lead haematoxylin and periodic-Schiff reagent. They occur predominantly in the rostral lobe.

(6) *Melanotropes*. As noted above, birds lack a pars intermedia, but it has long been known that they produce melanocyte-stimulating hormone (MSH) (Kleinholtz and Rahn, 1939; Mialhe-Voloss and Benoit, 1954). Melanotropes, which stain strongly with Herlant's tetrachrome and lead hematoxylin, occur in the rostral lobe, which is consistent with distribution of the hormone.

(7) *Somatotropes (= growth hormone cells)*. These are acidophil cells of the caudal lobe which are more numerous and active in young birds than in adults. Although they have properties similar to mammalian somatotropes, the evidence for identification in this scheme is entirely indirect.

Using tinctorial and cytochemical techniques different from those of Tixier-Vidal and her colleagues, Shin-Ichi Mikami developed an alternative system of putative cytophysiological functions for cells of the avian adenohypophysis (for summaries see Mikami, 1958, 1969, 1986; Mikami *et al.*, 1973a,b, 1975). Evidence was first derived from the domestic fowl, but later from White-crowned Sparrow and Japanese Quail. These investigations, like those of Tixier-Vidal and colleagues, involved extirpations of gonads and thyroid glands, but in addition, successful adrenalectomy. Also similar to those of Tixier-Vidal and colleagues, Mikami and associates ultimately included electron-microscopic investigations of the morphology of secretory granules and other ultrastructural features of the putative secretory cells. The major feature of the scheme of Mikami, together with some comparisons with that of Tixier-Vidal, are as follows (for reviews see Mikami, 1958, 1969, 1986; Mikami *et al.*, 1969, 1975):

(1) *Thyrotropes.* The putative thyrotropes resemble closely those of Tixier-Vidal and are probably identical with those that occur in the rostral lobe in her scheme. Mikami (1986) has summarized additional evidence that thyrotropes occur only in the rostral lobe.

(2,3) *Gonadotropes.* Tinctorial studies with light microscopy and electron-microscopic observation, both in correlation with reproductive state and following castration, lead to the conclusion that there are two types of basophilic gonadotropes, each type occurring in both the rostral and caudal lobes (Mikami, 1969; Mikami *et al.*, 1969, 1975). These conclusions seem to be at least partially sustained. However, the situation has become more complex with results of immunocytochemical studies on the Japanese Quail indicating that FSH and LH co-exist in the same cells. This argues that there well may be only a single type of gonadotrope in the avian pars distalis.

(4) *Lactotropes.* Although absolutely conclusive evidence is not available, it seems highly probable that the rostral lobe acidophils designated as lactotropes by Tixier-Vidal and Mikami are identical. Recent immunocytochemical evidence as reviewed by Mikami (1986) appears to be consistent with this identification.

(5) *Corticotropes.* The systems of Mikami and Tixier-Vidal agree that the amphophilic rostral lobe cell, designated as the V-cell by Mikami (1958; see also Matsuo *et al.*, 1969) and the *eta* cell by Tixier-Vidal (1963) is the corticotrope. This designation has been confirmed by its behavior after adrenalectomy and following administration of metapirone, which suppresses secretion of corticosterone by the adrenal cortex (Tixier-Vidal and Assenmacher, 1963; Tixier-Vidal *et al.*, 1968).

(6) *Melanotropes.* This type of cell has been described in the rostral lobe of the Pekin Duck (Herlant, 1960; Tixier-Vidal *et al.*, 1962) as the *kappa* cell. Since the evidence that it, in fact, secretes MSH is very indirect, because both Matsuo *et al.* (1969) and Mikami (1958) have raised the possibility that the corticotropes also secrete MSH, and because Mikami *et al.* (1973a,b) were unable to demonstrate *kappa* cells in *Z. l. gambelii* they have not been included in the scheme of Mikami. It is now known that ACTH and MSH can be derived from the same precursor molecule—proopiomelanocortin (POMC)—(see Gorbman *et al.*, 1983).

(7) *Somatotropes.* As in the scheme of Tixier-Vidal this cell type is a caudal lobe

acidophil. Its secretory function appears now to be adequately confirmed by combined tinctorial and immunocytochemical studies (Mikami, 1986). In the domestic fowl, somatotropes of growing chicks had larger secretory granules than those of older animals (Malamed *et al.*, 1988) suggesting a change in intracellular packaging of growth hormone with age.

As they have culminated in two somewhat different morphological schemes, this brief review of investigations on cytology of the avian pars distalis displays the difficulties and sources of differences in opinion in designation of the putative functions of adenohypophysial cells with classical tinctorial, cytochemical, and electron-microscopical techniques. Nevertheless, it is patently clear that such investigations were essential for subsequent correlations with endocrine state (Section E).

### 5. *Hypothalamus*

As in other classes of vertebrates, the avian hypothalamus is functionally an integral part of both nervous and endocrine systems. Indeed, the great strides of the past two decades in research on hypothalamic function now challenge the wisdom of the conventional division of the apparatus for internal communication and regulation into separate nervous and endocrine systems. Because of the function of this chapter, the hypothalamus is treated very largely from the aspect of its role in the control of reproduction, only a single set of its numerous functions.

There is surprisingly little agreement among published anatomical definitions of the hypothalamus. For reasons of both ontogenetic development and function our definition is broad. Although differing in terminology, it is similar in concept to those of Wingstrand (1951), Da Lage (1955), Dodd *et al.* (1971) and Oksche and Farner (1974). The hypothalamus *sensu lato* thus consists rostrally of the floor and lateral walls of the third ventricle; it extends caudally as the infundibular stalk and pars nervosa, which enclose the infundibular recess of the third ventricle.

Albeit somewhat arbitrarily, it is useful to recognize the following parts of the hypothalamus:

(1) *Hypothalamus* sensu stricto (e.g., Kuenzel and van Tienhoven, 1982; Mikami, 1986). This part consists of the floor and lateral walls of the third ventricle extending from the lamina terminalis (just rostral to the optic chiasm) to the supramammillary decussation. It contains most of the hypothalamic nuclei, fibers, and tracts associated there with glial and ependymal cells. Because avian hypothalamic nuclei are somewhat diffuse there has been considerable disagreement with respect to the number of them and hence also with nomenclature. It appears that there are of the order of 15–20,

only a few of which are known to be directly involved in control of reproductive function. With considerable logic the hypothalamus *sensu stricto* has been divided arbitrarily into preopticohypothalamic, tuberal, and mammillary regions (e.g., Kuenzel and van Tienhoven, 1982).

(2) *Median eminence*. The median eminence is a specialized ventral region of the floor of the hypothalamus, which emerges at the posterior slope of the optic chiasma and tapers into the infundibular stalk. It consists principally of tracts and axonal projections from hypothalamic nuclei and glial cell elements. It is characterized by specialized ependymal cells that terminate on primary portal capillaries in juxtaposition with terminals of axons. Kobayashi *et al.* (1970) have appropriately defined this organ exteriorly as the portion of the hypothalamus covered by the capillaries of the primary plexus of the hypophysial portal vessels, and interiorally as the basal portion of the hypothalamus containing processes of the basal ependymal tanycytes. It may contain some neurons of hypothalamic origin which are excluded from this definition. The median eminence is thus a neurohemal organ in which transfer of neurohormones from axons of hypothalamic nuclei into blood of the hypophysial portal system occurs. To varying extents, according to species, the median eminence and the primary portal capillaries are encased by the pars tuberalis of the adenohypophysis.

(3) *Infundibular stalk*. Caudally the median eminence narrows into the tubular infundibular stalk, which subsequently terminates in the pars nervosa. Although the stalk contains the supraoptico-paraventriculo-hypophysial tract, along which neurosecretory material passes to the pars nervosa, and retains some of the structural features of the median eminence, its organization is simpler and its innervation much sparser in comparison with the latter (see Wingstrand, 1951; Oksche and Farner, 1974 for details, discussion, and review).

(4) *Pars nervosa (= neurohypophysis)*. Anatomically this organ is the termination of the infundibulum of the hypothalamus, i.e., the infundibular recess of the third ventricle. Its lumen may be open or closed. It consists largely of branched terminals of fibers of the supraoptico-paraventriculo-hypophysial tract, specialized glial cells, pituicytes, and ependymal cells. Terminals of nerve fibers lie in close juxtaposition with systemic capillaries. The neurohypophysis is thus, like the median eminence, a neurohemal organ from which hypothalamic neurohormones can be transferred into circulating systemic blood. An early description of the avian neurohypophysis was that of Müller (1871) who, from studies on the Rock Dove, described its lumen as continuous with the third ventricle. An early, perhaps the earliest, description of this organ in a wild species was that of Haller (1898) on the Yellowhammer. Haller was apparently the first to recognize the cellular composition of the organ as described briefly in the paragraph above. Although not described in the text,

one of Haller's illustrations clearly shows hypothalamo-hypophysial vessels. The monograph of Wingstrand (1951) remains a very important publication on the anatomy and histology of this organ in wild species of birds, not only because of its discussion of the literature but also because of its comparative observations of the organ in species of no less than 14 orders.

Results of neuroanatomical and neurohistological investigations have provided the basis for recognition of the mode by which information is transferred in the form of neurohormones, and probably neurotransmitters and modulators into three compartments of body fluid: (1) from median eminence into blood of primary capillaries of the hypothalamo-adenohypophysial portal system; (2) from neurohypophysis to systemic capillaries; and (3) from other circumventricular organs of the hypothalamus and elsewhere (e.g., organum vascularum of the lamina terminalis, lateral septal organ, subfornical organ, paraventricular organ, subcommissural organ, and other partly investigated vascular beds, some of which appear to be highly specialized (see Kuenzel and van Tienhoven, 1982). Transfer of information in the reverse direction, i.e., from cerebrospinal fluid to hypothalamus, should remain in consideration as should transfers through morphologically unspecialized areas of the walls of the third ventricle and its recesses. On the basis of present knowledge, hypothalamic neurohormones released by the pars nervosa may be directly involved in reproduction only through a role of oxytocic hormones in oviposition, possibly by interaction with prostaglandins (for discussion and references see Shimada, 1980; Olson *et al.*, 1986). However, to our knowledge, this relationship has only been studied in two domesticated species, Japanese Quail and domestic fowl.

(5) *Hypothalamic nuclei and tracts.* Rationalization of the results of early investigations based on tinctorial methods has proved to be difficult. Among the reasons therefore are: (1) the diffuse nature of nuclei ("clusters" of perikarya) and many tracts in the avian hypothalamus in comparison with other tetrapod groups; (2) frequent, but seldom successful, attempts to apply mammalian nomenclature to avian nuclei and tracts; (3) real and specific differences among even closely related species; (4) difficulties in comparison of results from various investigations because of differences in fixation, tinctorial procedures, and especially topographic problems caused by differences in planes of section; (5) lack of tinctorial and histochemical methods for differentiation among types of nerve cells; and (6) lack of adequate silver-impregnation techniques for demonstration of fine nerve fibers. However, the introduction of immunocytochemical methods, and *in situ* hybridization techniques to indicate gene expression for protein and peptide hormones, has fundamentally improved this situation.

The earliest microanatomical studies of the avian hypothalamus were concerned primarily with cytoarchitecturally conspicuous "cell clusters", i.e., nuclei. Among these were those of Rehndahl (1924) on the domestic fowl, Craigie (1928, 1931) on three species of hummingbirds (Trochilidae), and on the Brown Kiwi, *Apteryx australis* (Craigie, 1930); Huber and Crosby (1929) on domestic fowl and House Sparrow, and Kurotsu (1935) on the "nucleus magnocellularis periventricularis" in seven species from three orders. Unfortunately, these classical studies have led to the development of two different (partly diverging) lines of nomenclature for the avian hypothalamus; one for passerine, and one for galliform birds.

With respect to hypothalamic tracts, Wingstrand (1951), and Boon (1938) reported that fibers from the supraoptic nucleus can be traced to the neurohypophysis in the Ostrich, *Struthio camelus*. For the domestic fowl, Drager (1945) described a hypothalamo-hypophysial tract with fibers that terminate mostly in the neurohypophysis; rostrally the fibers were traceable to a region dorsal to the optic chiasma. Green (1951), in a comparative study, was able to identify the origins of the tract as the nucleus supraopticus.

The important milestone in the development of our knowledge of the nuclei and tracts of the hypothalamus *sensu lato* was the extensive set of observations by Wingstrand (1951) in conjunction with his investigation of both the adeno- and neurohypophysis. The study of the latter, based primarily on the Rock Dove, but with extensive comparative observations on other species, employed silver-impregnation methods as well as the chromealum–hematoxylin procedure of Gomori. The latter identifies neurosecretory cells, nuclei, fibers, and tracts because of its affinity for sulfhydryl groups in the neurohormones and/or their carrier proteins.

Wingstrand (1951) recognized, among others, the following tracts in the hypothalamus of the Rock Dove: (1) tractus supraoptico-hypophyseus, which originates in the supraoptic nucleus and terminates in the neurohypophysis. It is a "Gomori-positive" tract; (2) tractus tuberohypophyseus with "Gomori-negative" fibers that pass from the nucleus tuberalis (= n. infundibularis) to an extensive area of the median eminence; (3) tractus hypophyseus posterior, which receives principally fibers from the basal tuberal nucleus and n. subdecussationis. It was Wingstrand's interpretation that this tract terminates in the pars nervosa; and (4) tractus hypophyseus anterior, formed by fibers from nucleus lateralis hypothalami, from paraventricular areas, and from the preoptic areas that terminate in the median eminence. This tract contains a mixture of Gomori-positive and -negative fibers. A reconsideration of the descriptions of Wingstrand in light of investigations of Oksche, Farner and numerous colleagues on *Z. l. gambelii* (see Oksche and Farner, 1974, for summary of observations and discussion), and on other wild species e.g., Zebra Finch, *Poephila guttata* (Oksche *et al.*, 1963; Oehmke, 1971), Long-tailed Grassfinch, *P. acuticauda* (Oehmke, 1971), Greenfinch (Oehmke, 1968; Oehmke *et al.*, 1969), House Sparrow (Oehmke, 1968; Oehmke *et al.*, 1969), Sharp-tailed Sandpiper, *Calidris acuminata* and Rufous-necked Sandpiper, *C. ruficollis* (Oehmke, 1971), and Mallard (Oehmke, 1969), suggested some extensions and revisions of the still basic interpretations of Wingstrand. These became possible largely because of development of new methods for demonstration of

"Gomori-type" neurosecretory material, such as use of aldehyde fuchsin, pseudo-isocyanin, and others; refined techniques for demonstration of biogenic amines at cellular levels, and electron microscopy (for discussions see Dodd *et al.*, 1971; Oksche and Farner, 1974). A further improvement of this analysis was introduced by the use of immunocytochemical methods.

Although we do not preclude the possibility of real differences between the Rock Dove, on one hand, and *Z. l. gambelii* and more recently investigated wild species on the other, Oksche and Farner (1974) suggested that axons of Wingstrand's tractus hypophyseus posterior do not enter the neurohypophysis. Rather it seemed preferable to consider this bundle as a part of the tractus tuberohypophyseus. Furthermore, they concluded that all nuclei contributing to this tract are more logically considered as parts of the tuberal or infundibular nuclear complex. In contrast with the interpretation of Wingstrand, these observations indicate that it is unnecessary to separate the rostral and caudal bundles of the tuberal connections to the median eminence from the principal part of the tuberohypophysial tract. A further clarification of this situation can be expected from the systematic use of immunocytochemical methods.

Contributions to our knowledge of hypothalamic tracts and nuclei have also allowed the preparation of stereotaxic atlases of avian brains. These atlases consist of precise illustrations of cross sections of the brain cut with the organ maintained in a standard position with respect to the stereotaxic instrument. A variety of classical tinctorial methods have been used to stain the sections. Stereotaxic atlases have been invaluable for experiments involving lesions and implantations at specific localized sites in the brain and hypothalamus. Although secondarily, they have enhanced understanding of three dimensional relationships among nuclei and tracts. Thus far, stereotaxic atlases have been prepared for brains, or parts thereof, of the domestic fowl (van Tienhoven and Juhász, 1962; Feldman *et al.*, 1973; Youngren and Phillips, 1978; Kuenzel and van Tienhoven, 1982), Japanese Quail (Baylé *et al.*, 1974), Rock Dove (Karten and Hodos, 1967), Ring Dove, *Streptopelia risoria* (Vowles *et al.*, 1975), Domestic Mallard (Zweers, 1971), and Canary, *Serinus canaria* (Stokes *et al.*, 1974). There are rather substantial differences in nomenclature among these atlases with respect to hypothalamic nuclei and tracts, caused by the above-mentioned historical developments. Although problems remain from the aspect of the various orders of birds, Kuenzel and van Tienhoven (1982) have made significant progress towards a general system of nomenclature for avian hypothalami. Again, new insight can be expected based on immunocytochemical investigations of hypothalamic nuclei and pathways.

### 6. *Hypothalamo-hypophysial Portal System*

This system consists of two networks of primary portal capillaries that arise, essentially independently and respectively, on the ventral surfaces of the anterior and posterior divisions of the median eminence of the hypothalamus. Anterior and posterior portal veins, usually enclosed as a bundle by the pars tuberalis of the adenohypophysis, course respectively to separate secondary capillary plexuses and sinuses of the rostral and caudal lobes of the pars distalis (e.g., Vitums *et al.*, 1964, 1966; Duvernoy *et al.*, 1969, 1970; Singh and Dominic, 1975). Hypothetically this arrangement could provide a basis for delivery of neurohormones from rather specific areas in the hypothalamus to specific areas in the rostral and caudal lobes of the pars distalis (Vitums *et al.*, 1966). However, it must be emphasized that there is as yet no physiologic evidence that supports this hypothesis.

Although knowledge of the anatomy of the avian hypothalamo-hypophysial portal system has been accumulated largely during the past four decades, incidental observations of elements thereof are much older. For example Haller (1898), in a communication on the adenohypophysis of the Yellowhammer, clearly illustrates portal veins without label or comment. Other earlier observations are cited by Green (1951) and Wingstrand (1951), each of whom, also by examination of injected systems from several species, contributed significantly to knowledge of this system in birds.

Demonstration of the role of daylength and encephalic photoreceptors in the functional development of testes in Pekin drakes by Benoit and colleagues (for references see Benoit, 1937, 1961, 1962, 1975; Assenmacher, 1963) was accompanied by a very extensive investigation of the entire vascularization—arterial, venous, and portal—of the hypothalamo-hypophysial complex (Assenmacher, 1952a,b). The description of the hypothalamo-hypophysial portal system conforms generally with that given above. Although not specifically described, the illustrations indicate that anterior and posterior portal vessels pass through the "tuberoportal zone" sleeve formed by the pars tuberalis to enter, respectively, the rostral and caudal lobes of the adenohypophysis. Because portal vessels designated by Assenmacher (1952a) as "posterior" do not pass through the tuberal sleeve, they are perhaps more appropriately designated as "accessory portal vessels" which are of irregular occurrence among individuals of several species (Vitums *et al.*, 1964; Singh and Dominic, 1975).

The first thorough investigation of the hypothalamo-hypophysial-portal system of a wild avian species was that of Vitums *et al.* (1964) on *Z. l. gambelii*, on which the brief description earlier in this section is based. It was emphasized that the morphologies of the anterior and posterior plexuses corresponded well with neuroanatomical differences between the anterior and posterior divisions of the median eminence. It was also emphasized that there appear to be only minor capillary anastomoses between the two primary plexuses.

Simultaneously with the investigations of Vitums *et al.*, Duvernoy and Koritké (1964) began investigations of the vascularization of the periventricular organs of selected species of mammals and birds which led to studies of the vascularization of the hypothalamo-hypophysial complex (Duvernoy and Koritké, 1968; Duvernoy *et al.*, 1969, 1970). Avian species studied included the Black-billed Magpie,

Muscovy Duck, *Cairina moschata*, domestic fowl, Rock Dove, and Long-eared Owl, *Asio otus*. Although observations are in general agreement with those of Vitums *et al.* (1964, 1966), there are two noteworthy differences: (1) the inclusion of a subependymal capillary network as a part of the primary portal capillary plexus. This may be due to differences in interpretation, or it may represent true specific differences as the observations of Singh and Dominic (1975) suggest; and (2) as in the interpretation of Assenmacher (1952a,b) on material from the Muscovy Duck, that posterior portal veins arise from the posterior median eminence or infundibular stalk and pass directly to the caudal lobe of the pars distalis, i.e., outside the "post-tuberal zone" and the sleeve of pars tuberalis. Figures 1 and 3 in Duvernoy *et al.* (1969) suggest the occurrence of posterior portal veins in the sense of Vitums *et al.* (1964). However, this interpretation is tenuous because the pars tuberalis is not visible and because of the interpretation in schematic Fig. 1 in this chapter. Nevertheless, the possibility remains open that the "posterior vessels" in the Muscovy Duck, like those in the Pekin Duck, are "accessory portal vessels" that occur irregularly in *Z. l. gambelii* (Vitums *et al.*, 1964) and in *Sturnus pagodarum* and the Jungle Crow, *Corvus macrorhynchos* (Singh and Dominic, 1975). It should be noted that for anatinine species, *Tadorna ferruginea*, Singh and Dominic (1975) have described distinct anterior and posterior plexuses and distinct anterior and posterior portal vessels even though some do not pass through the "porto-tuberal zone" *en route* to the rostral and caudal lobes of the pars distalis. It also may well be that variation is a matter of differences in development and of trivial physiologic significance.

Knowledge of the avian hypothalamo-hypophysial portal system has been broadened extensively by a series of investigations on about 50 wild Indian species of 11 orders (Dominic and Singh, 1969; Singh and Dominic, 1970, 1973, 1975; Singh, 1972). In all species studied these authors identified distinct groups of anterior and posterior portal vessels that supply respectively the cytologically distinctive rostral and caudal lobes of the pars distalis. However, in about 30% of the species, distinct anterior and posterior primary plexuses could not be detected; for these species it was reported also that the median eminence is not differentiated into two divisions. Detailed quantitative examination of the median eminence and primary portal plexus, including reconstructions from serial sections would be of very great interest.

On morphological bases, three generalizations concerning the functional relationship between hypothalamus and the pars distalis of the avian hypophysis can be made: (1) anterior and posterior portal veins conduct blood from the primary portal plexus, respectively, to the rostral and caudal lobes of the pars distalis; (2) the portal vessels are generally the sole vascular supply to the pars distalis; and (3) there is no unequivocal evidence of direct hypothalamic or other innervation of the pars distalis, although even recently such projections have been claimed to exist.

Defined hypothalamic nuclei contain cells of different functions despite their

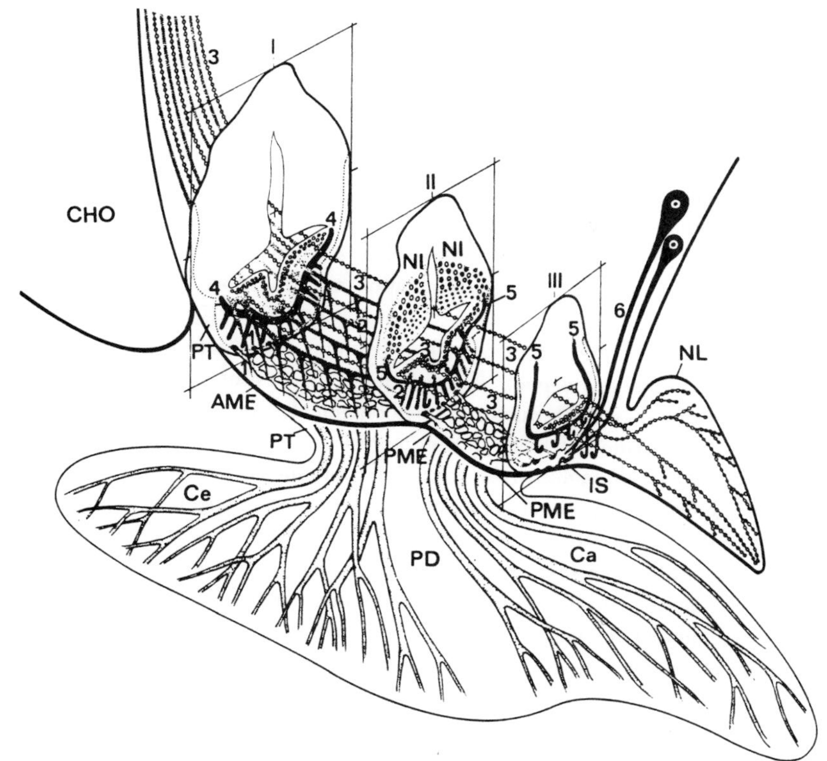


FIG. 1. Schematic diagram of the hypothalamo-hypophysial system in the White-crowned Sparrow, *Zonotrichia leucophrys gambelii*, with particular emphasis on the neurohemal areas of the median eminence. CHO, optic chiasma; AME, anterior median eminence; PME, posterior median eminence; IS, infundibular system; NL, neural lobe; PD, pars distalis; PT, pars tuberalis. AME contains Gomori-positive neurosecretory material. A system of fine Gomori-positive axons (1–2) passes from the anterior hypothalamus with the neurosecretory tr. suproptico-hypophyseus (3) and penetrates into the AME with rostral root-like bundles (1) or fine fibers (2) which subsequently leave the common Gomori-positive pathway in a cascade-like manner. This fiber system seems to lie beneath the coarser fiber bundles to the neural lobe (3). The characteristic granule of the neural-lobe system has a diameter of approximately 2,000 Å. The Gomori-positive fibers to the median eminence contain elementary granules 1,200–1,500 Å in diameter. At the border between the AME and PME this Gomori-positive system is nearly exhausted. Bilateral tubero-hypophysial (4, 5) penetrate into the median eminence. In our material they are composed of (Gomori-negative) aminergic and non-aminergic axons. The latter may be associated with different types of releasing hormones (factors). Within this system elementary granules, 800–1,000 Å in diameter, have been observed. Note that in the basal infundibular nucleus (NI) there are clusters of neurons of different types. These perikarya are embedded in a neuropil very rich in axo-somatic and axo-dendritic synapses. A strong bundle of tubero-hypophysial fibers (6) that are at least partly fluorescent extends from a higher level of the tuberal complex to the neurohemal contact area of the infundibular stem (IS). For functional considerations it is important to note that in *Z.l. gambelii* the cephalic (Ce) and caudal (Ca) lobes of the pars distalis (PD) are supplied by independent bundles of portal vessels. The capillaries leading to these vessels are in a spatial point-to-point relation with the endings of distinct divisions of hypothalamic tracts. From Oksche and Farner (1974).

near-uniform appearance when examined by classical cytological and cytochemical techniques. The same may hold for fiber systems within anatomically defined tracts. A crude analogy can be drawn with industrial cities, railways and highways, and destinations. Factories (perikarya) in several cities (nuclei) produce a variety of products (neurohormones) some of which are produced in more than one city. It then follows that railways and highways (axons and tracts) are involved in delivery of a variety of products (neurohormones, etc.) to different destinations (neurohemal organs, synapses (?), cerebrospinal fluid). The analogy is of course incomplete since nothing is included about communication and transport of materials to the factories.

Traditional morphology and topography of nuclei and tracts become less important from a functional aspect. Nevertheless, the results of neuroanatomical investigations with classical tinctorial, cytochemical, and electron-microscopic techniques have provided an essential framework for contemporary investigations of the role of the hypothalamus in the control of reproductive function. However, as interest in control of reproductive endocrinology in wild species under natural conditions expands from the present handful of passeriform, psittaciform, anseriform, and columbiform species, rapid progress in basic knowledge of the control of reproductive function can be expected. As mentioned above, there is no doubt that immunocytochemical methods for detection of neurohormones, modulators, and transmitters offer a new and functionally promising approach to peptidergic and associated aminergic systems of the avian hypothalamus.

### 7. *Pineal Body (Epiphysis Cerebri)*

As a circumventricular organ, the avian pineal body is an integral part of the brain. It is an organ of extraordinarily great morphological diversity (e.g., Studnika, 1905; Tilney and Warren, 1919; Krabbe, 1952, 1955; Quay and Renzoni, 1963; Renzoni, 1965), which, at least as yet, cannot be rationalized on phylogenetic relationships. In general the pineal body of adult birds consists of a stalk that arises from the roof of the diencephalon in the region between the habenular and posterior commissures from which it extends and expands dorsally to become the pineal organ proper or epiphysis cerebri *sensu stricto*. The latter lies close to the interior surface of the dorsal wall of the brain case in a triangular area delimited by the posterior margins of the cerebral hemispheres and the anterior margin of the cerebellum.

The diversity of development and adult morphology of the avian pineal body was first appreciated by Studnika (1905) largely from his own extensive studies, but also from a careful review of investigations published by others during the 19th century. He recognized three basic structural patterns: (1) simple elongated sac with thick walls (e.g., in the Hawfinch, *Coccothraustes coccothraustes*) and lateral outpocketings, but no follicles; stalk connected or discontinuous with diencephalon; (2) in most cases with follicles and short tubes, solid stalk (e.g., in the Domestic Turkey); and (3) solid organ, with narrow, often obliterated lumen (e.g., in the Domestic Fowl). Studnika also described several intermediate forms. For descriptive purposes his scheme still survives but with further intermediate forms and additional modifications (cf. Bargmann, 1943; Krabbe, 1952, 1955; Quay and Renzoni, 1963, 1966a,b; Oksche and Vaupel-von Harnack, 1965; Quay, 1965; Renzoni, 1965; Collin, 1966; Oksche, 1968). In addition to extensive variations in the morphology of the pineal body proper alluded to above, further anatomical and microanatomical variations involve a complete or interrupted stalk, a luminal or closed stalk, innervation (Ueck, 1979), and its relationships to habenular and posterior commissures (Renzoni, 1965), and the extent of development of secretory pinealocytes. The situation in the six species of owls studied by Renzoni (1968) is noteworthy. In the adults examined, the pineal body was either lacking or represented only by degenerate remnants (cf. Breucker, 1967).

All in all, pineal bodies of about 100 species of 13 avian orders have been studied embryologically and/or in adult form by classical cytological and/or histological techniques, histochemistry, and electron microscopy (for cytological details on some selected wild species see Quay and Renzoni, 1963, 1966a,b; Collin, 1966, 1967; Oksche and Kirschstein, 1969; Ueck, 1979). Notably deficient among these rather numerous investigations is attention to possible seasonal changes in microanatomy and ultrastructure. In the House Sparrow, Ralph and Lane (1969) detected no truly seasonal changes in the pineal body. Conversely, Barfus and Ellis (1971) found that the activity of hydroxyindole-*O*-methyl transferase (an important enzyme in the biosynthetic pathway of melatonin) in pineal bodies of the same species was lowest during the breeding season. More recently, Saxena *et al.* (1979) obtained similar results for pineal activity in the Baya Weaver, *Ploceus philippinus*. In the Indian Tree Pie, *Dendrocitta vagabunda*, nuclear diameter of pineal parenchymal cells decreased, whereas cell density and serotonin (a precursor of melatonin) content increased during gonadal maturation in both sexes. These trends were reversed during gonadal regression and the non-breeding season (Chandhuri and Maiti, 1989). Of additional importance to investigations of the avian pineal were their demonstrations of age-dependent changes. However, it is beyond the scope of this brief review to attempt to wring generalities from the results of these diverse investigations. Suffice it to indicate here that phylogenetic relationships with lower vertebrates suggest that the avian pineal body evolved from a photosensitive organ as first suggested by Studnika (1905). For additional comments and supporting evidence of this concept see Bargmann (1943), Oksche and Vaupel von-Harnack (1965), Collin (1967, 1971), Oksche (1968, 1971, 1989), Menaker and Oksche (1974), Hartwig and Oksche (1981, 1982), Hartwig (1987), and Binkley (1988).

From a phylogenetic aspect it has been assumed that (photo-) receptor cells of lower vertebrates evolved into secretory "pinealocytes" through loss of polarization, degeneration of outer segments, and further development of secretory apparatus. However, as it has become evident that these cells in lower vertebrates synthesize indolamines, including melatonin, it appears more realistic (e.g., Kappers, 1981; Oksche, 1989) to regard the evolution as one of change from primitive photo-neuroendocrine cells to secretory cells that have lost their primary photosensory function, (cf. Collin 1971; Oksche, 1971, 1989). There is no unequivocal electrophysiological evidence that the avian pineal body is a photoreceptive organ responding to light in the sense of the true photoreceptor pineals of lower vertebrates. In contrast to anurans and fish (e.g., Morita, 1975), direct illumination does not (at least under the conditions used) alter the electrical activity of the avian pineal body (Morita, 1966; Ralph and Dawson, 1968; Homma *et al.*, 1980). The above conclusion, however, is not to say that the pineal is unaffected by light. The secretion of melatonin for example, is an endogenous circadian function that is entrained by the environmental photoregime (e.g., Ralph *et al.*, 1967; Ralph, 1976; Binkley *et al.*, 1977; Menaker *et al.*, 1978; Cassone and Menaker, 1984). In cell culture, Deguchi (1982) was able to show that the chicken pineal gland contains a rhodopsin-like photoreceptor and an endogenous oscillator that controls the circadian rhythm of serotonin *N*-acetyltransferase activity, the key enzyme of melatonin synthesis in the pineal gland. However, in other species, whether pineal functions affected by light can be viewed as evidence that this organ is directly photosensitive has been questioned (cf. Oliver and Baylé, 1982). There is sufficient evidence now to suspect extensive interspecific differences (cf. Cassone and Menaker, 1984). In constructing generalizations and models one must not forget the adult owls (Renzoni, 1968) that apparently lack pineal bodies! Furthermore, the habenular ganglion of different vertebrates was shown to contain pinealocyte-like cells endowed with molecular markers (proteins) characteristic of photoreceptor cells (Korf *et al.*, 1989).

With respect to other putative functions of the avian pineal body there are numerous reports of varying quality of both anti- and progonadal effects which have been critically reviewed by Ralph (1970) and Binkley (1988). Consideration of his analyses and our examination of many of the papers involved led us to the conclusion that as yet there is little acceptable evidence of a significant direct role of the pineal body in regulation of gonadal function. In a well-designed experiment, Balasubramanian and Saxena (1973) demonstrated that pinealectomy of male Baya Weaver resulted in significantly enhanced rate of photoperiodically induced testicular growth. Furthermore, when birds in breeding condition were transferred to short days, the pinealectomized group underwent only partial testicular regression in comparison with an essentially normal regression in the sham operated group. The authors interpreted these results as effects of removal of an inhibitory effect of melatonin on the hypothalamo-hypophysial-gonadal axis. However, assuming that the pineal body is one of the components of the avian circadian system (cf. Cassone and Menaker, 1984), one could rationalize the results in terms of a shift in phase angle between a daily cycle in photosensitivity and the daily environmental photocycle. A similar explanation could be advanced for other reports of either positive or negative effects of pinealectomy on avian gonads (e.g., Gwinner *et al.*, 1981).

We have included this brief discussion of the pineal body in a treatise on the endocrinology of reproduction primarily because it, possibly through its hormone melatonin, is a component of a "biological clock" or "circadian system". Other components of the avian "circadian system" (Gaston and Menaker, 1968) include the suprachiasmatic nucleus, retina, and deep encephalic (hypothalamic) photoreceptors (e.g., Cassone and Menaker, 1984; Underwood and Siopes, 1985). It should be noted that all four of these components have a common developmental origin in the diencephalon. They are also either photoreceptive or receive photic information after transduction by associated structures (Menaker, 1982). The relative importances and inter-relationships of these four putative components of the avian circadian system must involve substantial interspecific variation; furthermore, it would be unwise at this time to preclude the existence of further components (cf. Cassone and Menaker, 1984; Underwood and Siopes, 1985). As first suggested for birds by Hamner (1964, 1966a), there is considerable evidence in support of a cycle in photosensitivity that is driven by the circadian system. Such a cycle could provide the basis of a system for measurement of daylength in photoperiodic responses proposed by Pittendrigh and Minis (1964) or by the general, but more complex model of Pittendrigh and Daan (1976; for reviews see Farner, 1975, 1986; Klein *et al.*, 1991).

It seems abundantly clear that the pineal body, as a component of the circadian system, has a role that varies, at least in photoperiodic species, intraspecifically in the measurement of daylength. But it is also abundantly clear that many more experimental investigations are essential before precise mechanisms are elucidated with respect to other components of the system.

## B. BIOASSAYS OF ENDOCRINE GLANDS

The histological and histochemical investigations described above generally left open the question of mechanisms that cause changes in secretion rates and plasma levels of hormones in relation to physiological adaptations to the environment, and in the course of reproductive cycles. In the 1950s and 1960s several investigations developed bioassay techniques that allowed measurements of glandular concentrations of endocrine secretions.

### *1. Hypothalamic and Hypophysial Hormones*

The concept of hypothalamic control of pituitary hormone secretion is relatively recent, and purification and identification of a spectrum of releasing and inhibiting hormones of neuroendocrine origin occurred at a time when radioimmunoassay techniques were well established. For example, gonadotropin-releasing hormone, GnRH (also called luteinizing hormone releasing hormone, LHRH), content of the avian hypothalamus was measured by determining the ability of hypothalamic extracts to release LH *in vitro*. In some cases the released LH was measured by a second bioassay (e.g., P-32 uptake in chick testis, e.g., Erickson, 1975 in White-crowned Sparrows, *Z. l. gambelii*), or by radioimmunoassay either in the incubation medium (e.g., Bicknell and Follett, 1975 in Japanese Quail), or in blood after injection of hypothalamic extracts *in vivo* (R. Hudson, J. C. Wingfield and D. S. Farner, in *Z. l. gambelii*, unpublished). Recently, it was found that the hypothalamus of the domestic fowl had two distinct GnRH molecules, designated chicken GnRH-I and GnRH-II that differed from mammalian GnRH by one and three amino acid substitutions (see Fig. 2, King and Millar, 1982a,b; Miyamoto *et al.*, 1984). It has since been shown that the two chicken GnRHs are also found in the hypothalamus of wild species (European Starling and Song Sparrow, *Melospiza melodia*) with c-GnRH-I predominating (Sherwood *et al.*, 1988). Both have potent effects on release of LH in male and female Song Sparrows in a more or less dose-dependent fashion (Figs 3 and 4).

Bioassays were used to estimate changes in pituitary levels of gonadotropins in the Pheasant, *Phasianus colchicus*, (Greely and Meyer, 1953) and in both captive and free-living *Zonotrichia leucophrys* (King *et al.*, 1966). In the latter study, highest levels of pituitary gonadotropin content were measured in May and June in both males and females sampled on their breeding grounds in Alaska. Lowest levels were assayed in photorefractory birds, although an increase in gonadotropic activity of pituitaries was noted in autumn on wintering grounds in Washington State. These changes correlated well with periods of gonadal growth and regression. However, this assay is based upon phosphorus-32 uptake by testes of very young domestic fowl chicks following injection of pituitary extract, and thus fails to distinguish between LH and FSH. In addition, this assay is not sufficiently sensitive to measure plasma levels of gonadotropin activity. Often, several pituitaries from small birds had to be pooled for a single determination. Other more

**GONADOTROPIN-RELEASING HORMONES IN MAMMALS AND BIRDS**

| | 1 | 2 | 3 | 4 | 5 | 6 | 7 | 8 | 9 | 10 | |
|---|---|---|---|---|---|---|---|---|---|---|---|
| MAMMALIAN | pGlu- | His- | Trp- | Ser- | Tyr- | Gly- | Leu- | Arg- | Pro- | Gly- | NH2 |
| CHICKEN-1 | pGlu- | His- | Trp- | Ser- | Tyr- | Gly- | Leu- | <u>Gln</u>- | Pro- | Gly- | NH2 |
| CHICKEN-2 | pGlu- | His- | Trp- | Ser- | <u>His</u>- | Gly- | <u>Trp</u>- | <u>Tyr</u>- | Pro- | Gly- | NH2 |

FIG. 2. Amino acid sequence of GnRH molecules. Dark bars indicate amino acid substitutions.

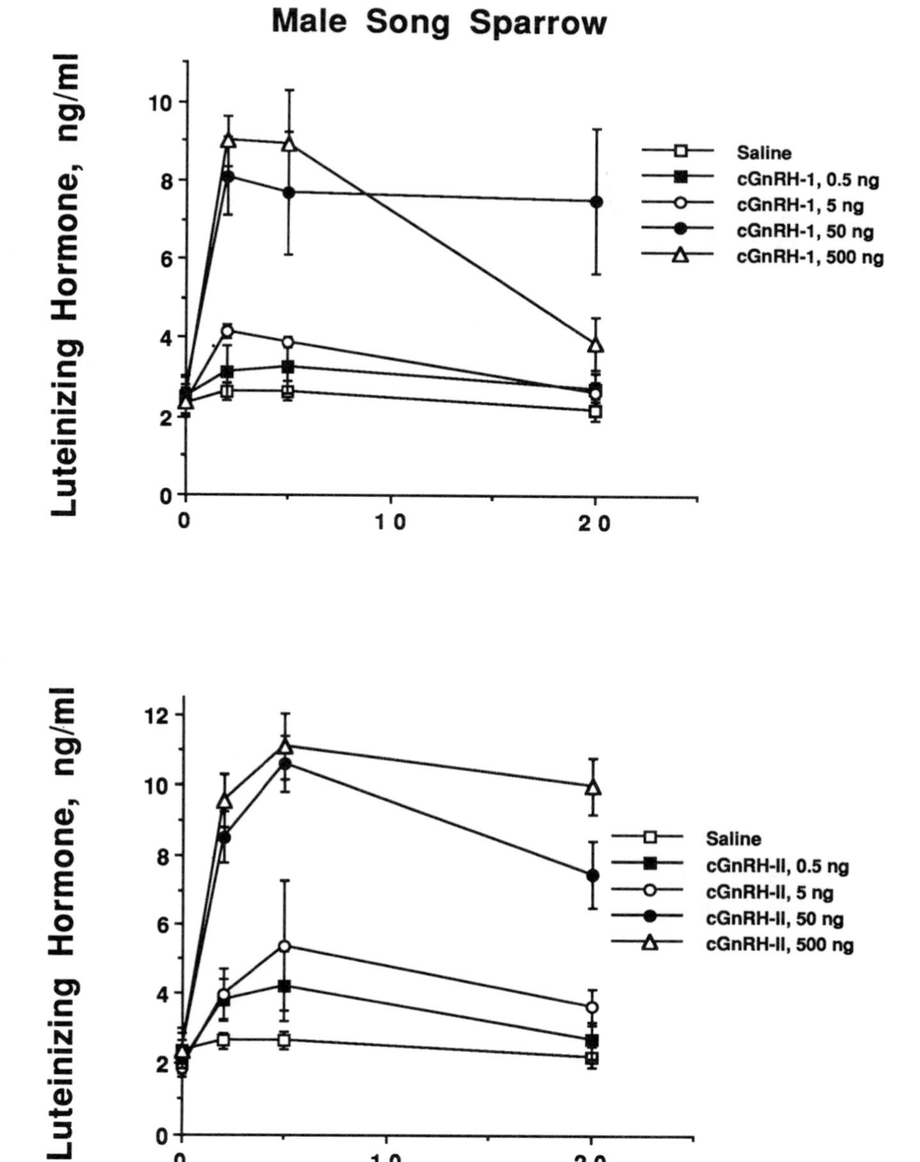


FIG. 3. Effects of intravenous injection of chicken gonadotropin-releasing hormones I and II (cGnRH-I and cGnRH-II) on circulating levels of luteinizing hormone in male Song Sparrows, *Melospiza melodia*. Vertical bars are standard errors of means, $n = 5$. From J. C. Wingfield, unpublished.

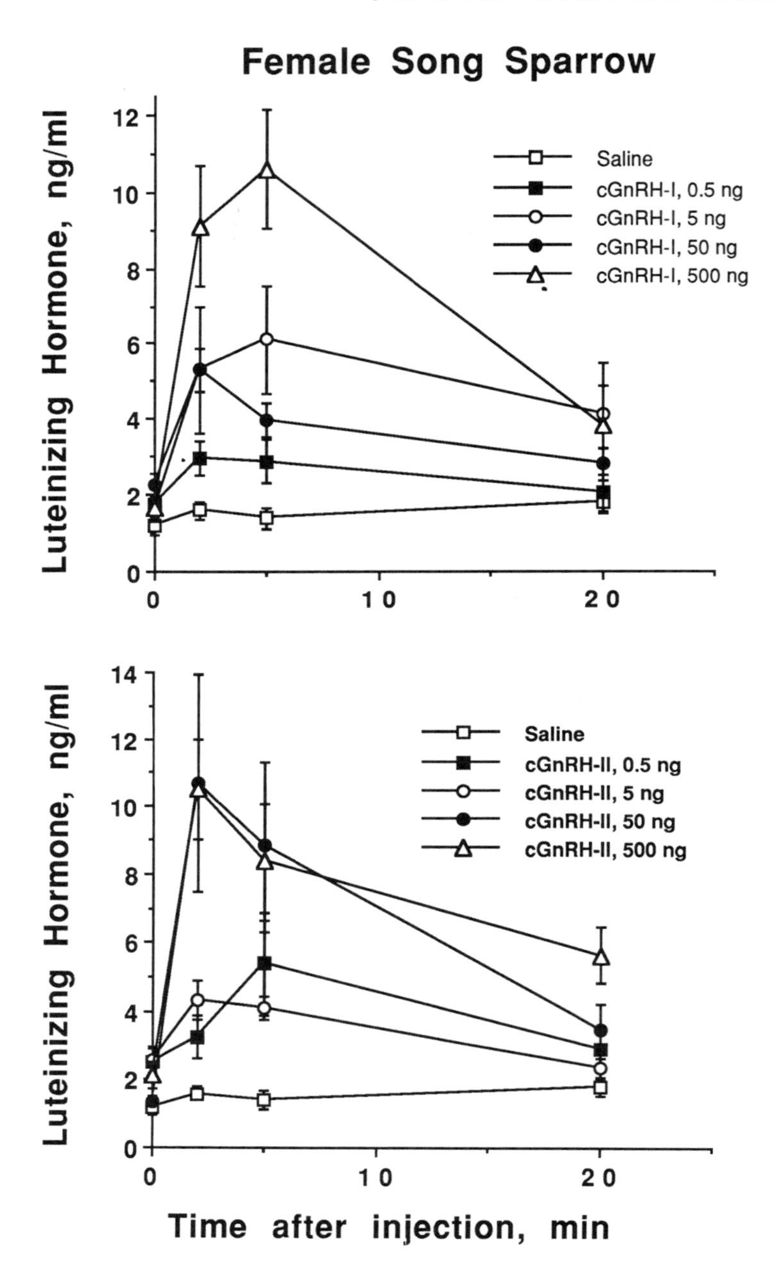


FIG. 4. Effects of intravenous injection of chicken gonadotropin-releasing hormones I and II (cGnRH-I and cGnRH-II) on circulating levels of luteinizing hormone in female Song Sparrows, *Melospiza melodia*. Vertical bars are standard errors of means, $n = 5$. From J. C. Wingfield, unpublished.

specific bioassays have been developed, e.g., the ovarian augmentation assay for FSH (Steelman and Pohley, 1953), and the weaver finch plumage bioassay for LH (Witschi, 1961). Once again, however, these assays cannot be used to measure plasma levels of gonadotropins.

Bioassays for other adenohypophysial hormones have also been developed. Pituitary levels of prolactin have been measured traditionally by the pigeon crop-sac bioassay. Prolactin stimulates hypertrophy of epithelial cells in the crop sac resulting in production of "crop-milk" that is used to feed young. Injection of pituitary extracts stimulates proliferation of the epithelial cells to varying degrees dependent upon the prolactin content. Thus an estimate of the concentration of prolactin is obtained if the response after injection of extract is compared with responses to known quantities of partially purified mammalian prolactin (Nicoll, 1967). This assay has been used by Meier *et al.* (1965) on *Zonotrichia leucophrys gambelii*; Meier *et al.* (1969) on the White-throated Sparrow, *Z. albicollis*; and by Dyachenko (1976, 1982) on the Chaffinch, *Fringilla coelebs*, and several other passerine species.

Other examples include the mouse tibia growth test for growth hormone (Greenspan *et al.*, 1949), a bioassay for ACTH based on ascorbic acid depletion of the adrenal gland (Woods, 1957), and an assay for TSH utilizing the stimulation of iodine-131 uptake by the thyroid gland (see Assenmacher, 1973 for review). However, as with gonadotropin assays, these bioassays were not suitable for measurements of circulating levels. Furthermore, since the pituitary concentrations of hormones are a function of synthesis, storage, and release, it is not always pertinent, or possible, to relate temporal patterns in these kinds of data to changes in circulating levels.

### 2. *Hormones of Peripheral Endocrine Glands*

The glandular concentrations of peripheral endocrine secretions were first determined by either bioassay or colorimetric techniques. Release of adrenocortical steroid hormones during, for example, stressful episodes was often estimated by changes in ascorbic acid content of adrenal tissue (Woods, 1957). More popular, however, was a fluorimetric technique that involved extracting steroids from gonadal or adrenal tissue and purifying the various fractions by thin layer silica gel, or descending paper chromatography (see Sandor and Idler, 1972). Purified steroid hormones were then reacted with concentrated sulphuric acid to produce a chromogen that could then be quantified colorimetrically or fluorimetrically. Very pure, and non-aromatic, steroid fractions could also be quantified by absorption of ultraviolet light.

A third technique utilized steroidogenic tissue incubated *in vitro* with fixed quantities of tritiated or carbon-14 precursors such as pregnenolone or progesterone. The percentage conversion of radio-labeled precursor to biologically active steroid (isolated by chromatography) gave some indication of the activity of the steroid secreting tissue. This technique has been used to show that testosterone production by testicular tissue of *Passer montanus* increased during gonadal recrudescence and was low during periods of gonadal regression and non-breeding (Chan and Lofts, 1974). In addition, Chan and Phillips (1973) showed that *in vitro* production of corticosterone by adrenocortical tissue of the Herring Gull, *Larus argentatus*, was greater in the winter, or non-breeding season, than in the reproductive season. Once again, however, this technique gave no indication of changes in circulating levels of steroid hormones.

The double isotope derivative assay (DIDA) is a more sensitive technique for measuring levels of steroid hormones that involves labeling C-14 metabolites, following *in vitro* incubation, with tritiated anhydride to give a doubly-labeled derivative. Recrystallization to constant isotope ratio gave not only an accurate measurement of concentration, but also a tentative identification of the hormone measured (Sandor and Idler, 1972). This was important since steroid hormones have still only been identified definitively in a few avian, mostly domesticated, species (see Sandor and Idler, 1972). DIDA, however, is a very tedious procedure and thus only a limited number of samples can be assayed.

Secretory activity of the thyroid gland was first determined by uptake and metabolism of injected radio-iodine (see Assenmacher, 1973). Extraction of radioiodinated hormones resulted in the identification of thyroxine (T4) and tri-iodothyronine (T3) in such wild species as Bobwhite Quail, *Colinus virginianus* (Morgan and Mraz, 1970); White-crowned Sparrow and the Brown Towhee, *Pipilo fuscus* (Fink, 1957); White-throated Sparrow and Golden Bishop, *Euplectes cafer* (Kobayashi *et al.*, 1960); and the Spotted Munia, *Lonchura punctulata* (Chandola, 1972), and also allowed an assessment of thyroid gland activity. Obviously, however, and as with all bioassays, these techniques leave open the question of changes in circulating levels of thyroid hormones.

### C. ISOLATION AND PURIFICATION OF AVIAN HORMONES AND DEVELOPMENT OF SPECIFIC RADIOIMMUNOASSAYS

Most hormones, both neuroendocrine and endocrine, circulate in blood to their target organs (exceptions are paracrine and autocrine actions of some hormones). Thus it is desirable to measure the circulating titers of hormones as an indication of the concentrations to which target organs respond. Changes in circulating levels of hormones would thus be a powerful means by which function and mechanisms could be determined. As mentioned above, the majority of bioassays were not sufficiently sensitive to detect plasma concentrations. The development of radioimmunoassay (RIA) systems in the 1960s and 1970s revolutionized endocrinology because these assays were incredibly sensitive and able to measure circulating levels of hormones. Many of the smaller hormones have very similar, or identical, structures regardless of phylogeny (e.g., for sex steroid hormones see Ozon, 1972a,b; for adrenocorticosteroid hormones see Huibregste *et al.*, 1973; Holmes and Phillips, 1976). RIA systems for these hormones in mammals were thus readily adaptable for use in birds (e.g., Wingfield and Farner, 1975). Similarly assays for thyroid hormones in mammals can also be applied to avian materials (e.g., Smith, 1979, 1982).

However, RIAs developed for mammalian pituitary hormones are rarely applicable to avian systems because the structure of respective avian pituitary hormones differs sufficiently from those of mammals so that there is only partial, or no, cross-reaction of the former with antibodies raised against the latter. Although it should be noted that Follett (1976) adapted an heterologous assay for FSH in Japanese Quail, and McNeilly *et al.*, (1978) similarly applied a mammalian prolactin RIA to avian materials, the development of RIAs for avian pituitary hormones by and large awaited their isolation and purification. Avian LH and FSH were first purified from adenohypophyses of the domestic fowl by Stockell-Hartree and Cunningham (1969), and later by Furuya and Ishii (1974), and Sakai and Ishii (1980). Bioassay studies show that as in mammals, LH acts primarily on the endocrine gonad and FSH on the gametogenic gonad (e.g., Brown *et al.*, 1975; Furuya and Ishii, 1976; Brown and Follett, 1977; Maung and Follett, 1977; Follett *et al.*, 1978). Using these purified gonadotropins, specific RIAs have been developed for chicken LH (Follett *et al.*, 1972), Turkey LH (Burke *et al.*, 1979); and for chicken FSH (Scanes *et al.*, 1977; Sakai and Ishii, 1985). With the exception of the FSH assay developed by Scanes *et al.* (1977) these gonadotropin RIAs have proved to be applicable to plasma samples from a wide range of avian species (see Follett *et al.*, 1978; Goldsmith and Follett, 1983; Sakai and Ishii, 1985, 1986). Isolation of LH and FSH from truly wild species appears to be restricted to the Ostrich, *Struthio camelus*, in which LH and FSH fractions have potent activity in both avian and mammalian bioassay systems (Bona-Gallo *et al.*, 1983). Furthermore, LH from the Ostrich is far more potent than FSH in stimulating secretion of testosterone from immature Mallard testes suggesting that as in domesticated species, increased androgen secretion is highly specific for LH (D. Chase, 1982).

Thyroid stimulating hormone (TSH) has only been partially purified from pituitaries of the domestic fowl and Ostrich. Both TSH preparations are active in stimulating thyroid activity as in mammals (MacKenzie, 1981; Papkoff *et al.*, 1982). Injections of chicken TSH resulted in a dose-dependent increase in plasma levels of T4 within 30 min in Japanese Quail (Goldsmith and Follett, 1980). As far as we are aware, a specific and widely applicable RIA for avian TSH has not, as yet, been developed.

Prolactin was first purified from adenohypophyses of domestic fowl by Scanes *et al.* (1975) and later from Domestic Turkey by Burke and Papkoff (1980) and

Proudman and Corkoran (1981). These prolactin preparations have potent biological activity in the pigeon crop-sac assay and have also been used to develop RIAs (Scanes *et al.*, 1975; Burke and Papkoff, 1980), although their validity in measuring plasma prolactin has been questioned (Nicoll, 1975). Nevertheless, it is of interest to note that application of a heterologous prolactin RIA using an antihuman prolactin serum and ovine prolactin as standard (see McNeilly *et al.*, 1978; Goldsmith and Hall, 1980), to a dilution series of prolactin content in pituitary extracts of European Starlings, gave identical results to those of a radioreceptor assay based on rat lymphoma cells that respond to lactogenic hormones (Tanaka *et al.*, 1980; see Wingfield *et al.*, 1989 for details). Moreover, the assay developed by McNeilly *et al.* (1978) and Goldsmith and Hall (1980) was used to measure changes in prolactin levels during a breeding cycle in the Spotted Sandpiper, *Actitis macularia*, and gave identical results to another prolactin RIA based on antisera to Turkey prolactin and standards (Oring *et al.*, 1986a,b).

Purified growth hormone (GH) has also been isolated from domestic fowl and Turkeys (Farmer *et al.*, 1974; Harvey and Scanes, 1977) and appears to have many biological actions on growth similar to GH in mammals (Scanes and Harvey, 1981; Papkoff *et al.*, 1982). Specific RIAs have also been developed for avian GH (ibid). Avian GH influences lipid metabolism in adult birds by increasing plasma levels of free-fatty acids and promoting lipolysis. In addition, plasma levels of GH increase during fasting in the domestic fowl (Scanes *et al.*, 1983). It is possible that GH may have an important role in regulating some metabolic function during the annual cycle of adult birds (see below).

Reports on the isolation of other pituitary hormones in birds are extremely sparse and appear to be restricted to domesticated species. Adrenocorticotropin (ACTH) and ß-lipotropin have been isolated and sequenced from pituitary extracts of the Domestic Turkey and Ostrich (Chang *et al.*, 1980; Yamashiro *et al.*, 1984) and mesotocin (Ile-8-oxytocin) and arginine vasotocin (AVT) have been demonstrated in extracts of posterior pituitaries in domestic fowl, Turkey, and Goose. In the domestic fowl, AVT appears to be antidiuretic, and mesotocin acts on contraction of smooth muscle (e.g., in oviduct), as does oxytocin in mammals (see George, 1980 for review). As far as we are aware, the isolation of these hormones in pituitary extracts of wild species has not been performed. As mentioned previously, these antibodies have also been used to visualize cells that contain specific hormones—immunocytochemistry.

### D. MEASUREMENT OF PLASMA LEVELS OF HORMONES—SHORTCOMINGS

Although the advent of RIAs allowed endocrinologists to measure plasma levels of hormones from minute quantities of blood thereby permitting studies of individuals in progressive changes of physiological (or pathological) state in response to changing environmental conditions, it has also become apparent that caution must be exercised in the interpretation of absolute levels of hormones in blood. Under some circumstances the metabolic clearance rate of a hormone may increase without a concomitant change in secretion rate. Thus plasma level of the hormone may decline despite no apparent change in secretory activity of the gland, e.g., a decline in plasma levels of testosterone despite high circulating levels of LH in *Anas platyrhynchos* (Jallageas *et al.*, 1974) and White-crowned Sparrow (Lam and Farner, 1976; Wingfield and Farner, 1978a,b; Wingfield, 1984a,b).

Plasma levels of hormones also do not reflect the concentration and distribution of receptors in target organs that may vary from season to season and/or between the sexes. For example, testosterone, and perhaps even progesterone, were metabolized to estradiol, 5α– or 5ß-dihydrotestosterone (DHT) in the brain of European Starlings, Ring Doves, Japanese Quail, and domestic fowl (Massa *et al.*, 1977, 1982, 1983; Sharp and Massa, 1980; Steimer and Hutchison, 1981). Peripheral conversion of testosterone to estradiol or 5α-DHT was important for many biological actions of testosterone, especially in the brain (Massa *et al.*, 1983), whereas conversion to 5ß-DHT appeared to be a deactivation shunt (Steimer and Hutchison, 1981). In the adenohypophysis and hypothalamus of the European Starling, there was an increase in 5ß-reductase activity as reproduction was terminated suggesting a biological role for the deactivation shunt (Bottoni and Massa, 1981).

Changes in receptor populations target organs may also play a vital role in the actions of hormones. In the Ring Dove, Balthazart *et al.* (1980) have shown that receptors for progesterone increase in the hypothalamus of males during courtship and copulation even though plasma levels of progesterone do not change. Thus, they suggest that progesterone may be important in regulating the onset of parental behavior of this species despite no change in circulating levels (see Balthazart, 1983 for further discussion).

A further problem that may confound interpretation of circulating levels of hormones in plasma concerns binding proteins. For thyroid hormones there are binding proteins in blood that vary considerably in binding affinity, specificity and capacity. In birds, thyroid hormones appear to circulate mostly bound to pre-albumin (transthyretin or thyroxine binding pre-albumin, e.g., Gorbman *et al.*, 1983). In the White Stork, *Ciconia ciconia*, thyroid hormones may, however, circulate bound to three plasma proteins: albumin, transthyretin, and a protein that appears to be intermediate between the two (Cookson *et al.*, 1988). Most steroid hormones also circulate bound to proteins that fall into three major classes: (1) albumins that bind most steroids with low affinity and very high capacity; (2) corticosteroid binding globulin (CBG) which binds primarily glucocorticosteroids and progesterone; and (3) sex hormone binding globulins which bind sex steroid hormones with affinities and specificities that vary greatly among vertebrate classes (e.g., Burton and Westphal, 1972; Anderson, 1974; Petra and Schiller, 1977; Wingfield, 1980a; Wingfield *et al.*, 1984). When bound to the high affinity binding systems, steroid hormones are generally thought to be biologically inactive since only free hormone can enter cells and encounter receptors. Although it has been demonstrated that some protein–steroid complexes may enter the cytoplasm of target cells in mammals (e.g., Bordin and Petra, 1980), the significance of this uptake remains to be assessed. In addition, steroid hormones bound to plasma proteins are cleared less rapidly, a process that may help to stabilize the unbound concentration of the steroid hormone thus providing for more precise assessment by the target organ (e.g., Burton and Westphal, 1972). Thus the ratio of bound to unbound hormone may have important implications for endocrine investigations, especially because assay techniques for plasma levels of steroid hormones tend to measure *total* concentrations.

The distribution of the different types of steroid binding proteins in blood are by no means uniform throughout the vertebrate classes. Unlike some species of mammals, birds appear to be entirely without specific, high affinity, binding proteins for androgens and estrogens (Corvol and Bardin, 1973; Wenn *et al.*, 1977; Wingfield, 1980a; Wingfield *et al.*, 1984). Of some 23 species from eight orders, none showed high affinity binding for testosterone or estradiol-17ß, although all species had CBG-like binding activity for progesterone and corticosterone (Wingfield *et al.*, 1984; Table 1). In some groups, however, testosterone may bind with low affinity to CBG, e.g., Columbidae and Alcidae (Wingfield *et al.*, 1984). Thus it appears that sex steroid hormones in birds circulate bound largely with low affinity to plasma albumins whereas corticosterone and progesterone circulate bound with high affinity to CBG. Clearly, interpretation of changes in plasma levels of corticosterone and progesterone may depend on changes in binding capacity of CBG (Wingfield *et al.*, 1984). Despite the cautions outlined above, there is, nevertheless, no doubt that RIAs of circulating concentrations of hormones have contributed enormously to the field of endocrinology in general.

### E. RECENT TECHNIQUES AND FUTURE DIRECTIONS

Although great advances have been made in the endocrinology of wild species over the last 25 years, it is anticipated that even greater developments lie ahead (see also, Silver and Ball, 1989). Techniques in molecular biology are now well established and are applicable to all biological systems. For example, the gene for the ß-subunit of chicken LH has been sequenced (Noce *et al.*, 1989) and used to demonstrate that mRNA for this subunit increases following photostimulation in the pituitary of Japanese Quail (Ando and Ishii, 1990). The gene for chicken prolactin has also been isolated (Shimada *et al.*, 1987). As genes for other hormones, and their receptors, are cloned it is likely that we will be able to study the endocrinology of wild species not only at the organismal level, but also at the level of regulation of gene expression.

TABLE 1

SUMMARY OF THE DISTRIBUTION OF STEROID HORMONE-BINDING PROTEINS IN AVIAN PLASMA

| Species | Binding Protein | | |
|---|---|---|---|
| | Corticosterone | Progesterone | Testosterone |
| Procellariiformes | | | |
| *Oceanodroma homochroa* | + | + | ND |
| Anseriformes | | | |
| *Anas platyrhynchos* | + | + | ND |
| Ciconiiformes | | | |
| *Bubulcus ibis* | + | + | ND |
| Galliformes | | | |
| *Gallus gallus* | + | + | ND |
| *Coturnix sp.* | + | + | ND |
| Charadriiformes | | | |
| *Larus occidentalis* | + | + | ND |
| *Rissa tridactyla* | + | + | PT |
| *Rissa brevirostris* | + | + | ND |
| *Thalasseus maximus* | + | + | ND |
| *Uria aalge* | + | + | PT |
| *Uria lomvia* | + | + | PT |
| *Endomychura hypoleuca* | + | + | ND |
| *Aethia pusilla* | + | + | ND |
| Columbiformes | | | |
| *Columba livia* | + | + | PT |
| *Zenaida macroura* | + | + | PT |
| Piciformes | | | |
| *Colaptes auratus* | + | + | ND |
| Passeriformes | | | |
| *Turdus merula* | + | + | ND |
| *Molothrus ater* | + | + | ND |
| *Agelaius phoeniceus* | + | + | ND |
| *Sturnella neglecta* | + | + | ND |
| *Passer domesticus* | + | + | ND |
| *Pooecetes gramineus* | + | + | ND |
| *Zonotrichia leucophrys* | + | + | ND |

NB. ND, not detectable; PT, possible trace; +, high affinity and specific binding was detected. Estradiol-17β was bound by the plasma of none of these species. From Wingfield *et al.* (1984), with permission.

Other techniques look equally promising. Another example is the application of quantitative autoradiography to avian tissues. Recently Ball *et al.* (1990) showed that muscarinic cholinergic receptors in the brain of Japanese Quail, European Starlings, and Song Sparrows could be investigated by quantitative autoradiographic techniques using the same agonists and antagonists as used in mammalian brain. Similarly, α-2 adrenergic receptors could also be demonstrated in brains from Japanese Quail (Ball *et al.*, 1989).

The future looks very bright indeed, especially since investigations of wild species allow thus far unexplored possibilities for the application of molecular techniques to organismal problems. Our background knowledge of the biology of wild species in both field and laboratory is perhaps the broadest of any vertebrate group, and thus provides a very extensive data base for molecular investigations.

## III. Reproductive Endocrinology

### A. GENERAL BACKGROUND—ANNUAL CYCLES OF REPRODUCTION AND ASSOCIATED EVENTS

There is a general consensus that reproduction in birds, and perhaps most vertebrates, is periodic and timed to occur when trophic resources are favorable for production and post-fledging survival of offspring as well as survival of the parents (e.g., Lack, 1968; Perrins, 1970; Murton and Westwood, 1977). The reproductive cycle itself is a complex sequence of distinct events that in most monogamous species includes: development of the gonad, establishment of a breeding territory, pair formation, nest construction, ovulation and oviposition, incubation, and finally feeding of young until they become independent. In many species that breed at mid- to low-latitudes, two or more broods may be raised in a single season. Additionally, storms or predators often destroy nests and renest attempts usually follow. Clearly these complex sequences of events require that individuals synchronize their cycles with the local phenological progression of the seasons. The repetition of parts of the reproductive cycle in multiple-brooded populations, and the readjustment of the cycle following loss of a nest also require a high degree of co-ordination of the reproductive effort between mates.

The regulation of annual or periodic cycles depends on information from the environment that allows an individual to predict the mean time for onset of reproductive effort, or nesting phase, and thus initiate gonadal development in advance of the ensuing breeding period (Farner and Lewis, 1971; Farner and Follett, 1979; Farner and Gwinner, 1980; Wingfield, 1980b, 1983; Wingfield and Farner, 1980a; Wingfield *et al.*, 1991). Since the phenology of "spring" varies from year to year, local environmental influences are important in fine-tuning final gonadal maturation and onset of the nesting phase. Behavioral interactions between mates, and among adults and young, are also crucial for synchronizing and integrating the sequence of events within the nesting phase (e.g., Lehrman, 1965; Hinde, 1965).

There are two categories of environmental factors. First, "ultimate environmental factors" select individuals that produce young at a favorable time for survival. Secondly, "proximate environmental factors" *regulate* gonadal development and the temporal progression of the reproductive effort so that production of young occurs at a time when the probability of survival is greatest (Baker, 1938; Thompson, 1950; Farner, 1964; Immelmann, 1971, 1973).

The vast literature on the numerous proximate environmental factors known to influence avian reproduction has been reviewed several times (e.g., Lofts and Murton, 1968; Immelmann, 1971, 1973; Lewis and Orcutt, 1971; Dolnik, 1975, 1976; Murton and Westwood, 1977; Farner and Follett, 1979). The complex array of proximate factors can, however, be subdivided into four major categories that allow a more precise analysis of the interactions of environmental information and reproduction (Wingfield, 1980b, 1983; Wingfield and Kenagy, 1991).

*Initial predictive information.* These factors initiate gonadal development and other vernal phenomena, and serve to bring the bird into the breeding area in a physiological state in which responses to local information are possible. Such predictive information is critical because of the time required for development of the reproductive system to a functional, or near functional, condition (Farner, 1970). Initial predictive information also sustains a physiological state that allows rapid responses to other environmental information throughout the breeding season, and finally induces termination of reproductive function, usually in mid- or late-summer. Perhaps the most widely used initial predictive information among non-equatorial species is the annual cycle of daylength which appears to act either directly as a "driver" or indirectly as a *Zeitgeber* for endogenous rhythms (e.g., Farner, 1964, 1970; Gwinner, 1975, 1981; Dolnik, 1976; Berthold, 1977; Farner and Wingfield, 1978; Meier and Ferrell, 1978; Farner and Follett, 1979; Farner and Gwinner, 1980; Follett and Robinson, 1980; Wingfield and Farner, 1980a). It should be noted, however, that these factors alone are not sufficient for culmination of the final reproductive effort (Immelmann, 1963b, 1967b; Farner, 1964; King *et al.*, 1966; Wingfield, 1980b, 1983).

*Essential supplementary information.* Generally this type of information supplements initial predictive information and promotes final gonadal development and onset of the nesting phase. Typical examples of these factors are inclement weather, temperature, availability of food, and access to nest sites (Wingfield, 1983).

*Synchronizing and integrating information.* The reproductive cycle involves a complex temporal sequence of events (e.g., territory establishment, pair bonding, nest building, parental behavior) and in addition, varying degrees of synchronization within a pair are important for reproductive success. Behavioral interactions between mates and among adults are well known examples of this type of information (e.g., Lehrman, 1965; Hinde, 1965; Harding, 1981; Wingfield and Marler, 1988).

*Modifying information.* This is a rather broad category that includes those factors that disrupt the normal temporal progression of the nesting phase. Examples are loss of the eggs or young because of such factors as storms or predators. In some cases readjustments of the cycle occur and renest attempts are initiated (Wingfield, 1988b; Wingfield and Kenagy, 1991).

Wingfield (1980b) has suggested that it is possible to assign any environmental factor to one of the above categories. It is conceded that this may represent an oversimplification of the diverse information utilized by an individual, but more recent investigations have supported this scheme (Wingfield, 1983; Wingfield and Kenagy, 1991), and raise the possibility that within a category different environmental factors can exert their effects by common pathways and mechanisms.

### B. RADIOIMMUNOASSAY OF REPRODUCTIVE AND ASSOCIATED HORMONES IN CAPTIVE WILD SPECIES

It was first demonstrated six decades ago (Rowan, 1925, 1926) that the annual cycle in daylength regulates gonadal development and regression in many avian species that breed in mid- to high-latitudes (for reviews see Lofts and Murton, 1968; Farner and Follett, 1979; Farner and Gwinner, 1980; Follett, 1984; Farner, 1985). In this respect, the annual photocycle acts as initial predictive information that induces gonadal maturation well in advance of the breeding season. Over the years since Rowan's experiments, the effects of daylength on gonadal development have been demonstrated in about 70 species. Since an exhaustive review of the mechanisms underlying these responses to daylength (e.g., driver versus *Zeitgeber*) is beyond the scope of this chapter, the reader is referred to recent detailed treatises on this topic (Gwinner, 1975; Dolnik, 1976; Berthold, 1977; Murton and Westwood, 1977; Farner and Follett, 1979; Aschoff, 1980; Farner and Gwinner, 1980; Follett and Robinson, 1980).

The endocrine basis of photoperiodically induced gonadal cycles in birds remained relatively unknown, except for inferences, until the development of RIAs, or similar competitive binding assays, for avian LH (Follett *et al.*, 1972), FSH (Follett, 1976; Sakai and Ishii, 1985), and steroid hormones (e.g., Temple, 1974; Wingfield and Farner, 1975). In general, wild avian species held on natural daylengths in captivity showed an increase in plasma levels of LH and testosterone in spring as gonadal maturation progressed. Plasma levels of LH remained high until mid-summer and then declined (e.g., *Zonotrichia leucophrys gambelii*, Mattocks *et al.*, 1976; Mallard, e.g., Donham, 1979; Green-winged Teal, *Anas crecca*, Assenmacher and Jallageas, 1980; Canvasback, *Aythya valisineria*, Bluhm *et al.*, 1983; Willow Ptarmigan, *Lagopus lagopus lagopus* and Red Grouse, *L. l. scoticus*, Stokkan and Sharp, 1980a, Sharp and Moss, 1981; and Herring Gull, Scanes *et al.*, 1974). In the Japanese Pheasant, *Phasianus colchicus versicolor*, Sakai and Ishii (1986) showed that plasma levels of LH and FSH increased through spring and were highest at egg-laying. Changes in circulating concentrations of estradiol were correlated with FSH levels, and elevated levels of progesterone with LH. In male Pheasants, FSH levels were correlated with testicular maturation and with increasing testosterone levels in spring. However, the temporal pattern of plasma LH titers was clearly bimodal with peaks in February and September. Curiously, the autumnal increase in LH levels was not accompanied by an increase in testosterone.

The correlation of increasing plasma levels of LH and testosterone with lengthening days in spring has been confirmed in experiments in which photosensitive adults were transferred from short to long days in the laboratory. In adult male White-crowned Sparrows transferred from 8L 16D to 20L 4D (L refers to hours of light and D to hours of dark), there was a rapid increase in plasma levels of LH to a maximum within 5 days of photostimulation (Fig. 5), where after these levels remained maximal until about day 50 of photostimulation when spontaneous gonadal regression, or photorefractoriness, terminated the reproductive cycle. Similar results have also been obtained for female White-crowned Sparrows (Fig. 5d), and for female American Tree Sparrows, *Spizella arborea* (Wilson and Follett, 1974). Plasma levels of FSH in males, however, showed a more gradual increase that paralleled testicular development (Figs 5a,c). Note that plasma levels of LH and FSH did not change in controls held on 8L 16D. It is also known that long day treatment resulted in an increase in the number of receptors for FSH as indicated by the specific binding of rat FSH by the testes of *Z. l. gambelii* sacrificed at intervals throughout the photoperiodically induced cycle (Fig. 6), from Ishii and Farner (1976). Note that the binding capacity increased in parallel with testicular development for the first 20–25 days and then remained level as the logarithmic phase of growth neared completion. Circulating testosterone and DHT did not increase in a manner parallel with LH. Maximum androgen levels were attained at about day 30 when testicular development was nearing completion (Fig. 5c).

In the House Sparrow, plasma levels of LH also increased immediately after photostimulation whereas testosterone levels in blood did not increase significantly until 21 days later (Fig. 7, from Donham *et al.*, 1982). In this case the testicular content of testosterone increased in parallel with LH reaching a plateau after 1 week. Plasma levels of testosterone did not increase until the main period of testicular growth was completed suggesting that the testis may sequester testosterone, presumably for spermatogenesis. A synergistic action of FSH and testosterone on spermatogenesis has been demonstrated in Japanese Quail (Ishii and Furuya, 1975; Brown and Follett, 1977). Thus it is possible to explain the initial dissociation of LH and testosterone patterns in plasma by a sequestering action of the seminiferous tubules during at least the early stages of spermatogenesis. As the testis neared maturity perhaps less testosterone was sequestered so that plasma levels began to increase (Fig. 7). It is also curious that plasma levels of testosterone declined soon after the maximum was reached, despite the fact that LH levels were still elevated. A similar dissociation of LH and testosterone secretion has been demonstrated in Mallards by Jallageas *et al.* (1974). These authors showed that

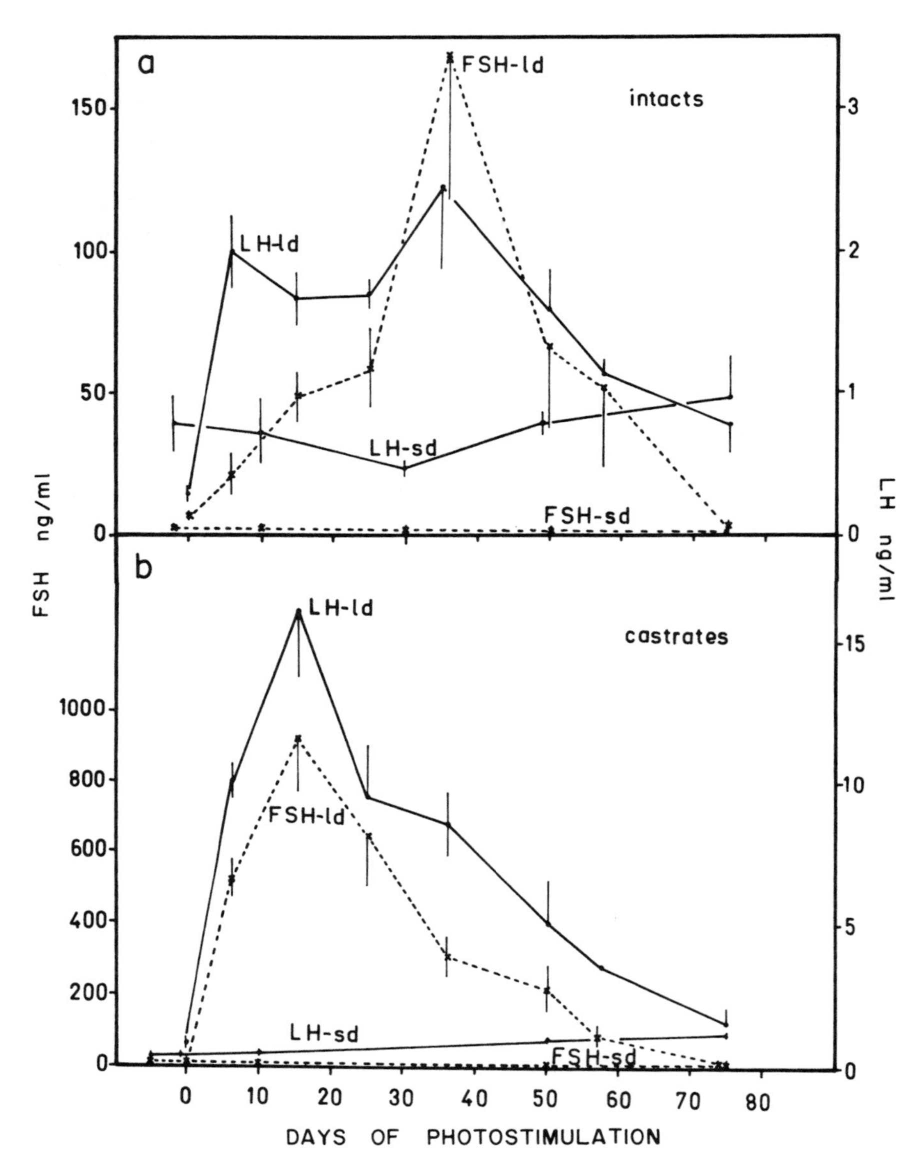


plasma levels of T4 also increased during the reproductive cycle resulting in an increase in the metabolic clearance rate of testosterone. Thus plasma levels of testosterone declined even though the secretion rate may not have changed.

Essentially similar patterns of LH, and in some cases FSH and testosterone, have been demonstrated during photoperiodically induced gonadal growth in Willow Ptarmigan and Red Grouse (Stokkan *et al.*, 1982, 1988), American Tree

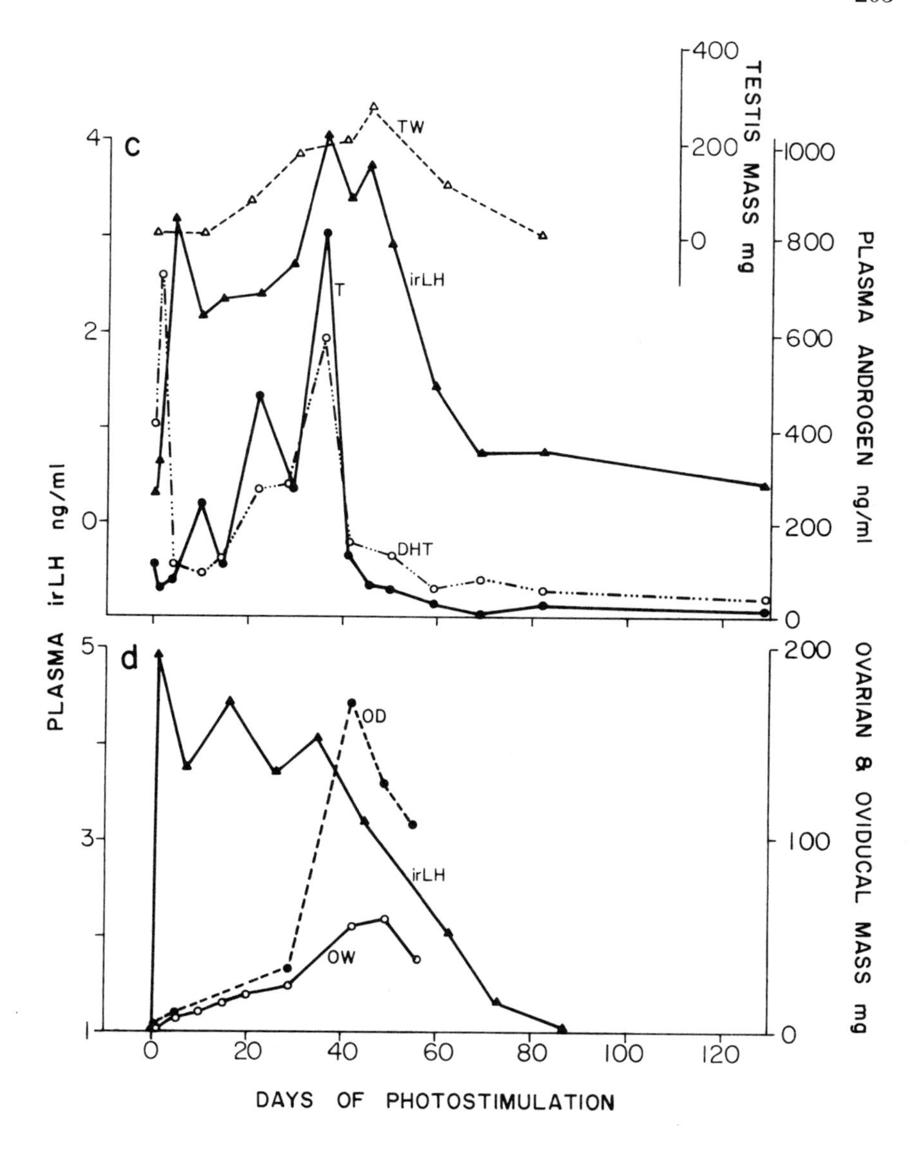


FIG. 5. Photoperiodically induced gonadal functions in White-crowned Sparrows, *Zonotrichia leucophrys gambelii.* (a) Changes in plasma levels of luteinizing hormone (LH), follicle-stimulating hormone (FSH) in intact adult males, (b) changes in plasma levels of LH and FSH in castrated males. ld, long days (20L 4D); sd, short days (8L 16D), (c) changes in plasma levels of LH, dihydrotestosterone (DHT), and testis mass (TW), (d) changes in plasma levels of LH, ovarian mass (OW) and oviduct (OD). From Follett *et al.* (1975), Lam and Farner (1976), Wingfield and Farner (1980a), courtesy of S. Karger AG, Basel, and Wingfield *et al.* (1980a).

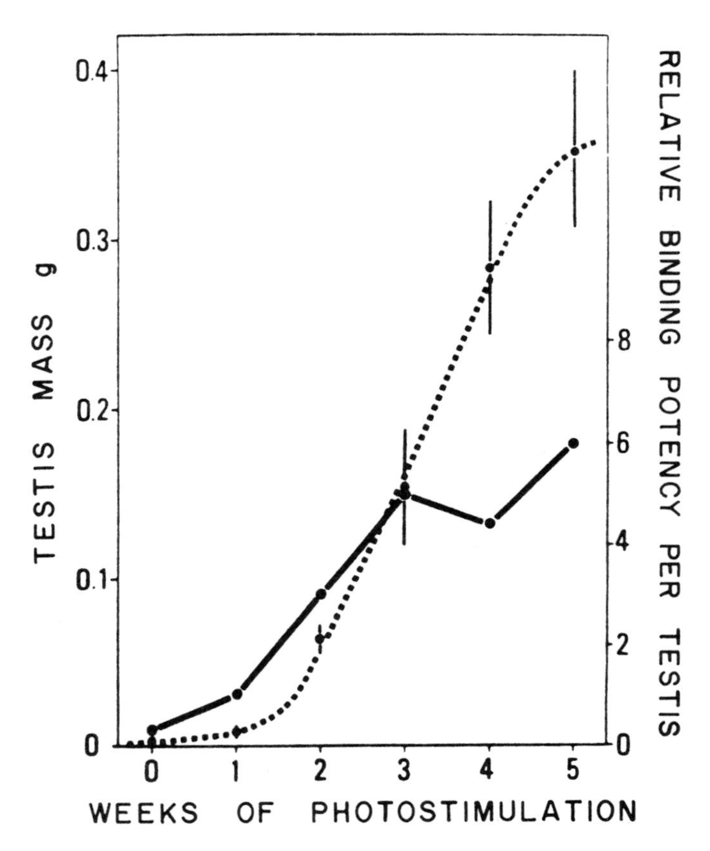


FIG. 6. Effect of long day treatment (20L 4D) on the mass of testes (solid circle, broken line) and relative capacity (open circles, solid line) of the specific binding of I-125 rat FSH by the testes of the White-crowned Sparrow, *Zonotrichia leucophrys gambelii.* Mean values and their standard errors are indicated by circles and vertical bars respectively. From Ishii and Farner (1976), with permission.

plasma levels of T4 also increased during the reproductive cycle resulting in an increase in the metabolic clearance rate of testosterone. Thus plasma levels of testosterone declined even though the secretion rate may not have changed.

Essentially similar patterns of LH, and in some cases FSH and testosterone, have been demonstrated during photoperiodically induced gonadal growth in Willow Ptarmigan and Red Grouse (Stokkan *et al.*, 1982, 1988), American Tree Sparrow (Wilson and Follett, 1974, 1978), European Starling (Ebling *et al.*, 1982; Dawson *et al.*, 1985), the Grey Partridge, *Perdix perdix* (Sharp *et al.*, 1986), Red-legged Partridge, *Alectoris rufa* (Creighton, 1988a,b), and the British race of the Bullfinch, *Pyrrhula pyrrhula pileata* (Storey and Nicholls, 1982a). Photoperiodically induced changes in androgen secretion are less well known, but Fevold and Eik-Nes (1962) showed that the conversion, *in vitro*, of C-14 progester-

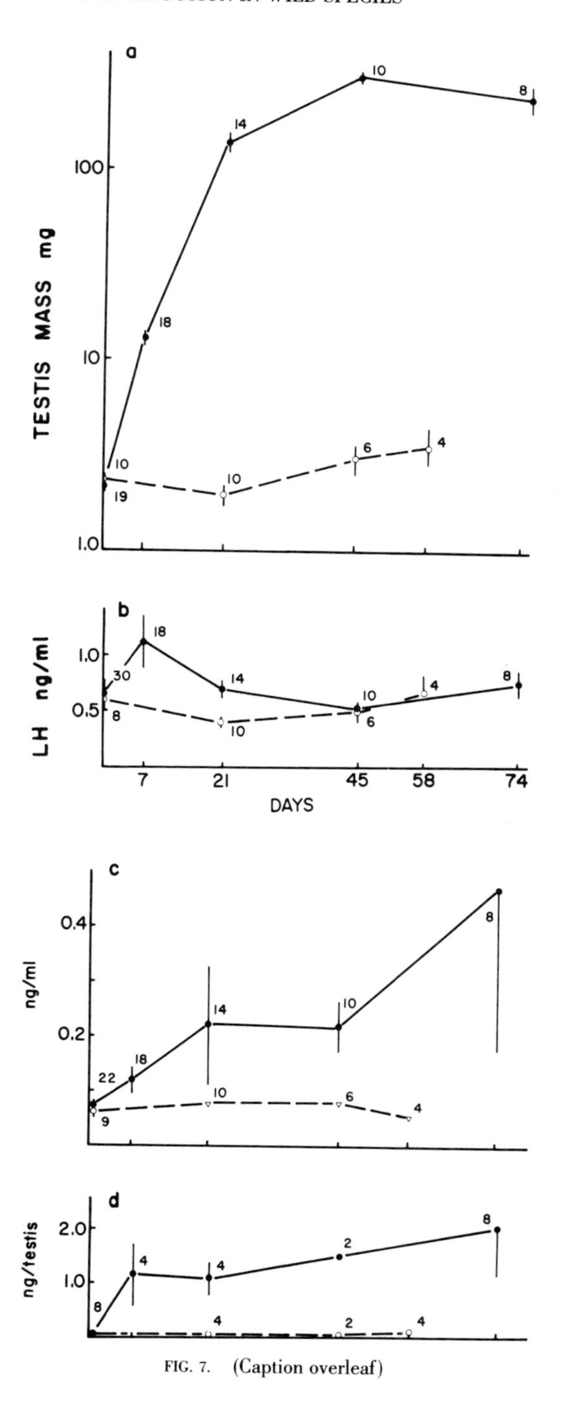


FIG. 7. (Caption overleaf)

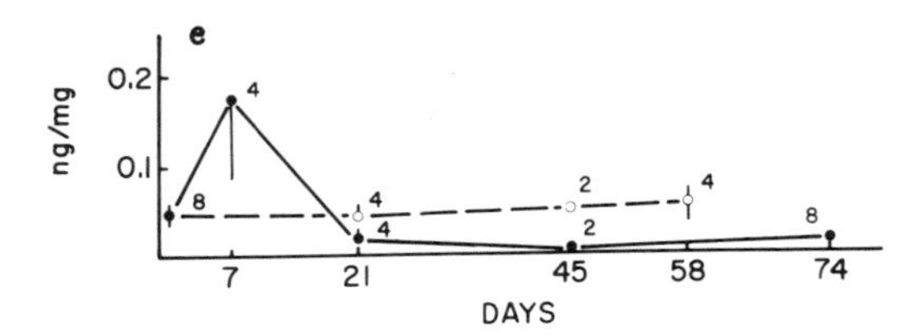


FIG. 7. Changes in testis mass (a), and plasma levels of luteinizing hormone (LH, b), of House Sparrows, *Passer domesticus*, exposed to 16L 8D (solid line) and 8L 16D (broken line). Note the logarithmic scale for testis mass. Changes in the plasma level (c), testicular content (d), and testicular concentration (e) of testosterone in House Sparrows exposed to 16L 8D (solid line) and 8L 16D (broken line). All data are means and standard errors. Adjacent numbers are sample sizes. From Donham *et al.* (1982), with permission.

but was basal in refractory birds. This trend was also paralleled by plasma levels of testosterone (Kerlan and Jaffe, 1974). Essentially similar patterns of *in vitro* testosterone production have been demonstrated in testicular material from European Tree Sparrows exposed to natural daylength (Chan and Lofts, 1974).

Photoperiod-induced elevations of reproductive hormones accompanying gonadal maturation are not ubiquitous. In at least one species, the Cockatiel, *Nymphicus hollandicus*, transfer to long days did not result in an increase of circulating LH in either sex. Other cues, such as presence of a nest box and mate, may be far more effective in stimulating LH release (Shields *et al.*, 1989). Whether the lack of a photostimulated LH response is an adaptation to a potentially opportunistic breeding strategy in the Australian deserts remains to be determined.

From the data presented in Fig. 5, and from other experiments cited above, it is clear that the temporal patterns of LH and FSH are different suggesting separate controls of these two gonadotropins, but presumably by a single gonadotropin-releasing factor from the hypothalamus. Note, however, that in Fig. 5b the temporal patterns of LH and FSH secretion in photostimulated and castrated male White-crowned Sparrows were identical (from Wingfield *et al.*, 1980a) suggesting that feedback from circulating androgens is important in regulating not only the absolute levels of gonadotropin, but also the temporal pattern. Very similar data have been presented for the castrated Canary by Storey *et al.* (1980), and castrated American Tree Sparrows (Wilson and Follett, 1974). In the Bullfinch, plasma levels of LH were also elevated after castration and then slowly declined to basal values 40–50 weeks after photostimulation (Storey and Nicholls, 1982a).

In mammals it is well known that there are marked daily, or diel, rhythms of circulating levels of reproductive hormones, and that secretion of at least LH is pulsatile or episodic. Pulsatile secretion of LH has been demonstrated in the domestic fowl (Wilson and Sharp, 1975), and Japanese Quail (Gledhill and Follett, 1976), although the extent to which this pattern of hormone secretion is found in wild species remains to be demonstrated. However, there is extensive evidence for diel rhythms of hormone secretion in domesticated species that are supported by a few studies of wild species. In the Bobwhite Quail, plasma levels of LH became maximal late in the day and were low early in the morning in reproductively active females (Rattner *et al.*, 1982). In adult male White-crowned Sparrows, there were marked changes in the diel rhythms of plasma LH levels in relation to daylength and reproductive state (Fig. 8). In photosensitive males on short days, and in photorefractory males held on both long and short days there was a marked elevation of LH late in the day and at night (similar to female Bobwhite Quail), and low levels were measured 1–2 h after "lights on" (Fig. 8b,c,d). In refractory birds there appears to be a slight elevation of LH in the middle of the day also (Fig. 8b,d). However, in contrast there was no discernable rhythm of plasma LH levels in photosensitive males on long days (i.e., undergoing testicular development, Fig. 8a). In the latter group LH levels were uniformly high. It is curious that plasma

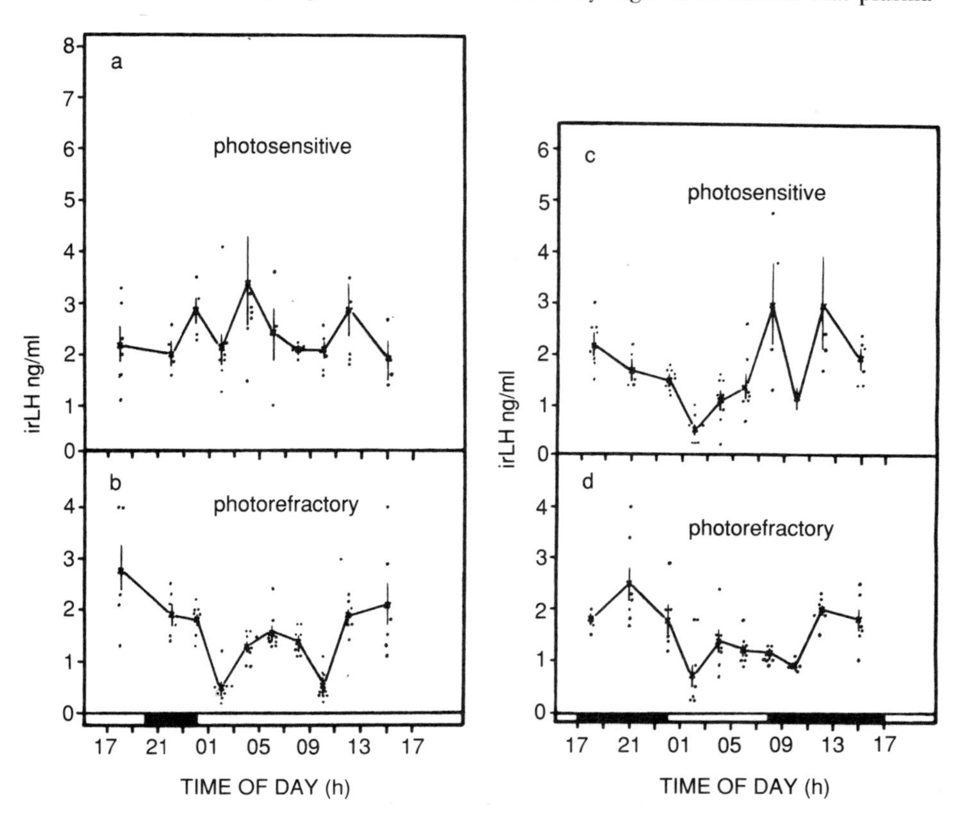


FIG. 8. Diel changes in plasma levels of immunoreactive luteinizing hormone (irLH) in photosensitive and photorefractory adult male White-crowned Sparrows, *Zonotrichia leucophrys gambelii*, exposed to (a) 20L 4D and (b) 8L 16D. Each cross is the mean with standard error, small dots are individual values within each group. From Wingfield *et al.* (1981), courtesy of Wiley-Liss Inc.

levels of LH in photorefractory males sampled at night were as high as in reproductively active males on long days. Thus it is possible that photoperiodically induced gonadal growth requires high levels of LH throughout the 24-h period (see also Wingfield *et al.*, 1981).

There is also evidence for photoperiodically induced changes in GnRH content in the hypothalamus of wild species. In the White-crowned Sparrow, GnRH activity (as measured by the ability of hypothalamic extracts to release LH from rooster pituitaries *in vitro*) was highest during photostimulation and low when birds became photorefractory and the testes regressed (Fig. 9, from Erickson, 1975). Furthermore, if hypothalamic extracts from male White-crowned Sparrows were injected into mature males, then only extracts from photosensitive birds resulted in an increase in plasma levels of LH (Fig. 10). Extracts of cerebrum, and extracts of both hypothalamus and cerebrum from photorefractory males were without effect. In other species, photostimulation increased GnRH content (as measured by radioimmunoassay) only in female Red-legged Partridges (Creighton, 1988a,b), whereas in the European Starling there was no change in GnRH content following transfer of males to long days (Dawson *et al.*, 1985). However, when birds became photorefractory there was a significant decrease in hypothalamic GnRH content. In the Baya Weaver, GnRH (measured as the ability of hypothalamic extracts to release mammalian GnRH from rat pituitaries *in vitro*), content of the hypothalamus increased during testicular maturation (in parallel with plasma LH concentrations), was maximal during breeding, and minimal when the testes regressed (Narula and Saxena, 1981). Curiously, similar changes in GnRH content also occurred in the pineal (Saxena and Narula, 1981).

### C. FIELD ENDOCRINOLOGY TECHNIQUES

Although experiments with domesticated and semi-domesticated species have enhanced our knowledge of how the nesting phase, once initiated, is synchronized and integrated (e.g., Murton and Westwood, 1977; Hinde and Steel, 1978; Silver, 1978; Cheng, 1979) and are ideal for investigations of fundamental endocrine mechanisms, these species have lost, to varying degrees, the ability to respond to the environmental information used by their wild ancestors. Therefore, the elucidation of the mechanisms involved in the fine temporal adjustment of all reproductive and associated functions ultimately depends on investigations of truly wild populations. In the past such studies were impeded by the failure of most wild avian species to breed in captivity. Females often undergo ovarian development to the pre-yolk deposition phase in captivity, but stimulation of vitellogenesis and final maturation of the ovary in captive wild species is unusual without elaborate aviary space and diets (King *et al.*, 1966).

For this reason, techniques have been devised whereby blood samples, and

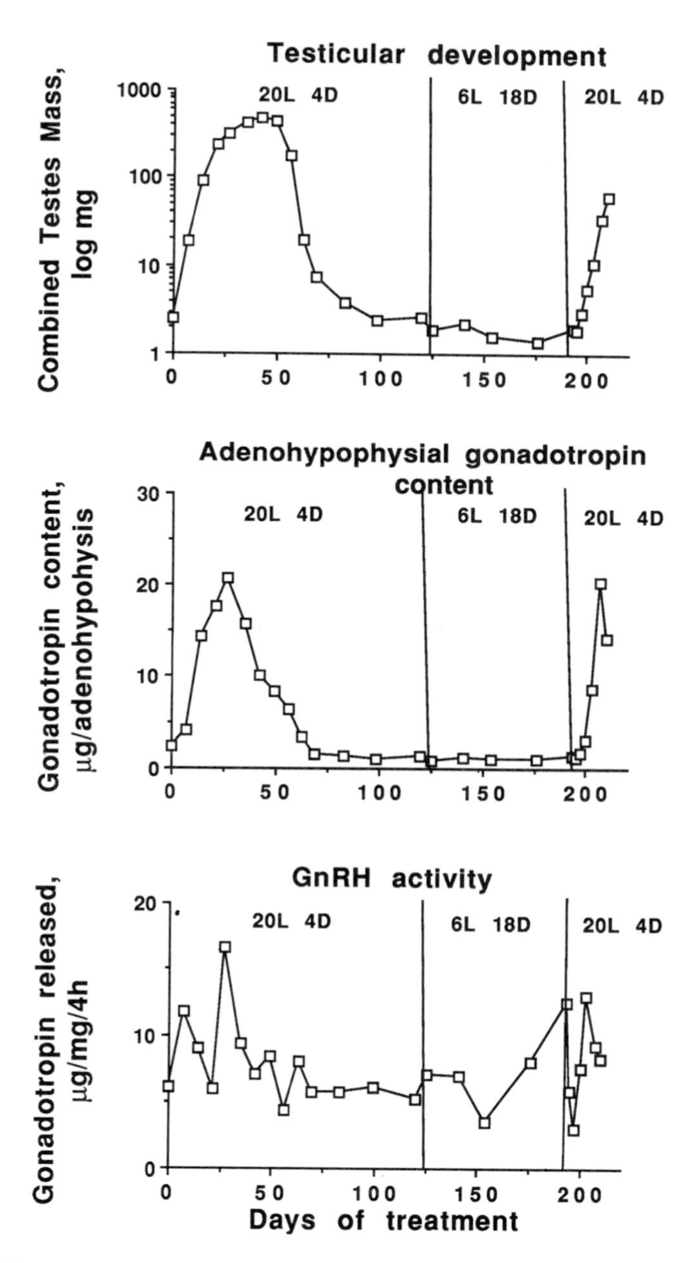


FIG. 9. Changes in testicular development, gonadotropin content of the adenohypophysis and GnRH activity (as measured by ability to release luteinizing hormone from rooster adenohypophyses *in vitro*) in the hypothalamus of adult male White-crowned Sparrows, *Zonotrichia leucophrys gambelii*, during a photoperiodically induced gonadal cycle. From Erickson (1975).

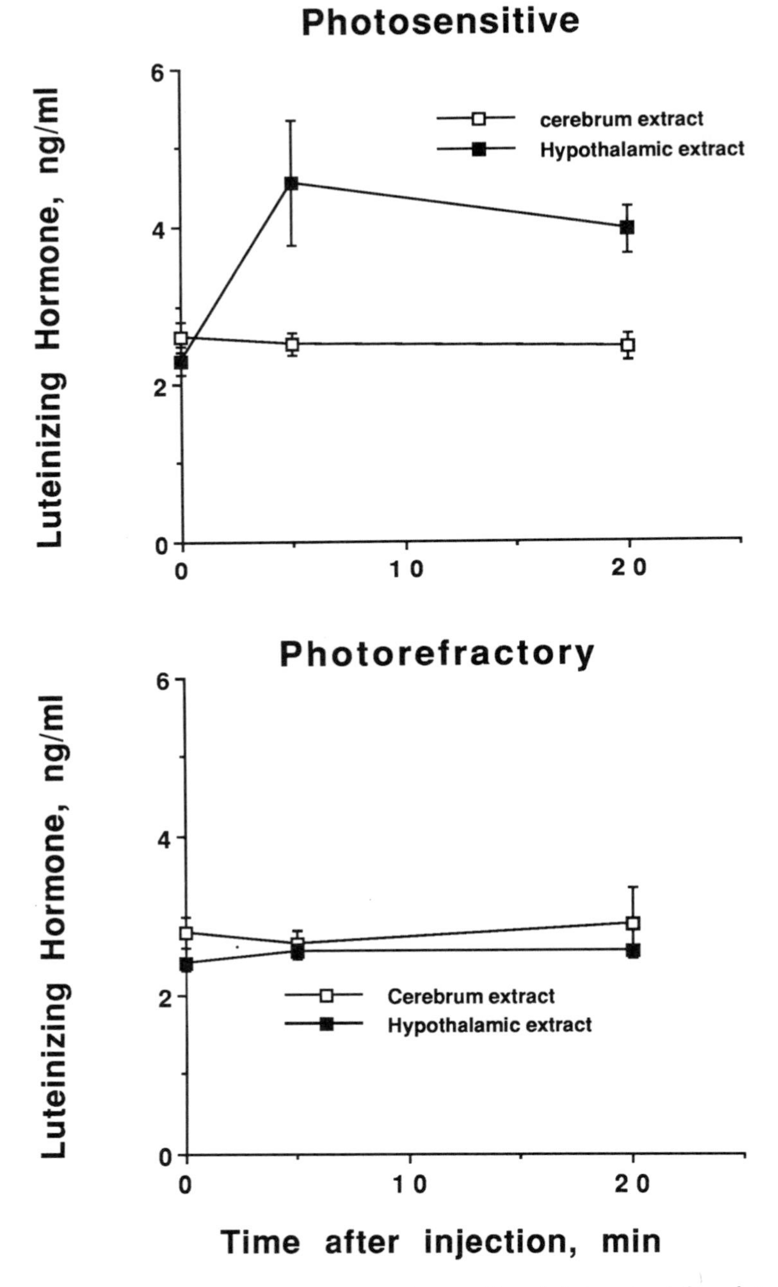


FIG. 10. Effects of intravenous injections of cerebrum and hypothalamus extracts from photosensitive and photorefractory male White-crowned Sparrows, *Zonotrichia leucophrys gambelii*, on circulating levels of luteinizing hormone. Vertical bars are standard errors of the means. From R. Hudson and J. C. Wingfield, unpublished.

often serial samples, can be collected from free-living birds immediately after capture in Japanese mist nets, or traps (Wingfield and Farner, 1976). After withdrawal of blood, each bird is color banded, and data such as body mass, fat score, and, by laparotomy, state of the reproductive organs, gathered. The bird is then released for subsequent observation under natural conditions. Recapture, re-examination, and procurement of further blood samples is then possible. Data obtained from analyses of hormone levels in the blood samples can be related directly to stages of the annual cycle, behavior, weather and other variables (Wingfield and Farner, 1976). This type of information provides not only valuable bases for designing appropriate laboratory or field experiments to elucidate mechanisms, and for evaluation of the significance of results obtained under rigidly controlled conditions, or from domesticated species in the laboratory, but also has provided new research directions and concepts that could probably not be approached in the laboratory (e.g., Wingfield *et al.*, 1990).

### D. ENDOCRINOLOGY OF NATURAL POPULATIONS OF THE WHITE-CROWNED SPARROW

The first detailed investigations on temporal changes in reproductive hormones in repeatedly captured and free-living birds were performed on two taxa of White-crowned Sparrow, the short distance migrant, double brooded *Z.l. pugetensis*, which breeds in, for example, the region of Puget Sound, Washington (48° N) and winters in central and northern California (Blanchard, 1941; Cortopassi and Mewaldt, 1965); and the long distance migrant, single brooded *Z.l. gambelii*, which breeds, for example, in central Alaska (64° N) and winters from central Washington to northern Mexico (Blanchard and Erickson, 1949; Cortopassi and Mewaldt, 1965). These data are summarized in Figs 11, 12, and 13 (from Wingfield and Farner, 1977, 1978a,b, 1980a), and are arranged as to stage in the reproductive cycle and spaced according to the average duration of the stages. For example, the incubation period is usually 14 days, and young birds become independent at about 30 days of age (see Wingfield and Farner, 1976, 1978a). Reasons for this arrangement become clear when one considers that owing to asynchrony among pairs once the nesting phase is underway, it is meaningless to arrange the data purely by calendar time because on any one day some birds might be preparing to lay, others in lay, and some incubating. Since plasma levels of hormones change dramatically between sexual and parental phases, it is crucial to record the stage of breeding whenever possible.

As daylength increased in spring there was a rapid recrudescence of the gonads in both the short- and long-distance migrants (Fig. 11). Note, however, that maximum testicular mass and ovarian stage were attained about 1 month earlier in *Z.l. pugetensis* which breeds at mid-latitudes where summers begin earlier and last longer. Additionally, it can be seen that *pugetensis* produced at least two broods

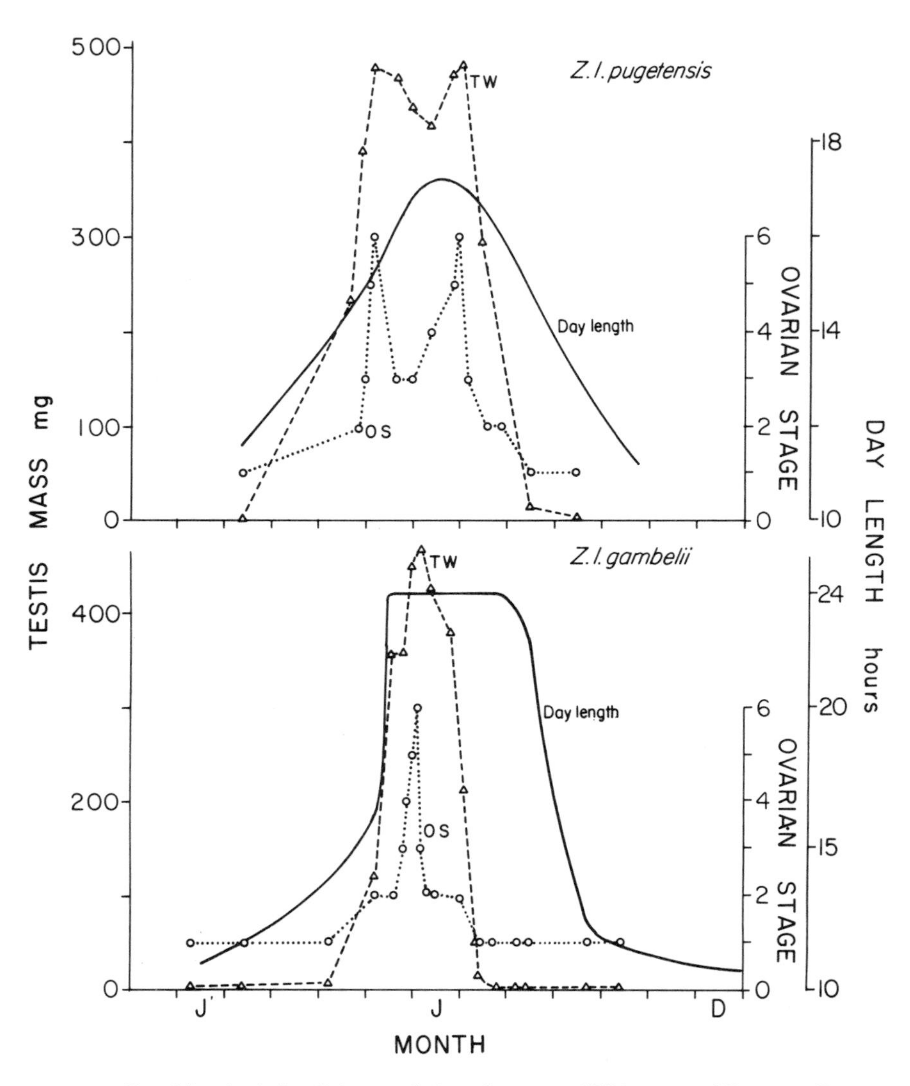


FIG. 11. Gonadal cycles in free-living populations of two taxa of White-crowned Sparrows, *Zonotrichia leucophrys gambelii*, and *Z.l. pugetensis*, in relation to the annual photocycle. From Wingfield and Farner (1977, 1978a,b, 1980a), courtesy of S. Karger AG, Basel.

per summer whereas *Z.l. gambelii* raised only a single brood in the short subarctic summer. Plasma levels of LH and testosterone also increased in the spring reaching maximum levels as males arrived on the breeding grounds, established territories, and attracted mates (Fig. 12). These data are consistent with laboratory findings that testosterone regulates both aggressive territorial behavior and sexual

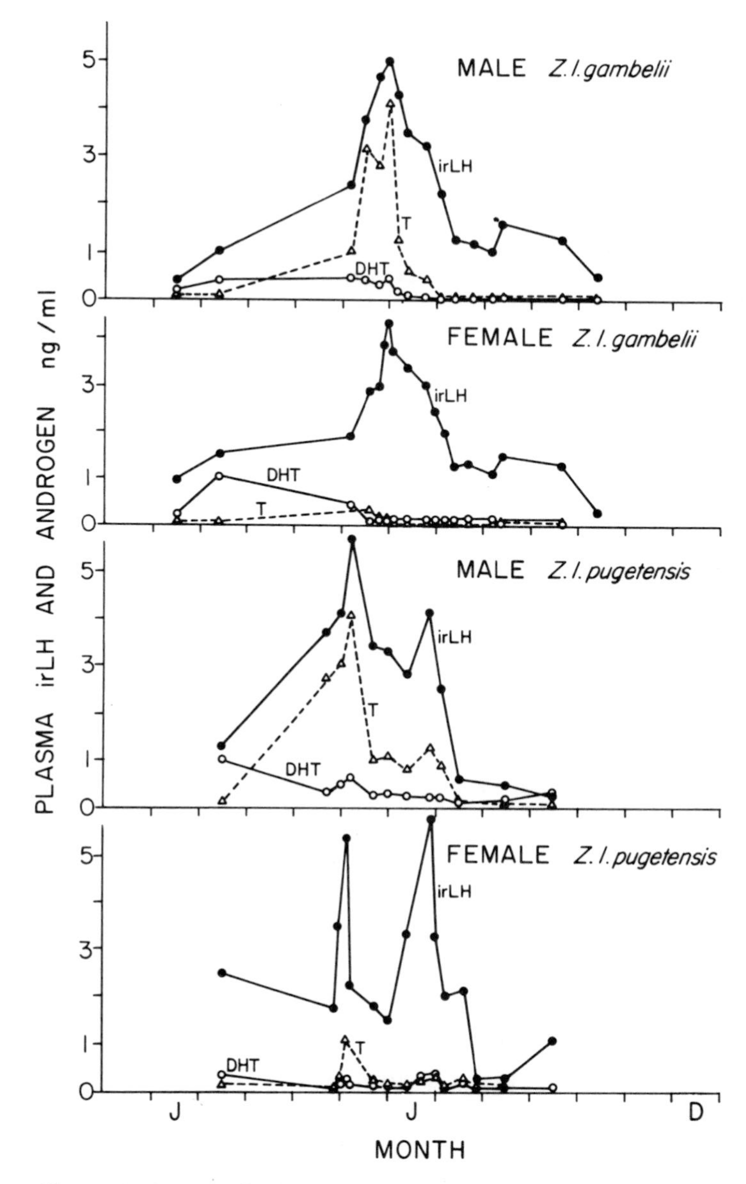


FIG. 12. Changes in plasma levels of immunoreactive luteinizing hormone (irLH), testosterone (T) and dihydrotestosterone (DHT) in free-living populations of two taxa of White-crowned Sparrows, *Zonotrichia leucophrys gambelii*, and *Z.l. pugetensis*. From Wingfield and Farner (1977, 1978a,b, 1980a), courtesy of S. Karger AG, Basel.

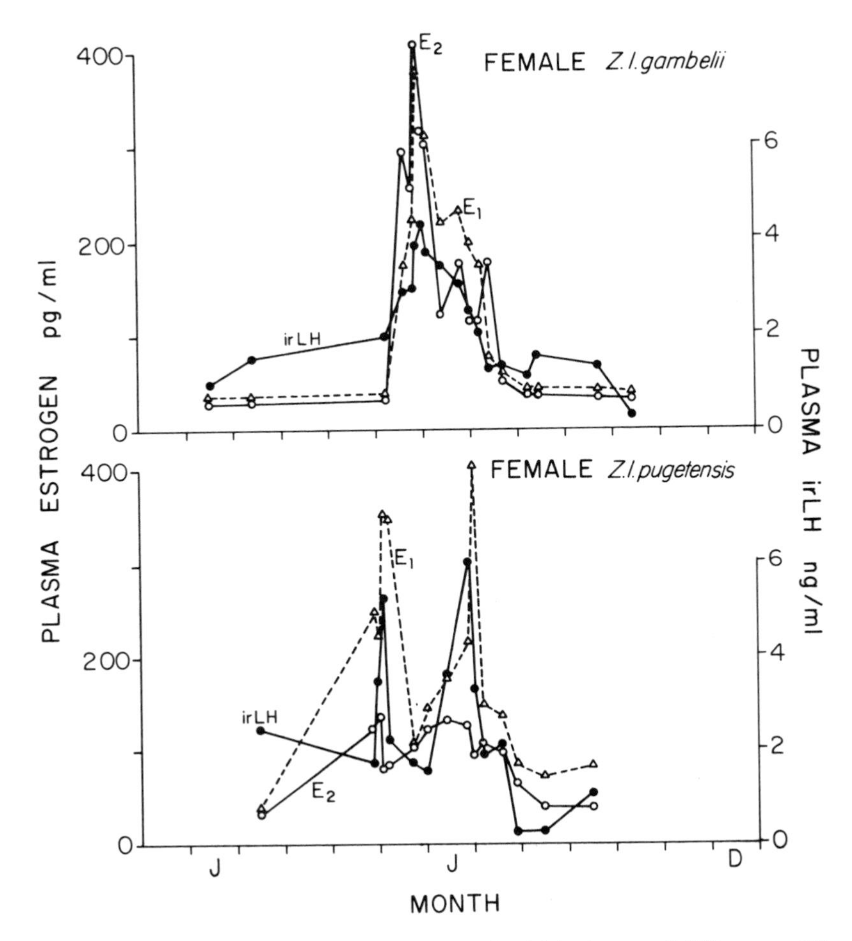


FIG. 13. Changes in plasma levels of immunoreactive luteinizing hormone (irLH), estrone (El) and estradiol-17β (E2) in free-living populations of two taxa of White-crowned Sparrows, *Zonotrichia leucophrys gambelii*, and *Z.l. pugetensis*. From Wingfield and Farner (1977, 1978a,b, 1980a), courtesy of S. Karger AG, Basel.

behavior (e.g., Cheng and Lehrman, 1975; Adkins and Pniewski, 1978; Adkins-Regan, 1981; Harding, 1981, 1983; Harding *et al.*, 1988; see also Balthazart, 1983 for review). Note also that circulating levels of testosterone were even higher in males when females were laying. At this time males copulated and became very aggressive toward conspecific males as they mate-guarded to protect their paternity of the young (see also Wingfield and Moore, 1987). High circulating levels of testosterone, and to a lesser extent DHT (Fig. 12) were also correlated with development of secondary sex characters such as the cloacal protuberance (see also Adkins and Pniewski, 1978; Wingfield and Farner, 1978a,b).

In females there were slight but significant vernal increases in plasma levels of testosterone and DHT, just before and during egg-laying (Fig. 12). This is consistent with field observations that female White-crowned Sparrows are territorial and may accompany the male during agonistic interactions over boundaries (Blanchard, 1941; Wingfield, 1984a). As soon as the female began to incubate (males of these populations did not incubate) there was a dramatic decline in plasma levels of LH and testosterone in male *gambelii* that reached basal values by the time the young fledged. At this time, there was a tendency to abandon defense of territorial boundaries and breeding activity was terminated soon after. The gonads then regressed very rapidly, where after the pre-basic (post-nuptial) molt began. A basically similar pattern of circulating testosterone levels has been observed in the Mountain White-crowned Sparrow (*Z. l. oriantha*), although levels peaked before clutch initiation (Morton *et al.*, 1990). In contrast, plasma levels of LH and testosterone in male *pugetensis* did not decline to basal but stabilized at a lower maintenance level (Fig. 12). As females prepared for the second clutch there was a resurgence of plasma levels of LH, but not testosterone. When the second parental phase developed, circulating levels of LH and testosterone declined to basal. Termination of the reproductive effort in July was followed by the onset of pre-basic molt (Fig. 12). Note also that plasma levels of FSH remained high throughout the breeding period in male *Z. l. pugetensis*, whereas in females FSH peaked after each laying period (Hiatt *et al.*, 1987).

In females, plasma levels of LH and estrogens increased to maxima coincident with the egg-laying periods (Fig. 13). Note the decline in LH and estrogen levels between broods in *pugetensis* females. The increase in plasma levels of estrogen soon after arrival in the breeding area presumably had an important role in the development of the female reproductive tract and in egg formation. Hypercalcemia, often associated with vitellogenesis, shell formation, and egg-laying, has been induced by injections of estradiol in female *Z. l. gambelii* subjected to both short (8L 16D) and long (16L 8D) days (Kern *et al.*, 1972). Estrogen is also known to induce synthesis of yolk proteins by the liver of the domestic fowl (Heald and MacLachlan, 1964; O'Malley *et al.*, 1969), possibly in synergy with testosterone (Yu and Marquardt, 1973a). Estrogen and testosterone also had synergistic effects in promoting growth of the oviduct and synthesis of albumin by the magnum (Yu and Marquardt, 1973b,c). Note that plasma levels of DHT and testosterone increased at the time when the oviduct was developing, and during egg-laying (Fig. 12). After breeding, plasma levels of LH and sex steroid hormones in females declined to basal, the ovaries regressed, and the pre-basic molt began.

Subsequent investigations of free-living populations, and captive populations breeding in near natural environments, have accumulated information on temporal patterns of circulating levels of reproductive and associated hormones for approximately 40 wild species. As one might expect, the techniques used vary considerably so that in some investigations stages of the breeding cycle were not recorded, whereas in other studies birds were shot rather than captured and released. In some instances birds were recaptured so that serial samples were obtained. Despite these disparities in approach, we have attempted here to review data from most of the wild species studied thus far. Since this array of species covers different mating systems and breeding strategies, we have, for convenience, considered monogamous and polygamous species separately.

### E. CYCLES IN GONADOTROPINS AND SEX HORMONES OF MONOGAMOUS SPECIES

Common themes in most species studied thus far are vernal increases in plasma levels of FSH, LH and/or sex steroids that are more or less parallel with gonadal development. These species are listed below:

Emperor Penguin, *Aptenodytes forsteri* and Adelie Penguin, *Pygoscelis adeliae*, (Groscolas, 1982; Groscolas *et al.*, 1985, 1986, 1988); Wandering Albatross, *Diomedea exulans*, (Hector *et al.*, 1986a); Black-browed and Gray-headed Albatrosses, *D. melanophris* and *D. chrysostoma*, (Hector *et al.*, 1986b); Canada Goose, *Branta canadensis*, (John *et al.*, 1983; Akesson and Raveling, 1981); Hawaiian Goose, *B. sandwichensis*, (R. S. Donham, P. Banko, and D. S. Farner unpublished); Spur-winged Goose, *Plectropterus gambensis*, (Halse, 1985); Bar-headed Goose, *Anser indicus*, (Dittami, 1981; Dittami *et al.*, 1985); Snow Goose, *A. caerulescens caerulescens*, (Campbell *et al.*, 1978); Canvasback, *Aythya valisinaria*, (Bluhm *et al.*, 1983); wild stocks of Mallard, (Donham, 1979; Haase and Sharp, 1975; Haase *et al.*, 1982); Wood Duck, *Aix sponsa*, (Hepp *et al.*, 1991); Black Swan, *Cygnus atratus*, (Goldsmith, 1982a); Rock Ptarmigan, *Lagopus mutus*, (Stokkan *et al.*, 1986); Gray Partridge, (Bottoni *et al.*, 1984; Fraissinet *et al.*, 1987); both sexes of the White Stork, (Hall *et al.*, 1987); male and female Rose-ringed Parakeets, *Psittacula krameri* (Krishnapradan *et al.*, 1988; Sailaja *et al.*, 1988); European Kestrel, *Falco tinnunculus*, (Meijer and Schwabl, 1989); Collared Dove, *Streptopelia decaocto*, (Péczely and Pethes, 1979); Lesser Sheathbill, *Chionis minor*, (Burger and Millar, 1980); Semi-palmated Sandpiper, *Calidris pusilla*, (Gratto-Trevor *et al.*, 1990a); Australian Magpie, *Gymnorhina tibicen*, (Schmidt *et al.*, 1991); European Starling, (Temple, 1974; Dawson and Goldsmith, 1982; Dawson, 1983); Rook, (Lincoln *et al.*, 1980; Péczely and Pethes, 1982); Great Tit, (Röhss and Silverin, 1983); Willow Tit, *Parus montanus*, (Silverin *et al.*, 1986); Song Sparrow, (Wingfield, 1984a); Stonechat, *Saxicola torquata axillaris*, (Dittami and Gwinner, 1985); House Sparrow, (Hegner and Wingfield, 1986a,b); Indian Baya Weaver, *Ploceus philippinus*, (Narasimhacharya *et al.*, 1988); Blue-eared Glossy Starling, *Lamprotornis chalybaeus*, and Rüpell's Long-tailed Glossy Starling, *L. purpuropterus*, (Dittami, 1987); and European Blackbird, *Turdus merula* (Schwabl *et al.*, 1980).

There are some notable exceptions, however. In the female Capercaillie, *Tetrao urogallus*, and Black Swan, plasma levels of FSH appear not to change throughout the year (Goldsmith, 1982a; Hissa *et al.*, 1983), or only slightly as in female Bar-headed Geese (Dittami *et al.*, 1985), although this may be an artifact of the heterologous assay system used for FSH. Both LH and testosterone levels in the blood of Fiscal Shrikes, *Lanius collaris*, were highly variable with little obvious correlation with gonadal development (Dittami and Knauer, 1986). In the Cape Cormorant, *Phalacrocorax capensis*, there were no significant changes in plasma levels of LH throughout the season in either males or females, although in males circulating testosterone reached a maximum in November and December of the austral spring. Once again, the lack of a change in LH may be a reflection of poor sensitivity of the heterologous assay system for this species (Berry *et al.*, 1979). Lack of strong correlations of gonadal development with reproductive hormones were also obtained in Reichenow's Weaver, *Ploceus baglafecht reichenowi*, and rufous sparrow, *Passer motitensis rufocinctus*, studied in equatorial East Africa (Dittami, 1986). Nevertheless, in most species, the maximum levels of LH in males in spring were accompanied by significant elevations of circulating testosterone, and were coincident not only with development of the testes and secondary sex characters, but also with periods of heightened aggression as territories were formed, and with periods of heightened sexual behavior (see also Dittami and Reyer, 1984 for an in depth discussion of the correlation of hormone levels and behavior).

In European Starlings sampled in northeastern United States, plasma levels of testosterone rose in late autumn and remained elevated throughout the winter and breeding season (Temple, 1974). In contrast, investigations of a British population of this species revealed only very slightly elevated levels of testosterone during the winter and a sharp increase in April as breeding commenced. During the parental phase, plasma testosterone declined rapidly (Dawson, 1983). Since Temple (1974) did not record phase of the breeding cycle in his study, it is possible that a distinct vernal increase in testosterone level was overlooked. On the other hand, since the North American population is double-brooded, it is possible that these and other differences from the British population are real. However, Ball and Wingfield (1987) working with a population from the mid Hudson Valley of New York State, found a marked increase of testosterone in free-living males in April as breeding commenced. Both pre- and post-breeding levels of testosterone were basal. It should also be borne in mind that differences and contradictions in patterns of testosterone levels among populations may be related to seasonal changes in enzyme activity that metabolizes testosterone in the target tissues as has been shown in the Great Tit (Silverin and Deviche, 1991).

In the Common Eider, males showed clear seasonal cycles of sexual displays that were tightly correlated with high levels of testosterone in blood. However, these displays also occurred in daily cycles in phase with the tides—maximum activities at flood tide and low water. Levels of neither testosterone nor DHT showed short-term correlations with the tidal cycle (Gorman, 1977). Further, there were no changes in circulating concentrations of androgens during copulation. The author concluded that although seasonal cycles in plasma androgens undoubtedly regulate expression of sexual behavior in spring, there was no quantitative correlation between overt sexual activity and androgen concentrations. It should be pointed out, however, that plasma levels of androgens in displaying males were not compared with those of sexually inactive, or unpaired males.

Circulating levels of sex steroid hormones in females of monogamous species are, for the most part, similar to those described above for female White-crowned Sparrows (Figs 12 and 13). In Wood Ducks, circulating LH and estradiol were elevated in February prior to onset of breeding (Hepp *et al.*, 1991). Plasma levels of LH and estradiol tend to be highest during periods of yolk formation and deposition, and development of the oviduct, e.g., Brown Kiwi (Cockrem and Potter, 1991), Emperor Penguin (Groscolas, 1982; Groscolas *et al.*, 1985, 1986); Canvasback (Bluhm *et al.*, 1983); Canada Goose (Akesson and Raveling, 1981); Bar-headed Goose (Dittami, 1981); Mallard (Donham, 1979); American Kestrel, *Falco sparverius* (Rehder *et al.*, 1986); Collared Dove (Péczely and Pethes, 1979); European Starling (Dawson, 1983); European Blackbird (Schwabl *et al.*, 1980); and Rook, (Péczely and Pethes, 1982). Plasma levels of estrogens were below the sensitivity of the assay throughout the year in some of the species, e.g., Cape Cormorant (Berry *et al.*, 1979). Plasma levels of estrogens in males also tended to remain basal (Péczely and Pethes, 1979, 1982; Wingfield *et al.*, 1982a).

Progesterone levels have been measured in the blood of relatively few wild species. In the female Canvasback, circulating progesterone is high at egg-laying, and again during incubation (Bluhm *et al.*, 1983), whereas in the female European Starling, Collared Dove, and Rook, progesterone was found to be elevated only during egg-laying (Péczely and Pethes, 1979, 1982; Dawson, 1983). It has also been suggested that progesterone is involved in development of the oviduct and in ovulation (Sharp, 1980). In other species, high circulating progesterone is correlated with deferred sexual maturity and with "non-breeding" in species that reproduce biennially (e.g., *Diomedia* albatrosses, Hector *et al.*, 1986a; Hector, 1988). In male Collared Doves and Rooks, and male and female Brown Kiwi, there were no changes in plasma levels of progesterone throughout the breeding season (Péczely and Pethes, 1979, 1982; Cockrem and Potter, 1991).

With the exception of circulating LH in incubating male Emperor Penguins (Groscolas, 1982; Groscolas *et al.*, 1986) and the possible exception of plasma levels of testosterone in male European Starlings sampled in northeastern United States (Temple, 1974), there are distinct declines in plasma levels of LH, FSH, and sex steroid hormones during the parental phase of the breeding cycle in all monogamous species (Gorman, 1977; Campbell *et al.*, 1978; Wingfield and Farner, 1978a,b, 1980b; Berry *et al.*, 1979; Péczely and Pethes, 1979, 1982; Burger and Millar, 1980; Lincoln *et al.*, 1980; Schwabl *et al.*, 1980; Akesson and Raveling, 1981; Dittami, 1981; Goldsmith, 1982a; Bluhm *et al.*, 1983; John *et al.*,

1983; Röhss and Silverin, 1983; Wingfield, 1984a; Hegner and Wingfield, 1986a,b; Cockrem and Potter, 1991; Schoech *et al.*, 1991; Mays *et al.*, 1991; Fornasari *et al.*, 1992). In the double-brooded European Blackbird, as in *Z. l. pugetensis*, there were resurgences of circulating levels of LH and estrogens in females as they prepared to lay a second clutch, whereas in males there was no second increase in plasma testosterone despite a significant increase in LH (Schwabl *et al.*, 1980; see also Section III–I for full discussion of multiple brooding).

There are several exceptions to the general temporal pattern of hormone secretion cited above, and one has been studied in detail. Investigations of the Western Gull, *Larus occidentalis wymani*, on Santa Barbara Island off southern California, revealed that changes in plasma levels of LH, DHT, and testosterone were slight in both males and females with highest concentrations tending to occur just prior to egg-laying (Fig. 14, Wingfield *et al.*, 1980b, 1982a). However, these increases were not significantly higher than levels measured earlier in spring. Nevertheless, there was a significant decline of testosterone during incubation in both males and females as appears to be typical of monogamous species. Plasma levels of estrogens reached maxima coincidentally with final development of the ovary and egg-laying as described for other species. Changes in progesterone were also consistent with those of other species with a maximum during egg-laying in females, although in males progesterone titers declined steadily throughout the breeding season. Of additional interest in this species were the virtually identical concentrations of DHT and testosterone in the blood of males and females. The ratio of androgen in male plasma to that in the female exceeded 2 only slightly. In contrast, this ratio may vary between 3 and 20 in the White-crowned Sparrow (Wingfield and Farner, 1977, 1978a,b, 1980a) and other species (see Table 2).

Such specific differences in male–female androgen ratios (Table 2) may be related to differences in the degree of sexual dimorphism in reproductive behavior and possibly also in plumage. For example, the male White-crowned Sparrow establishes and defends the territory, does not incubate, but does feed young. Females also defend the territory but less so than males. Furthermore, there are well-defined dimorphisms in sexual behaviors in this species (Blanchard, 1941; Blanchard and Erickson, 1949). In the Ring Dove, males and females share incubation and feeding of young, although there is a well documented dimorphism in sexual behavior in which males and females perform distinct "bow-coo" displays (e.g., Silver, 1978; Cheng, 1979). Unlike the White-crowned Sparrow and Ring Dove, there is a marked dimorphism in plumage in the Mallard in which the male plays no part in incubation or care of young (e.g., Donham, 1979). In the Western Gull, parental duties are more or less equally divided between members of the pair (Wingfield *et al.*, 1982a), although males defend territory more than females, the differences between the sexes are less when the mate is absent at sea (Hunt *et al.*, 1984). Moreover, some females are capable of establishing their own territories.

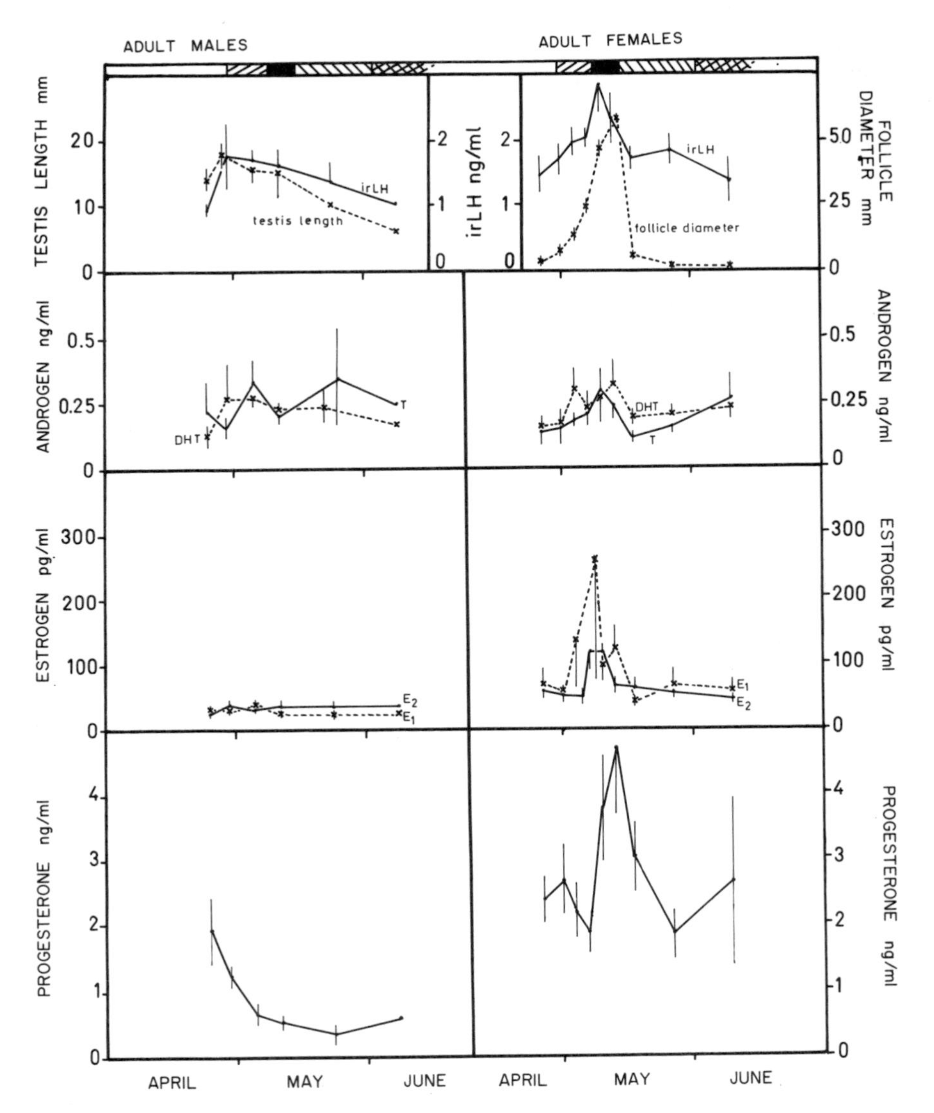


FIG. 14. Changes in length of testis, diameter of ovarian follicles and plasma levels of immunoreactive luteinizing hormone (irLH), testosterone (T), dihydrotestosterone (DHT), estrone (El), estradiol-17β (E2), and progesterone in adult male and female Western Gulls, *Larus occidentalis wymani*, during the breeding season on Santa Barbara Island. Stages in the reproductive cycle are indicated at the top of the figure. The cross hatched bar, up from left to right, represents the period of yolk deposition; solid bar, the period of egg-laying; cross hatching, down from left to right, incubation period; and the double hatching, the period during which chicks are being fed. Vertical bars are standard errors of the means. From Wingfield *et al.* (1982a), with permission.

TABLE 2

COMPARISON OF PLASMA LEVELS OF ANDROGENS IN MALES AND FEMALES DURING THE BREEDING SEASON AMONG SPECIES WITH DIFFERING LEVELS OF SEXUAL DIMORPHISM

| | Level of sexual dimorphism | | |
|---|---|---|---|
| Species | External morphological differences | Behavioral | Ratio of androgens M/F |
| *Anas platyrhynchos* | Great | Great | 2 |
| *Turdus merula* | Great | Great | 10–20 |
| *Molothrus ater* | Great | Great | 10 |
| *Passer domesticus* | Great | Great | 10–20 |
| *Agelaius phoeniceus* | Great | Great | 10–20 |
| *Ficedula hypoleuca* | Great | Great | 5 |
| *Apteryx australis* | Moderate | Great | 19 |
| *Aphelocoma coerulescens* | Moderate | Great | 13 |
| *Actitis macularia* | Slight | Great | 10 |
| *Sturnella neglecta* | Slight | Great | 5–10 |
| *Poocetes gramineus* | Slight | Great | 5 |
| *Melospiza melodia* | Slight | Great | 5–10 |
| *Parus montanus* | Slight | Great | 4 |
| *Zonotrichia leucophrys* | Slight | Great | 10–20 |
| *Aethia pusilla* | Polymorphic | Moderate | 10 |
| Phalaropus tricolor | Moderate | Moderate | 7 |
| *P. lobatus* | Moderate | Moderate | 10–20 |
| *Parabuteo unicinctus* | Moderate | Moderate | 5 |
| *Larus ridibundus* | Slight | Moderate | 15 |
| *Streptopelia risoria* | Slight | Moderate | 3–5 |
| *Anser indicus* | Slight | Moderate | 5 |
| *Rissa tridactyla* | Slight | Moderate | 5 |
| *Uria lomvia* | Slight | Moderate | 4 |
| *Calidris pusilla* | Slight | Moderate | 4 |
| *Psittacula krameri* | Slight | Moderate | 1–2 |
| *Larus occidentalis wymani* | Slight | Slight | 1–2 |
| *Sula capenis* | Slight | Slight | 1 |
| *Diomedea exulans* | Slight | Slight | 2 |
| *D. melanophris* | Slight | Slight | 1–2 |
| *D. chrysostoma* | Slight | Slight | 1–2 |

Compiled from: Beletsky *et al.* (1989, and unpublished), Cockrem and Potter (1991), Dittami (1981), Donham (1979), Dufty and Wingfield (1986a,b), Feder *et al.* (1977), Fivizzani and Oring (1986), Fivizzani *et al.* (1986), Gratto-Trevor *et al.* (1990a), Groothuis and Meeuwissen (1992), Hall (1986), Hector *et al.* (1986a,b), Hegner and Wingfield (1986a,b), Krishnaprasadan *et al.* (1988), Mays *et al.* (1991), Sailaja *et al.* (1988), Schoech *et al.* (1991), Silverin *et al.* (1986), Silverin and Wingfield (1982), Vleck *et al.* (1991), Wingfield and Farner (1977, 1978a,b, 1980b and unpublished data), J. C. Wingfield and G. L. Hunt Jr (unpublished).

There are less well-defined differences in sexual behavior between males and females. Males do mount and courtship feed more than females, although the latter do mount other females and occasionally males. There is no dimorphism in plumage; although males average larger than females, there is at least a 10% overlap in body size.

In other species of seabirds there is also very little dimorphism of plumage but the ratio of plasma testosterone in males and females ranges between 3 and 10 for the Black-legged Kittiwake, *Rissa tridactyla*, Thick-billed Murre, *Uria lomvia*, and Least Auklet, *Aethia pusilla* (Table 2). However, the latter three species were sampled on their breeding grounds on St George Island in the Bering Sea (J. C. Wingfield and G. L. Hunt Jr, unpublished), where seasonal changes in climate are marked. Most of the breeding population winters at sea well away from the breeding colony. In spring, males return to the colonies and establish a breeding territory before the females arrive. As in most colonial species, females do, however, defend the breeding territory as reflected in their significant circulating levels of testosterone.

As in many other avian groups, breeding plumage and territorial aspects of reproductive behavior of both male and female gulls appear to be under the control of androgens. Castrated male Black-headed gulls, *Larus ridibundus*, failed to develop the adult color of the bill, legs, and plumage (van Oordt and Junge, 1930). Replacement therapy with testicular tissue resulted in development of normal summer plumage in spring (van Oordt and Junge, 1933). In gonadectomized Laughing Gulls, *L. atricilla*, failure to develop adult color of head feathers, eye rings, and legs during the breeding season was reversed by injections of testosterone propionate in both sexes (Noble and Wurm, 1940). Treatment with testosterone also induced long-call vocalizations and postures common to both sexes, whereas injections of estradiol appeared to affect the characteristic sexual behavior of females only (Terkel *et al.*, 1976). In the Herring Gull, premature development of adult beak color and plumage as well as increased aggressiveness, territorial behavior and adult vocalizations were induced by androgen treatment. Estradiol was without effect (Boss and Witschi, 1941; Boss, 1943). Similarly in the Black-headed Gull, implants of testosterone into chicks induced premature darkening of the head plumage and precocial development of long call vocalizations (Groothuis and Meeuwissen, 1992). However, the plasma level of testosterone in May (period of maximum sexual behavior and mate guarding behavior) was much higher in males (Table 2, Groothuis and Meeuwissen, 1992). Since these birds were held in captivity, it is possible that male–male interactions in a confined area may have resulted in high secretion of testosterone (see also below). In a field study in Hungary, Péczely (1986) found that the ratio of testosterone in males to females was much lower (about 1) when the dark head plumage had developed and gonads were recrudescing.

A second interesting physiological feature of reproduction in the Western Gull is the lack of a marked seasonal cycle in plasma levels of LH, DHT, and testosterone (Fig. 14), despite marked changes in gonad size. Sexual maturity was attained by late April and early May and the amplitude of gonadal development was similar to that observed in the Mew Gull (Thybusch, 1965), and the California Gull, *L. californicus* (Johnston, 1956a,b). The low amplitude cycle of LH in the Western Gull, and also in the Cape Cormorant (Berry *et al.*, 1979), contrasts conspicuously with the 10- to 100-fold seasonal changes in Herring Gulls (Scanes *et al.*, 1974) and other monogamous species (see above).

The low amplitude cycles of DHT and testosterone in Western Gulls may be similar to those of the California Gull and the Fulmar Petrel since in these species the cells of Leydig were active from January to July (Marshall, 1949a; Johnston, 1956a). This suggests relatively high levels of secretory activity (and thus androgen level) in winter as is the case in Western Gulls, at least when compared with circulating concentrations during the breeding season. Western Gulls are also sedentary with many individuals remaining in the vicinity of the breeding colony throughout the year. Although they appear not to visit their breeding territories from August through early December, territorial birds are frequent on Santa Barbara Island from December through late July. In contrast, other marine species, e.g., the Cape Cormorant and Lesser Sheathbill, maintain basal levels of testosterone throughout the non-breeding season (Berry *et al.*, 1979; Burger and Millar, 1980).

Clearly the relationships among territorial and sexual behavior, and reproductive cycles, are complex and undoubtedly related directly to the mating system and breeding strategy of each species. Nevertheless, there is a distinct trend for a "monogamy pattern" with high levels of LH and testosterone in males only early in the breeding season (the Western Gull notwithstanding).

## F. CYCLES IN GONADOTROPINS AND SEX HORMONES OF POLYGAMOUS SPECIES

### 1. *Polygyny*

Plasma levels of LH and/or testosterone reached maxima in spring during the most intense period of courtship in two polygynous galliform species, the Capercaillie (Hissa *et al.*, 1983), and the wild Turkey, *Meleagris gallopavo* (Lissano and Kennamer, 1977). However, males of these species take no part in raising of the young and, after copulation, females may not cohabit with the male for the rest of the breeding season. Thus plasma levels of LH and testosterone were elevated only during the period that females were available for copulation.

In more detailed studies, Silverin and Wingfield (1982) and Silverin and Goldsmith (1983), showed that plasma levels of LH, FSH, and testosterone in polygy-

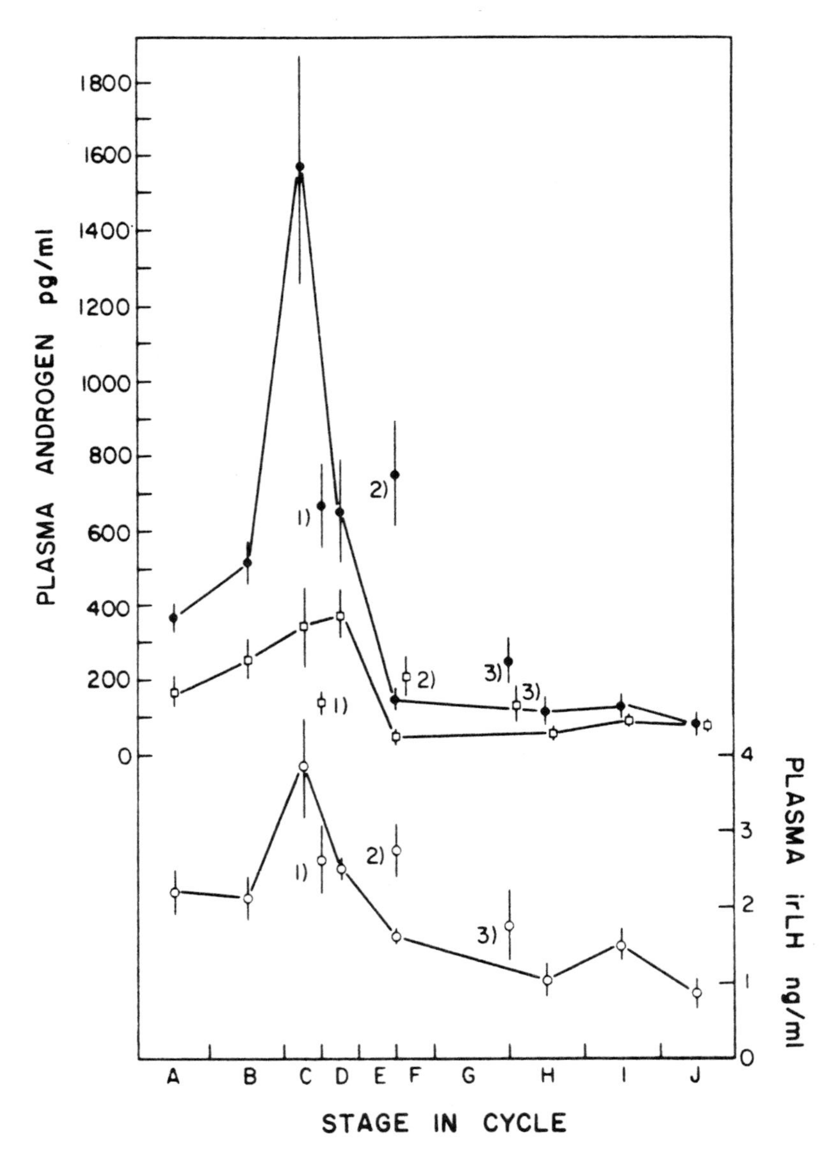


FIG. 15. Plasma levels of immunoreactive luteinizing hormone (irLH), testosterone (solid circles, and dihydrotestosterone (DHT, open squares) in male Pied Flycatchers, *Ficedula hypoleuca*, during the breeding season. Each point is the mean with standard error. Stages are as follows: A, B, territorial males; C, D, nest building; E, F, egg-laying period; G, H, incubation; I, J, feeding nestlings. (1), territorial unmated males from the average nest-building period; (2), paired males on secondary territories while primary female is laying eggs; (3), paired males on secondary territories while primary female is incubating. From Silverin and Wingfield (1982), courtesy of the Zoological Society of London.

nous Pied Flycatchers remained elevated for longer than in monogamous males in the same population (Fig. 15). Polygynous males left their first mate as soon as she began incubation and attempted to attract another female to a second nest site. Throughout this period, plasma levels of LH, FSH, and testosterone remained higher than in monogamous males. As soon as the second female was incubating, her mate abandoned her and returned to his first mate. At this time eggs of the first female were hatched and the male assisted with provisioning of nestlings. By this time plasma levels of LH and testosterone had declined to basal, as in monogamous

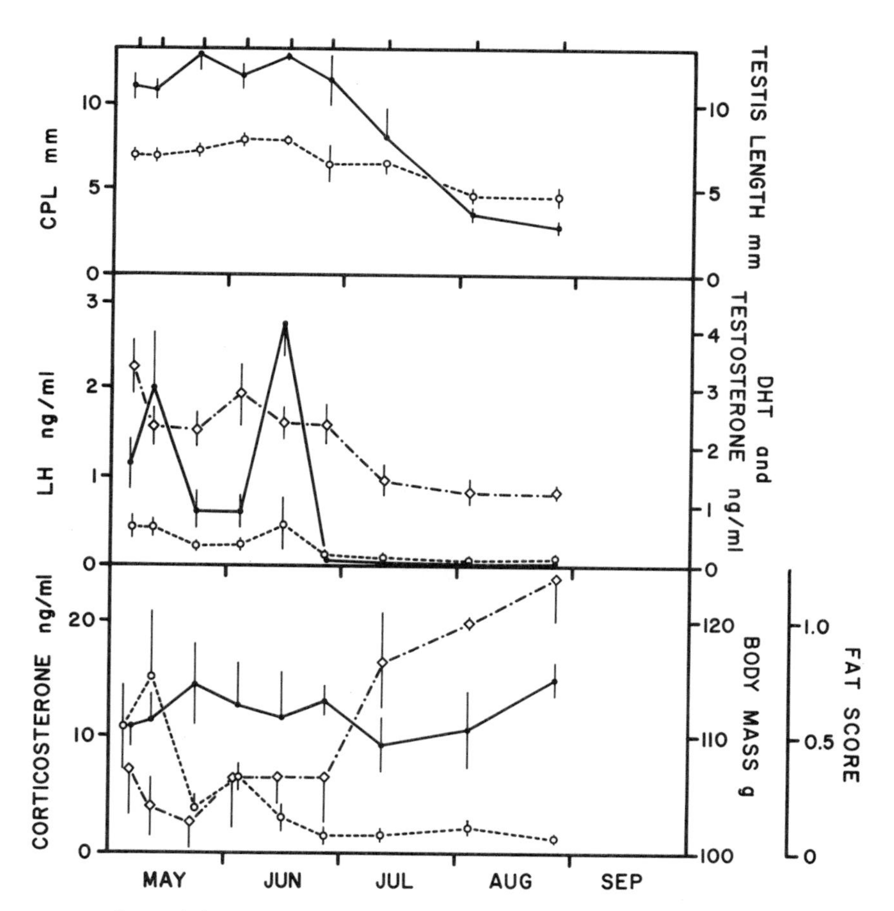


FIG. 16. Seasonal changes in (a), upper panel: testis length (solid circles), length of cloacal protuberance (CPL, open circles); (b), middle panel: plasma levels of luteinizing hormone (LH, open diamonds), testosterone (solid circles), dihydrotestosterone (DHT, open circles); and (c), lower panel: corticosterone (open circles), body mass (closed circles), and fat depot (open diamonds) in free-living populations of adult male Western Meadowlarks, *Sturnella neglecta*, sampled in southeastern Montana. Each point is the mean with standard error. From Wingfield and Farner (1980b).

males (Silverin and Wingfield, 1982; Silverin, 1983a,b; Silverin and Goldsmith, 1983).

In the red-winged blackbird, which displays simultaneous polygyny, plasma levels of testosterone remained high throughout the breeding period, or at least well into the parental phase (Beletsky *et al.*, 1989), in contrast with monogamous species in which plasma levels of testosterone declined precipitously as the parental phase began. Investigations of the partially polygynous Western Meadowlark, *Sturnella neglecta*, sampled on the high plains of southeastern Montana, showed that plasma levels of LH and testosterone were characteristically high during the

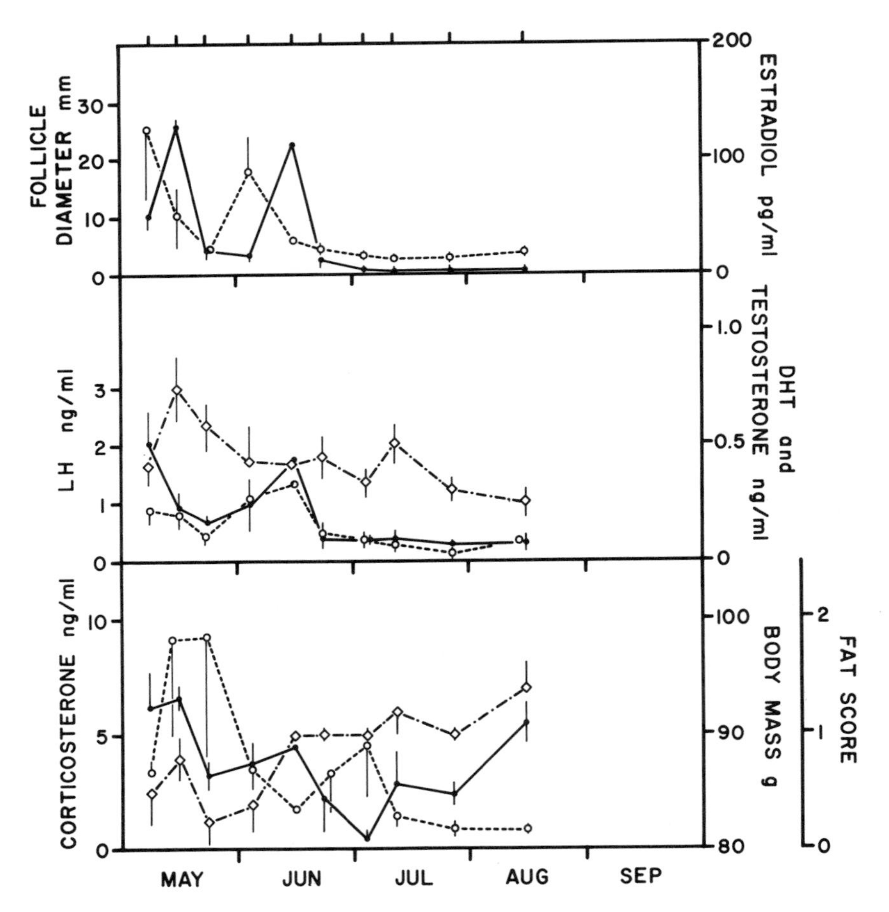


FIG. 17. Seasonal changes in (a), upper panel: ovarian follicle diameter (solid circles), plasma levels of estradiol-17β (E2 open circles); (b), middle panel: luteinizing hormone (LH, open diamonds), testosterone (solid circles), dihydrotestosterone (DHT, open circles); and (c), lower panel: corticosterone (open circles), body mass (closed circles), and fat score (open triangles) in adult female Western Meadowlarks, *Sturnella neglecta*, sampled in southeastern Montana. From Wingfield and Farner (1980b).

courtship period for the first brood and then declined as the parental phase began (Fig. 16; Wingfield and Farner, 1980b). However, unlike primarily monogamous species, male Western Meadowlarks showed a highly significant increase in LH and testosterone during the courtship phase of the second brood (Fig. 16) as they competed for females once again (Lanyon, 1957). However, in female Western Meadowlarks (Fig. 17), the cycles in plasma levels of LH, DHT, testosterone, and estradiol were essentially identical to those of females of monogamous species. In the Yellow-headed Blackbird, *Xanthocephalus xanthocephalus*, plasma levels of testosterone were high in males during the period of territory establishment and the first part of the nesting phase. Males of this species, although polygynous, feed

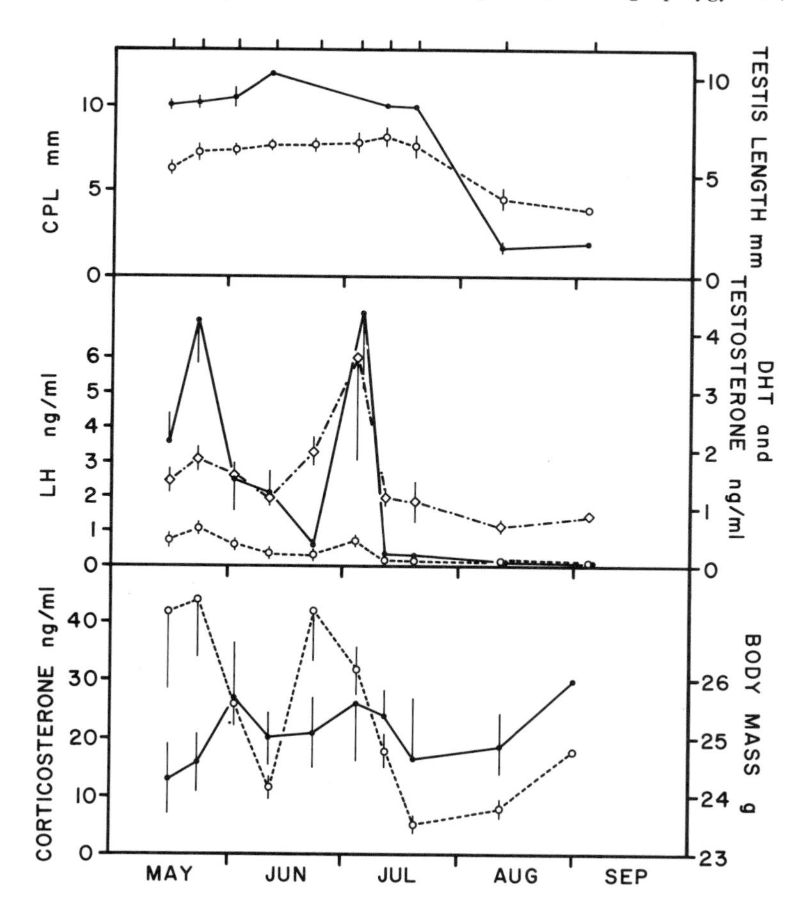


FIG. 18. Seasonal changes in (a), upper panel: testis length (solid circles), length of cloacal protuberance (CPL, open circles); (b), middle panel: plasma levels of luteinizing hormone (LH, open diamonds), testosterone (solid circles), dihydrotestosterone (DHT, open circles); and (c), lower panel: corticosterone (open circles), and body mass (solid circles), in free-living populations of adult male Vesper Sparrows, *Pooecetes gramineus*, sampled in southeastern Montana. From Wingfield and Farner (1980b).

nestlings in the nest of the primary female. By the time males were feeding young, plasma testosterone levels had declined to near basal (Beletsky *et al.*, 1990).

Thus it appears that in males of polygynous species there is a trend for plasma levels of testosterone to remain elevated for longer periods, and to overlap with the parental phase, or to attain second maxima during the sexual phase of a subsequent brood. Interestingly, cycles of LH and testosterone in male Vesper Sparrows, *Poocetes gramineus*, also attained secondary maxima coincident with the courtship phase of second broods (Fig. 18, Wingfield and Farner, 1980b). Owing to a lack of behavioral data we cannot be certain, however, that the Vesper Sparrows sampled

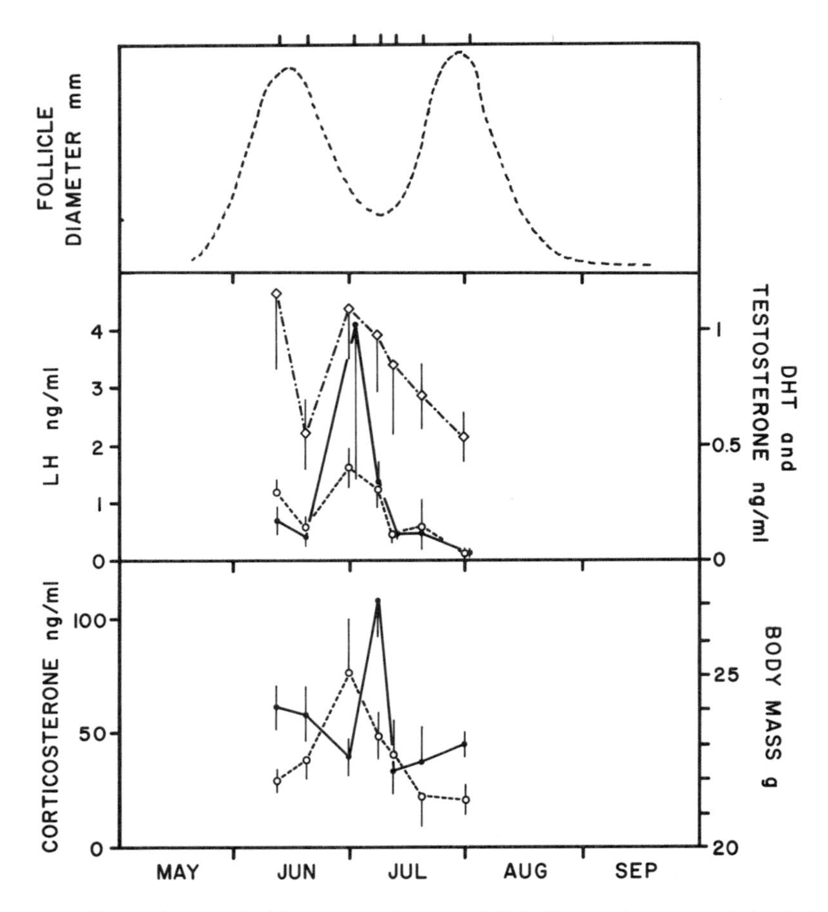


FIG. 19. Seasonal changes in (a), upper panel: ovarian follicle diameter (not measured, but deduced from breeding stage); (b), middle panel: plasma levels of luteinizing hormone (LH, open diamonds), testosterone (solid circles), dihydrotestosterone (DHT, open circles); and (c), lower panel: corticosterone (open circles), and body mass (solid circles), in free-living populations of adult female Vesper Sparrows, *Pooecetes gramineus*, sampled in southeastern Montana. From Wingfield and Farner (1980b).

in southeastern Montana were polygynous as appears to be the case in Western Meadowlarks. However, it is tempting to predict that this species was at least weakly polygynous given the temporal pattern of gonadal hormones in males. Again, females showed temporal patterns of LH and testosterone (Fig. 19; Wingfield and Farner, 1980b) that were typical of those of both polygynous and monogamous species described above.

### 2. *Polyandry*

Several recent studies have clarified the endocrine basis of polyandry in birds. Earlier investigations by Höhn and Cheng (1967) reported that in the Wilson's Phalarope, *Phalaropus tricolor*, the testosterone content of ovaries from mature females was greater than that of testicular material from mature males. Conversely, testosterone content was highest in testicular tissue of mature Killdeers, *Charadrius vociferus*, a species that does not show "sex-role reversal". However, since testosterone is a precursor of estradiol, it should not be concluded that this androgen is secreted into the peripheral blood in concentrations greater than in males. A later investigation by Rissman and Wingfield (1984) found that plasma levels of testosterone were about two- to four-fold higher in males than in females of the polyandrous Spotted Sandpiper, *Actitis macularia*. More detailed studies of this species showed that LH was high when establishing a territory and in the sexual phase and decreased during incubation in males (Oring *et al.*, 1986a). Circulating levels of testosterone were higher in males than in females during the sexual phase and decreased dramatically when incubation began. However, plasma levels of both testosterone and estradiol were higher in breeding than in non-breeding females (Fivizzani and Oring, 1986). The Wilson's Phalarope is not territorial, but none the less, plasma levels of testosterone were much higher in males than in females. Levels of sex steroids and LH declined precipitously as the parental phase began (Fivizzani *et al.*, 1986; Oring *et al.*, 1988b). In another polyandrous species, the Red-necked Phalarope, *P. lobatus*, testosterone was again highest in males, but as in Wilson's Phalarope, both progesterone and estradiol concentrations were greatest in females (Oring *et al.*, 1988b; Gratto-Trevor *et al.*, 1990a). Note also that in another shorebird species that does not show "sex-role reversal", the Semi-palmated Sandpiper, *Calidris pusilla*, testosterone levels were highest in males as expected (Gratto-Trevor *et al.*, 1990a). Hormonal correlates of polyandry in other orders of birds are virtually non-existent. However, in the Harris' Hawk, *Parabuteo unicinctus*, polyandry is not infrequent. Again, there was no sex reversal in the pattern of circulating testosterone levels (Mays *et al.*, 1991).

These data clearly indicate that there is no sex reversal of sex steroid hormone secretions in polyandrous species studied to date. Investigations cited above also suggest that in females of monogamous birds with no sex-role reversal, that testosterone may regulate aggression and development of nuptial plumage in females as well as males (see Section III-E). It is possible that there are differences in hormone secretion related to parental care (see below) since male Phalaropes provide all the parental care and females none. However, it is curious that although males have high levels of testosterone they do not develop the bright nuptial plumage (known to be androgen dependent; Johns, 1964) as do females. Recent work of Schlinger *et al.* (1989) indicated that enzymes which convert testosterone to biologically active metabolites showed sexual dimorphisms in target tissues. The skin of a dimorphically colored area of Wilson's Phalaropes (the scapular area) had higher levels of 5α- and 5ß-reductase activity in females. This may explain why females, but not males, develop a bright nuptial plumage under the influence of testosterone. Interestingly there were no sex differences in content of aromatase or 5α- and 5ß-reductases in several brain areas, but there was a dimorphism in pattern of distribution. In the anterior preoptic area and the posterior hypothalamus, aromatase activity was higher in males (Schlinger *et al.*, 1989). Thus the "sex-role reversal" seen in polyandrous shorebirds has little correlation with sex steroid hormones except in the activity of testosterone metabolizing enzymes in dimorphic areas of skin. A possible ontogenetic basis, in which sex steroids may play an important organization role for development of behavior, requires further investigation (Fivizzani *et al.*, 1990).

### G. ANDROGENS AND MATING SYSTEMS: TESTOSTERONE-INDUCED POLYGYNY IN NORMALLY MONOGAMOUS BIRDS

From the previous sections it appears that circulating levels of LH and testosterone in males of polygynous species remain elevated for longer during the reproductive cycle than in males of monogamous species. In recent experiments it has been found that if the hormonal profile of polygynous males was mimicked by administering a subcutaneous implant of testosterone to males of monogamous species, thus maintaining plasma levels of this androgen at the vernal maximum, then territorial aggression and courtship were maintained so that a substantial fraction of these males became polygynous (Wingfield, 1984c). In one experiment performed on free-living populations of White-crowned Sparrows, *Z. l. pugetensis*, known to be primarily monogamous (Blanchard, 1941; Lewis, 1975), subcutaneous implants of testosterone were given to territorial males, and empty implants were administered to a control group of territorial males. Blood samples were collected at intervals and measurements of plasma testosterone levels indicated that experimental males had significantly higher concentrations than controls during the parental phase of the reproductive cycle in May and June (Wingfield, 1984c). Unexpectedly, some of the males were found to have two or even three mates. These data suggest that there was indeed a relationship between prolonged elevation of circulating levels of testosterone in males and the incidence of polygyny. This hypothesis has been tested in greater detail in free-living populations of the

Song Sparrow in the mid-Hudson valley of New York State. This species has a natural history that is very similar to *Z. l. pugetensis*. It is monogamous, although occasional incidences of polygyny have been reported (Nice, 1943; Smith and Roff, 1980). In this investigation experimental male Song Sparrows given implants of testosterone to maintain high circulating levels throughout the breeding season showed a highly significant trend toward polygyny (Wingfield, 1984c). Of eight male Song Sparrows implanted with testosterone, six had two mates, one had a single mate, and one had three females, whereas all controls were monogamous.

Since female Song Sparrows and White-crowned Sparrows also defend territory, especially early in spring, and since these agonistic encounters usually involve interactions with other females, it is puzzling that two or even three females would tolerate each other on a territory unless the size of that territory was sufficiently large to accommodate all females. Watson and Parr (1981) have shown that an implant of testosterone given to a single male Red Grouse (this species shows varying degrees of polygyny), not only resulted in an increase in size of the territory, but also in the attraction of two mates. Similarly with the Song Sparrow, testosterone-implanted males had territories that averaged twice as large as those of controls (Wingfield, 1984c). This strengthens the intriguing possibility that experimental modification of the temporal pattern of testosterone secretion from a monogamous to a polygynous type results in a change in mating system in the same direction. Since polygynous matings in normally monogamous systems have the potential for increasing the reproductive success of males, the question arises as to why polygyny is not more widespread in both White-crowned and Song Sparrows. Clearly other restraints act to conserve monogamy in these species, although the nature of these restraints remains obscure (Wingfield, 1990a).

The defense of a larger than normal territory, and feeding young in more than one nest did not appear to be stressful to polygynous male Song Sparrows, at least as reflected by changes in body mass, fat depots, and plasma levels of corticosterone. Furthermore, over-winter survival of testosterone-implanted males was not impaired (Wingfield, 1984c). In Dark-eyed Juncos, *Junco hyemalis*, implants of testosterone early in spring decreased fat depots and body mass and increased circulating levels of corticosterone indicating a possible elevated susceptibility to stress. Implants later in the season were less effective in decreasing fat and mass, but still increased corticosterone levels (Ketterson *et al.*, 1991).

It was possible that heightened levels of aggression in testosterone-implanted males might reduce parental behavior, or even result in males driving away fledglings before they were fully independent. Support for this hypothesis comes from Silverin (1980) who administered testosterone to free-living male Pied Flycatchers during the parental phase of a breeding cycle when endogenous levels of testosterone were basal (Silverin and Wingfield, 1982). Treated males spent most of their time singing and patrolling the territory and failed to feed young thus resulting in a dramatic reduction of reproductive success. More recently it has been shown that high levels of testosterone during the parental phase decreased food provisioning of nestlings by male House Sparrows and increased aggression related to defense of a nest box (Hegner and Wingfield, 1987). Conversely, an anti-androgen, flutamide, maintained high levels of parental care by males (Hegner and Wingfield, 1987). In the Spotted Sandpiper, implants of testosterone into males during the egg-laying period dramatically reduced incubation behavior (Oring *et al.*, 1989).

If the habitat for a given breeding population were sufficiently rich so that females were able to raise young to independence without the help of a male, as appears to be the case in naturally polygynous species (Orians, 1969; Wittenberger, 1976), it is perhaps reasonable to speculate that selection would act on the neuroendocrine and endocrine systems by favoring males that tend to have protracted high levels of testosterone. The lack of parental care in these males would allow high levels of testosterone especially if male–male aggression was high. In other words, the temporal patterns of testosterone in males may be a "trade-off" between the degree of parental care by males (that would tend to depress testosterone secretion) and male–male aggression (that would result in high testosterone secretion, see Wingfield *et al.*, 1990). Another example of this is provided by a brood parasite, the brown-headed cowbird, *Molothrus ater*, that shows no parental care. Males have high levels of testosterone throughout much of the breeding season even though they are not always polygynous (Dufty and Wingfield, 1986a). Also in the House Sparrow and European Starling, there were second increases in testosterone secretion accompanying the second brood (Hegner and Wingfield, 1986a; Ball and Wingfield, 1987) similar to the polygynous Western Meadowlark (Fig. 16, Wingfield and Farner, 1980b). Further, in the polygynous Yellow-headed Blackbird and Pied Flycatcher, males showed a marked decrease in plasma levels of testosterone as parental behavior was expressed (Silverin and Wingfield, 1982; Beletsky *et al.*, 1990). Thus it appears that the relationship of the degree of male parental care and male–male aggression is a more meaningful basis for the patterns of testosterone secretion than simple mating system since males of polygynous species that show parental care, and males of monogamous species that have high levels of male–male aggression between broods, show similar patterns of testosterone secretion. The regulation of temporal patterns of hormone secretion may play a more important role in determining the nature of avian mating systems and breeding strategies than has been realized before (see Silverin, 1988; Wingfield, 1990a,b; Wingfield *et al.*, 1990a).

### H. CYCLES IN PLASMA LEVELS OF PROLACTIN

It is well known that prolactin has been implicated in the regulation of parental behavior (e.g., Lehrman, 1965; Silver, 1978; Cheng, 1979; Scanes and Harvey, 1981; Goldsmith, 1983, 1991), and also may play a role in osmoregulation especially in marine species that excrete excess sodium chloride through the nasal salt glands (e.g., Ensor, 1975; Scanes and Harvey, 1981). Until recently, however, there was no suitable assay for plasma levels of this hormone. Since this problem has been at least partly resolved, there is now a wealth of information concerning temporal patterns of plasma levels of prolactin in relation to reproductive function. Many of the detailed investigations have been conducted on domesticated or semi-domesticated species. Some of these are included in this section because of their interest to the endocrinology of wild species in general.

In altricial species, such as the European Starling, there was an increase in plasma levels of prolactin during the breeding season reaching a maximum in females during egg-laying and incubation. Prolactin concentrations were basal in June after breeding had terminated (Dawson and Goldsmith, 1982; see also Fig.

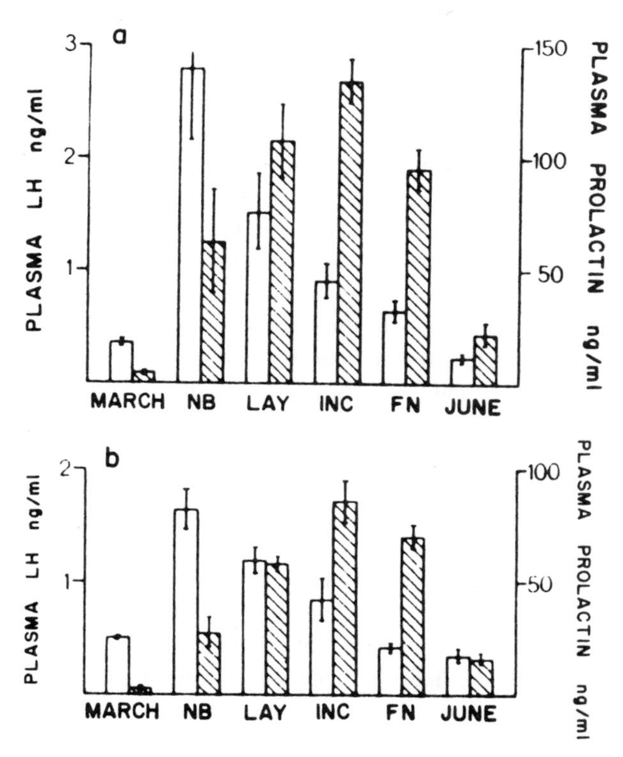


FIG. 20. (a), Changes in plasma concentrations of LH (open bars) and prolactin (cross hatched bars) in female European Starlings, *Sturnus vulgaris*, during the breeding cycle: March ($n$ = 5), nest building ($n$ = 6), laying ($n$ = 9), incubating ($n$ = 15), feeding nestlings ($n$ = 12), and after breeding in June ($n$ = 6); (b), changes in plasma concentrations of LH (open bars) and prolactin (cross hatched bars) in male European Starlings during the breeding cycle: March ($n$ = 10), nest building ($n$ = 7), laying ($n$ = 3), incubating ($n$ = 6), feeding nestlings ($n$ = 11), and after breeding in June ($n$ = 6). Each point represents the mean and standard error. From Dawson and Goldsmith (1982), with permission.

20). Males showed a very similar pattern even though they generally do not incubate (Fig. 20), nor developed a brood patch (Dawson and Goldsmith, 1982; but see also Lloyd, 1965). In Pied Flycatchers, circulating levels of prolactin increased during incubation reaching a maximum as the eggs hatched. Thereafter, prolactin declined rapidly (Silverin and Goldsmith, 1983). In males on their home territories (i.e., monogamous males) prolactin levels increased during incubation, even though males did not incubate, and declined after the young hatched. In polygynous males, however, this increase in prolactin was delayed until after the second female had begun incubating (Silverin and Goldsmith, 1983; Silverin, 1983a). Also in the Rook, Cockatiel, Dark-eyed Junco, White-crowned Sparrow, Wandering, Grey-headed and Black-browed Albatrosses, plasma levels of prolactin in males began to increase during egg-laying and were high during the parental phase and then declined sharply (Lincoln *et al.*, 1980; Hector *et al.*, 1984; Hiatt *et al.*, 1987; Myers *et al.*, 1989; Ketterson *et al.*, 1990). Note that increases may occur before egg laying as in European Kestrels (Meijer *et al.*, 1990).

In captive European Starlings, elevation of prolactin levels appeared to be at least partly induced by increasing daylength and to lag behind photoperiodically induced increases in LH. Additionally, on natural photoperiods LH began to rise after daylength had reached 11.5 h whereas prolactin levels in blood did not rise until daylength had reached 14.5 h (Ebling *et al.*, 1982). However, in addition to daylength it is quite clear that there is a close relationship between stimuli from the nest and eggs, or young, and secretion of prolactin (see also Lehrman, 1965; Dawson and Goldsmith, 1985). In the Cape Gannet, *Sula capensis*, which incubates its eggs with its feet rather than via a brood patch, prolactin levels also increased in both males and females during incubation. Further, prolactin levels were highest in breeders versus non-breeders (Hall, 1986). In Pied Flycatchers, plasma levels of prolactin remained low in females if the eggs failed to hatch. On the other hand, if the eggs hatched late following experimental switching of eggs, then prolactin levels decreased prematurely. If incubation was shortened by 2 days (by switching eggs at different stages of development) then prolactin concentrations remained high for longer (Silverin, 1983a,b; Silverin and Goldsmith, 1984). These data suggest there may be an endogenous period that regulates a high level of prolactin secretion during incubation that is only partly influenced by stimuli from the nest and eggs. When Pied Flycatchers were brooding nestlings prolactin also remained high. If nestlings at varying stages of development were switched to prolong the brooding period for 12 days, then prolactin remained higher than in controls initially, but then declined despite continued brooding behavior. Conversely, if the brooding period was shortened by substituting older nestlings, then prolactin levels decreased earlier than in controls (Silverin and Goldsmith, 1990). These data provide further support for an interaction of an endogenous period (regulated by daylength?) and cues from the eggs and young in the control of prolactin secretion related to parental care.

In species that produce precocial young, plasma levels of prolactin also increased in females during incubation, but then declined as the young hatched (e.g. in Ruffed Grouse, *Bonasa umbellus*, Etches *et al.*, 1979; Mallard, Hall and Goldsmith, 1983, see also Fig. 21; Capercaillie, Hissa *et al.*, 1983; Bar-headed Goose, Dittami, 1981; Black Swan, Goldsmith, 1982a,b; Semi-palmated Sandpiper, Gratto-Trevor *et al.*, 1990b; and Canvasback, Bluhm *et al.*, 1983). In males that also incubated, there was a parallel increase in circulating prolactin (e.g., Black Swan, Goldsmith, 1982a,b), whereas in males that did not incubate there was no increase in prolactin during incubation (e.g., Mallard, Hall and Goldsmith, 1983), or an increase late in the season that may have been concerned more with gonadal regression than parental behavior (e.g., Bar-headed Goose, Dittami, 1981). Note, that in the polyandrous Wilson's Phalarope, prolactin levels increased in males as they began incubation. The increase of prolactin secretion in females was less marked (Gratto-Trevor *et al.*, 1990b). Similarly in polyandrous Spotted Sandpipers, prolactin concentrations were greatest in the blood of incubating males compared with females and non-incubating males (Oring *et al.*, 1986b).

The mechanisms by which stimuli from the nest, eggs, and young may influence prolactin secretion appear to involve tactile and possibly thermal sensory modes. In the Mallard, Hall (1987) showed that if the brood patch was anesthetized or denervated, then prolactin levels declined within 24 h compared with controls

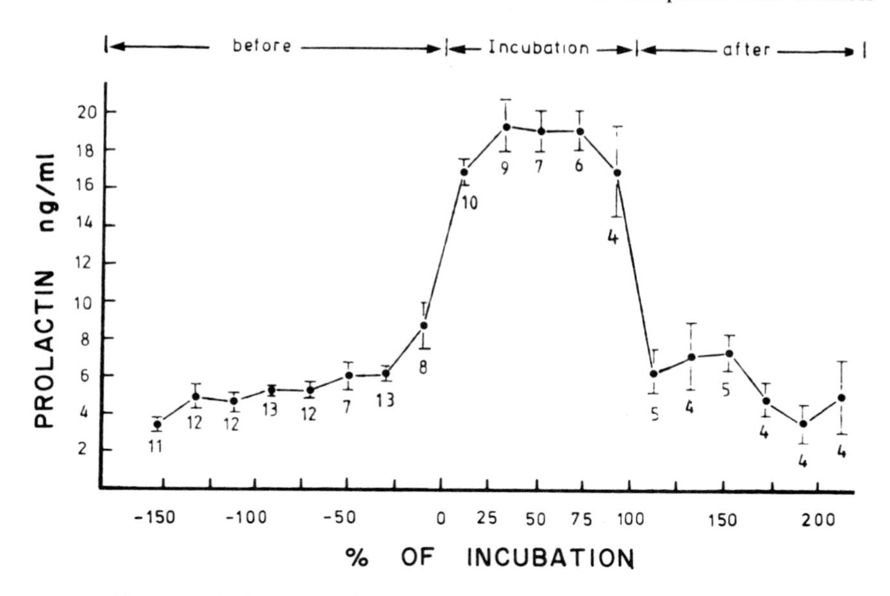


FIG. 21. Plasma prolactin concentration (mean and standard error) in breeding female Mallards, *Anas platyrhynchos*. The incubation period varied from 28 to 36 days in different individuals, the results were standardized by plotting the values in relation to length of the incubation period. Numbers of samples are shown under each point. From Hall and Goldsmith (1983), with permission.

which presumably were still receiving tactile stimuli from the eggs through the brood patch. Mallards were also given artificial eggs connected to pipes so that by pumping water of different temperatures through the eggs the clutch could be artificially heated or cooled. Both heating to 41°C or cooling to 12°C resulted in a decrease of prolactin secretion (Hall, 1987). It is also interesting to note that plasma levels of prolactin did not increase in female Canvasbacks that failed to incubate (Bluhm *et al.*, 1983) or were generally much lower in female Bar-headed Geese that were unsuccessful in breeding (Dittami, 1981). Additionally, young female geese that were not breeding or were sexually immature had much lower levels of prolactin (Dittami, 1981; Campbell *et al.*, 1981). In female Ruffed Grouse there was only a slight increase in circulating prolactin following photostimulation, but there was a dramatic further rise during ovulation and oviposition reaching a maximum when the last egg was laid and as incubation began (Etches *et al.*, 1979). Females that failed to incubate showed a dramatic decline of circulating prolactin after the last egg was laid, or if the eggs were experimentally removed, there was also a decrease in prolactin (Etches *et al.*, 1979). However, the role of stimuli from the nest and its contents in the regulation of prolactin secretion may not be universal. For example, in the Wandering Albatross, incoming adults arriving at the colony to take over an incubation bout had as high a level of prolactin as the mate that had been incubating for several days. Even if the incoming bird was experimentally prevented from touching or seeing the egg, prolactin levels were not affected (Hector and Goldsmith, 1985).

There are several interesting variants on the prolactin–parental care relationship. There may be increases in prolactin secretion in species that show no parental care. In the brood parasitic Brown-headed Cowbird, prolactin showed marked increases throughout the breeding season in both sexes (Dufty *et al.*, 1987). These data concur with an earlier investigation by Höhn (1959) who showed that pituitary prolactin content (measured by pigeon crop-sac bioassay) of Brown-headed Cowbirds was similar to that of nesting species in the same family. However, this species appears to have lost the ability to express parental care even when prolactin levels are high. Injection of prolactin into females did not activate nest-building behavior or incubation (Selander and Yang, 1966; but see also Robinson and Warner, 1964). Furthermore, treatment of either sex with prolactin alone or in combinations with estradiol and progesterone failed to stimulate development of a brood patch (Selander, 1960; Selander and Kuich, 1963).

Most avian species showed pronounced aggression and other behavior toward potential nest predators. Prolactin may also increase nest defense behavior as shown in free-living female Willow Ptarmigan following injections of ovine prolactin. Incubation behavior was not affected (Pedersen, 1989).

Of interest here are the Columbiiformes because of the secretion of "crop milk" by the specialized epithelium of the crop (crop gland), the development of which has long been known to be controlled by prolactin (Riddle *et al.*, 1935). In Ring

Doves, plasma levels of prolactin remained low through egg-laying and began to increase in mid-incubation in both males and females (both of which incubate). The crop gland of both sexes also began development at this time so that a fully functional state was attained coincident with highest concentrations of prolactin as the eggs hatched (Goldsmith *et al.*, 1981; see also Fig. 22). Thereafter, prolactin levels declined through the nestling phase. Exposure of Ring Doves to squabs (especially after nest deprivation for several days) tended to increase prolactin secretion and decrease LH. Females appeared to be more sensitive to the effects of squabs than males (Lea and Sharp, 1991). Passive immunization of Ring Doves to vasoactive intestinal peptide (a potent secretogogue of avian prolactin) decreased plasma levels of prolactin and crop sac development, but did not appear to influence incubation and brooding behavior (Lea *et al.*, 1991). However, injections of prolactin both systemically and intracranially increased feeding regurgitations and "crouching" in the nest (Buntin *et al.*, 1991). The latter authors conclude that

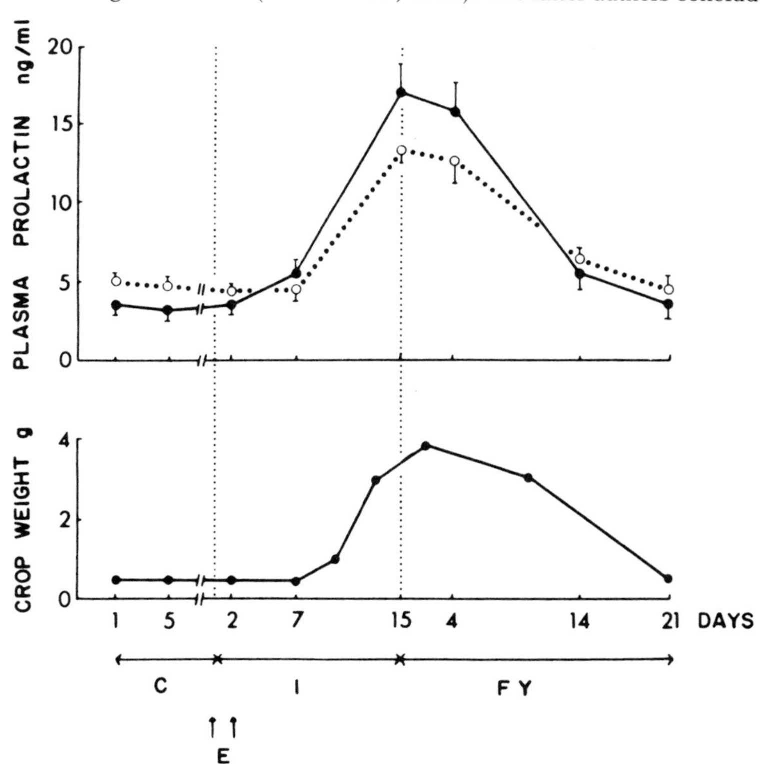


FIG. 22. Plasma concentrations of prolactin (mean and standard error) in female (solid circles) and male (open circles) Ring Doves, *Streptopelia* sp., and changes in crop weight (sexes combined) during the breeding cycle. From Goldsmith *et al.* (1981), with permission of author and *Journal of Endocrinology*.

prolactin may have facilitative actions on parental behavior by acting directly on the brain, but is not essential for activation of this behavior.

From the extensive literature cited above it is clear that cycles of prolactin secretion are complex. Although part of this complexity may be due to the diverse roles prolactin has on physiology in general, there do appear to be trends in relation to reproductive function and in particular the regulation of parental care. The relationship of prolactin and its control to the behavioral ecology of different species could be a rewarding avenue of future research (see also Goldsmith, 1983).

### I. MULTIPLE BROODS

Many species that breed at mid- to low latitudes raise more than one brood within a single breeding season (e.g., Wingfield and Farner, 1978a, 1980a). In females there were distinct cycles in plasma levels of LH and estrogens with maxima at each egg-laying period (Fig. 13). Males were somewhat more complex; despite a second maximum in circulating LH during the second egg-laying period, there was no parallel increase in circulating testosterone (Fig. 12; Wingfield and Farner, 1978a, 1980a; Schwabl *et al.*, 1980; Wingfield, 1984a). However, of the species studied—White-crowned Sparrow, Song Sparrow, and European Blackbird—males remained paired to the same mate and remained on the same territory for the second brood. Thus the high levels of testosterone associated with territory establishment and attraction of a mate apparently became unnecessary for the second brood (Wingfield and Farner, 1978a; Schwabl *et al.*, 1980; Wingfield, 1984a; Wingfield and Moore, 1987; Wingfield and Goldsmith, 1990). This can be rationalized by the possibility that high circulating levels of testosterone could result in aggression directed at fledglings that are at least partly dependent on the male for food (Nice, 1943) since females generally begin a second clutch soon after young of the first brood fledge (Wingfield and Farner, 1980a; Wingfield and Moore, 1987; Wingfield and Goldsmith, 1990). Also, high levels of testosterone may interfere with parental behavior (see Section III-G; Silverin, 1980; Hegner and Wingfield, 1987). In other species, e.g., House Sparrow, there were equal intensity peaks of LH and testosterone with each brood (Hegner and Wingfield, 1986a). In this species it appeared that males were susceptible to nest box takeover by other males at the time young fledge. Increasing levels of testosterone between broods appeared to be closely correlated with aggression related to nest box defense as well as mate-guarding. Parental behavior by the male decreased and the female provided most care for the fledged young. As a result, a second brood was not initiated until the young became independent (Hegner and Wingfield, 1986a,b). Similarly in the European Starling, there was an increase of LH and testosterone during initiation of the second clutch although the second peak of testosterone was significantly less than the first (Ball and Wingfield, 1987). Again, the second clutch was not initiated until young of the first brood were independent. Additionally, only about 30% of the population initiated a second clutch and it is thought that greatly reduced competition for nest holes (boxes) may have been the reason for a smaller second peak of testosterone (Ball and Wingfield, 1987).

The mechanisms by which testosterone levels remain low despite elevated LH are by no means clear, but an antigonadal role for prolactin has been proposed (Wingfield and Farner, 1978a, 1980a; Moore, 1983; Wingfield and Moore, 1987). After fledgling, males may feed young for a further 2 weeks, presumably under the influence of increased plasma levels of prolactin (see above), which in turn could exert an antigonadal effect (Meier and Dusseau, 1968; Camper and Burke, 1977a,b), although under certain conditions in some mammalian species, prolactin may have no anti-gonadal, or even pro-gonadal effects (e.g., Bartke *et al.*, 1977, 1982). In contrast, stimuli from the young that induce prolactin secretion in females (Lehrman, 1965; Goldsmith *et al.*, 1981; Goldsmith, 1983) would be lost at fledging. Thus the inhibition of prolactin on ovarian function (Camper and Burke, 1977a,b; Scanes and Harvey, 1981; Dorrington and Gore-Langton, 1981) would be removed and a second period of yolk deposition and egg-laying initiated. It should be noted, however, that although there was no increase in circulating levels of testosterone in males during the egg-laying period of the second clutch, these levels were not basal and were sufficient to maintain territorial and sexual behavior (Wingfield and Farner, 1978a,b, 1980a; Moore, 1983).

There is considerable evidence for an antagonistic relationship between reproduction and prolactin in birds. In multiple brooded Canaries, plasma levels of prolactin in females increased during incubation, became maximal around hatching and then declined through the nestling stage to basal values during the subsequent egg-laying phase (Goldsmith, 1982b; see also Fig. 23). These data fit well

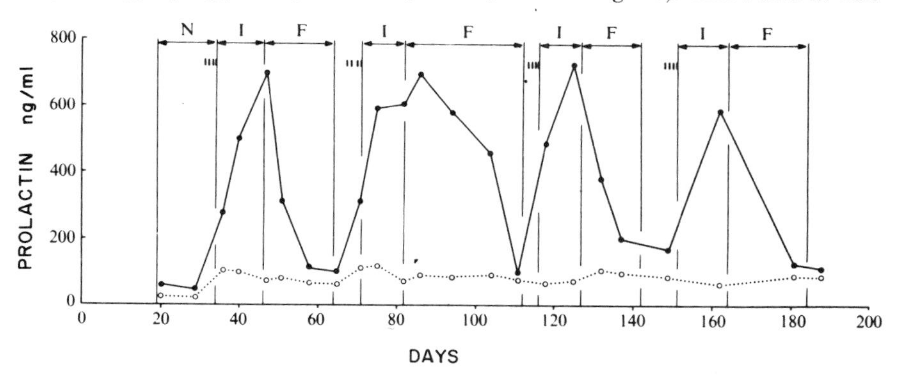


FIG. 23. Plasma concentrations of prolactin in female (solid circles), and male (open circles) Canaries, *Serinus canarius*. The vertical lines delineate the periods during which nest building (N), incubation (I) and parental feeding of young (F) occurred. The vertical bars beneath these notations indicate the days on which eggs were laid. From Goldsmith (1982b), with permission of author and *Journal of Endocrinology*.

with the general scheme for transition of behavior from the parental to sexual phase, and then again to parental behavior. However, in males, changes in prolactin during four broods were slight even though males feed young (Fig. 23). Initially, plasma levels of prolactin were low and then increased slightly during incubation of the first clutch and remained elevated through subsequent clutches. Whether or not this slight increase in circulating prolactin was sufficient to depress plasma levels of testosterone remains to be determined. In the Ring Dove, removal of the nest and eggs resulted in a marked decrease in prolactin and an elevation of plasma LH levels. Both LH and prolactin were unaffected by removal of the mate (Ramsey *et al.*, 1985). However, although infertile eggs did not maintain prolactin secretion in males, they did in females if the mate was present (Ramsey *et al.*, 1985). In other species there is strong evidence that if the nest and eggs were removed then prolactin declined and LH increased to precipitate a renest attempt (e.g., Turkey, El-Halawani *et al.*, 1980; Mallard, Hall, 1987; Canary, Goldsmith, 1982b). Additionally, in free-living Dark-eyed Juncos, removal of the male depressed prolactin levels (that could also be a result of stress of capture in addition to removal of stimuli from the nest) but had no effect of prolactin levels in the female (Ketterson *et al.*, 1990).

The data summarized above appear to support an antagonistic relationship of prolactin and reproductive hormones in the regulation of transitions between sexual and parental behavior. However, it must be pointed out that this mechanism may not be universal. In the Song Sparrow, and White-crowned Sparrow, plasma levels of prolactin increased in males and females during egg-laying and showed no major fluctuations until the end of the breeding season. In other words there were no significant correlations with multiple brooding and transitions between sexual and parental phases (Hiatt *et al.*, 1987; Wingfield and Goldsmith, 1990). In the Song Sparrow, removal of the nest to induce renesting did not result in a decline of prolactin in either males or females, despite significant resurgences of LH and sex steroid hormones (Wingfield and Goldsmith, 1990). Furthermore, implants of estradiol into females to maintain them in a sexual receptive state had no effects on prolactin levels and females continued to show parental care (Wingfield *et al.*, 1989). Male Song Sparrows mated to estrogen-treated females had higher levels of testosterone than males mated to controls, but there was no difference in circulating concentrations of prolactin (Wingfield *et al.*, 1989). Thus in this species it appears that high prolactin does not suppress secretion of sex steroid hormones and high levels of testosterone and estradiol do not inhibit prolactin secretion.

### J. CYCLES IN PLASMA LEVELS OF GROWTH HORMONE

It has now been well established that plasma levels of growth hormone are elevated in young of domesticated species (see Scanes and Harvey, 1981 for review).

However, the role of growth hormone in adults, especially in relation to reproduction is unclear. The lipolytic actions of growth hormone are perhaps of importance in adults since, in the Mallard, annual fluctuations in circulating levels of growth hormone appeared to be directly related to changes in serum levels of free-fatty acids (Assenmacher and Jallageas, 1980). In the Green-winged Teal and Canada Goose, plasma levels of growth hormone were variable but tended to be highest in July–September when birds were in molt (Scanes *et al.*, 1980; John *et al.*, 1983; see also Fig. 24). In the Canada Goose high levels of growth hormone also coincided with high serum levels of free fatty acids and possibly reflected increased metabolic requirements during molt and hypertrophy of leg muscles during this flightless period (John *et al.*, 1983).

Other investigations find different patterns. Circulating concentrations of growth

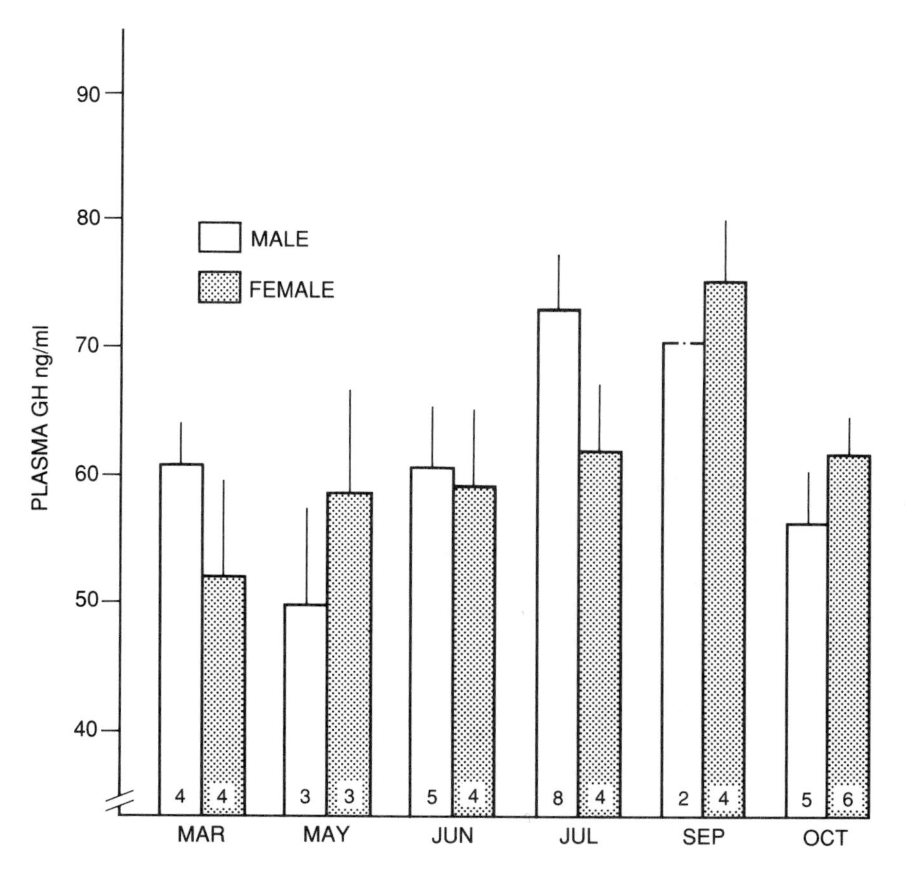


FIG. 24. Seasonal changes in levels of plasma growth hormone (GH) in the Canada Goose, *Branta canadensis*. Values are expressed as means with standard errors. Numbers within columns denote sample sizes. From John *et al.* (1983).

hormone in captive male and female Red Grouse on natural daylengths were highest in spring and early summer with lowest levels in autumn and early winter (Harvey *et al.*, 1982). Since this species is non-migratory, high vernal levels may represent increased metabolic requirements of the onset of the breeding season. In a population of Rock Ptarmigan at extreme high latitude (Svalbard), growth hormone levels were highest in winter and low in May and September, thus correlated with a period of fat mobilization (Stokkan *et al.*, 1985). Similarly in the Ring Dove, high circulating growth hormone occurred in winter when birds were held outside under ambient conditions (Lea *et al.*, 1986), whereas in the Great Tit, fluctuations in growth hormone were variable with no obvious trends (Silverin *et al.*, 1989). In contrast, growth hormone titers were greatest in winter and spring in the Willow Tit (Silverin *et al.*, 1986). The roles of growth hormone in metabolism and reproduction remain to be clarified and will await further elucidation as to its spectrum of function in adult birds.

## K. CYCLES IN PLASMA LEVELS OF THYROID HORMONES

Seasonal changes in thyroxine (T4) and tri-iodothyronine (T3) in wild species are complex, but two major trends appear to emerge. In one group of species, circulating levels of T4 and sometimes T3 increased during the breeding season reaching maxima as reproductive function was terminated, and as pre-basic molt began (e.g., Emperor Penguin, Groscolas, 1982; Groscolas and Leloup, 1986; Groscolas *et al.*, 1985; Adelie Penguin, Groscolas and Leloup, 1986; King Penguin, *Aptenodytes patagonica*, Cherel *et al.*, 1988; Ruffed Grouse, Garbutt *et al.*, 1979; Bar-headed Goose, Dittami 1981; Green-winged Teal, Scanes *et al.*, 1980; Jallageas *et al.*, 1978; Mallard, Jallageas *et al.*, 1978; Collared Dove, Péczely and Pethes, 1980; and Spotted Munia, Pathak and Chandola, 1983). However, Dittami and Hall (1983) examined T4 data for Bar-headed Geese and found an increase at the beginning of molt in males. T4 remained elevated throughout the molt. Conversely, in females, T4 decreased at onset of molt suggesting more complex mechanisms possibly at the level of thyroid hormone receptors in skin, and binding proteins in blood (see Cookson *et al.*, 1988). In a population of rooks sampled in Hungary, plasma levels of T3 reached a maximum during egg-laying in females, whereas in males maximum levels were attained during the pre-basic molt. On the other hand, T4 reached high levels during egg-laying and again during molt in females, whereas in males T4 concentrations were maximal from the courtship period through the molting phase (Péczely and Pethes, 1982). This is unlike a British population of Rooks in which plasma levels of T4 remained low throughout the reproductive period (Lincoln *et al.*, 1980). In some species there is good evidence for an anti-gonadal role for thyroid hormones in termination of the breeding season (Jallageas *et al.*, 1974; Chandola and Bhatt, 1982; Pathak and Chandola, 1983; see also Nicholls *et al.*, 1988 and Section III-R).

In other species there were no seasonal changes in circulating levels of T3, e.g., White-crowned Sparrows and House Sparrows (Smith, 1979, 1982; see also Fig. 25). On the other hand plasma levels of T4 were highest during the pre-basic molt, but did not increase until after gonadal regression had begun (Fig. 25). Similar results have been obtained in the Red Grouse (Klandorf *et al.*, 1982). In the Snow Goose, circulating T4 increased during gonadal recrudescence and declined well before onset of molt (Campbell and Leatherland, 1980).

It also appears that thyroid hormones may have a role in migratory functions in some wild species. In White-crowned Sparrows, T4 levels increased in April at the height of vernal migration, whereas there were no changes in T4 during spring in the sedentary House Sparrow (Smith, 1979, 1982; Fig. 25). Also in the Canada Goose, T4 was lowest in post-migratory period in autumn, and highest in the vernal post-migratory period. Plasma levels of T3, on the other hand, were maximal during spring migration. More detailed investigations in the Red-headed Bunting, *Emberiza bruniceps*, showed an elevation of the T3/T4 ratio prior to vernal migration (Chandola and Pathak, 1980), and thyroidectomy decreased the migratory disposition (Pathak and Chandola, 1982a), although this result should be treated with caution since thyroidectomy may result in general suppression of metabolism and activity independent of any effect on migration. Apparently T3 may be the important hormone regulating migratory behavior in the Red-headed Bunting. Furthermore, there was an increase in the activity of a peripheral deiodinating enzyme that catalyzes the conversion of T4 to T3 during migration. Pharmacological blockade of this enzyme significantly decreased pre-migratory fattening (Pathak and Chandola, 1982b).

In the Snow Goose, thyrotropin-releasing factor (TRF) increased plasma levels of T4 within 20 min of injection, and TSH injections were effective within 15 min. Generally, elevated circulating T3 lagged behind that of T4 (Campbell and Leatherland, 1979). It is possible that the role of thyroid hormones on migration may be associated with growth hormone. TRF administration also resulted in an elevation of growth hormone in the domestic fowl (Harvey *et al.*, 1978; Pethes *et al.*, 1979); also in the Green-winged Teal there was a significant correlation of plasma levels of T4 and growth hormone with highest concentrations of both hormones in August just prior to onset of autumnal migration (Scanes *et al.*, 1980).

### L. CYCLES IN PLASMA LEVELS OF ADRENOCORTICAL HORMONES

Investigations of temporal patterns of corticosteroid hormones in wild species are relatively few, possibly due to problems of sampling stress that result in rapid elevations of circulating corticosterone (e.g., Siegel, 1980; Wingfield *et al.*, 1982b; Harvey *et al.*, 1984). In European Starlings, Dawson and Howe (1983) found that following capture, plasma levels of corticosterone increased rapidly after 1 min of handling. Wingfield *et al.* (1982b) also found that plasma levels of

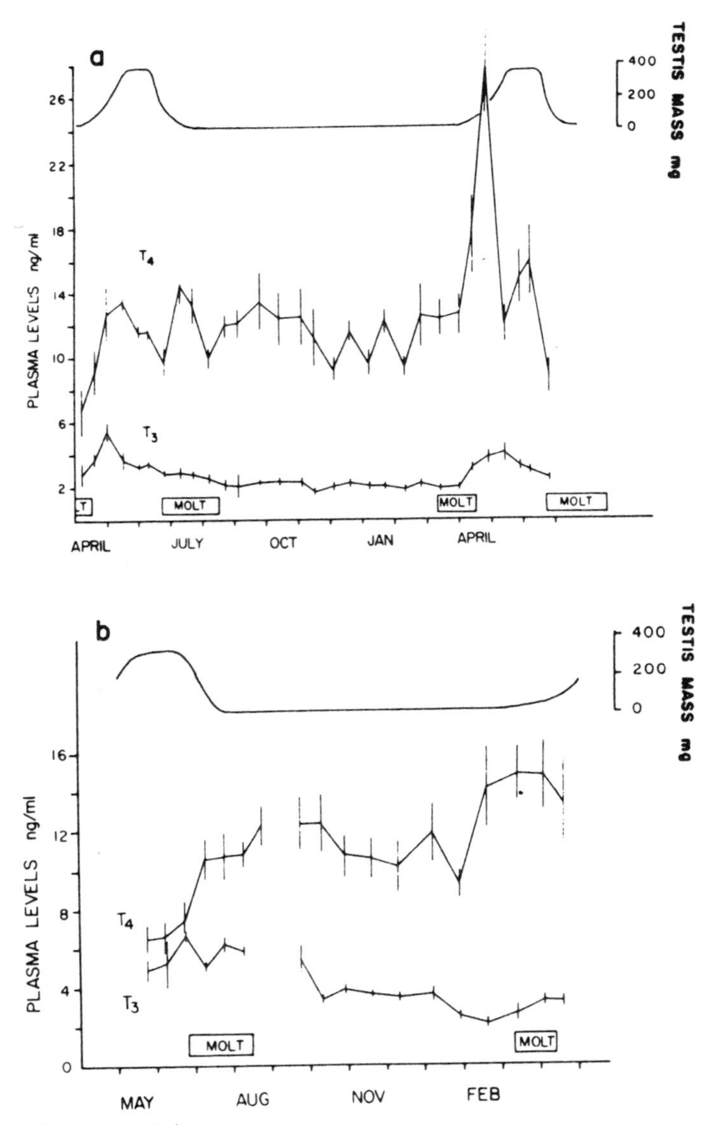


FIG. 25. (Caption opposite)

corticosterone were elevated within minutes of capture in free-living White-crowned Sparrows, but only in the non-breeding season. During the reproductive phase of the annual cycle, the increase in plasma corticosterone following capture was delayed by about 15–20 min in males, whereas in females there was no change over a period of 1 h post capture (Wingfield *et al.*, 1982b, Fig. 26). In contrast, circulating levels of LH and sex hormones were not affected or tended to decline

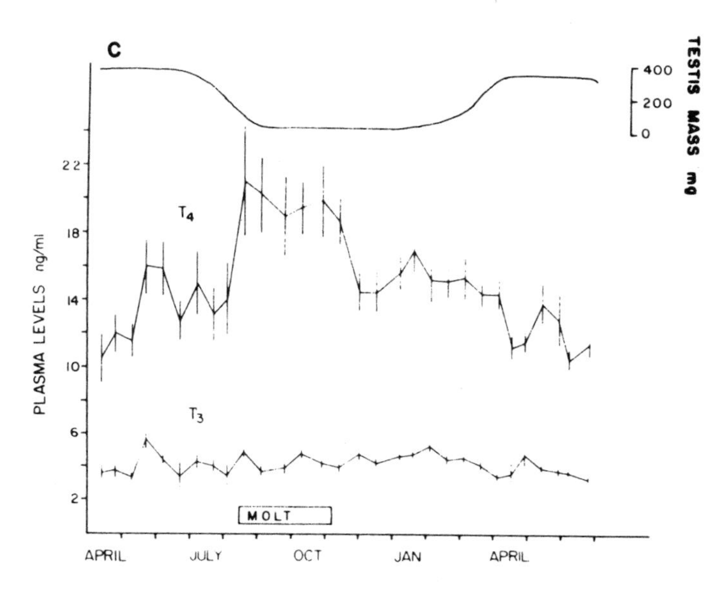


FIG. 25. (a), Plasma levels of thyroxine (T4) and triiodothyronine (T3) in captive male and female White-crowned Sparrows, *Zonotrichia leucophrys gambelii* (10 of each sex), together with estimated testicular mass and molt periods; (b), plasma levels of T4 and T3 in free-living White-crowned Sparrows (eight of each sex); (c), plasma levels of T4 and T3 in captive male and female House Sparrows, *Passer domesticus* (10 of each sex), together with estimated testicular mass and molt periods. Hormone levels are expressed as means and standard errors. From Smith (1979, 1982), with permission.

(Fig. 26). The functional significance of such changes in adrenocortical responses to acute stress remains unknown, but it has been suggested that breeding strategies in severe environments (e.g., high latitude and altitude) may provide an ecological basis (see Wingfield, 1988a,b for discussion). Nevertheless, the general rapid response of the hypothalamo-adenohyophysial-adrenocortical axis to stress requires that investigators minimize possible effects of capture stress.

Bearing these procedural problems in mind, it nevertheless appears that seasonal changes in plasma levels of corticosterone in wild species exhibit diverse temporal patterns. In White-crowned Sparrows, plasma levels of corticosterone increased gradually throughout the breeding season in male *Z.l. pugetensis*, whereas they increased abruptly in male *Z.l. gambelii* (Fig. 27). In females, high levels of plasma corticosterone were correlated with egg-laying periods (Wingfield and Farner, 1978a,b; see also Fig. 27). In male Western Meadowlarks and Vesper Sparrows, corticosterone levels were high in spring and then declined (Figs 16 and 18), whereas in females of these species highest levels were associated with egg-laying (Figs 17 and 19). Very similar patterns of corticosterone secretion have been observed in free-living European Starlings (Dawson and Howe, 1983), and Pied

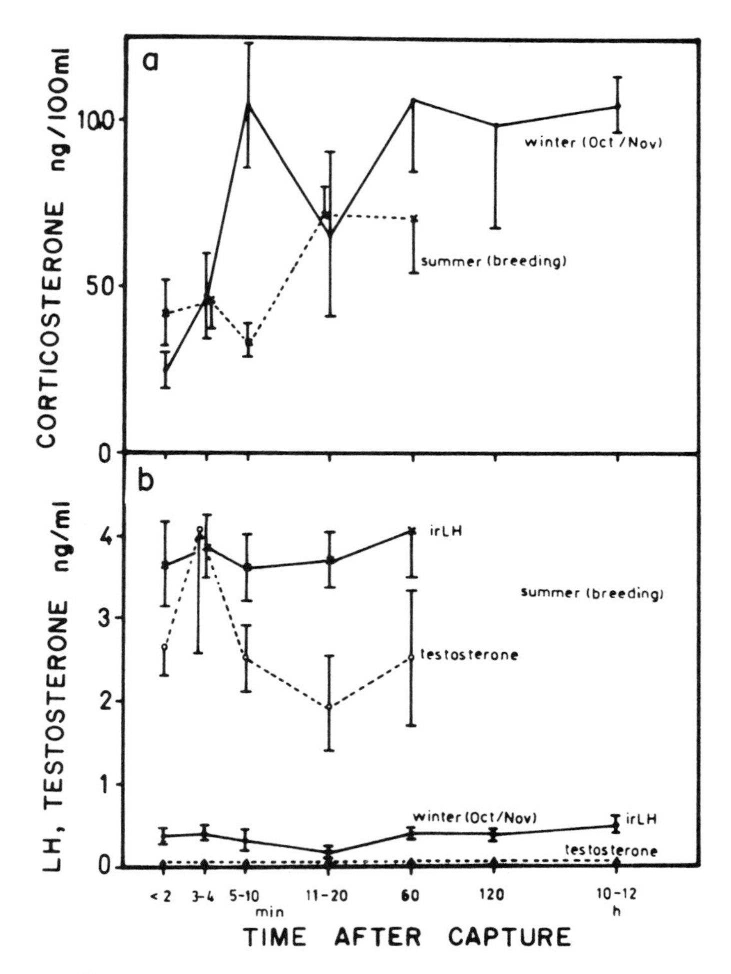


FIG. 26. (Caption opposite)

Flycatchers (Silverin and Wingfield, 1982). A common trend in most of these species was a dramatic decline to basal as the reproductive phase was terminated (e.g., Fig. 27). Changes in the capacity of corticosterone-binding globulin (CBG) are also marked. In Pied Flycatchers there was a decrease in CBG as the breeding season progressed (Silverin, 1986), whereas in contrast there was an increase in CBG in breeding White-crowned Sparrows. However, in the latter species, CBG levels dropped dramatically as breeding was terminated and pre-basic molt ensued (Wingfield and Farner, 1980b).

These data raise the possibility that changes in corticosterone secretion may promote gonadal regression, although the experimental evidence does not always support such a statement. Wilson and Follett (1975) showed that implants of

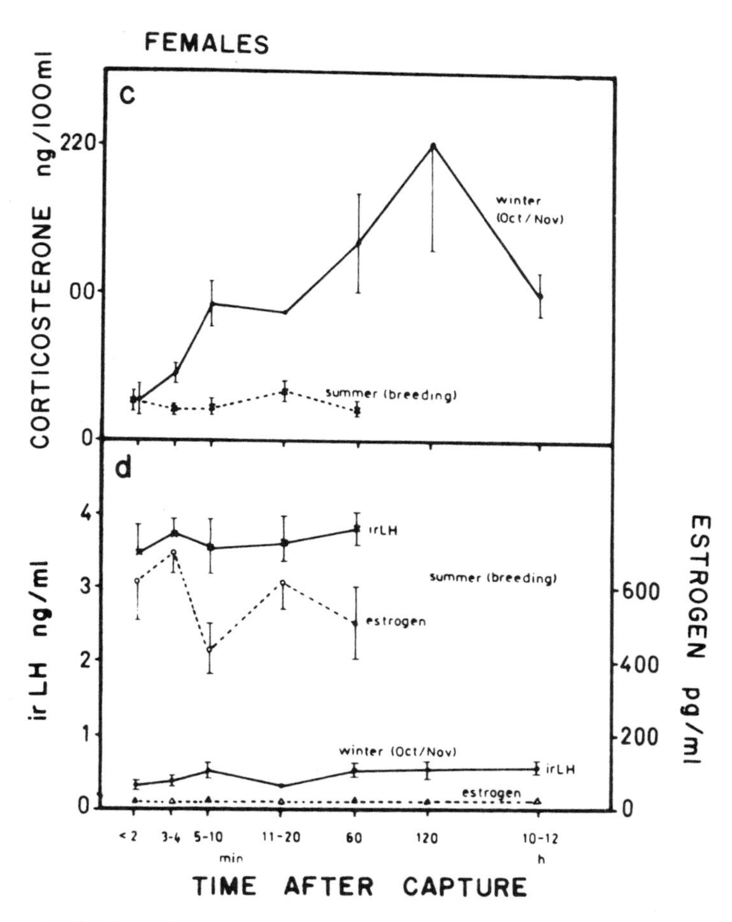


FIG. 26. (a, b), Plasma levels of corticosterone, immunoreactive luteinizing hormone (irLH), and testosterone in male White-crowned Sparrows, *Zonotrichia leucophrys*, as functions of time after capture in the field; (c, d), plasma levels of corticosterone, irLH and estrogen (estradiol-17β and estrone combined) in female White-crowned Sparrows as functions of time after capture in the field. Vertical lines represent standard errors of the means. From Wingfield *et al.* (1982b).

corticosterone placed in the basal hypothalamus of photostimulated American Tree Sparrows blocked testicular growth and depressed plasma levels of LH. Chaturvedi and Suresh (1990) showed in the Red-headed Bunting, that daily injections of corticosterone for 30 days had no effect on testis growth, but may enhance regression. On the other hand, Silverin (1979) demonstrated in Pied Flycatchers that there was an increase in adrenocortical activity during the parental phase, and when gonads regress, even though plasma levels of corticosterone declined after breeding, and were also high during autumnal migration. However, injections of

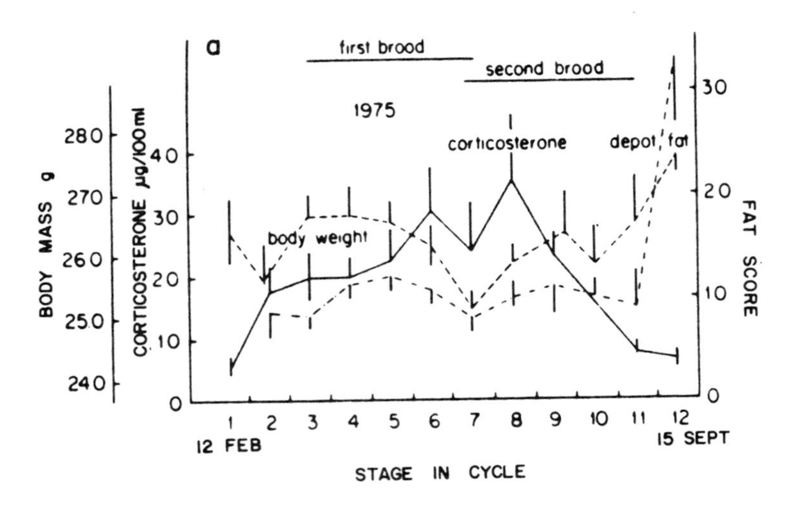


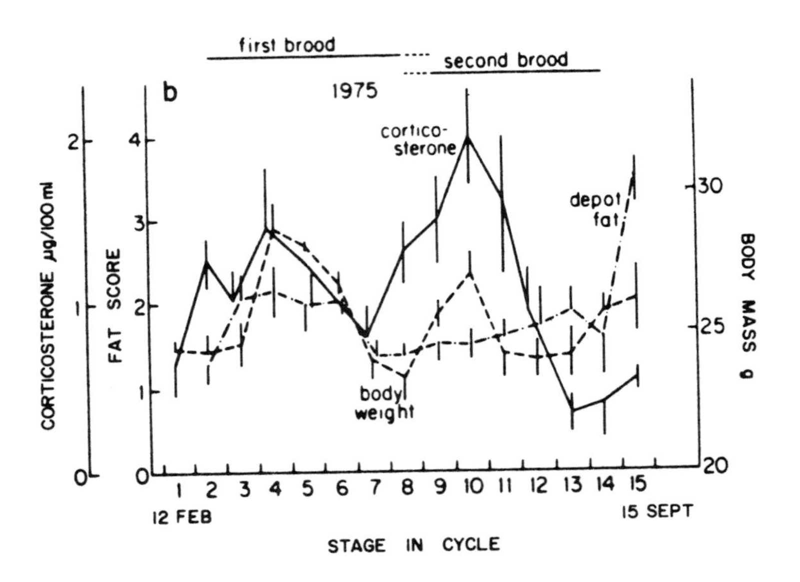


FIG. 27. Body mass, depot fat, and plasma corticosterone in free-living populations of White-crowned Sparrows, *Zonotrichia leucophrys*. (a), male and (b), female *Z.l. pugetensis*; (c), male and (d), female *Z.l. gambelii*. Vertical lines are standard errors of the means. From Wingfield and Farner (1978a,b), with permission of the Society for the Study of Reproduction and © 1978 by the University of Chicago. Stages for panel (a) are as follows: 1–3, pre-breeding; 4, courtship and egg-laying for first brood; 5–7, parental phase; 8, courtship and egg-laying for second brood; 9–10, second parental phase; 11–12, post-breeding. Stages for panel (b) are as follows: 1–2, pre-breeding; 3–5, courtship and

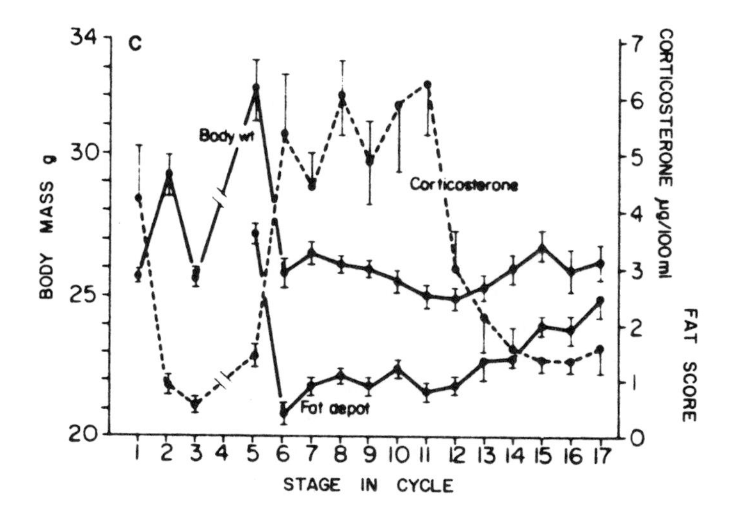


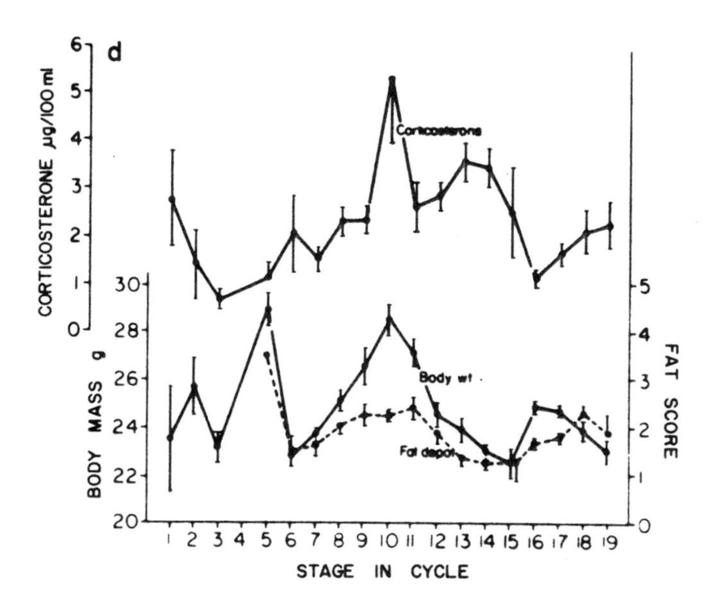


egg-laying for first brood; 6–8, parental phase; 8, courtship and egg-laying for second brood; 9–10, second parental phase; 11–12, post-breeding. Stages for panel (c) are as follows: 1–7, pre-breeding; 8, courtship and egg-laying; 9–12, parental phase; 13–17, post-breeding. Stages for panel (d) are as follows: 1–7, pre-breeding; 8–10, courtship and egg-laying; 11–14, parental phase; 15–19, post-breeding.

ACTH earlier in the season did not affect spermatogenesis, testis size, or development of the seminal vesicles (Silverin, 1979).

In the Pied Flycatcher, there was also an increase in plasma levels of corticosterone if the brood was experimentally enlarged suggesting a possible stress response since the parents must find more food for the enlarged brood. As a result, the nestlings of enlarged broods weighed significantly less at fledging than those from normal, or reduced broods (Silverin, 1982). Furthermore, secondary females, who had been abandoned by their mate, had higher circulating levels of corticosterone than primary females who were assisted by the male (Silverin, 1982). It should be noted, however, that even higher levels of corticosterone, as induced by stress, can have deleterious effects. Implants of corticosterone into Pied Flycatchers resulted in reduced feeding of nestlings and fewer fledged. Extremely high levels of corticosterone induced by open implants resulted in abandonment of the nest and territory (Silverin, 1986). Similarly in male Song Sparrows, implants of corticosterone suppressed the aggressive territorial response to a challenge (Wingfield and Silverin, 1986).

Investigations of other species have produced additional patterns of adrenocortical function. In the Canada Goose, plasma levels of corticosterone declined throughout the breeding season in immature or non-breeding males, and showed no change in immature or non-breeding females. In reproductively active geese, there were no changes in circulating corticosterone, except in females in which corticosterone levels actually declined at egg-laying (in contrast with White-crowned and Vesper Sparrows, Western Meadowlarks and others, Figs 17, 19, 27). During incubation, corticosterone levels in Canada Geese increased again only to decline once more after the eggs hatched (Akesson and Raveling, 1981). Note, adults of this species do not feed young directly but rather escort them. Since this captive experimental population was provided with food, it is perhaps not pertinent to compare seasonal changes in corticosterone with those of species sampled under free-living conditions. In the Collared Dove and Rook, plasma levels of corticosterone not only increased during the breeding season but also remained elevated throughout the pre-basic molt (Péczely and Pethes, 1980, 1982). Another variant cycle comes from Chan and Phillips (1973) who found that production of corticosterone and deoxycorticosterone *in vitro* by adrenals of Herring Gulls were highest in winter and lowest in summer. Clearly much more work is needed before these apparently diverse and confusing temporal patterns of adrenocortical activity can be evaluated further.

There is considerable evidence that adrenocortical secretions may influence migration. It has been shown for the White-crowned Sparrow that corticosterone and prolactin had marked effects on pre-migratory fattening and even the direction of migratory activity (e.g., Meier *et al.*, 1965, 1980; Meier and Ferrel, 1978). There appeared to be diurnal cycles in plasma levels of corticosterone the phase of which changed with season and with another diurnal cycle of circulating concentrations of prolactin (Dusseau and Meier, 1971). In constant light these cycles apparently disappeared so that levels became arrhythmic (Joseph and Meier, 1973). The phase angle between the diurnal rhythms of corticosterone and prolactin have been suggested to regulate migratory behavior and perhaps other phases of the annual cycle in the White-throated Sparrow (e.g., Meier *et al.*, 1980), although investigations of other species have failed to find a clear role for corticosterone and prolactin in migratory behavior e.g., the European Robin (Ieromnimon, 1977, 1978), and White-crowned Sparrow (Vleck *et al.*, 1980). In fact in the former species, prolactin may even inhibit migratory activity whereas, in agreement with the section on thyroid hormone above, small doses of TSH actually increased night activity (Ieromnimon, 1977, 1978). A re-analysis of the White-throated Sparrow data cited above raises a number of problems concerning interpretation that must be resolved before a synergistic role of corticosterone and prolactin in the regulation of migration and reproductive cycles can be clarified (Rankin, 1991).

Péczely (1976) has also presented extensive correlative evidence for a relationship between adrenocortical activity and migration. In these experiments he measured corticosteroid production *in vitro* using adrenal glands collected from wild species in the field. In non-migrants, such as the House Sparrow, Yellowhammer, Linnet, *Carduelis cannabina*, and Great Tit, corticosterone production was lowest in spring and high during the post-breeding period (unlike in the North American species described above), autumn and winter. In migratory species, such as the Stonechat, Whitethroat, *Sylvia curruca*, Red-backed Shrike, *Lanius collurio*, and Brambling, *Fringilla montifringilla*, corticosterone production was also high, but even higher levels occurred during the migratory periods. Given these correlations it would be of great interest to explore the seemingly complex relationship of corticosterone, and other hormones such as T3 and T4, and growth hormone (see above) with migration.

## M. NON-PHOTOPERIODIC ENVIRONMENTAL INFORMATION AND REGULATION OF REPRODUCTIVE CYCLES

The temporal patterns of circulating hormone levels in free-living populations are often markedly different from those obtained under artificial experimental conditions in the laboratory. In the Song Sparrow, the temporal patterns in plasma levels of LH and testosterone throughout the reproductive cycle in the field differed markedly from those in the laboratory (Fig. 28, and Wingfield, 1984a,c; Wingfield and Moore, 1987). Under natural conditions, males arrived in the breeding area in March followed 1–2 weeks later by females. The establishment of territory and attraction of mate were characterized by high circulating levels of LH and testosterone whereas testis mass was low and growth of the cloacal protuberance (CPL) was just beginning. Curiously, plasma levels of LH and testosterone increased for a

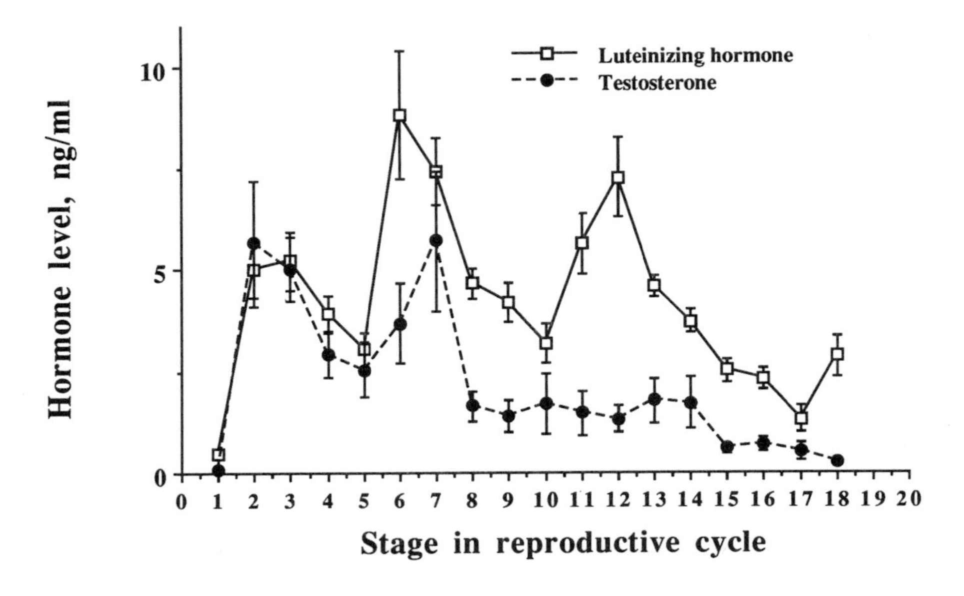


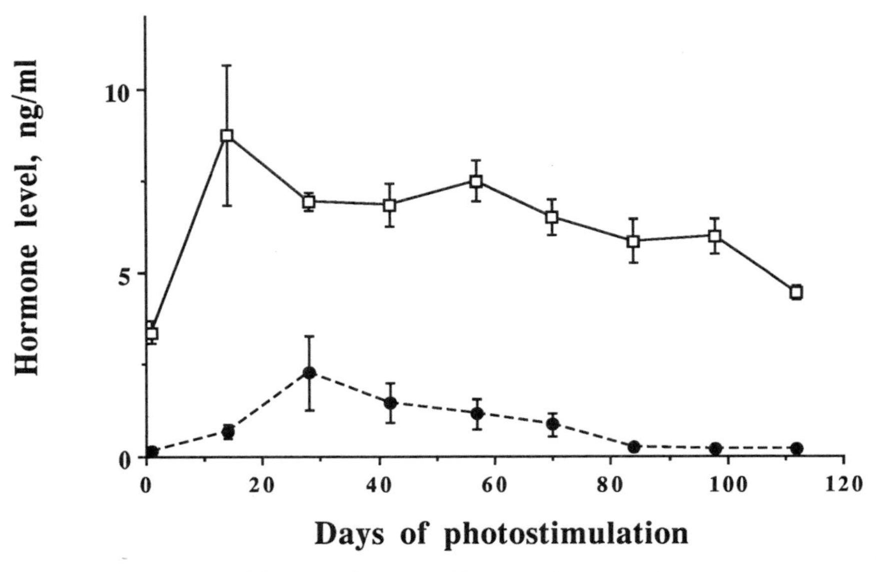


FIG. 28. A comparison of the temporal patterns of luteinizing hormone and testosterone levels in the plasma of free-living (top panel) and captive Song Sparrows, *Zonotrichia melodia* exposed to long days (bottom panel). From Wingfield, 1984a, with permission, and submitted. Stages in cycle on top panel are as follows: 1–3, territory establishment and pairing; 4–7, courtship and egg-laying for first brood; 8–10, parental phase; 11–12, courtship and egg-laying for the second brood; 13–15, second parental phase; 16–18, post-breeding.

second time, however, in late April and early May coincident with the egg-laying period. At this time males "guard" their sexually receptive mates. Territory establishment and pair formation occurred more than a month before the nesting phase began and plasma levels of LH and testosterone declined in the interim. In April, testes and CPL developed rapidly to a maximum in early May when females were laying eggs and when most copulations occurred. Agonistic interactions over territories were much less intense at this time (Nice, 1943; J. C. Wingfield unpublished). As soon as females were incubating the first clutch, circulating levels of LH and testosterone in males declined, although not to basal, remaining at a lower

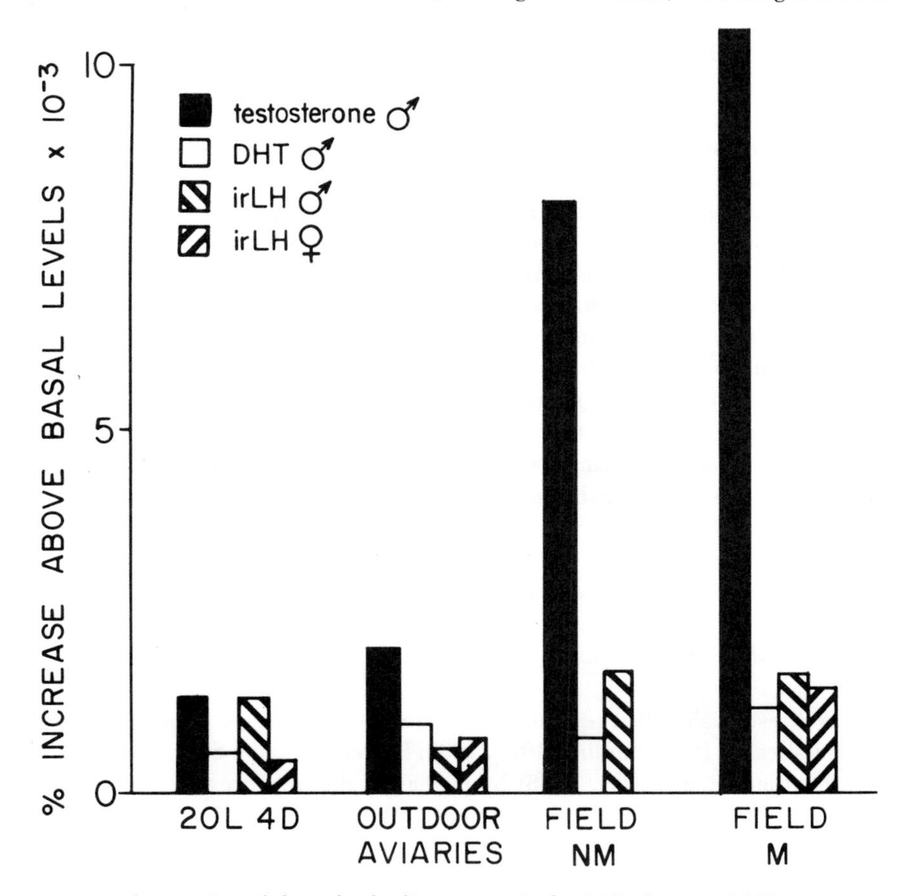


FIG. 29. A comparison of plasma levels of immunoreactive luteinizing hormone (irLH), testosterone and dihydrotestosterone (DHT) in the White-crowned Sparrows, *Zonotrichia leucophrys gambelii*, sampled in captivity and the field. Data are presented as the percentage increase from basal values to the maximum level measured under each treatment. With birds sampled in the field, (M) denotes mated and territorial; and (NM) denotes unmated birds that are establishing a territory. From Wingfield and Farner (1980a,c), with permission, S. Karger AG, Basel.

level throughout the rest of the breeding cycle (Fig. 28). This pattern is typical of males of monogamous species that provide parental care (see also Section III-E).

Clearly, there is a strong positive correlation of increased aggression and elevated levels of LH and testosterone in male Song Sparrows in agreement with laboratory studies that demonstrate a role for these hormones in the regulation of agonistic behavior (e.g., Adkins, 1977; Murton and Westwood, 1977; Adkins and Pniewski, 1978; Searcy and Wingfield, 1980; Adkins-Regan, 1981; Harding, 1981, 1983; Balthazart, 1983). Generally, expression of aggression is at a low frequency when males are providing parental care and plasma levels of LH and testosterone are correspondingly low (Fig. 28). These temporal patterns of hormone secretion and behaviors differed from those induced by long days in the laboratory (Fig. 28). In the laboratory, circulating LH and testosterone had a single maximum during testis maturation and then declined. Aggressive behavior and sexual behavior were expressed under captive conditions but not in the predictable patterns seen in the field. These data indicate that environmental information in addition to daylength may play an important role in regulation of the reproductive cycle.

In addition to differences in temporal patterns of hormone secretion between laboratory and field investigations, there were major disparities in the absolute circulating concentrations attained (Fig. 29, Wingfield and Farner, 1980a,c). The maximum levels of testosterone reached in the field were almost 10-fold higher than those attained in the laboratory, suggesting further that environmental factors other than daylength have pronounced influences on endocrine secretion. As discussed earlier, these factors may include weather, behavioral interactions, availability of food and many others. Below some of the better characterized responses to non-photoperiodic environmental information are outlined.

## N. BEHAVIORAL INTERACTIONS AND REGULATION OF REPRODUCTIVE HORMONES

### *1. Male–Male Interactions*

It is well established that testosterone can regulate aggressive behavior in birds (Adkins-Regan, 1981; Harding, 1981, 1983; Balthazart, 1983; Silver *et al.*, 1973; Wingfield and Ramenofsky, 1985) by increasing the frequency and intensity of aggressive displays (Selinger and Bermant, 1967; Adkins and Pniewski, 1978; Searcy and Wingfield, 1980). In some cases, treatment with testosterone not only increased aggressive behavior but also resulted in an elevation of social status that persisted after testosterone administration ceased (e.g., *Z. l. gambelii*—Baptista *et al.*, 1987). In other species, testosterone raised the frequency of aggression but did not result in a change in dominance status (e.g., Ramenofsky, 1982, 1984). The effects of testosterone on aggression appear to be restricted to reproductive contexts as has been shown in species that show territorial behavior year round. In

European Robins and Northern Mockingbirds, *Mimus polyglottos*, autumnal territorial behavior was not influenced by testosterone (Logan and Carlin, 1991; Schwabl and Kriner, 1991).

Seasonal changes of plasma testosterone levels also correlate positively with periods of heightened aggression (e.g., Wingfield and Farner, 1977, 1978a,b). In Fig. 28 it can be seen that circulating levels of testosterone were high in late March and early April, and again in early May coincident with two periods of intense aggression. However, in April, plasma levels of LH and testosterone declined despite the fact that daylength was increasing rapidly at this time. These data indicate very strongly that environmental information, in addition to the stimulatory effects of daylength, is important for regulation of plasma levels of LH and testosterone. It has been suggested that the agonistic interactions among males as they compete for territories and mates may act as synchronizing and integrating information (Wingfield, 1980b, 1983; Wingfield *et al.*, 1987; Wingfield and Marler, 1988; Wingfield and Kenagy, 1991) to stimulate secretion of testosterone (Harding, 1981). On the other hand, many more investigations have failed to confirm this relationship (e.g., Lumia, 1972; Balthazart *et al.*, 1979; Harding and Follett, 1979; Tsutsui and Ishii, 1981; Ishii and Tsutsui, 1982; Mench and

TABLE 3

CORRELATIONS OF CIRCULATING CONCENTRATIONS OF TESTOSTERONE WITH DOMINANCE STATUS OR LEVEL OF TERRITORIAL AGGRESSION

| Species | Relationship positive (+) none (-) | Reference |
|---|---|---|
| *Anser indicus* | +/- | Dittami and Reyer (1984) |
| *Anas platyrhynchos* | + | Balthazart and Hendrick (1976) |
| *Coturnix* sp. | - | Balthazart *et al.* (1979) |
| | - | Tsutsui and Ishii (1981) |
| | + | Ramenofsky (1984) |
| *Larus occidentalis* | - | Wingfield *et al.* (1982a) |
| *Agelaius phoeniceus* | - | Harding and Follett (1979) |
| | + | Beletsky *et al.* (1989) |
| *Xanthocephalus xanthocephalus* | +/- | Beletsky *et al.* (1990) |
| *Molothrus ater* | - | Dufty and Wingfield (1986a) |
| | + | Dufty and Wingfield (1986c) |
| *Zonotrichia leucophrys* | - | Wingfield *et al.* (1982b) |
| | + | Wingfield (1985b) |
| *Z. querula* | - | Rohwer and Wingfield (1981) |
| *Z. albicollis* | - | Schwabl *et al.* (1988) |
| | + | Schlinger (1987) |
| *Melospiza melodia* | + | Wingfield (1984b) |
| *Passer domesticus* | + | Hegner and Wingfield (1987) |
| *Sturnus vulgaris* | + | Ball and Wingfield (1987) |

Ottinger, 1991). These conflicting studies are summarized in Table 3. Some field studies have also produced results to the contrary. Rohwer and Wingfield (1981) found that plasma levels of testosterone and position in a dominance hierarchy were not correlated in flocks of Harris' Sparrows, *Zonotrichia querula*. Furthermore, Wingfield *et al.* (1980b, 1982a) could find no consistent seasonal changes in circulating testosterone in relation to aggressive territorial behavior in Western Gulls.

There are several hypotheses that attempt to explain this disparity (e.g., Leshner, 1978), but none give satisfactory discussions of the good correlation of testosterone and aggression in some species, and the lack thereof in others. Possibilities are social inertia (Guhl, 1964) or other forms of individual recognition (Rohwer, 1981; Boag, 1982; I.D. Chase, 1982) that may maintain social relationships in stable situations and independently of hormonal regulation (Ramenofsky, 1982, 1984; Mench and Ottinger, 1991). Indeed, most of the investigations that failed to correlate plasma levels of testosterone and aggressiveness were conducted on individuals in established and stable groups. Even in the field, flocks of Harris' Sparrows were often stable for most of the winter period (Rohwer and Wingfield, 1981); and the Western Gull usually pairs for life and is territorial for prolonged periods before and after breeding (Wingfield *et al.*, 1980b, 1982a; Hunt *et al.*, 1980). On the other hand, investigations that revealed positive correlations between testosterone and aggressiveness, involved some form of social "challenge" to the individual or group, or were conducted during the establishment of agonistic relationships. In some of these studies a novel individual, or individuals, were introduced to a stable group. In field studies, birds were sampled as they arrived in the breeding area and established territories. It is thus possible that testosterone increases the frequency and intensity of aggression when stable groups or relationships are challenged, and that there may often be an increase in testosterone levels of all individuals involved regardless of position in a hierarchy, although in most cases dominants tend to have higher levels of testosterone. In contrast, once groups become stable, or relationships are established, the correlation with testosterone is lost and relationships are maintained by social inertia or similar mechanisms (Schuurman, 1980; Ramenofsky, 1982, 1984; Wingfield *et al.*, 1987).

This point is illustrated further in an experiment in which castrated adult White-crowned Sparrows were given implants of testosterone, or control implants, and aggressive behavior monitored (Fig. 30). Over a period of 14 days after implantation there were no differences in frequency of total aggressive behaviors exhibited by either group, despite the wide differences in circulating levels of testosterone (Wingfield, 1985b, 1988a). However, since these males had been housed together for over 6 months it was probable that social relationships had been established for some time. If this stable social system was then disrupted by placing a novel male in a cage alongside each of the subject's cages, there was an immediate increase in total aggression and males with high levels of testosterone demonstrated more

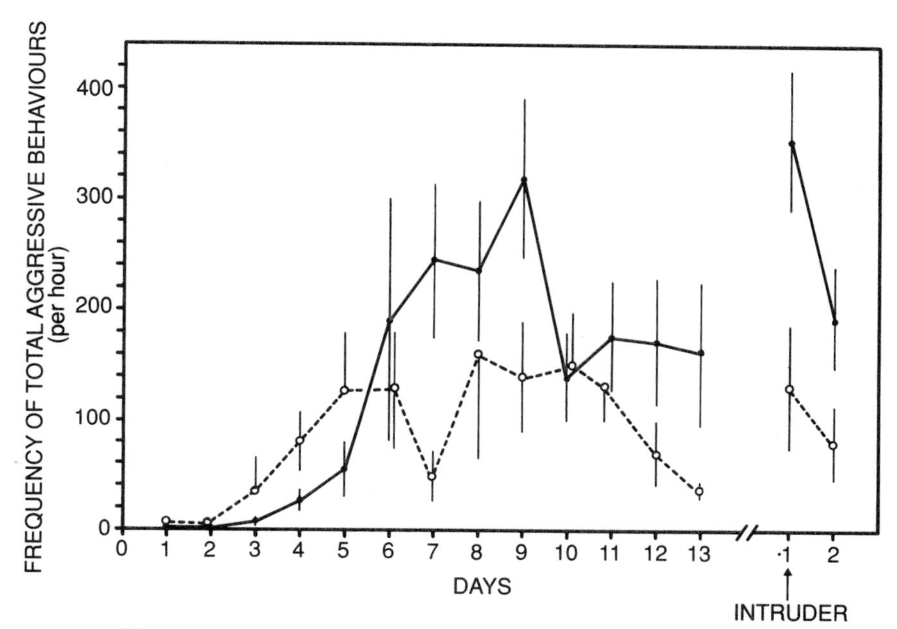


FIG. 30. Effects of testosterone on total aggressive behaviors in castrated and photostimulated White-crowned Sparrows, *Zonotrichia leucophrys gambelii*. Testosterone-implanted birds, solid line; control-implanted birds, broken line. After 13 days of implantation, a novel male White-crowned Sparrow (intruder) was introduced into the chamber and behavior recorded for a further 2 days. Vertical bars are standard errors of the means. From Wingfield (1985b), with permission of Springer-Verlag.

intense aggressive behavior than controls (Wingfield, 1985b). Thus, following a simulated "challenge" by introducing a novel male, there was a clear relationship between level of circulating testosterone and intensity of aggressive behavior (see also Wingfield *et al.*, 1987).

Given this plasticity of testosterone secretion, what are the environmental stimuli for increased secretion of testosterone above that induced simply by exposure to long days of spring? Since increased plasma levels of LH and testosterone coincide with establishment of a breeding territory, it has been proposed that stimuli associated with gaining and defending a territory may induce secretion of these hormones. To test this hypothesis, adult male Song Sparrows were removed from their territories and the replacement males sampled as they claimed vacant territories. It was also possible to sample the neighboring males as they established a new territory boundary with the replacement male. Plasma levels of testosterone were higher in the replacement males than in those with established territories in control areas (Fig. 31). Furthermore, circulating testosterone was also higher in neighboring males that had already occupied a territory, but were establishing a boundary with the new male (Wingfield, 1984a, 1985a). These data suggested that the territory *per se* was not the stimulus for increased secretion of testosterone, but

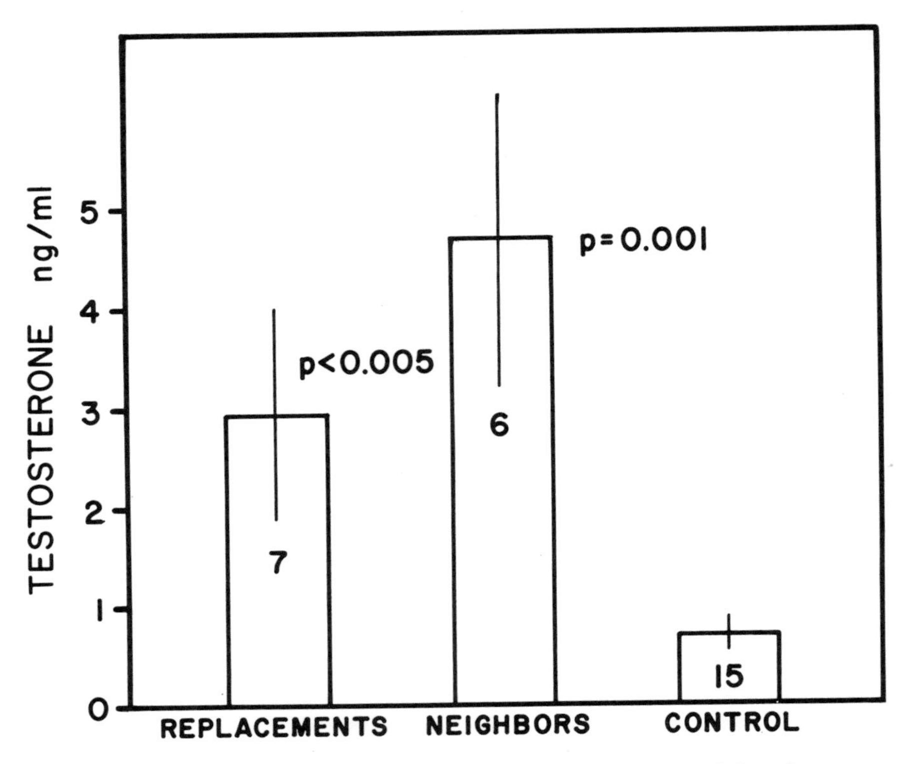


FIG. 31. Plasma levels of testosterone in male Song Sparrows, *Melospiza melodia*, taking over a territory (replacement) after removal of the resident, compared with neighbors and territorial control males. Vertical bars are standard errors and numbers within bars are sample sizes. From Wingfield (1985a), with permission.

rather that stimuli emanating from agonistic interactions as boundaries were formed appear to be more important.

To test this hypothesis further, territorial intrusions were simulated by placing a caged male in the center of the territory of a free-living male and by playing tape-recorded song through a speaker placed alongside the cage. Territorial male Song Sparrows responded vigorously to the apparent intruder and attempted to drive him away (see also Searcy *et al.*, 1981 for details of the experimental procedure). In control males, sampled as they foraged, there was no relationship between day and hormone titer, thus eliminating any possible confusion in the results owing to diel rhythms in plasma hormone levels (Wingfield, 1985a). An analysis of plasma levels of LH and testosterone against time after onset of simulated intrusion revealed a significant increase in LH by about 1 h, but no such relationship with testosterone was apparent. However, the mean testosterone level of all males exposed to simulated territorial intrusion was significantly higher than in control males (Fig. 32, Wingfield, 1985a).

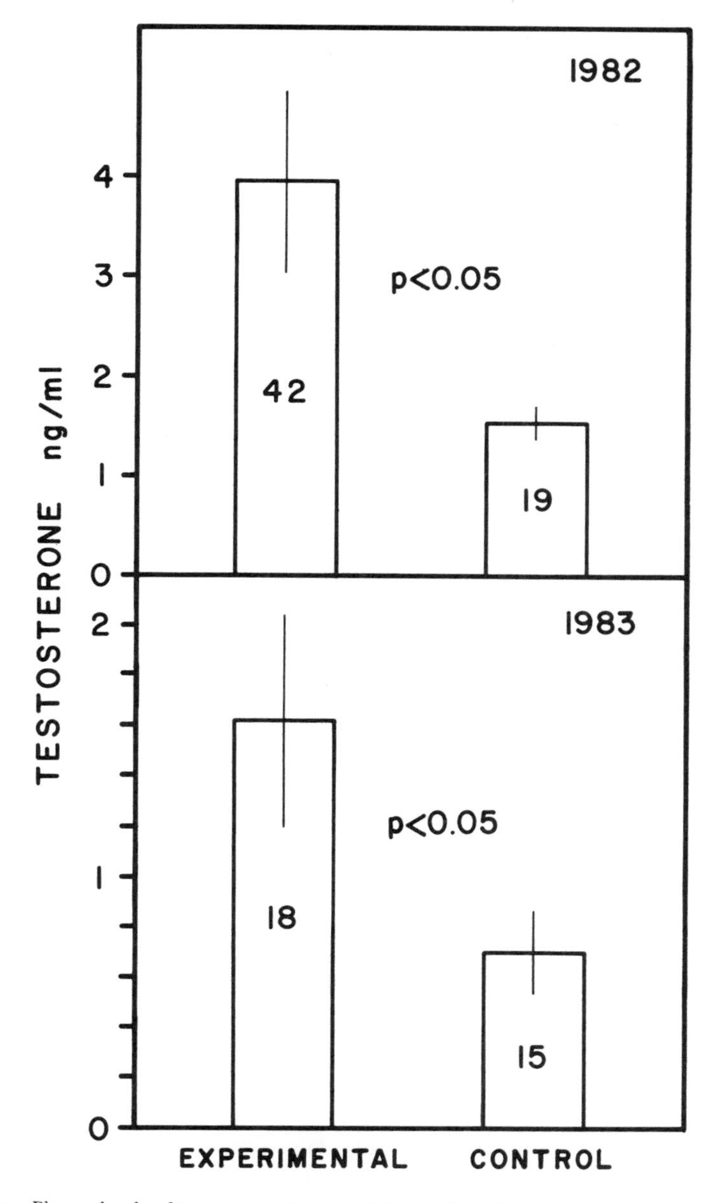


FIG. 32. Plasma levels of testosterone in territorial male Song Sparrows, *Melospiza melodia*, after exposure to a simulated territorial intrusion (experimental) and controls. Vertical bars are standard errors of the means and numbers within bars are sample sizes. From Wingfield (1985a), with permission.

The lack of a correlation of testosterone with time after onset of intrusion may be due to effects of other stimuli such as sexually receptive females, or presence of young. Thus some males may have had high levels and others low levels of testosterone just prior to onset of simulated intrusion which may have introduced variability into the results. Nevertheless, the significant correlation between LH and time after intrusion, and the overall higher levels of testosterone in males responding to simulated intruders suggested that stimuli from an aggressive male induced secretion of LH and testosterone within 1 h. To test this interpretation, the experiment was repeated in April when all males were in the same stage of the breeding cycle. Once again there was no relationship between circulating testosterone level or even LH and time after onset of simulated territorial intrusion, although mean levels of testosterone in all males exposed to intrusions were again higher than those of controls (Fig. 32, Wingfield, 1985a). This was confirmed by Wingfield and Wada (1989) using both captive and free-living male Song Sparrows exposed to a simulated challenge. Furthermore, they found that plasma levels of LH and testosterone did not increase until at least 10 min after onset of intrusion.

In a similar experiment, Harding and Follett (1979) also could find no consistent increase in plasma levels of DHT, testosterone and LH after simulated territorial intrusion in male Red-winged Blackbirds. However, a later analysis indicated that testosterone levels in males that attacked the intruder more aggressively were higher than those of males that approached the challenge indirectly (Harding, 1981). In sum, these data suggest that testosterone levels do increase in response to agonistic interactions, although the mechanisms involved require more research (see Wingfield *et al.*, 1987).

A further experiment in which groups of free-living male Song Sparrows were implanted with testosterone, or control tubes, testosterone-treated males were more aggressive than controls throughout the breeding season (Wingfield, 1984b), as tested by their responses to the simulated territorial intrusions described above. Given the results of the challenge experiments, it is reasonable to suggest that testosterone-implanted, aggressive, males may induce an endocrine response in their immediate neighbors who are otherwise untreated. Plasma levels of testosterone in males with territories next to control males were found to decline in April and remain low throughout the rest of the season, whereas testosterone levels in males with territories next to males implanted with testosterone did not decline (Wingfield, 1984b, see also Fig. 33). Levels of circulating testosterone decreased in the latter males during May, possibly due to some degree of habituation to testosterone-implanted males, or because other environmental information from mates or young took precedence at that time. However, concentrations of testosterone began to increase again in June, and were significantly higher than controls in July. Curiously, plasma levels of LH did not differ leaving open the question of the mechanism by which behavioral information from agonistic interactions resulted in an increase in testosterone (but see Wingfield and Wada, 1989).

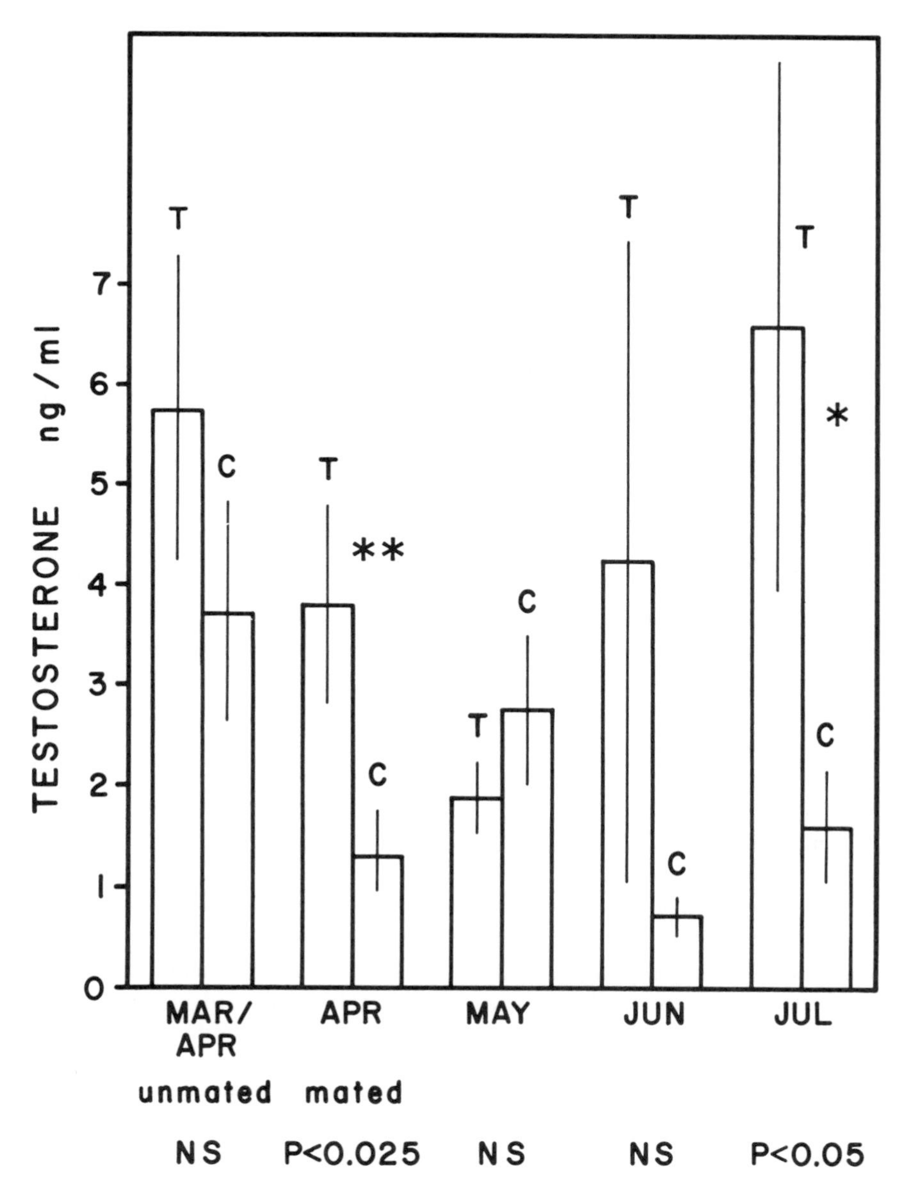


FIG. 33. Seasonal changes in plasma levels of testosterone in territorial male Song Sparrows, *Melospiza melodia*, with territories next to testosterone-implanted males (T) and control implanted (C). NS, not significant. Vertical bars are standard errors, and numbers within bars are sample sizes. From Wingfield (1984b), with permission.

Restriction of periods of elevated testosterone, and thus heightened aggression, to a minimum may be important for integration of other reproductive events. Continual singing, threat displays, and territorial vigilance are energetically demanding, expose an individual to predation or may even attract predators to the nest (see Dufty, 1989; Wingfield, 1990a). Additionally, parental behavior is impaired in aggressive males resulting in reduced nesting success (Silverin, 1980; Hegner and Wingfield, 1987; Oring *et al.*, 1989; Wingfield, 1990a).

The mechanisms by which male–male aggression elevates secretion of LH and testosterone appears to involve specificity of the challenge. Male Song Sparrows challenged with another male Song Sparrow showed an increase in testosterone levels, but males challenged with a male House Sparrow did not (Wingfield and Wada, 1989). This endocrine response to a conspecific male required a combination of auditory and visual stimuli since visual cues alone (a devocalized male) or tape recorded songs (auditory stimulus only) were not as effective as both simultaneously, i.e., an intact male or a tape recording plus a devocalized male (Wingfield and Wada, 1989). In Japanese Quail, deafening reduced the area of the cloacal protuberance suggesting hormonal changes indicative of low gonadotropin secretion (Kerlan *et al.*, 1991). These authors concluded that auditory stimuli may be important in regulating male reproductive function. The elucidation of the sensory modalities important for mediating endocrine responses to male–male interactions has not been well studied. However, given that we now have a broad behavioral and hormonal base to male aggression, the stage is set for new developments in this area.

### 2. *Effects of Male on Female*

A number of investigations have established that a mature male provides environmental information for the female, although the mechanisms have yet to be elucidated fully. The pioneering studies of Lehrman (e.g., 1965) and Hinde (e.g., 1965); Hinde and Steel (1978); and Brockway (1965) have demonstrated that females show more intense nesting activity if paired with a male. In addition, tape-recorded songs of males played to intact female canaries elevated plasma levels of LH over controls (Hinde *et al.*, 1974) and hence accelerated ovarian development and nesting activity (Kroodsma, 1976; Hinde and Steel, 1978). Similar data on ovarian development have been obtained for female Gambel's White-crowned Sparrows exposed to tape-recorded songs of males (Morton *et al.*, 1985). In female Ring Doves there was an increase in plasma levels of LH (Cheng and Follett, 1976) and estrogen (Korenbrot *et al.*, 1974) following pairing with a male. Similarly in the White-crowned Sparrow, Western Gull, Canada Goose, and Bar-headed Goose, only mated females had elevated levels of LH and estrogen (Wingfield and Farner, 1977; Wingfield, 1980b; Akesson and Raveling, 1981; Dittami, 1981;

Wingfield *et al.*, 1982a). In the Canvasback, females whose mates were switched experimentally thus disrupting the pair bond, there was a decline in plasma levels of LH in the female and they failed to breed (Bluhm *et al.*, 1983). Clearly pair bond and courtship with a chosen mate were important for maintenance of reproductive function in females of at least some species. For more details and further discussion see Wingfield and Marler (1988).

The need for a pair bond with a male is not a necessity for onset of breeding in females of all avian species (domesticated species notwithstanding), since female Western Gulls on Santa Barbara Island formed homosexual pairs that comprised up to 14% of breeding pairs in the population. This colony was unusual in that there was a substantial bias in the sex ratio toward females (Hunt *et al.*, 1980). Some of the excess females formed homosexual pair bonds. Both members of the pair laid eggs in the same nest, and also defended a breeding territory. Plasma levels of estrogen in homosexually paired females were elevated over those of unpaired, non-breeding females (Wingfield *et al.*, 1982a). It cannot be precluded, however, that these females failed to gain some stimulation from males because this species breeds in fairly dense colonies so that sexually active males were always present. Furthermore, some of the eggs from female–female pairs were fertile indicating that promiscuous matings did occur. These kinds of associations could provide considerable stimulation for breeding in female Western Gulls without permanent male mates.

### 3. *Effects of Female on Male*

Sexually mature female birds have pronounced stimulatory effects on the endocrine state of the male, including maturation of the testes (e.g. Burger, 1953). In Ring Doves and Rock Doves there was an increase in plasma levels of LH and testosterone in males exposed to females (Haase, 1975; Haase *et al.*, 1976; Feder *et al.*, 1977). Laboratory and field experiments with White-crowned Sparrows also indicated that implantation of females with estradiol, to induce copulation-soliciting displays, resulted in increases in plasma levels of LH, testosterone, and DHT in photostimulated male cage-mates (Moore, 1983, see also Table 4). Furthermore, if sexual activity of free-living female White-crowned Sparrows was prolonged by estradiol implants then the decline of circulating testosterone in their mates (as the parental phase began), was prevented (Moore, 1982; see also Fig. 34). In Brown-headed Cowbirds, exposure of males to estrogen-treated females accelerated the rate of testicular development, although the maximum levels of testosterone attained were not higher than control males (Dufty and Wingfield, 1986c; see also Wingfield and Marler, 1988 for detailed discussion).

The effects of such interactions on endocrine state are, presumably, mediated by visual and auditory cues. Deafened male Ring Doves exposed to sexually mature

TABLE 4

THE EFFECTS OF SEXUALLY RECEPTIVE AND NON-RECEPTIVE FEMALES ON PLASMA LEVELS OF LUTEINIZING HORMONE, TESTOSTERONE AND DIHYDROTESTOSTERONE (DHT) IN MALE WHITE-CROWNED SPARROWS, *ZONOTRICHIA LEUCOPHRYS GAMBELII*

| | Males caged alone | Males caged with cholesterol-implanted female | Males caged with estradiol-implanted female |
|---|---|---|---|
| Luteinizing hormone | 2.44 ± 0.46 | 2.59 ± 0.47 | 3.93 ± 0.24* |
| Testosterone | 0.33 ± 0.10 | 0.29 ± 0.07 | 1.16 ± 0.19** |
| DHT | 0.11 ± 0.006 | 0.25 ± 0.08 | 0.70 ± 0.27** |

Data from Moore (1983), with permission of author, and the Zoological Society of London. All samples were taken at day 32 of photostimulation (20L 4D) a time of high androgen levels in blood. Concentrations expressed as means ± standard errors, and levels of significance were determined by Student's t-test.

* $p < 0.05$

** $p < 0.01$.

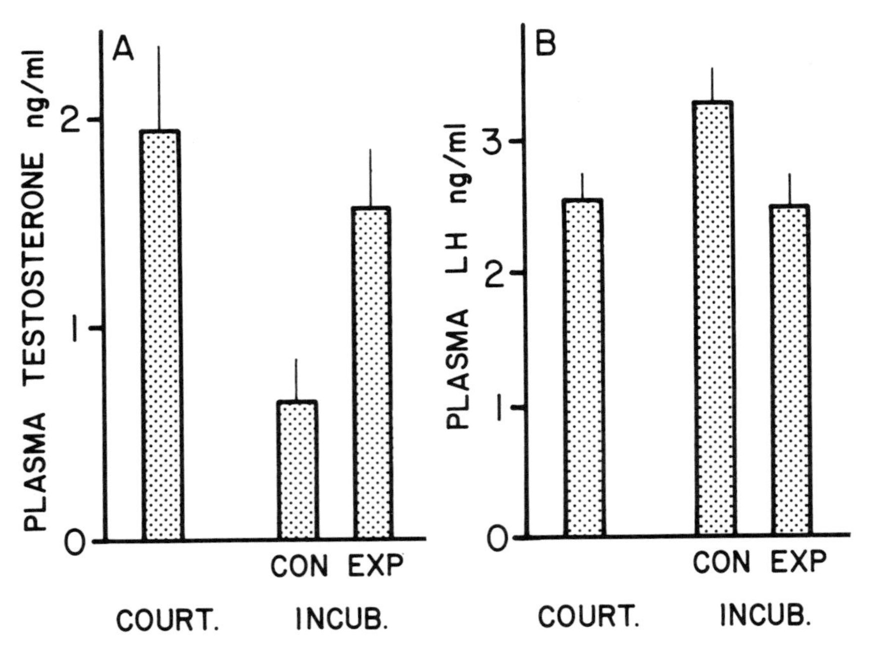


FIG. 34. (a), Plasma levels of testosterone (T) and dihydrotestosterone (DHT) of male White-crowned Sparrows, *Zonotrichia leucophrys pugetensis*, during courtship (COURT.), and incubation (INCUB.), including comparison of incubation levels in control males (CON), whose untreated mates had ceased soliciting copulations, and in experimental males (EXP) whose estradiol-implanted mates continued to solicit copulations; (b), plasma levels of luteinizing hormone (LH) from the same males. All values are means and standard errors. From Moore (1982b), with permission.

females had lower plasma levels of androgen than control males paired with females. In addition, plasma levels of androgens in males separated from females by a glass partition were higher than those of males held in isolation suggesting that visual cues were also important and physical contact less so (O'Connell *et al.*, 1981). Moreover, in female White-crowned Sparrows, non-photoperiodic information perceived by the eyes apparently caused an inhibition of gonadotropin secretion since bilaterally enucleated females held on a stimulatory daylength (20L 4D) had plasma levels of LH greatly elevated over those of controls (Yokoyama and Farner, 1976). Although the mechanisms by which both visual and auditory information influence endocrine section remain to be determined, there is interesting evidence that self-feedback control may be involved in female Ring Doves (Cheng, 1983). Female doves respond to courtship from males by increasing ovarian development. However, experimentally muted females exposed to their own coo vocalizations showed even greater ovarian development than muted females who heard males coo. These data suggested that the female's own cooing mediated the ovarian response to male courtship (Cheng, 1986).

### O. WEATHER AS ESSENTIAL SUPPLEMENTARY INFORMATION

Weather in early spring can have marked effects on gonadal recrudescence and onset of the nesting phase (e.g., Marshall, 1949b, 1959) in part, at least, by influencing the availability of food. It has been suggested that qualitative and quantitative changes in food availability must reach a certain threshold before egg-formation will begin (e.g., Lack, 1968; Harris, 1969; Perrins, 1970; Drent and Daan, 1980; Moore, 1980) and that this threshold varies with environmental temperature and rainfall. For example, in the Chaffinch, in which onset of breeding is dependent on a threshold temperature being reached, about 5 days were required for reaction to a favorable ambient temperature (Newton, 1964). It has yet to be determined, however, whether the birds were responding to ambient temperature *per se* or to food supply which may vary as a consequence of temperature and other weather factors.

Those species that feed on invertebrates, such as insects and their larvae in spring and early summer, are particularly susceptible to changes in weather. In warm springs, the emergence of insects and other invertebrates is generally greater than in cold springs. In addition, fair weather reduces the energy needs of the birds themselves, especially for overnight survival, thus making more energy and nutrients available for feeding young (Newton, 1973; O'Connor, 1978; Dhondt and Eyckerman, 1979). Certain cardueline finches such as the Linnet, Bullfinch, and Greenfinch were also sensitive to weather even though they did not feed primarily on invertebrates. These finches appeared to time onset of breeding in relation to effects of weather on food supply rather than directly to temperature (or other weather factors), since the herbaceous plants from which they feed also flowered and fruited earlier in warm springs (see Newton, 1973).

Marshall (1949b) found in the Chaffinch, European Robin, Great Tit and Blue Tit, *Parus caerulescens*, that testicular development was retarded considerably after the severe European winter of 1946–47 compared with an unusually mild winter of 1947–48. He concluded that this delay was probably due to a combination of very low environmental temperature and restricted food supply. However, egg-laying dates in the species investigated were only slightly delayed (compared with normal) after the severe winter suggesting that gonadal recrudescence was accelerated once underway, whereas after the very mild winter, oviposition occurred up to 1 month earlier than normal. In a more recent field study on the Song Sparrow, it was also found that gonadal recrudescence was delayed after the severe winter of 1981–82 in northeastern United States (Wingfield, 1985c,d; Figs 35 and 36). However, gonadal recrudescence accelerated in late April resulting in sexual maturity at more or less the normal time. The mode egg-laying date in spring 1982 was only 1 week later than in spring 1981 (Wingfield, 1985c,d). In 1982, plasma levels of testosterone were depressed in free-living male Song Sparrows (Fig. 37) although circulating LH, and both LH and estradiol in females, were not affected (Wingfield, 1985c,d). However, Song Sparrows did not appear to be stressed by the severe winter because plasma levels of corticosterone were not elevated, and fat depots in early 1982 were actually greater than at the same time in 1981 (Wingfield, 1985c,d; Fig. 38). Thus, in this case the delay of gonadal recrudescence may possibly be a result of direct effects of weather, or of more subtle indirect effects involving increased energetic requirements for thermoregulation (see also O'Connor, 1978) that are independent of a typical stress-response (see also Wingfield, 1984d, 1988b).

The mechanisms by which weather in early spring can supplement photoperiodic information and regulate gonadal recrudescence are still obscure, but there is now considerable evidence for domesticated species that low environmental temperature can depress testis growth and disrupt ovulatory cycles despite free access to food and water. Such treatment also depressed circulating levels of gonadal hormones and elevated plasma levels of corticosterone (Riddle and Honeywell, 1924; Flickinger, 1959; Assenmacher, 1973; Edens and Siegel, 1975; Huston, 1975; Nir *et al.*, 1975; Etches, 1976; Holmes and Phillips, 1976; Siegel, 1980). Responses of wild species to low environmental temperature had only a marginal effect on photoperiodically induced testicular growth (Lewis and Farner, 1973) and failed to depress plasma levels of LH and testosterone, or to elevate circulating levels of corticosterone (Wingfield *et al.*, 1982b; Table 5). Similarly Rowan (1925) showed that Dark-eyed Juncos exposed to artificial long days in winter, but under ambient temperatures of central Alberta, Canada, underwent normal gonadal recrudescence even though temperature dropped to almost −40 °C.

Some field investigations concur with these results. European Siskins, *Carduelis*

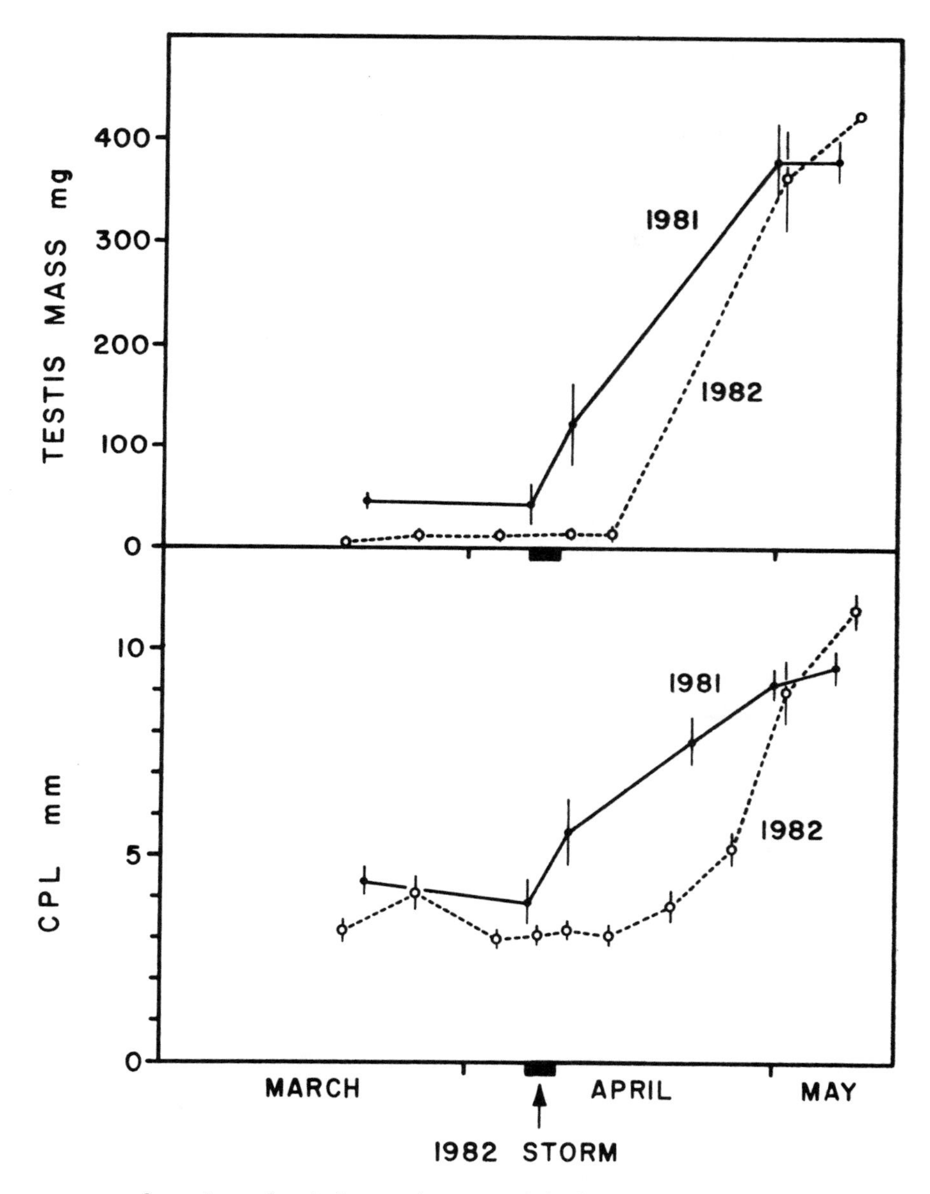


FIG. 35. Comparisons of testicular recrudescence and development of the cloacal protuberance (CPL) of free-living male Song Sparrows, *Melospiza melodia*, in early spring 1981 and 1982. Note the aseasonal storm on 6 April, 1982. Data expressed as means and standard errors. From Wingfield (1985c), courtesy of the Zoological Society of London.

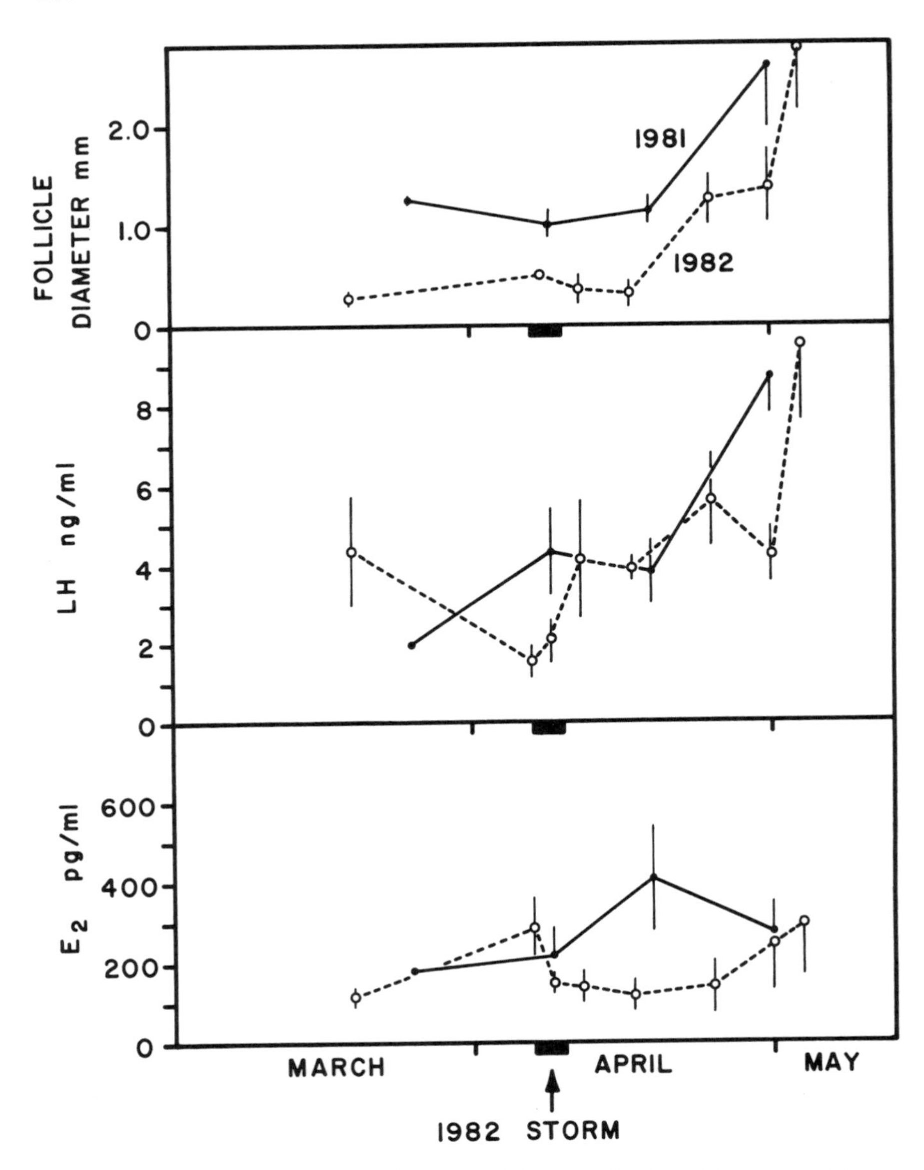


FIG. 36. Comparisons of ovarian recrudescence and plasma levels of luteinizing hormone (LH) and estradiol-17β (E2) of free-living female Song Sparrows, *Melospiza melodia*, in early spring 1981 and 1982. Note the aseasonal snow storm of 6 April, 1982. Data expressed as means and standard errors. From Wingfield (1985d), courtesy of the Zoological Society of London.

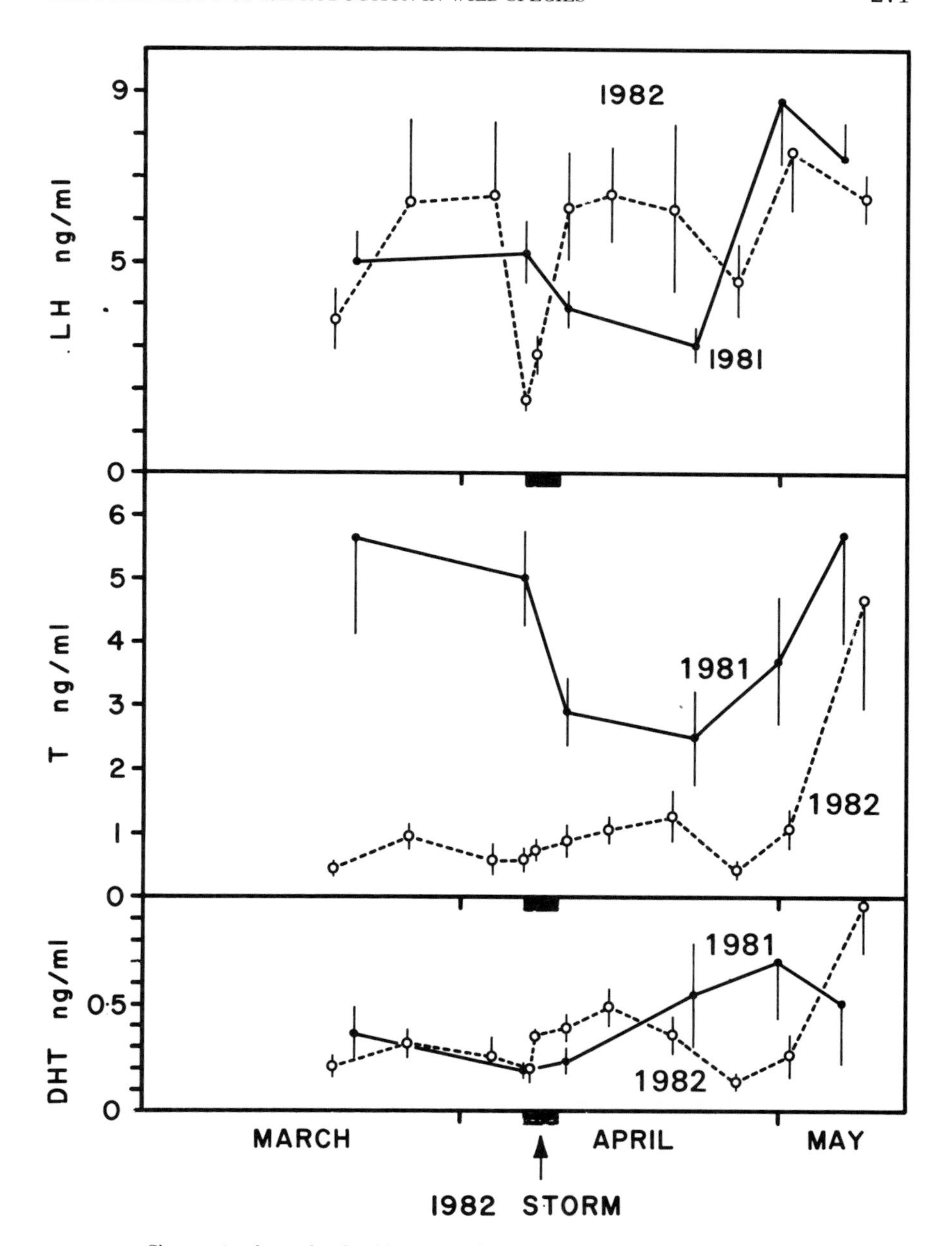


FIG. 37. Changes in plasma levels of luteinizing hormone (LH), testosterone (T), and dihydrotestosterone (DHT) of free-living male Song Sparrows, *Melospiza melodia*, during the early spring periods of 1981 and 1982. Note the aseasonal snow storm on 6 April, 1982. Circles are the means, and vertical bars the standard errors. From Wingfield (1985c), courtesy of the Zoological Society of London.

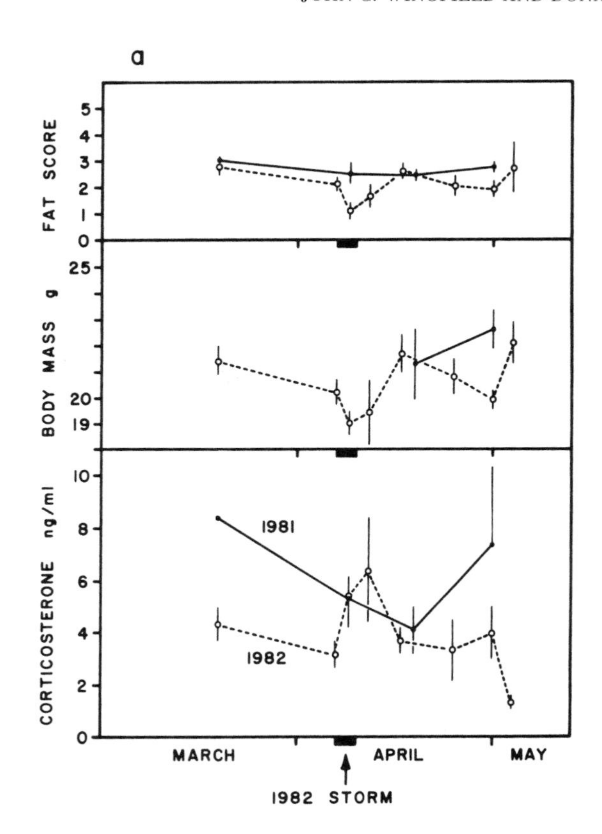


TABLE 5

EFFECTS OF LOW ENVIRONMENTAL TEMPERATURE ON PLASMA LEVELS OF LUTEINIZING HORMONE, TESTOSTERONE AND CORTICOSTERONE IN PHOTOSTIMULATED MALE WHITE-CROWNED SPARROWS

| Group | Day 31 | | | Day 35 | | | Day 39 | | |
|---|---|---|---|---|---|---|---|---|---|
| | LH | T | B | LH | T | B | LH | T | B |
| 5 °C | 2.3± | 0.2± | 13.5± | 2.5± | 0.7± | 13.1± | 2.4± | 0.3± | 13.3± |
| $n = 7$ | 0.5 | 0.02 | 1.8 | 0.3 | 0.3 | 2.5 | 0.5 | 0.1 | 2.2 |
| 23 °C | 1.9± | 0.4± | 12.2± | 2.3± | 1.0± | 12.0± | 1.7± | 0.3± | 17.1± |
| $n = 6$ | 0.2 | 0.3 | 1.4 | 0.2 | 0.3 | 2.3 | 0.2 | 0.1 | 3.7 |

LH, luteinizing hormone; T, testosterone; B, corticosterone. All plasma levels are expressed as mean ng/ml ± standard errors. From Wingfield *et al.* (1982b).

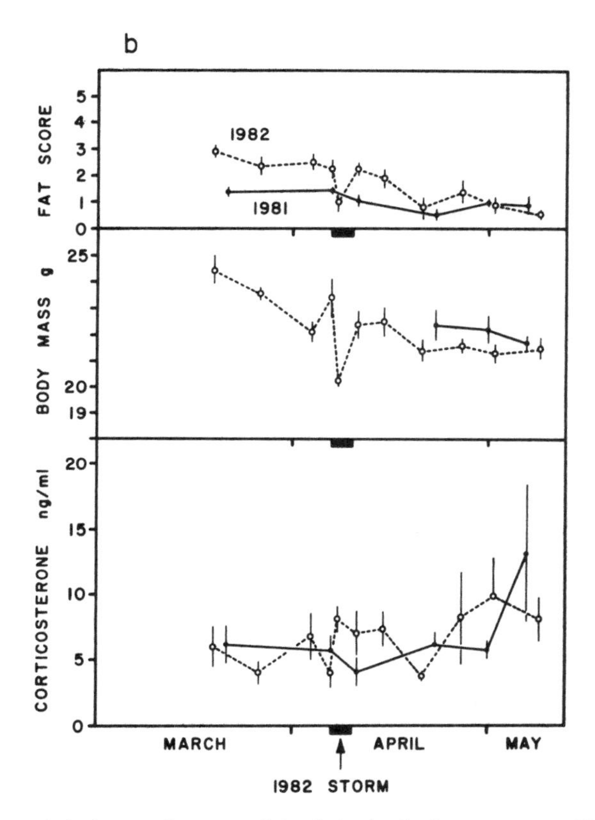


FIG. 38. Changes in body mass, fat score and circulating levels of corticosterone of free-living female (a) and male (b) Song Sparrows, *Melospiza melodia*, during the early spring of 1981 and 1982. Note the aseasonal snow storm of 6 April, 1982. Circles are means and vertical bars are standard errors. From Wingfield (1985c,d), courtesy of the Zoological Society of London.

*spinus* and Redpolls, *Acanthis flamea*, of Fennoscandia will breed regardless of weather so long as food, primarily seeds of coniferous trees, is abundant. In one year when the spruce (*Picea* sp.) crop was unusually abundant, nesting Redpolls were found as early as March and April when snow was 0.5 m in depth, and ambient temperatures were as low as −20 °C (Witt-Strømer *et al.*, 1956). In normal years, these species do not breed until May. Additionally, finches of the genus *Loxia* may breed at any time of year dependent upon seed crops of coniferous trees. Low environmental temperature does not seem to inhibit reproduction in these species as long as food is abundant (Bailey *et al.*, 1953; Tordoff and Dawson, 1965; Newton, 1973). In one nest in Russia, the eggs of a Red Crossbill hatched on a day when air temperature fell to −35 °C (Newton, 1973). Clearly the influences of weather as supplementary information and differences among species are complex. Low environmental temperature appears to be ineffective in delaying gonadal maturation in some species, and highly effective in others. The endocrine bases of these different responses remain, as yet, unknown.

The response of Song Sparrows to the snow storm in April 1982 were markedly different from those of male White-crowned Sparrows to a prolonged period of low environmental temperatures and heavy precipitation in April and May 1974. In this instance, onset of the nesting phase was delayed by almost a full month (Wingfield *et al.*, 1983; Fig. 39). However, this storm occurred later in the reproductive cycle when gonadal recrudescence was almost complete, and when plasma levels of LH and testosterone were approaching the vernal maximum. During this spring, plasma levels of LH and testosterone in male White-crowned Sparrows remained elevated for a month longer than in 1975. Since females delayed oviposition for almost 30 days, presumably the high levels of LH and testosterone in males were at least partly maintained by stimuli from the female, and possibly by other males (Wingfield *et al.*, 1983; see also Section III-N). The storm of April 1974 involved prolonged rain storms and cool temperatures but was not as severe as temperatures of $-15\,^{\circ}C$ and snow cover that accompanied the storm of 6 April, 1982 in New York. Furthermore, the White-crowned Sparrows investigated by Wingfield *et al.* (1983) were short-distance migrants that winter in California where winters are considerably milder than those of northeastern United States. Additionally, the storm of 6 April occurred relatively earlier in the reproductive cycle of Song Sparrows, i.e., at the early stages of recrudescence, compared with late stages of development in the investigations of White-crowned Sparrows. Thus the severity of the storm, and the stage in the reproductive cycle are likely important determinants of the mechanisms by which the endocrine system responds to inclement weather (Wingfield *et al.*, 1983; Wingfield, 1985c, d).

### P. WEATHER AS MODIFYING INFORMATION

Storms can also influence reproduction of birds by disrupting the normal temporal progression of the breeding season after it has been initiated. Weather may regulate rate of gonadal development leading up to the egg-laying stage, but subsequently, storms, etc. interrupt breeding which then requires radical readjustment often leading to a renest attempt. Thus it appears clear that the hormonal mechanisms underlying regulation of gonadal development by weather are different from those underlying disruption of breeding after egg-laying (see Wingfield, 1988b for full discussion). In male White-crowned Sparrows and Song Sparrows, the endocrine responses to storms during the parental phase in May and June did not affect plasma levels of LH and testosterone (Wingfield *et al.*, 1983; Wingfield, 1985c,d), possibly because all males were feeding young at the time so that plasma levels of LH and testosterone were lower than vernal levels, but above

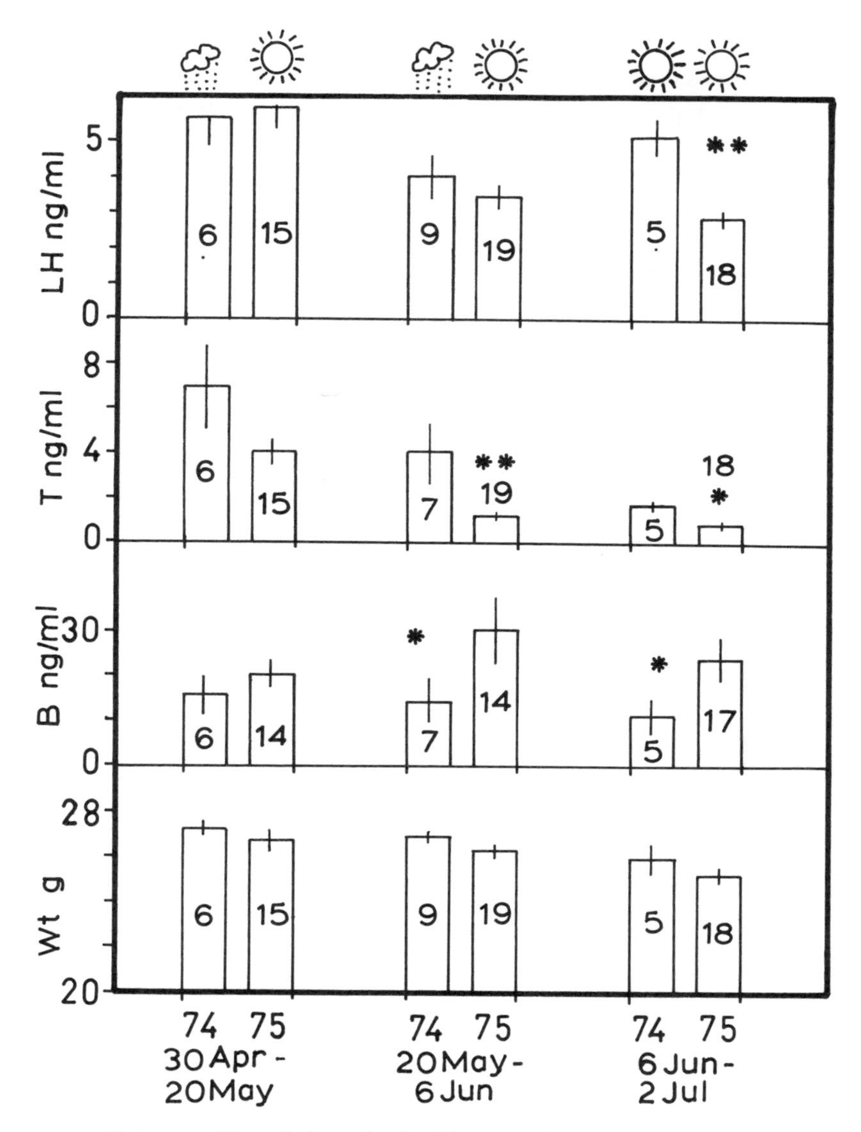


FIG. 39. Body mass (M), and plasma levels of luteinizing hormone (LH), testosterone (T), and corticosterone (B) in free-living male White-crowned Sparrows, *Zonotrichia leucophrys pugetensis*, sampled during spring 1974 and 1975. The stylized rain cloud indicates periods of stormy weather and the sun indicates fair weather. Vertical lines are standard errors of means, and numbers within open bars indicate sample sizes. Asterisks indicate significant difference between 1974 and the same time in 1975, * = $p < 0.05$; ** = $p < 0.02$. From Wingfield *et al.*, 1983, with permission of the American Ornithologist's Union.

basal (Wingfield and Farner, 1977, 1978a,b). However, fat depots in male White-crowned Sparrows (Fig. 40), and body mass in male Song Sparrows (Fig. 41) were depressed during the stormy period compared with levels measured at the same time the previous year. In both species, plasma levels of corticosterone were elevated during the storms (Figs 40 and 41). However, female Song Sparrows,

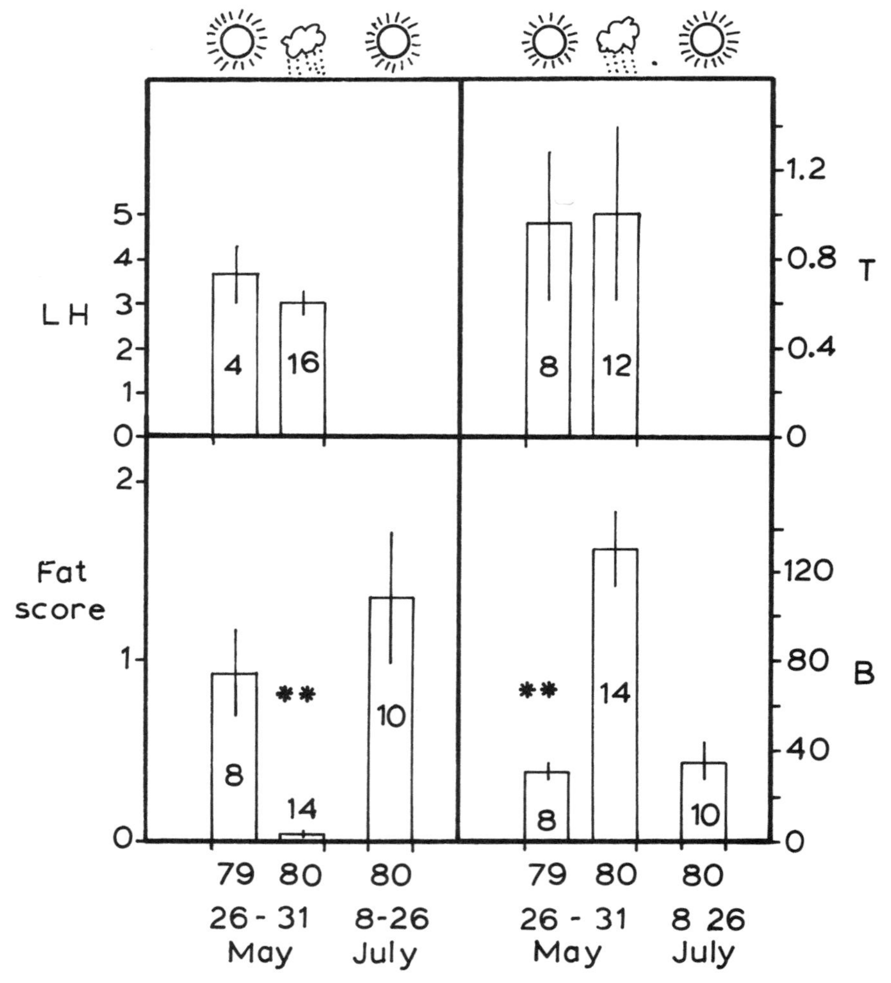


FIG. 40. Fat score and plasma levels of luteinizing hormone (LH), testosterone (T), and corticosterone (B) in free-living male White-crowned Sparrows, *Zonotrichia leucophrys pugetensis*, sampled during spring and summer 1979 and 1980. The stylized rain cloud indicates the period of stormy weather, and the sun indicates periods of fair weather. Vertical lines are standard errors of means and numbers within bars denote sample sizes. Asterisks indicate significant differences between 1979 and the same period in 1980. ** = $p < 0.001$. From Wingfield *et al.* (1983), with permission of the American Ornithologist's Union.

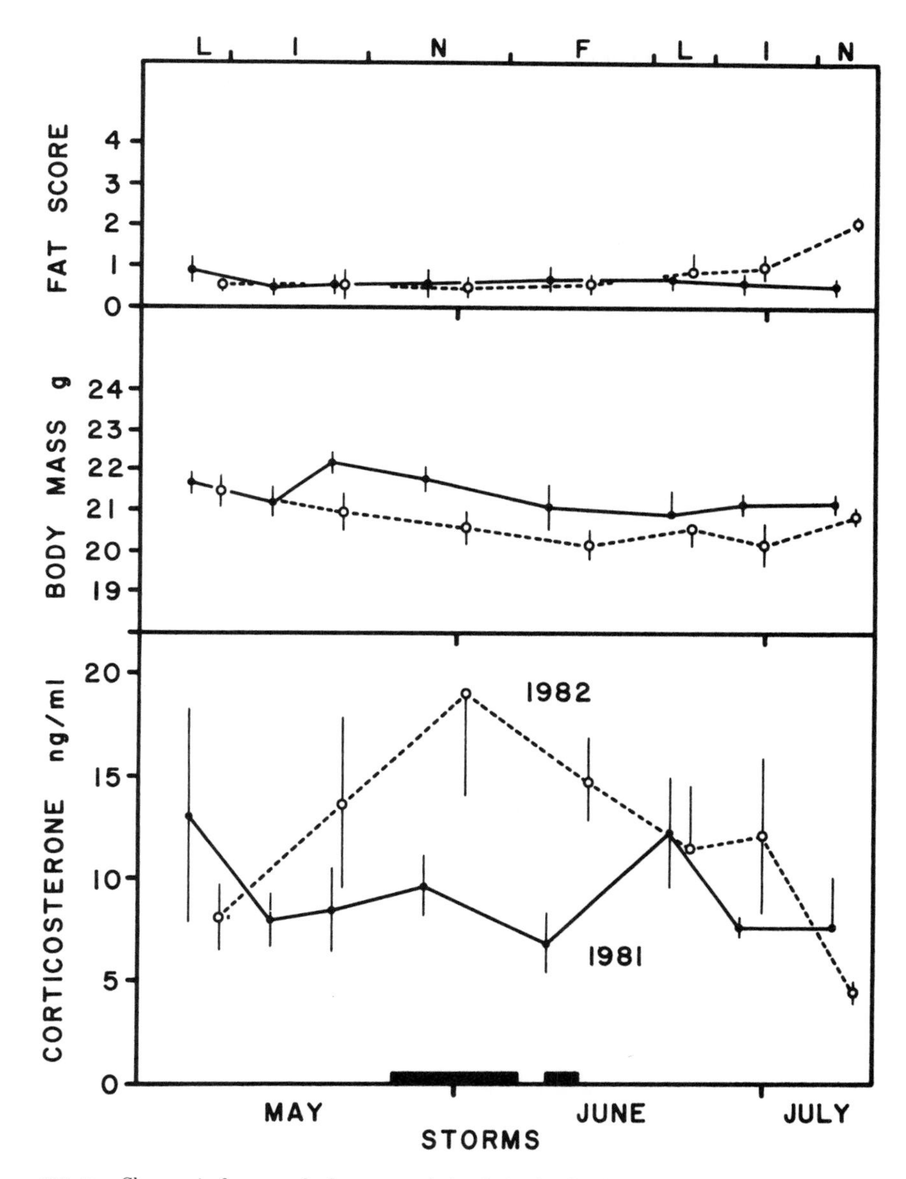


FIG. 41. Changes in fat score, body mass, and circulating levels of corticosterone during the period May–July 1981 and 1982 in free-living male Song Sparrows, *Melospiza melodia*. Note the stormy period in late May and early June 1982. Circles are means and vertical bars are standard errors. From Wingfield (1985c), with permission of the Zoological Society of London.

which did not lose weight or undergo a decrease in fat depots, did not appear to be stressed by the storms in May and June 1982 (Wingfield, 1985d). Since at the time the storms struck, most Song Sparrows were feeding fledglings, it is possible that females were about to begin a second clutch and were not feeding young as regularly as males, thus possibly were not as stressed. However, more investigation is needed to tease apart the complex mechanisms by which wild species respond to inclement weather (see also Wingfield, 1988b for detailed review).

## Q. HORMONAL CHANGES DURING RENESTING AFTER LOSS OF CLUTCH OR BROOD

Individuals that have lost the nest and young to a storm or predator frequently will renest. As a result, reproductive success increases and is critical, especially in those species that breed at high latitudes where the short summers allow successful raising of one brood only. In these instances, if the nest is lost sufficiently early in the season, renesting makes the difference between total reproductive failure for that year, or at least some success. Since loss of the nest disrupts the normal temporal progression of the reproductive cycle, an extensive reorganization of

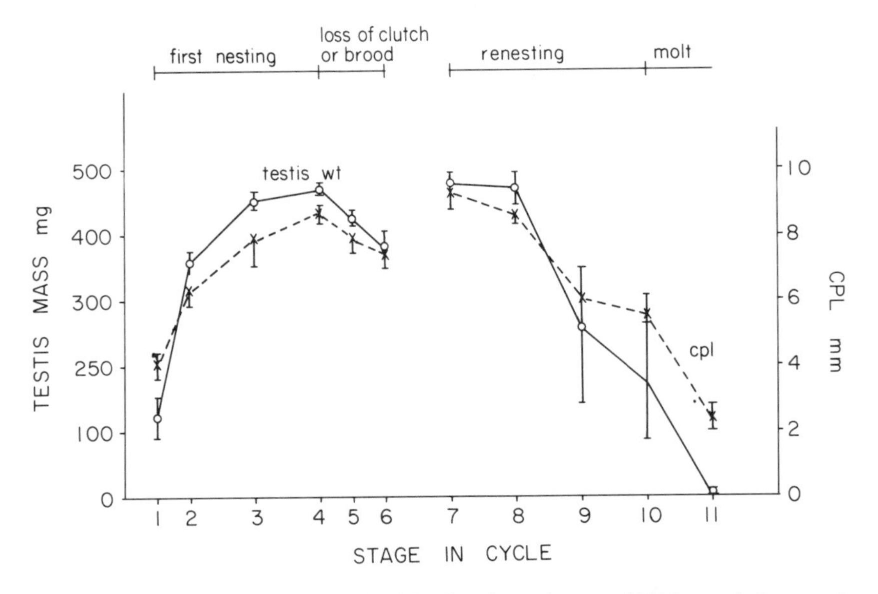


FIG. 42. Estimated testis mass and length of the cloacal protuberance (CPL) in mm during renesting in male White-crowned Sparrows, *Zonotrichia leucophrys gambelii*. Each point is the mean and vertical bars are the standard errors. Stages 1 and 2, migration; 3, courtship period; 4 and 5, incubation; 6, feeding nestlings. Nests with eggs or young were lost between 2 and 20 June; stage 7, renesting, courtship period; 8, incubation of second clutch; 9 and 10, feeding of young of second brood; 11, post-nuptial (pre-basic) molt. From Wingfield and Farner (1979), with permission.

endocrine functions is required. The endocrine changes associated with renesting have been investigated thus far in three wild species.

In White-crowned Sparrows, there were marked resurgences of LH, testosterone and testis mass after loss of the nest (Wingfield and Farner, 1979; Figs 42 and 43), and coincident with resurgences of LH, and estrogens in females that culminated in production of a replacement clutch (Wingfield and Farner, 1979; Fig. 44). As soon as the clutch was complete, and incubation began, plasma levels of LH and sex hormones declined rapidly as was observed after production of the first clutch (Wingfield and Farner, 1977, 1978a,b, 1979; Figs 12, 13, 42–44). Essentially similar results have been obtained in female Mallards after experimental removal of the eggs. Plasma levels of LH increased significantly within 12 h of loss of clutch (Donham *et al.*, 1976). Similarly in Song Sparrows that lost nests to extensive flooding in 1982, there was a resurgence of plasma levels of LH and sex hormones in both males and females when renesting (Wingfield, 1985c,d). This second increase in levels of testosterone when producing a replacement clutch was unlike multiple brooding in which there is no second rise in testosterone during the egg-laying period of the second clutch after successful raising of the first brood (see

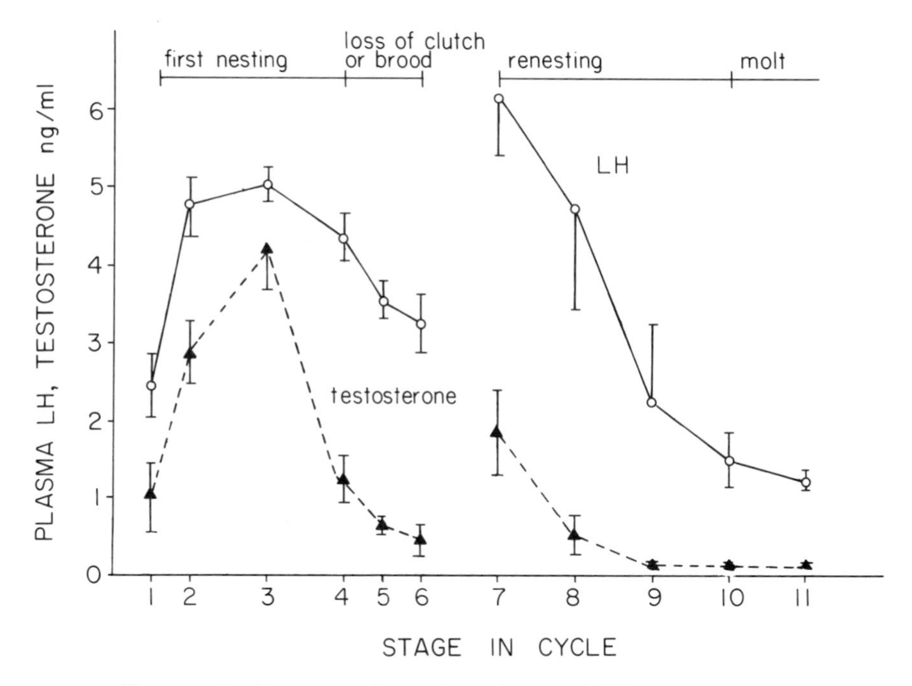


FIG. 43. Plasma levels of immunoreactive luteinizing hormone (irLH), testosterone and dihydrotestosterone (DHT) during renesting in male White-crowned Sparrows, *Zonotrichia leucophrys gambelii*. Each point is the mean and vertical bars are the standard errors. See legend for Fig. 42 for explanation of stages. From Wingfield and Farner (1979), with permission.

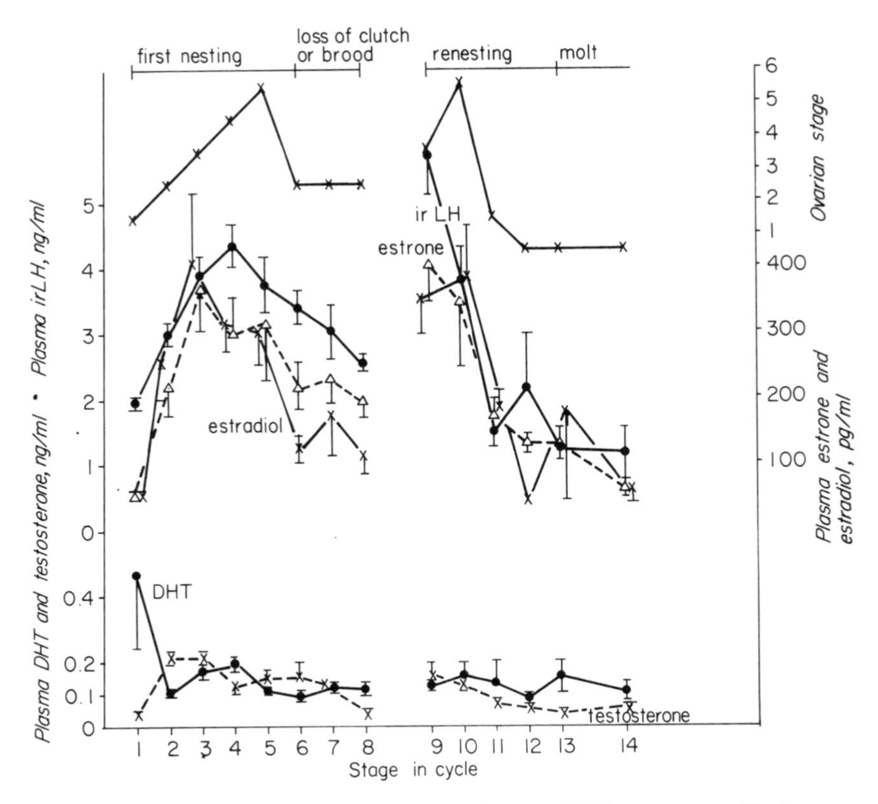


FIG. 44. Plasma levels of immunoreactive luteinizing hormone (irLH), testosterone, dihydrotestosterone (DHT), estrone and estradiol during renesting in female White-crowned Sparrows, *Zonotrichia leucophrys gambelii*. Ovarian stages are: (1), ovary in winter condition; (2), follicles 0.5–1.0 mm in diameter; (3), follicles 3–4 mm in diameter; (5), follicles 5–10 mm in diameter, about to ovulate; (6), egg in oviduct. Stages in reproductive cycle are: 1 and 2, spring migration; 3, courtship; 4 and 5, egg-laying; 6 and 7, incubation. Nest with eggs or young were lost between 2 and 20 June; stage 8, courtship for renesting attempt; 9, ovulating second clutch; 10, incubation of second clutch; 11 and 12, feeding young of second brood; 13, post-nuptial (pre-basic) molt. Each point is the mean and vertical bars are standard errors. From Wingfield and Farner (1979), with permission.

Wingfield and Moore, 1987; Wingfield and Goldsmith, 1990; see also Section III-I). It has been suggested that if the nest and young are lost to a predator or storm, the stimulus for prolactin secretion (from the nest and eggs/young) is lost, and plasma levels decline. This allows a second increase in circulating levels of LH and sex hormones, especially testosterone in males (Wingfield and Moore, 1987). However, in Song Sparrows, if the nest was removed experimentally then prolactin levels did not decline in either males or females. Sex hormone concentrations increased despite continued high titers of prolactin (Wingfield and Goldsmith,

1990). This is also unlike other species in which surges of sex hormones and prolactin do appear to be mutually exclusive (see Section III-I).

The functional significance of the elevation of testosterone titer during renesting, but not during the egg-laying period of a normal second brood, is unclear. One possible suggestion involves survival of young. High levels of testosterone accompanied by increased territorial and "mate-guarding" aggression might result in reduced parental behavior when males are feeding fledglings (Silverin, 1980; Hegner and Wingfield, 1987). Presumably fitness of the male is enhanced if he invests more time ensuring independence of fledglings from the first brood (whose chances of survival are greater than for young from subsequent broods, e.g., Perrins, 1970), rather than maximizing chances of his paternity of later clutches. This explanation is speculative but could be tested. On the other hand, if the nest and eggs are lost, it would be of no disadvantage to the male if plasma levels of testosterone rise thus increasing the levels of aggression as he mate-guards and protects his paternity of the replacement clutch (Wingfield, 1985d; Wingfield and Moore, 1987; Wingfield and Goldsmith, 1990).

### R. PHOTOREFRACTORINESS AND TERMINATION OF REPRODUCTION

Reproduction in all avian species investigated to date is timed to coincide with periods of favorable trophic resources for feeding young (Lack, 1968; Perrins, 1970). Equally important are control systems that terminate the nesting phase before trophic resources decline. In many species that breed during the brief summer of high latitudes and altitudes, the nesting phase must also be terminated in sufficient time for pre-basic molt and preparations for autumnal migration (Marshall, 1960; Lofts and Murton, 1968; Farner and Lewis, 1971; Dolnik, 1976; Murton and Westwood, 1977; Farner and Follett, 1979; Wingfield and Farner, 1980a). On the other hand, reproduction in avian species that breed during the longer summer of mid-latitudes is often terminated in July and August, sometimes earlier. In these species the pre-basic molt and preparations for migration are completed long before the first inclement weather in October and November. Such an apparent "premature" termination of the nesting phase may be linked to availability of food for survival of young. For example, the Great Tit feeds its young primarily on caterpillars that are available for only a limited time in spring (Perrins, 1970, 1973). The energetic requirements for molt and, when applicable, premigratory fat deposition, are satisfied by other foods that may not be suitable for the production and survival of young. Thus in many species, mechanisms to terminate reproduction have reached a complex stage of development owing to the need for precise timing in environments that undergo profound and predictable seasonal changes. These mechanisms include a refractory state in which individuals are no longer able to respond to the environmental stimuli that initiated the reproductive cycle in spring.

Refractory periods may not, however, be universal in avian breeding systems. A high level of internal nutrient reserves, especially protein, is apparently a major proximate factor timing the breeding seasons of several tropical avian species of southeast Asia and Indonesia (Fogden, 1972), e.g., the yellow-vented bulbul, *Pycnonotus goiavier*, (Ward, 1969) and two African species, the Red-billed Quelea, *Quelea quelea*, (Jones and Ward, 1976), and the Gray-backed Camaroptera, *Camaroptera brevicauda* (Fogden and Fogden, 1979). Breeding is terminated when these reserves become depleted. Whether this represents a refractory period or just withdrawal of appropriate stimuli is open to question.

In other tropical species, perhaps as many as 10–20% exhibit some degree of overlap of reproduction and molt (Snow, 1962; Stresemann and Stresemann, 1966; Immelmann, 1967a; Foster, 1975; Wingfield and Farner, 1980a). This coincidence of otherwise mutually exclusive events may be a reflection of less precise and less predictable fluctuations of environmental conditions in some tropical habitats. In many there may be no true refractory state since the non-breeding period, including molt, can be interrupted and breeding commence immediately if conditions permit (Foster, 1975). In addition, opportunistic breeders of arid regions maintain gonads at a near functional level so that breeding may begin as soon as conditions allow, e.g., some populations of the Zebra Finch (Immelmann, 1963a,b; Farner and Follett, 1979; Sossinka, 1980). A few species of northern latitudes may also lack a true refractory period. The gonads of the Red Crossbill, which may be able to breed over much of the year depending upon crops of their principal food—seeds in cones of coniferous trees—show a limited response to daylength (Tordoff and Dawson, 1965). Payne (1969) was able to stimulate gonadal growth in Tri-colored Blackbirds in autumn by providing insect food rich in protein. Similarly, in Pinyon Jays, *Gymnorhinus cyanocephalus*, autumnal gonadal regression can be reversed by exposure to cones of the Pinyon Pine, *Pinus edulis*, a major item in the diet (Ligon, 1974).

Photoperiodic control of refractory periods is by far the most intensively investigated environmentally induced function that terminates reproduction (e.g., Nicholls *et al.* 1988). It is well established that the vernal increase in daylength initiates gonadal growth in anticipation of the ensuing breeding season (see Section III-B). Conversely, decreasing daylength in late summer could act as predictive information for termination of reproduction, as appears to be the case in pigeons of the genus *Columba* (Lofts *et al.*, 1966; Lofts and Murton, 1968) and possibly also in the Baya Weaver (Thapliyal and Saxena, 1964). However, most passerine species, and many others, that breed in northern latitudes undergo spontaneous gonadal regression in mid-summer despite continuing long days which in spring were responsible for the stimulation of gonadal development. This photorefractory condition can also be induced by long days, and when in this state no known photoregime will stimulate gonadal recrudescence. The number of long days required to induce photorefractoriness is a function of daylength (Dolnik, 1976;

Farner and Gwinner, 1980; Farner *et al.*, 1983). In the White-crowned Sparrow, *Z.l. gambelii*, the gonads remained regressed for several years when maintained on permanently long days (Sansum and King, 1976; Farner and Follett, 1979; D. S. Farner, R. A. Lewis and R. S. Donham unpublished). Recovery of photosensitivity usually occurred only after birds have been exposed to short days for 40–60 days (Wolfson, 1952; Dolnik, 1964, 1976; Lofts and Murton, 1968; Turek, 1972, 1978; Farner and Follett, 1979; Farner *et al.*, 1983). Under natural conditions, photosensitivity was regained in late October and early November when daylength is decreasing. Gonadal recrudescence is thus prevented until days lengthen the following spring (Farner and Mewaldt, 1955).

There is much evidence for an involvement of an endogenous circadian rhythm in the measurement of daylength, although the underlying mechanisms remain somewhat obscure (for reviews see Farner and Follett, 1979; Farner and Gwinner, 1980; Follett and Robinson, 1980; Follett, 1981; Farner *et al.*, 1983). Furthermore, it has also been demonstrated that *Zonotrichia* and *Sturnus* continued to measure daylength in relation to recovery of photosensitivity (Turek, 1972, Schleussner and Gwinner, 1988). The photoreceptors in the photoperiodic control system were hypothalamic rather than retinal, eyes do not appear to be necessary (Benoit, 1964; Menaker and Keats, 1968; Yokoyama and Farner, 1976), although the pigments involved in hypothalamic photoreception may be rhodopsin-like (Foster *et al.*, 1985). In the Chukar Partridge, *Alectoris chukar*, and American Tree Sparrow, eyes were also not necessary for induction of photorefractoriness or recovery of photosensitivity (Siopes and Wilson, 1978; Wilson, 1989, 1990a,b).

This type of photorefractoriness is by no means common to all species investigated. In some recovery of photosensitivity may not be dependent upon short days, e.g., European Starlings (Schwab, 1971; Rutledge, 1974), and House Sparrow (Farner *et al.*, 1977). In the House Finch, *Carpodacus mexicanus*, photosensitivity was recovered by exposure to a daylength shorter than that used to induce gonadal growth and photorefractoriness, but not necessarily shorter than threshold daylength (Hamner, 1966b). In other species, such as Mallards (Lofts and Coombs, 1965), Japanese Quail (Follett and Robinson, 1980), and Northern Bobwhite (Kirkpatrick, 1959), spontaneous gonadal regression occurred after exposure to long days, or as daylength began to decrease after mid-summer. Subsequent exposure to even longer days can induce gonadal recrudescence. In the Red-billed Quelea, continual exposure to long days eventually resulted in gonadal regression that was followed by spontaneous gonadal development (Lofts, 1964). However, as this species lives in equatorial Africa where there is virtually no annual photocycle, these results are difficult to interpret (see also Farner *et al.*, 1983).

There is now considerable evidence that suggests the occurrence of circannual rhythms of gonadal growth and regression in several avian species (e.g., several taxa of *Phylloscopus* and *Sylvia*, and *Sturnus vulgaris* and others) which appear to be entrained by daylength into cycles of exactly 1 year (e.g., Gwinner, 1975,

1981, 1986; Berthold, 1977; Farner and Gwinner, 1980). This mechanism is particularly attractive for migratory species that winter in equatorial regions where photoperiodic information is absent or minimal, or for transequatorial migrants that are exposed to long days on their wintering grounds, e.g., Short-tailed Shearwater, *Puffinus tenuirostris*, (Marshall, 1959, 1960; Marshall and Serventy, 1959). In another transequatorial migrant, the Bobolink, *Dolichonyx oryzivorus*, photosensitivity was recovered on the shortening days during autumnal migration. On their wintering grounds in South America, Bobolinks are exposed to daylengths of up to 14 h during the austral summer. This daylength should be stimulatory in the absence of photorefractoriness, but laboratory experiments have shown that if photosensitive Bobolinks are subjected to 14 h of light per day, gonadal growth does not begin for at least 10 weeks. However, exposure to daylengths of 16 h or longer induced immediate gonadal recrudescence (Engels, 1962, 1969). The delayed response to a 14-h day thus prevented premature gonadal maturation on the wintering grounds until about the end of March when the natural northward migration began. Whether or not this delay in response to 14 h of light represents some circannual component remains to be determined.

A rigid test for the occurrence of endogenous circannual rhythms is to hold subjects under strictly constant conditions, including a constant photoregime, i.e., either continuous light or continuous dark. The only cases known to us in which evidence of an endogenous cycle has been demonstrated are reports by Chandola *et al.* (1983) for the spotted munia, and by Holberton (1991) for the Dark-eyed Junco. The facts that the testicular cycles of birds held under continuous light or dark, and 12-h days were rather similar, both with periods somewhat shorter than a cycle on natural daylength, adds credence to the occurrence of endogenous circannual cycles in at least some of the species cited above (Farner, 1985).

An additional hypothesis suggests that at least in the White-throated Sparrow and House Sparrow, the annual cycle of migration, gonadal growth, refractoriness, and molt is regulated by the phase angle between circadian rhythms of circulating corticosterone and prolactin which are themselves entrained by daylength, and possibly other factors such as ambient temperature (for reviews, see Meier and Ferrell, 1978; Meier *et al.*, 1980; see also Section III-L).

Changes in the endocrine system that accompany photoperiodically induced gonadal growth and regression may also play some role in development of photorefractoriness. In Willow Ptarmigan, Red Grouse, and Mallard, there is evidence for an increase in the sensitivity of the hypothalamus to gonadal steroid feedback so that even very low circulating levels of sex steroid hormones are sufficient to suppress secretion of gonadotropins (Sharp and Moss, 1977; Sharp, 1980; Stokkan and Sharp, 1980b; Haase *et al.*, 1982). A similar hypothesis has been proposed for *Z. l. gambelii* (Matt, 1980) and the Baya Weaver (Singh and Chandola, 1981), although the latter authors suggest that gonadal steroids may also increase the threshold daylength required for a photoresponse thus resulting in gonadal regression. Supporting data come from Cusick and Wilson (1972) who observed that implants of an androgen blocker, cyproterone acetate, into the basal hypothalamus prevented spontaneous gonadal regression in photostimulated American Tree Sparrows. However, it should be borne in mind that plasma levels of testosterone and dihydrotestosterone were basal in photorefractory *Z. l. gambelii* (Lam and Farner, 1976; Wingfield and Farner, 1978a,b, 1980a, see also Section III-B and -C), and were similar to those of castrated males (McCreery and Farner, 1979). If the hypothalamus of *Z. l. gambelii* did become sensitive to very low levels of sex steroid hormones, then these levels were below the sensitivity of our assay systems (20–40 pg/ml) and, in castrates, were presumably of adrenal origin. Furthermore, unlike the situation in *Lagopus* and *Anas*, castration of photorefractory Canaries and White-crowned Sparrows did not result in increases in plasma LH until photosensitivity was regained (Nicholls and Storey, 1976; P. W. Mattocks, Jr unpublished). Indeed, if photorefractoriness was dependent on negative feedback of gonadal hormones, the system should oscillate when birds are held continuously on long days. For example, *Z. l. gambelii* held on 12L 12D for 46.5 months did not molt but underwent four cycles of testicular growth and regression. When transferred to 20L 4D all molted and showed the typical signs of photorefractoriness (Farner *et al.*, 1980). These data suggest that photorefractoriness and molt require a daylength longer than 12L, and if held on 12L then the testes oscillate between development and regression possibly as a function of feedback. Changing sensitivity to feedback may interact with other direct effects of daylength to "fine-tune" gonadal regression, or may represent an adjustment of the hypothalamo-gonadotroph axis from a state of high activity during reproduction to one of much lower activity in autumn and winter.

Other components of the endocrine system may have direct or indirect effects on the induction of photorefractoriness, although the mechanisms involved are still unclear. Adrenocorticosteroids are classically regarded as antigonadal and have been implicated as a possible factor inducing photorefractoriness. However, the literature does not support this view wholly. Wilson and Follett (1975) and Deviche *et al.* (1979) have shown that administration of corticosterone, whether systemically or indirectly into the basal hypothalamus, depressed plasma levels of LH and FSH in American Tree Sparrows and Mallards. On the other hand, Deviche *et al.* (1980) were unable to suppress circulating levels of LH and FSH following injections of ACTH, despite the fact that such treatment did result in a significant elevation of plasma levels of corticosterone. Similarly, stress-induced rises in corticosterone levels of free-living White-crowned and Song Sparrows did not necessarily result in depression of LH and sex steroid secretion (Wingfield, *et al.*, 1983; Wingfield, 1985c,d). Moreover, circulating corticosterone levels in free-living populations of White-crowned and Song Sparrows show a precipitous decline as photorefractoriness developed (Wingfield, 1984a; Wingfield and Farner, 1977, 1978a,b). Injections of high doses of corticosterone (250 μg/day) also failed to modify the onset of photorefractoriness and prebasic molt in photostimulated male *Z. l. gambelii* (Wingfield and Farner, 1980b).

A role for the thyroid gland has also been implicated since thyroidectomy appears to block the onset of photorefractoriness in the European Starling (Woitkewitsch, 1940a,b; Wieselthier and van Tienhoven, 1972), Baya Weaver (Chandola *et al.*, 1974), and Spotted Munia (Pandha and Thapliyal, 1964). In addition, plasma levels of thyroid hormones rise just as photorefractoriness develops in Mallards and Green-winged Teal (Jallageas *et al.*, 1974, 1978). On the other hand, thyroidectomy did not influence development of photorefractoriness in the Chukar (Siopes and Wilson, 1978), and circulating levels of thyroxine and triiodothyronine did not rise until after gonadal regression in photostimulated and free-living populations of *Z.l. gambelii* (Smith, 1979, 1982). However, more recent work on the European Starling has indicated clearly that thyroid hormones may have a major role in regulating the termination of reproduction. Starlings maintained on 11 h of light per day underwent testicular recrudescence and remained sexually mature indefinitely. Administration of T4 to these birds resulted in rapid testicular involution and onset of a prebasic-type molt as is typical following natural onset of photorefractoriness. Furthermore, these birds failed to show gonadal recrudescence following transfer to 18 h of light per day suggesting strongly that they were indeed photorefractory (Goldsmith and Nicholls, 1984). Thyroidectomized Starlings also maintained high hypothalamic contents of GnRH compared with controls, and plasma levels of prolactin (that rise concomitantly with development of photorefractoriness) remained low (Goldsmith and Nicholls, 1984; Dawson *et al.*, 1985). Only short periods of T4 injections, or even only a single injection, are needed just after photostimulation to provoke a spontaneous regression in thyroidectomized birds several weeks later (Nicholls *et al.*, 1988).

Thyroid hormone also appears to be involved in the regulation of refractoriness in juvenile birds. It is well known in temperate latitudes that most birds raised in summer do not breed until the following summer despite being exposed to long days during development in the nest and at fledging. Castration of juvenile European Starlings did not increase plasma levels of LH, and exposure to short days (8L 16D) resulted in a spontaneous increase in LH after about 4–6 weeks—a treatment that signals development of photosensitivity in adults (Williams *et al.*, 1987). If Starlings were thyroidectomized when 8 days old in the wild and then hand reared in the laboratory on 16L 8D, all became sexually mature even though somatic development remained at the 3-week level for at least 23 weeks. Testicular growth began at 8 weeks and by 23 weeks both males and females had well-developed gonads (Dawson *et al.*, 1987). In the Chukar, young birds were thyroidectomized at 6.5 weeks and then exposed to either 20L 4D or 8L 16D. Thyroidectomized birds on short days showed no gonadal growth and T4 injections had no further effect. However, on long days, significant testicular development occurred at 18 weeks and was enhanced by injection of T4. None of the euthyroid controls showed any gonadal development (Creighton, 1988b). These experiments indicate that at least in Starlings and Chukars, juvenile photorefractoriness may have a similar basis to this phenomenon in adults. Note, however, that although it is apparent that juvenile Starlings are photorefractory, it appears that the effects of short days on recovery of photosensitivity only occur after the individual is full grown. Exposure of juveniles to short days during the nestling stage did not advance recovery of photosensitivity compared with young birds raised on long days and then transferred to short days after attaining adult size (McNaughton *et al.*, 1992).

It is now over five decades since photorefractoriness was first described in the European Starling (Bissonette and Wadlund, 1932), House Sparrow (Riley, 1936), and the European Redstart, *Phoenicurus phoenicurus* (Schildmacher, 1938), and yet it is still not clear which component or components of the central nervous and neuroendocrine control systems are involved. The possibility that the gonads become refractory to the actions of gonadotropins has been ruled out since injections of mammalian LH and FSH stimulated gonadal growth in photorefractory White-crowned Sparrows (Stetson *et al.*, 1973); House Sparrows (Vaugien, 1954); and others (Riley and Witschi, 1938; Schildmacher, 1938; Miller, 1949). Moreover, Novikov (1955) demonstrated that testes of photorefractory House Sparrows developed normally if transplanted into photostimulated hosts.

More recently it has been shown that plasma levels of LH in castrated *Z.l. gambelii* declined spontaneously in mid-summer in the same manner as those of intact controls (Mattocks *et al.*, 1976). Similarly in American Tree Sparrows subjected to artificial long days, plasma levels of LH decline spontaneously in castrates at the same time as intact controls (Wilson and Follett, 1974). In photostimulated *Z. l. gambelii* and Canaries, plasma levels of LH and FSH decreased in castrates up to 2 weeks before those of intact birds (Nicholls and Storey, 1976; Wingfield *et al.*, 1980a; Storey *et al.*, 1980), although in the former species, basal levels of LH and FSH in castrates were not attained until day 75 of photostimulation, the same time as in intact sparrows. These data suggest that although the presence of a gonad is not essential for the development of photorefractoriness, testicular factors may modify its timing. This led to a revival of the "exhaustion hypothesis" first suggested by Miller (1954) and Dolnik (1964), which asserts that during photostimulation some component of the hypothalamo-hypophysial unit becomes depleted. In the absence of negative feedback from sex steroid hormones, levels of gonadotropins in castrates are greatly elevated over those of controls thus leading to exhaustion of the gonadotrophs in the adenohypophysis, or hypothalamic neurones that secrete GnRH. Wingfield *et al.* (1979) demonstrated that intravenous injections of mammalian GnRH into male *Z. l. gambelii* elicited significant and approximately equal increases in plasma levels of LH in both photosensitive and photorefractory birds. McCarthy (1979) has confirmed and extended these findings to both male and female *Z. l. gambelii*. Essentially similar data have been presented for Domestic Mallard (Balthazart *et al.*, 1980b) and Snow Goose (Campbell *et al.*, 1978). Thus it appears that the gonadotrophs of the anterior pituitary do not become refractory to GnRH and, furthermore, these data argue against the exhaustion hypothesis, at least at the level of the pituitary. Moreover, when castrated male Canaries (Storey and Nicholls, 1981), *Z. l. gambelii* (Matt and Farner, 1979), and Willow Ptarmigan (Stokkan and Sharp, 1980b) were given implants of testosterone in Silastic tubing, photoperiodically induced rises in plasma levels of LH were suppressed. If at intervals after transfer to long days the implants were removed and circulating levels of LH measured for several weeks further, then LH levels increased rapidly after removal of the implant, but declined at about the same time as controls regardless of when the implants were withdrawn.

Additional evidence comes from Erickson (1975) using an *in vitro* technique to measure release of gonadotropins from adenohypophyses of roosters by hypothalamic extracts of adult male *Z. l. gambelii* during photoperiodically induced testicular growth and onset of photorefractoriness. The hypothalamic content of gonadotropin-releasing activity rose after transfer to long days, began to decline 30 days later, and reached a nadir as photorefractoriness developed. Gonadotropin-releasing activity remained low until photosensitivity was regained. Erickson (1975) used extracts of cerebral cortex as control, and although this extract did have some gonadotropin-releasing activity, there were no temporal changes during photostimulation.

More recently, R. A. Hudson, J. C. Wingfield, and D. S. Farner (unpublished) repeated these experiments *in vivo*. Hypothalami, and portions of the cerebral cortex as control, were collected from photosensitive and photorefractory male *Z. l. gambelii* and the tissue homogenized and extracted in 70% methanol at 0°C. Extracts were dried, taken up in 0.85% saline, and injected intravenously into photosensitive adult male *Z. l. gambelii* at a dose of four hypothalami per bird. Blood samples were collected at 0, 5, and 20 min after injection and plasma levels of LH measured. Neither saline alone or extracts of cerebral cortex induced an increase in plasma LH (Fig. 10). Hypothalamic extract from photosensitive birds injected into photosensitive sparrows increased circulating LH from 1.2 to 4.6 ng/ml within 5 min in a manner identical to that induced by injection of mammalian GnRH (Wingfield *et al.*, 1979). However, hypothalamic extract from photorefractory birds injected into photosensitive sparrows did not induce any change in levels of LH (Fig. 10) consistent with the findings of Erickson (1975) that the hypothalamic content of gonadotropin-releasing activity was low in photorefractory birds. In the Starling, radioimmunoassay of GnRH content in hypothalamus showed a decrease in assayable peptide in photorefractory males. When transferred to short days, the levels of GnRH in the hypothalamus increased as photosensitivity was regained (Dawson *et al.*, 1986). Clearly these investigations support the hypothesis that photorefractoriness lies at the hypothalamic level or higher in the central nervous system, although the precise site still awaits discovery.

### S. NON-PHOTOPERIODIC REGULATION OF TERMINATION OF REPRODUCTION

Although there appears to be no clear demonstration that birds become refractory to environmental stimuli other than daylength, there is evidence that the onset of photorefractoriness can be delayed by non-photoperiodic information. Investigations of naturally breeding populations of *Z. l. gambelii* reveal that the apparent timing of spontaneous gonadal regression is variable (Table 6). Unmated males showed the effects of photorefractoriness by the fourth week of June, breeding males and females in the first week of July, and finally renesting birds, that had lost their first brood and initiated a second clutch in mid-June, were able to delay gonadal regression and onset of molt until the young were fledged in the third week of July (Wingfield and Farner, 1978b, 1979, 1980a, Table 6). Clearly it would be maladaptive for the parental phase to be terminated abruptly by photorefractoriness since the young would not survive and the reproductive effort would be wasted. Rather, it appears that natural selection favors mechanisms that prevent initiation of clutches beyond a certain date, but allow completion of a nesting phase that was initiated before this time. In the case of *Z. l. gambelii*, this period appears to be the third and fourth weeks of June (Wingfield and Farner, 1979).

The mechanisms by which photorefractoriness is delayed are by no means clear. It is possible that elevated circulating levels of sex steroid hormones during the initiation of the second clutch somehow delay gonadal regression and molt since testosterone has been shown to maintain a functional testis in Japanese Quail (Brown and Follett, 1977), and prevented spontaneous gonadal involution in photostimulated House Sparrows (Turek *et al.*, 1976). Moreover, both testosterone and estradiol delayed onset of molt in Canaries (Kobayashi, 1952); European Starling (Schleussner *et al.*, 1985); and *Z. l. gambelii* (M. C. Moore, K. S. Matt, J. C. Wingfield and D. S. Farner, unpublished). Stimuli from the nest and eggs, or interactions between mates may also play a role, although data supporting such hypotheses are sparse. Runfeldt and Wingfield (1985) have shown that free-living

TABLE 6

DURATION OF MAXIMUM SIZE (> 300 MG) OF TESTES IN CAPTIVE AND FREE-LIVING WHITE-CROWNED SPARROWS, *ZONOTRICHIA LEUCOPHRYS GAMBELII*

| Treatment | Duration in days |
|---|---|
| Indoor cages (20L 4D) | 10 |
| Outdoor aviaries | 12 |
| Free-living, non-breeding | 25 |
| Free-living, breeding | 31 |
| Free-living, renesting | 44 |

From Wingfield and Farner (1979, 1980a), with permission.

female Song Sparrows given subcutaneous implants of estradiol remained sexually receptive and delayed onset of prebasic molt until at least the beginning of October. Control females given empty implants became photorefractory and began molt by mid-August, and all but two had vacated the breeding area by the end of August. Males mated to control females also were photorefractory by mid-August and most had left by early September. In contrast, males mated to estradiol-implanted females did not become photorefractory and failed to molt until late-August to mid-September, up to one full month later than controls (Table 7). These males also had higher plasma levels of testosterone (Fig. 45). Curiously, untreated females mated to males that had been given implants of testosterone became photorefractory and began molt at the same time as females mated to control males

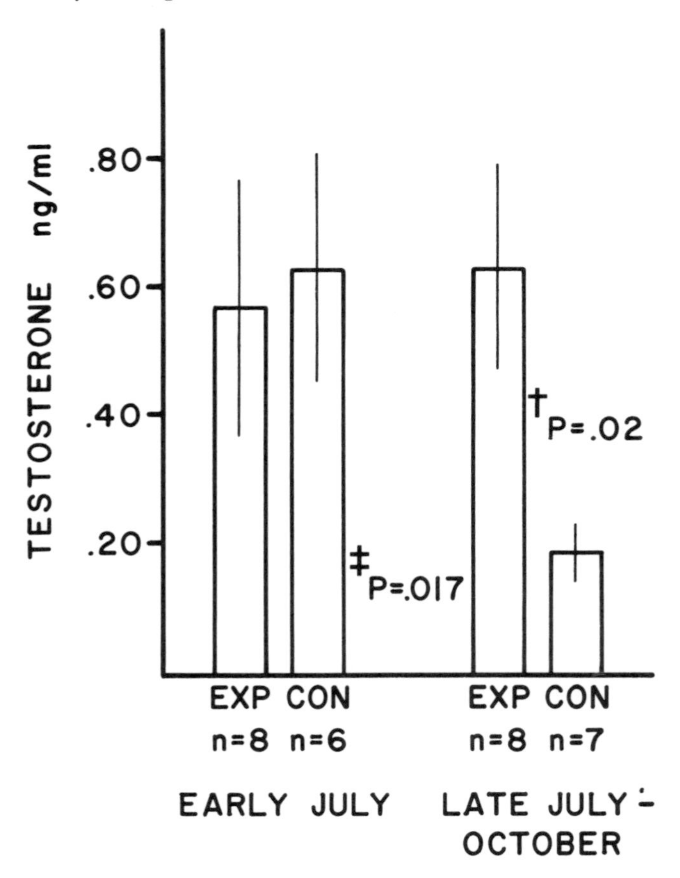


FIG. 45. Mean levels (± SE) of circulating testosterone from control male Song Sparrows, *Melospiza melodia* (mated to control implanted females) and experimental males (mated to estrogenized females) for the periods 2–22 July and 23 July–30 October: P values from Mann-Whitney U tests, $n$ = sample sizes. From Runfeldt and Wingfield (1985), with permission.

TABLE 7

EFFECT OF SEXUALLY RECEPTIVE FEMALE SONG SPARROWS ON THE TERRITORIAL BEHAVIOR AND ONSET OF MOLT IN MALES

| | Number on territory | | | |
|---|---|---|---|---|
| Female treatment | Aug | Sep | Oct | Initiation of molt |
| E2 implant | | | | |
| female, $n = 8$ | 8 | 8 | 8 | Sep–Oct |
| male, $n = 8$ | 8 | 8 | 6 | 24 Aug–10 Sep |
| Control implant | | | | |
| female, $n = 8$ | (9–3)* | 0 | 0 | 10 Aug–25 Aug |
| male, $n = 8$ | 6 | 1 | 0 | 24 July–26 Aug |

* Nine females on territory in early August, but only three remained by late August.
From Runfeldt and Wingfield (1985), with permission.

(mid-August) even though testosterone-implanted males remained on territory and delayed onset of molt until at least the beginning of October (Table 8). These data suggest that the male "fine-tunes" onset of photorefractoriness to the reproductive state of his mate, and thus it might be expected that unmated males would undergo gonadal regression before mated males (c.f. Wingfield and Farner, 1979, Table 6). On the other hand, females do not appear to fine-tune gonadal regression to the reproductive state of the males (Runfeldt and Wingfield, 1985).

In the Japanese Quail, recent evidence suggests that low ambient temperature may accelerate gonadal regression and suppress circulating levels of LH in late

TABLE 8

EFFECT OF SEXUALLY ACTIVE MALE SONG SPARROWS ON THE TERRITORIAL BEHAVIOR AND ONSET OF MOLT IN FEMALES

| | Number on territory | | | |
|---|---|---|---|---|
| Male treatment | Aug | Sep | Oct | Initiation of molt |
| T implant | | | | |
| female, $n = 9$ | (9–1)* | 0 | 0 | 24 Aug–10 Sep |
| male, $n = 9$ | 9 | 8 | 8 | Sep–Oct |
| Control implant | | | | |
| female, $n = 11$ | (11–3)** | 0 | 0 | 10 Aug–25 Aug |
| male, $n = 11$ | 11 | 1 | 0 | 24 July–25 Aug |

* Nine females on territory in early August, but only one remained by late August.
** Eleven females on territory in early August, but only three remained by late August.
From Runfeldt and Wingfield (1985), with permission.

summer or when daylength is reduced from 16L 8D to 8L 16D (Wada *et al.*, 1990a, b; 1992). Furthermore, it appears that alternation of high and low temperature during the day was also sufficient to suppress LH and precipitate gonadal regression (Tsuyoshi and Wada, 1991). Adjustments to termination of breeding induced by supplementary factors may be widespread.

## IV. Conclusions

The aim of this review is to provide a historical as well as current account of endocrinology in feral avian species with an emphasis on reproductive endocrinology and associated hormone secretions. Papers cited refer to approximately 150 species of 16 orders. It is hoped that this approach will provide a useful base for graduate students, postdoctoral associates and others entering basic research in the field of avian endocrinology. Although the list of approximately 750 references cited here is by no means comprehensive, we have tried to be representative of the field in general. However, that such a bibliography does not cover the entire topic is an indication of the breadth, depth, and vigor of avian endocrinology. It is also noteworthy that investigations of feral species have had a long history of bridging fields such as ecology, behavior, and physiology, and have repeatedly made fundamental contributions to biology as a whole (Konishi *et al.*, 1989). In recent years the application of cell and molecular techniques has expanded avian biology so as to allow many unique integrations of organismal and molecular biology within one group of vertebrates. Given the broad base of avian biology, and the tractability of avian species whether in the laboratory or free-living, the future looks very bright for integrative studies combining organismal and molecular studies of endocrinology as well as to interact more effectively with applied issues—especially the emerging discipline of conservation biology. To predict future directions and goals would be pointless here since current techniques and ideas become obsolete so quickly, and new ideas and technology spring up at rapid rates. One thing is certain, however, the next few decades will be at least as surprising and exciting as the last.

ACKNOWLEDGMENTS

Preparation of this chapter was facilitated by grant numbers DCB-8616189, DCB-9005081, and DPP-9023834 from the National Science Foundation to J. C. W. I am very grateful to Professor Dr Andreas Oksche and Professor M. Wada who provided many valuable comments on earlier drafts.

REFERENCES

Adkins, E. K. (1977). Effects of diverse androgens on the sexual behavior and morphology of castrated male quail. *Horm. Behav.* **8**, 201–207.

Adkins, E. K., and Pniewski, E. E. (1978). Control of reproductive behavior by sex steroids in male quail. *J. Comp. Physiol. Psychol.* **92**, 1169–1179.

Adkins-Regan, E. (1981). Hormone specificity, androgen metabolism and social behavior. *Amer. Zool.* **21**, 257–272.

Akesson, T. R., and Raveling, D. G. (1981). Endocrine and body weight changes of nesting and non-nesting Canada geese. *Biol. Reprod.* **25**, 792–804.

Anderson, D. C. (1974). Sex hormone-binding globulin. *Clin. Endocrinol.* **3**, 69–96.

Ando, H., and Ishii, S. (1990). Avian gonadotropins: chemical structures and evolution. *In* "Endocrinology of Birds: Molecular to Behavioral" (M. Wada, S. Ishii, and C. G. Scanes, eds), pp. 3–15, Jap. Sci. Soc. Press, Tokyo, and Springer-Verlag, Berlin.

Aschoff, J. (1980). Biological clocks in birds. *In* "Acta XII Cong. Int. Ornithol." (R. Nöhring, ed.), pp. 113–136. Deutschen Ornithologen-Gesellschaft, Berlin.

Assenmacher, I. (1952a). La vascularization du complex hypophysaire chez le Canard domestique. I. La vascularization du complexe hypophysaire adulte. *Arch. Anat. Microsp. Morphol. Exp.* **41**, 69–106.

Assenmacher, I. (1952b). La vascularization du complexe hypophysaire chez le Canard domestique. II. Le développement embrylogique de l'appareil vasculaire hypophysaire. *Arch. Anat. Microsc. Morphol. Exp.* **41**, 107–152.

Assenmacher, I. (1958). La mue des oiseaux et son déterminism endocrinien. *Alauda* **26**, 241–289.

Assenmacher, I. (1963). Les régulations hypothalamiques de la fonction gonadotrope. *Acta Neuroveget.* **25**, 339–382.

Assenmacher, I. (1973). The peripheral endocrine glands. *In* "Avian Biology" (D. S. Farner, and J. R. King, eds), Vol. 3, pp. 183–286, Academic Press, New York.

Assenmacher, I. and Jallageas M. (1980). *In* "Avian Endocrinology" (A. Epple, and M. H. Stetson, eds), pp. 391–411. Academic Press, New York.

Bailey, A. M., Niedrach, R. J., and Bailey, A. L. (1953). The red crossbills of Colorado. *Publ. Denver Mus. Nat. Hist.* **9**, 1–64.

Baker, J. R. (1938). The evolution of breeding seasons. *In* "Evolution" (G. R. deBeer, ed.), pp. 161–177. Oxford University Press, London.

Balasubramanian, K. S., and Saxena, R. N. (1973). Effect of pinealectomy and photoperiodism in the reproduction of Indian weaver birds, *Ploceus phillipinus*. *J. exp. Zool.* **185**, 333–340.

Ball, G. F., and Wingfield, J. C. (1987). Changes in plasma levels of sex steroids in relation to multiple broodedness and nest site density in male starlings. *Physiol. Zool.* **60**, 191–199.

Ball, G. F., Foidart, A., and Balthazart, J. (1989). A dorsomedial subdivision within the nucleus intercollicularis identified in the Japanese Quail (*Coturnix coturnix japonica*) by means of alpha 2-adrenergic receptor autoradiography and estrogen receptor immunocytochemistry. *Cell Tiss. Res.* **257**, 123–128.

Ball, G. F., Nock, B., Wingfield, J. C., McEwen, B. S., and Balthazart, J. (1990). Muscarinic cholinergic receptors in the song bird and quail brain: a quantitative autoradiographic study. *J. Comp. Neurol.* **298**, 431–442.

Balthazart, J. (1983). Hormonal correlates of behavior. *In* "Avian Biology" (D. S. Farner, J. R. King, and K. C. Parkes, eds), Vol. 7, pp. 221–365. Academic Press, New York.

Balthazart, J., and Hendrick, J. C. (1976). Relationships between the daily variations of social behavior and of plasma FSH, LH and testosterone levels in the domestic duck, *Anas platyrhynchos* L. *Behav. Proc.* **4**, 107–128.

Balthazart, J., Massa, R., and Negri-Cesi, P. (1979). Photoperiodic control of testosterone metabolism, plasma gonadotropins, cloacal gland growth and reproductive behavior in the Japanese quail. *Gen. Comp. Endocrinol.* **39**, 222–235.

Balthazart, J., Blaustein, J. D., Cheng, M.-F., and Feder, H. H. (1980a). Hormones modulate the concentration of cytoplasmic progestin receptors in the brain of male ring doves, (*Streptopelia risoria*). *J. Endocrinol.* **86**, 251–261.

Balthazart, J., Willems, J., and Hendrick, J. C. (1980b). Changes in pituitary responsiveness to luteinizing hormone-releasing hormone during an annual cycle in the domestic duck, *Anas platyrhynchos*. *J. exp. Zool.* **211**, 113–123.

Baptista, L. F., De Wolfe, B. B., and Avery-Beausoleil, L. (1987). Testosterone, aggression, and dominance in Gambel's white-crowned sparrows. *Wilson Bull.* **99**, 86–91.

Barfus, D. W., and Ellis, L. C. (1971). Seasonal cycles in melatonin synthesis by the pineal gland as related to testicular function in the house sparrow, (*Passer domesticus*). *Gen. Comp. Endocrinol.* **17**, 183–193.

Bargmann, W. (1943). Die Epiphysis cerebri. *In* "Handbuch der Mikroskopischen Anatomie des Menschen" (W. von Möllendorff, ed.), Vol. 4, pp. 309–502. Springer-Verlag, Berlin.

Barker-Jørgensen, C. (1971). John Hunter, A. A. Berthold and the origins of endocrinology. *Acta Histor. Sci. Natur. Med.* **24**, 7–54.

Bartke, A., Smith, M. S., Michael, S. D., Peron, F. G., and Dalterio, S. (1977). Effects of experimentally-induced chronic hyperprolactinemia on testosterone and gonadotropin levels in male rats and mice. *Endocrinology* **100**, 182–186.

Bartke, A., Steger, R. H., Klemcke, H. G., Siler-Khodr, T. M., and Goldman, B. D. (1982). Effects of experimentally induced hyperprolactinemia on the hypothalamus, pituitary and testis in the golden hamster. *J. Andrology* **3**, 172–177.

Baylé, J.-D., Ramade, F., and Oliver, J. (1974). Stereotaxic topography of the brain of the quail (*Coturnix coturnix japonica*). *J. Physiol.* (*Paris*) **68**, 219–241.

Beletsky, L., Wingfield, J. C., and Orians, G. H. (1989). Relationships of hormones and polygyny to territorial status, breeding experience and reproductive success in male red-winged blackbirds. *Auk* **106**, 107–117.

Beletsky, L., Orians, G. H., and Wingfield, J. C. (1990). Steroid hormones, polygyny and parental behavior in male yellow-headed blackbirds. *Auk* **107**, 60–68.

Benoit, J. (1923). Sur les modifications cytologiques des cellules interstitielles du testicule chez les oiseaux à activité sexuelle périodique. *C. R. Soc. Biol.* **88**, 202–205.

Benoit, J. (1937). Facteurs externes et internes de l'activité sexuelle. II. Étude du mecanisme de la stimulation par la lumière de l'activité testiculaire chez le canard domestique. Rôle de l'hypophyse. *Bull. Biol. France Belg.* **71**, 393–437.

Benoit, J. (1961). Opto-sexual reflex in the duck: physiological and histological aspects. *Yale J. Biol. Med.* **34**, 97–116.

Benoit, J. (1962). Hypothalamo-hypophyseal control of the sexual activity in birds. *Gen. Comp. Endocrinol. Suppl.* **1**, 254–274.

Benoit, J. (1964). The role of the eye and of the hypothalamus in the photostimulation of gonads in the duck. *Ann N.Y. Acad Sci.* **117**, 204–215.

Benoit, J. (1975). My research in neuroendocrinology: study of the photo-sexual reflex in the domestic duck. *In* "Pioneers in Neuroendocrinology" (J. Meites, B. T. Donovan, and S. M. McCann, eds), pp. 47–60. Academic Press, New York.

Berry, H. H., Millar, R. P., and Louw, G. N. (1979). Environmental cues influencing the breeding biology and circulating levels of various hormones and triglycerides in the Cape cormorant. *Comp. Biochem. Physiol.* **62A**, 879–884.

Berthold, A. A. (1849a). Über die Transplantation der Hoden. Nachrichten von der Georg-Augusts Universität und der Königl. Gesellschaft der Wissenschaften zu Göttingen, **19 Feb. 1849**, pp. 1–6.

Berthold, A. A. (1849b). Transplantation der Hoden. *Arch. Anat. Physiol*, **1849**, 42–46.

Berthold, P. (1977). Endogene Steuerung des Vogelzugs. *Vogelwarte* **30**, 4–15.

Bicknell, R. J. and Follett, B. K. (1975). A quantitative bioassay for luteinizing hormone releasing hormone using dispersed pituitary cells. *Gen. Comp. Endocrinol.* **26**, 141–152.

Bigalke, R. (1956). Uber die zyklischen Veränderungen der Schildrüse und des Körpergewichtes bei einigen Singvögeln im Jahresablauf. Inaug. Diss., Univ. Frankfurt am Main.

Binkley, S. (1988). "The Pineal. Endocrine and Neuroendocrine Function." Prentice Hall, Englewood Cliffs, New Jersey.

Binkley, S., Stephens, J. L., Riebman, J. B., and Reilly, K. B. (1977). Regulation of pineal rhythms in chickens: Photoperiod and dark-time sensitivity. *Gen. Comp. Endocrinol.* **32**, 411–416.

Bissonette, T. H., and Wadlund, A. P. (1932). Duration of testis activity of *Sturnus vulgaris* in relation to type of illumination. *J. exp. Zool.* **9**, 339–350.

Blanchard, B. D. (1941). The white-crowned sparrows (*Zonotrichia leucophrys*) of the Pacific seaboard: Environment and annual cycle. *Univ. Calif. Publ. Zool.* **46**, 1–178.

Blanchard, B. D. and Erickson, M. M. (1949). The cycle in the Gambel sparrow. *Univ. Calif. Publ. Zool.* **47**, 225–318.

Bluhm, C. K., Phillips, R. E., and Burke, W. H. (1983). Serum levels of luteinizing hormone (LH), prolactin, estradiol, and progesterone in laying and non-laying canvasback ducks (*Aythya valisinaria*). *Gen. Comp. Endocrinol.* **52**, 1–16.

Bluhm, C. K., Phillips, R. E., Burke, R. E., and Gupta, G. N. (1983). Effects of male courtship and gonadal steroids on pair formation, egg-laying, and serum LH in canvasback ducks (*Aythya valisinaria*). *J. Zool. Lond.* **204**, 185–200.

Boag, D. A. (1982). How dominance status of adult Japanese quail influences the viability and dominance status of their offspring. *Can. J. Zool.* **60**, 1885–1891.

Bona-Gallo, A., Licht, P., and Papkoff, H. (1983). Biological and binding activities of pituitary hormones from the ostrich, *Struthio camelus*. *Gen. Comp. Endocrinol.* **51**, 50–60.

Boon, A. A. (1938). Comparative anatomy and physiopathology of the autonomic hypothalamic centers. Haarlem.

Bordin, S., and Petra, P. H. (1980). Immunological localization of sex steroid binding protein of plasma and tissues of the adult monkey, *Macaca nemestrina*. *Proc. Natl. Acad. Sci. U.S.A.* **77**, 5678–5682.

Boss, W. R. (1943). Hormonal determination of adult characters and sex behavior in herring gulls (*Larus argentatus*). *J. exp. Zool.* **94**, 181–209.

Boss, W. R., and Witschi, E. (1941). Male sex hormones inducing adult characters in juvenile herring gulls (*Larus argentatus*). *Anat. Rec.* **81** suppl., 27–28.

Bottoni, L., and Massa, R. (1981). Seasonal changes in testosterone metabolism in the pituitary gland and central nervous system of the European starling (*Sturnus vulgaris*). *Gen. Comp. Endocrinol.* **43**, 532–536.

Bottoni, L., Lucini, V., Massa, R., Sharp, P. J., and Trocchi, W. (1984). Annual cycle of plasma LH in the grey partridge, *Perdix perdix*. *Gen. Comp. Endocrinol.* **53**, 462–463.

Breucker, H. (1967). Vergleichende histologische Studien an der Zirbel der Vögel. *Verhandlungen der Anatomischen Gesellschaft* **120**, 177–183.

Brockway, B. F. (1965). Stimulation of ovarian development and egg laying by male courtship vocalization in budgerigars (*Melopsittacus undulatus*). *Anim. Behav.* **13**, 575–578.

Brown, N. L., and Follett, B. K. (1977). Effects of androgen on the testis of intact and hypophysectomized Japanese quail. *Gen. Comp. Endocrinol.* **33**, 267–277.

Brown, N. L., Baylé, J.-D,, Scanes, C. G., and Follett, B. K. (1975). Chicken gonadotropins: their effects on the testes of immature and hypophysectomized Japanese quail. *Cell Tiss. Res.* **156**, 499–520.

Buntin, J. D., Becker, G. M., and Ruzycki, E. (1991). Facilitation of parental behavior in ring doves by systemic or intracranial injections of prolactin. *Horm. Behav.* **25**, 424–444.

Burger, A. E., and Millar, R. P. (1980). Seasonal changes of sexual and territorial behavior and plasma testosterone levels in male lesser sheathbills (*Chionis minor*). *Z. Tierpsychol.* **52**, 397–406.

Burger, J. W. (1938). Cyclic changes in the thyroid and adrenal cortex of the male Starling, *Sturnus vulgaris*, and their relation to the sexual cycle. *Amer. Nat.* **72**, 562–570.

Burger, J. W. (1953). The effect of photic and psychic stimuli on the reproductive cycle of the male starling, *Sturnus vulgaris*. *J. exp. Zool.* **124**, 227–239.

Burke, W. H., and Papkoff, H. (1980). Purification of turkey prolactin and the development of a homologous radioimmunoassay for its measurement. *Gen. Comp. Endocrinol.* **40**, 297–307.

Burke, W. H., Licht, P., Papkoff, H., and Bona-Gallo, A. (1979). Isolation and characterization of luteinizing hormone and follicle-stimulating hormone from pituitary glands of the turkey (*Meleagris galloparvo*). *Gen. Comp. Endocrinol.* **37**, 508–520.

Burton, R. M., and Westphal, U. (1972). Steroid hormone-binding proteins in blood plasma. *Metabolism* **21**, 253–276.

Campbell, R. R., and Leatherland, J. F. (1979). Effect of TRH, TSH, and LHRH on plasma thyroxine and triiodothyronine in the lesser snow goose *(Anser caerulescens caerulescens)* and plasma thyroxine in the Rouen duck (*Anas platyrhynchos*). *Can. J. Zool.* **57**, 271–274.

Campbell, R. R., and Leatherland, J. F. (1980). Seasonal changes in thyroid activity in the lesser snow goose (*Anser caerulescens caerulescens*). *Can. J. Zool.* **58**, 1144–1150.

Campbell, R. R., Ashton, S. A., Follett, B. K., and Leatherland, J. F. (1978). Seasonal changes in plasma concentrations of LH in the lesser snow goose (*Anser caerulescens caerulescens*). *Biol. Reprod.* **18**, 663–668.

Campbell, R. R., Etches, R. J., and Leatherland, J. F. (1981). Seasonal changes in plasma prolactin concentration and carcass lipid levels in the lesser snow goose (*Anser caerulescens caerulescens*). *Comp. Biochem. Physiol.* **68**, 653–658.

Camper, P. M., and Burke, W. H. (1977a). The effects of prolactin on the gonadotropin induced rise in serum estradiol and progesterone of the laying turkey. *Gen. Comp. Endocrinol.* **32**, 72–77.

Camper, P. M. and Burke, W. H. (1977b). The effect of prolactin on reproductive function in female Japanese quail (*Coturnix coturnix japonica*). *Poultry Sci.* **56**, 1130–1134.

Cassone, V. M., and Menaker, M. (1984). Is the avian circadian system a neuroendocrine loop? *J. Exp. Zool.* **232**, 539–549.

Chan, K. M. B., and Lofts, B. (1974). The testicular cycle and androgen biosynthesis in the tree sparrow, *Passer montanus saturatus*. *J. Zool. Lond.* **172**, 47–66.

Chan, S. W. C., and Phillips, J. G. (1973). Variations in the *in vitro* production of corticosteroids by herring gull adrenals. *Gen. Comp. Endocrinol.* **20**, 283–290.

Chandola, A. (1972). Thyroid in reproduction: reproductive physiology of *Lonchura punctulata* in relation to iodine metabolism and hypothyroidism. PhD Thesis, Banaras Hindu University, Varanasi.

Chandola, A., and Bhatt, D. (1982). Tri-iodothyronine fails to mimic gonado-inhibitory action of thyroxine in spotted munia: Effects of injections at different times of day. *Gen. Comp. Endocrinol.* **48**, 499–503.

Chandola, A., and Pathak, V. (1980). Premigratory increase in circulating triiodothyronine/thyroxine ratio in the red-headed bunting (*Emberiza bruniceps*). *Gen. Comp. Endocrinol.* **42**, 39–46.

Chandola, A., Thapliyal, J. P., and Pavnaskar, J. (1974). The effects of thyroidal hormones on the ovarian response to photoperiod in a tropical finch *(Ploceus phillipinus)*. *Gen. Comp. Endocrinol.* **24**, 437–441.

Chandola, A., Bhatt, D., and Pathak, V. K. (1983). Environmental manipulation of seasonal reproduction in spotted munia, *Lonchura punctulata*. *In* "Avian Endocrinology: Environmental and Ecological Perspectives" (S.-I. Mikami, K. Homma, and M. Wada, eds), pp. 229–242. Jap. Sci. Soc. Press, Tokyo, and Springer-Verlag, Berlin.

Chang, W. C., Chung, D., and Li, C. H. (1980). Isolation and characterization of ß-lipotropin and adrenocorticotropin from turkey pituitary glands. *Int. J. Peptide Prot. Res.* **15**, 261–270.

Chase, I. D. (1982). Dynamics of hierarchy formation – the sequential development of dominance relationships. *Behaviour* **80**, 218–240.

Chandhuri, S., and Maiti, B. R. (1989). Pineal activity during the seasonal gonadal cycle in a wild avian species, the tree pie (*Dendrocitta vagabunda*). *Gen. Comp. Endocrinol.* **76**, 346–349.

Chaturvedi, C. M., and Suresh, P. K. (1990). Effects of corticosterone, metapyrone, and ACTH on testicular function at different stages of the breeding cycle in migratory red-headed bunting, *Emberiza bruniceps. Gen. Comp. Endocrinol.* **78**, 1–11.

Cheng, M.-F. (1979). Progress and prospectus in ring dove research: a personal view. *In* "Advances in the Study of Behavior" (J. S. Rosenblatt, R. A. Hinde, E. Shaw, and C. Beer, eds), Vol. 9, pp. 97–129. Academic Press, New York.

Cheng, M.-F. (1983). Behavioral "self feedback" control of endocrine states. *In* "Hormones and Behavior in Higher Vertebrates" (J. Balthazart, E. Pröve, and R. Gilles, eds), pp. 408–421. Springer-Verlag, Berlin.

Cheng, M.-F. (1986). Female cooing promotes ovarian development in ring doves. *Physiol. Behav.* **37**, 371–374.

Cheng, M.-F., and Follett, B. K. (1976). Plasma luteinizing hormone during the breeding cycle of the female ring dove. *Horm. Behav.* **7**, 199–206.

Cheng, M.-F., and Lehrman, D. S. (1975). Gonadal hormone specificity in the sexual behavior of ring doves. *Psychoneuroendocrinol.* **1**, 95–102.

Cherel, Y., Robin, J. P., Walch, O., Karmann, H., Netchitailo, P. and Le Maho, Y. (1988). Fasting in king penguin. I. Hormonal and metabolic changes during breeding. *Amer. J. Physiol.* **23**, R170–R177.

Cockrem, J. F., and Potter, M. A. (1991). Reproductive endocrinology of the North Island brown kiwi, *Apteryx australis mantelli. In* "Acta XX Cong. Intern. Ornithol." (B. D. Bell, ed.), pp. 2092–2101. New Zealand Ornithological Congress Trust Board, Wellington.

Collin, J.-P. (1966). Étude préliminaire des photorécepteurs rudimentaires de l'epiphyse de *Pica pica* L. pendant la vie embryonnaire et post embryonnaire. *C.R. Acad. Sci. Ser D* **265**, 660–667.

Collin, J.-P. (1967). Sir l'évolution des photorécepteurs rudimentaires epiphysiares chez la pie (*Pica pica* L.). *C.R. Seanc. Soc. Biol.* **160**, 1876–1886.

Collin, J.-P. (1971). Differentiation and regression of the cells of the sensory line in the epiphysis cerebri. *In* "The Pineal Gland" (G. E. W. Wolstenholme, and J. Knight, eds), pp. 79–125. Churchill Livingstone, Edinburgh.

Cookson, E. J., Hall, M. R., and Glover, J. (1988). The transport of plasma thyroxine in white storks (*Ciconia ciconia*) and the association of high levels of plasma transthyretin (thyroxine-binding prealbumin) with moult. *J. Endocrinol.* **117**, 75–84.

Cortopassi, A. J., and Mewaldt, L. R. (1965). The circumannual distribution of white-crowned sparrows. *Bird Banding* **36**, 141–169.

Corvol, P., and Bardin, C. W. (1973). Species distribution of testosterone-binding globulin. *Biol. Reprod.* **8**, 277–282.

Craigie, E. H. (1928). Observations on the brain of the hummingbird (*Chrysolampis mosquitus* L. and *Chlorostilbon caribaeus* Lawr.). *J. Comp. Neurol.* **45**, 377–481.

Craigie, E. H. (1930). Studies on the brain of the kiwi *(Apteryx australis). J. Comp. Neurol.* **49**, 223–357.

Craigie, E. H. (1931). The cell masses in the diencephalon of the hummingbird. *Proc. Sci. Koninklijke Akad. Wettenschnappen Amsterdam* **34**, 1038–1050.

Creighton, J. A. (1988b). Thyroidectomy and the termination of juvenile refractoriness in the red-legged partridge *(Alectoris graeca chukar). Gen. Comp. Endocrinol.* **72**, 204–208.

Cusick, E. K., and Wilson, F. E. (1972). On control of spontaneous testicular regression in tree sparrows *(Spizella arborea). Gen. Comp. Endocrinol.* **19**, 441–456.

Dahl, E. (1970). Ultrastructure of ovarian interstitial tissue in the domestic fowl. Part 1. Fixation and normal fine structure. PhD Thesis, University of Oslo.

Da Lage, C. (1955). Innervation neurosécrétoire de l'adenohypophyse chez l'hippocampe. *C.R. Ass. Anat.* **85**, 161.

Davis, J., and Davis, B. S. (1954). The annual gonad and thyroid cycles of the English sparrow in southern California. *Condor* **56**, 328–345.

Dawson, A. (1983). Plasma gonadal steroid levels in wild starlings (*Sturnus vulgaris*) during the annual cycle and in relation to the stages of breeding. *Gen. Comp. Endocrinol.* **49**, 286–294.

Dawson, A., and Goldsmith, A. R. (1982). Prolactin and gonadotropin secretion in wild starlings (*Sturnus vulgaris*) during the annual cycle and in relation to nesting, incubation and rearing of young. *Gen. Comp. Endocrinol.* **48**, 213–221.

Dawson, A., and Goldsmith, A. R. (1985). Modulation of gonadotrophin and prolactin secretion by day length and breeding behavior in free-living starlings (*Sturnus vulgaris*). *J. Zool. Lond.* **206**, 241–252.

Dawson, A., and Howe, P. D. (1983). Plasma corticosterone in wild starlings (*Sturnus vulgaris*) immediately following capture and in relation to body weight during the annual cycle. *Gen. Comp. Endocrinol.* **51**, 303–308.

Dawson, A., Follett, B. K., Goldsmith, A. R., and Nicholls, T. J. (1985). Hypothalamic gonadotropin-releasing hormone and pituitary and plasma FSH and prolactin during photostimulation and photorefractoriness in intact and thyroidectomized starlings *(Sturnus vulgaris). J. Endocrinol.* **105**, 71–77.

Dawson, A., Goldsmith, A. R., Nicholls, T. J., and Follett, B. K. (1986). Endocrine changes associated with the termination of photorefractoriness by short daylengths and thyroidectomy in starlings (*Sturnus vulgaris*). *J. Endocrinol.* **110**, 73–79.

Dawson, A., Williams, T. D., and Nicholls, T. J. (1987). Thyroidectomy of nestling starlings appears to cause neotenous sexual maturation. *J. Endocrinol.* **112**, 125–126.

Deguchi, T. (1982). Endogenous oscillator and photoreceptor for *N*-acetyltransferase rhythm in chicken pineal gland. *In* "Vertebrate Circadian Systems" (J. Aschoff, S. Daan, and G. Groos, eds), pp. 164–172. Springer-Verlag, Berlin.

Deviche, P., Heyns, W., Balthazart, J., and Hendrick, J.-C. (1979). Inhibition of LH plasma levels by corticosterone administration in the male duckling (*Anas platyrhynchos*). *I.R.C.S. Med. Sci.* **7**, 622.

Deviche, P., Balthazart, J., Heyns, W., and Hendrick, J.-C. (1980). Endocrine effects of castration followed by androgen replacement and ACTH injections in the male domestic duck (*Anas platyrhynchos* L.). *Gen. Comp. Endocrinol.* **41**, 53–61.

Dhondt, A. A. and Eyckerman, R. (1979). Temperature and date of laying by tits *Parus* sp. *Ibis* **121**, 329–331.

Dittami, J. P. (1981). Seasonal changes in the behavior and plasma titers of various hormones in bar-headed geese, *Anser indicus*. *Z. Tierpsychol.* **55**, 289–324.

Dittami, J. P. (1986). Seasonal reproduction, moult and their endocrine correlates in two tropical Ploceidae species. *J. Comp. Physiol. B* **156**, 641–647.

Dittami, J. P. (1987). A comparison of breeding and molt cycles and life histories in two tropical starling species: the blue-eared glossy starling, *Lamprotornis chalybaeus*, and Rüppell's long-tailed glossy starling, *L. purpuropterus*. *Ibis* **129**, 69–85.

Dittami, J. P., and Gwinner, E. (1985). Annual cycles in the African stonechat, *Saxicola torquata axillaris*, and their relationship to environmental factors. *J. Zool. Lond.* **207**, 350–357.

Dittami, J. P., and Hall, M. R. (1983). Molt, T4, and testosterone in adult male and female bar-headed geese, *Anser indicus*. *Can J. Zool.* **61**, 2695–2697.

Dittami, J. P., and Knauer, B. (1986). Seasonal organization of breeding and molting in the fiscal shrike (*Lanius collaris*). *J. Orn.* **127**, 79–84.

Dittami, J. P., and Reyer, H.-U. (1984). A factor analysis of seasonal behavioral hormonal and body weight changes in adult male bar-headed geese, *Anser indicus*. *Behaviour* **90**, 114–124.

Dittami, J. P., Goldsmith, A. R., and Follett, B. K. (1985). Seasonal changes in follicle-stimulating hormone in a breeding population of bar-headed geese, *Anser indicus*. *Gen. Comp. Endocrinol.* **57**, 195–197.

Dodd, J. M., Follett, B. K., and Sharp, P. J. (1971). Hypothalamic control of pituitary function in submammalian vertebrates. *Adv. Comp. Physiol. Biochem.* **4**, 113–223.

Dolnik, V. R. (1964). O mekhanizime fotoperiodicheskovo kontrolya endogennovo ritma polovoi tsiklichnosti ptits. *Zoologicheskii Zh.* **43**, 720–733.

Dolnik, V. R. (1975). Fotoperiodicheskii kontrol sezonnykh tsiklov beca tela linki i polovoi aktivnosti u zyablekov (*Fringilla coelebs*). *Zoologicheskii Zh.* **54**, 1048–1056.

Dolnik, V. R. (1976). Fotoperiodizm u ptits. *In* "Fotoperiodizm Zhivotnykhi Rastenii" (L. Zaslavsky, ed.), pp. 47–81. Akademiya Nauk SSSR, Leningrad.

Dominic, C. J., and Singh, R. M. (1969). Anterior and posterior groups of portal vessels in the avian pituitary. *Gen. Comp. Endocrinol.* **13**, 22–26.

Donham, R. S. (1979). The annual cycle of plasma luteinizing hormone and sex hormones in male and female mallards (*Anas platyrhynchos*). *Gen. Comp. Endocrinol.* **29**, 152–155.

Donham, R. S., Dane, C. W., and Farner, D. S. (1976). Plasma luteinizing hormone and the development of ovarian follicles after loss of clutch in female mallards (*Anas platyrhynchos*). *Gen. Comp. Endocrinol.* **29**, 152–155.

Donham, R. S., Wingfield, J. C., Mattocks, P. W. Jr, and Farner, D. S. (1982). Changes in testicular and plasma androgens with photoperiodically induced increase in plasma LH in the house sparrow. *Gen. Comp. Endocrinol.* **48**, 342–347.

Dorrington, J., and Gore-Langton, R. E. (1981). Prolactin inhibits oestrogen synthesis in the ovary. *Nature* **290**, 600–602.

Drager, G. A. (1945). The innervation of the avian hypophysis. *Endocrinology* **36**, 124–129.

Drent, R. H., and Daan, S. (1980). The prudent parent: Energetic adjustments in avian breeding. *Ardea* **68**, 224–252.

Dufty, A. M. Jr (1989) Testosterone and survival: a cost of aggressiveness? *Horm. Behav.* **23**, 185–193.

Dufty, A. M. Jr, and Wingfield, J. C. (1986a). Temporal patterns of circulating LH and steroid hormones in a brood parasite, the brown-headed cowbird, *Molothrus ater*. I. Males. *J. Zool. Lond.* **208**, 191–203.

Dufty, A. M. Jr, and Wingfield, J. C. (1986b). Temporal patterns of circulating LH and steroid hormones in a brood parasite, the brown-headed cowbird, *Molothrus ater*. II. Females. *J. Zool. Lond.* **208**, 205–214.

Dufty, A. M. Jr, and Wingfield, J. C. (1986c). Influence of social cues on the reproductive endocrinology of male brown-headed cowbirds: field and laboratory studies. *Horm. Behav.* **20**, 222–234.

Dufty, A. M. Jr, Goldsmith, A. R., and Wingfield, J. C. (1987). Prolactin secretion in a brood parasite, the brown-headed cowbird, *Molthus ater*. *J. Zool. Lond.* **212**, 669–675.

Dusseau, J. W. and Meier, A. H. (1971). Diurnal and seasonal variations of plasma adrenal steroid hormone in the white-throated sparrow, *Zonotrichia albicollis*. *Gen. Comp. Endocrinol.* **16**, 399–408.

Duvernoy, H., and Koritké, J. G. (1964). Contributions à l'étude de l'angioarchitectonie des organes circumventriculaires. *Arch Biol. Suppl.* **75**, 693–748.

Duvernoy, H., and Koritké, J. G. (1968). Les vaisseaux sous-épendymaires du recessus hypophysaire. *J. Hirnforschung* **10**, 227–245.

Duvernoy, H., Gainet, F., and Koritké, J. G. (1969). Sur la vascularization de l'hypohyse des oiseaux. *J. Neuro-visceral Rel.* **31**, 109–127.

Duvernoy, H., Koritké, J. G., and Monnier, G. (1970). Architecture du plexus primaire du système porte hypophysaire. *Coll. Nat. Cent. Nat. Rech. Sci.* **927**, 137–144.

Dyachenko, V. P. (1976). Rol sezonnykh izmenenii sekretsii prolaktina i chustvitelnosti k nemu v regulyastsii migratsionnovo zhirootlozheniya u nekotorykh vitov ptits. *In* "Fiziologicheskie Osnovy Migratsionnovo Zoologichicheskii Institut", pp. 211–212. AN SSSR, Leningrad.

Dyachenko, V. P. (1982). The roles of seasonal variations in prolactin secretion and in prolactin responsiveness in the regulation of migratory fattening in some bird species. *In* "Ornithological Studies in the USSR" (V. M. Gavrilov, and P. L. Potapov, eds), Vol. 2, pp. 306–327. USSR Acad. Sci. Moscow.

Ebling, F. J. P., Goldsmith, A. R., and Follett, B. K. (1982). Plasma prolactin and luteinizing hormone during photoperiodically induced testicular growth and regression in starlings, (*Sturnus vulgaris*). *Gen. Comp. Endocrinol.* **48**, 485–490.

Edens, F. W., and Siegel, H. S. (1975). Adrenal responses in high and low ACTH response lines of chickens during acute heat stress. *Gen. Comp. Endocrinol.* **25**, 64–73.

El-Halawani, M. E., Burke, W. H., and Dennison, P. T. (1980). Effect of nest deprivation on serum prolactin level in nesting female turkeys. *Biol. Reprod.* **23**, 118–123.

Engels, W. L. (1962). Day length and the termination of photorefractoriness in the annual testicular cycle of the bobolink, (*Dolichonyx oryzivorous*). *Biol. Bull.* **123**, 94–104.

Engels, W. L. (1969). Photoperiodically induced testicular recrudescence in the transequitorial migrant *Dolichonyx oryzivorous* relative to natural photoperiods. *Biol. Bull.* **137**, 256–264.

Ensor, D. M. (1975). Prolactin and adaptation. *In* "Avian Physiology" (M. Peaker, ed.), pp. 129–148. Academic Press, New York.

Epple, A., and Stetson, M. H. (1980). "Avian Endocrinology". Academic Press, New York.

Erickson, J. E. (1975). Hypothalamic gonadotropin-releasing hormone and the photoperiodic control of the testes in the white-crowned sparrow, *Zonotrichia leucophrys gambelii*. PhD Thesis, University of Washington, Seattle.

Erpino, M. J. (1968). Aspects of thyroid histology in black-billed magpies. *Auk* **85**, 397–403.

Erpino, M. J. (1969). Seasonal cycle of reproductive physiology in the black-billed magpie. *Condor* **71**, 267–279.

Etches, R. J. (1976). A radioimmunoassay for corticosterone and its application to the measurement of stress in poultry. *Steroids* **28**, 763–773.

Etches, R. J., Garbutt, A., and Middleton, A. L. (1979). Plasma concentrations of prolactin during egg-laying and incubation in the ruffed grouse (*Bonasa umbellus*). *Can. J. Zool*, **57**, 1624–1627.

Etzold, F. (1891). Die Entwicklung der Testikal von *Fringilla domestica* von der Winteruhe bis zum Eintritt der Brunst. *Z. Wissenschaft. Zool.* **52**, 46–84.

Farmer, S. W., Papkoff, H., and Hyashida, T. (1974). Purification and properties of avian growth hormone. *Endocrinology* **95**, 1560–1565.

Farner, D. S. (1964). The photoperiodic control of reproductive cycles in birds. *Amer. Sci.* **52**, 137–156.

Farner, D. S. (1970). Predictive function in the control of annual cycles. *Environ. Res.* **3**, 119–131.

Farner, D. S. (1975). Photoperiodic controls in the secretion of gonadotropins in birds. *Amer. Zool.* **15**, 117–135.

Farner, D. S. (1983). Some recent advances in avian physiology. *J. Yamashina Inst. Ornithol.* **15**, 97–140.

Farner, D. S. (1985). Annual rhythms. *Ann. Rev. Physiol.* **47**, 65–82.

Farner, D. S. (1986). Generation and regulation of annual cycles in migratory passerine birds. *Amer. Zool.* **26**, 493–501.

Farner, D. S. and Mewaldt, L. R. (1955). The natural termination of the refractory period in the white-crowned sparrow. *Condor* **57**, 112–116.

Farner, D. S., and Lewis, R. A. (1971). Photoperiodism and reproductive cycles in birds. *In* "Photoperiodicity" (A. C. Giese, ed.), Vol. 6, pp. 325–370. Academic Press, New York.

Farner, D. S., and Follett, B. K. (1979). Reproductive periodicity in birds. *In* "Hormones and Evolution" (E. J. W. Barrington, ed.), pp. 829–872. Academic Press, New York.

Farner, D. S., and Gwinner, E. (1980). Photoperiodicity, circannual and reproductive cycles. *In*

"Avian Endocrinology" (A. Epple, and M. H. Stetson, eds), pp. 331–366. Academic Press, New York.

Farner, D. S., Donham, R. S., Lewis, R. A., Mattocks, P. W. Jr, Darden, T. R., and Smith, J. P. (1977). The circadian component in the photoperiodic mechanism of the house sparrow, *Passer domesticus*. *Physiol. Zool.* **50**, 247–268.

Farner, D. S., Donham, R. S., Moore, M. C., and Lewis, R. A. (1980). The temporal relationship between the cycle of testicular development and molt in the white-crowned sparrow, *Zonotrichia leucophrys gambelii*. *Auk* **97**, 63–75.

Farner, D. S., Donham, R. S., Matt, K. S., Mattocks, P. W. Jr, Moore, M. C., and Wingfield, J. C. (1983). The nature of photorefractoriness. *In* "Avian Endocrinology – Environmental and Ecological Perspectives" (S.-I. Mikami, K. Homma, and M. Wada, eds), pp. 149–166. Jap. Sci. Soc. Press, Tokyo, and Springer-Verlag, Berlin.

Feder, H. H., Storey, A., Goodwin, D., Reboulleau, C., and Silver, R. (1977). Testosterone and 5-alpha-dihydrotestosterone levels in peripheral plasma of male and female ring doves (*Streptopelia risoria*). *Biol. Reprod.* **16**, 666–677.

Feldman, S. E., Snapir, N., Yasuda, M., Treuting, F., and Lepkovsky, S. (1973). Physiological and nutritional consequences of brain lesions: a functional atlas of the chicken hypothalamus. *Hilgardia* **41**, 605–629.

Fevold, H. R., and Eik-Nes, K. B. (1962). Progesterone metabolism by testicular tissue of the English sparrow (*Passer domesticus*) during the annual reproductive cycle. *Gen. Comp. Endocrinol.* **2**, 506–515.

Fink, B. A. (1957). Radioiodine, a method for measuring thyroid activity. *Auk* **74**, 487–493.

Firket, J. (1920). Recherches sur l'organogénése des glandes sexuelles chez les oiseux. *Arch. Biol.* **30**, 393–516.

Fivizzani, A. J., and Oring, L. W. (1986). Plasma steroid hormones in the polyandrous spotted sandpiper, *Actitis macularia*. *Biol. Reprod.* **35**, 1195–1201.

Fivizzani, A. J., Colwell, M. A., and Oring, L. W. (1986). Plasma steroid hormone levels in free-living Wilson's phalaropes, *Phalaropus tricolor*. *Gen. Comp. Endocrinol.* **62**, 137–144.

Fivizzani, A. J., Oring, L. W., El-Halawani, M. E., and Schlinger, B. A. (1990). Hormonal basis of parental care and female intersexual competition in sex-role reversed birds. *In* "Endocrinology of Birds: Molecular to Behavioral" (M. Wada, S. Ishii, and C. G. Scanes, eds), pp. 273–286. Jap. Sci. Soc. Press, Tokyo, and Springer-Verlag, Berlin.

Flickinger, D. P. (1959). Adrenal responses of California quail subjected to various physiologic stimuli. *Proc. Soc. exp. Biol. Med.* **100**, 23–25.

Fogden, M. P. L. (1972). The seasonality and population dynamics of equitorial forest birds in Sarawak. *Ibis* **114**, 307–343.

Fogden, M. P. L., and Fogden, P. M. (1979). The role of fat and protein reserves in the annual cycle of the grey-backed camaroptera in Uganda (Aves: Sylviidae). *J. Zool. Lond.* **189**, 233–258.

Follett, B. K. (1976). Follicle-stimulating hormone during photoperiodically induced sexual maturation in male Japanese quail. *J. Endocrinol.* **69**, 117–126.

Follett, B. K. (1981). The stimulation of luteinizing hormone and follicle-stimulating hormone secretion in quail with complete and skeleton photoperiods. *Gen. Comp. Endocrinol.* **45**, 306–316.

Follett, B. K. (1984). Birds. *In* "Marshall's Physiology of Reproduction I. Reproductive Cycles of Vertebrates" (G. E. Lamming, ed.), pp. 283–350. Churchill-Livingstone, Edinburgh.

Follett, B. K., and Robinson, J. E. (1980). Photoperiod and gonadotropin secretion in birds. *Prog. Reprod. Biol.* **5**, 39–61.

Follett, B. K., Scanes, C. G., and Cunningham, F. J. (1972). A radioimmunoassay for avian luteinizing hormone. *J. Endocrinol.* **52**, 359–378.

Follett, B. K., Farner, D. S., and Mattocks, P. W. Jr (1975). Luteinizing hormone in the plasma of white-crowned sparrows, *Zonotrichia leucophrys gambelii*, during artificial photostimulation. *Gen. Comp. Endocrinol.* **26**, 126–134.

Follett, B. K., Davies, D. T., Gibson, R., Hodges, K. J., Jenkins, N., Maung, S. L., Maung, Z. W., Redshaw, M. R., and Sumpter, J. P. (1978). Avian gonadotropins – their purification and assay. *Ind. J. Ornithol.* (*Pavo*) **16**, 34–55.

Follett, B. K., Ishii, S., and Chandola, A. (1985). "The Endocrine System and the Environment". Jap. Sci. Soc. Press, Tokyo, and Springer-Verlag, Berlin.

Fornasari, L., Bottoni, L., Schwabl, H., and Massa, R. (1992). Testosterone in the breeding cycle of the male red-backed shrike, *Lanius collurio*. *Ethol. Ecol. Evol.* **4**, 193–196.

Foster, M. S. (1975). The overlap of molting and breeding in some tropical birds. *Condor* **77**, 304–314.

Foster, R. G., Follett, B. K., and Lythgoe, J. N. (1985). Rhodopsin-like sensitivity of extra-retinal photoreceptors mediating the photoperiodic response in quail. *Nature* **313**, 50–52.

Fraissinet, M., Varriale, B., Pierantoni, R., Caliendo, M. F., Di Matteo, L., Bottoni, L. and Milone, M. (1987). Annual testicular activity in the grey partridge (*Perdix perdix* L.). *Gen. Comp. Endocrinol.* **68**, 28–32.

Fromme-Bouman, H. (1962). Jahresperiodik Untersuchungen am der Amsel (*Turdus merula*). *Vogelwarte* **21**, 188–198.

Furuya, T., and Ishii, S. (1974). Separation of chicken adenohypophysial gonadotropins. *Endocrinol. Jap.* **21**, 329–334.

Furuya, T., and Ishii, S. (1976). Effects of follicle-stimulating hormone and luteinizing hormone on the incorporation of P-32 into testis of immature Japanese quail. *Gen. Comp. Endocrinol.* **29**, 556–559.

Gal, G. (1940). A madarak pajsmirigyszerkezetének ciklikus változásai. *Math. Termeszettudomanyi Ertesito* (*Budapest*) **59**, 360–378.

Garbutt, A., Leatherland, J. F., and Middleton, A. L. A. (1979). Seasonal changes in serum thyroid hormone levels in ruffed grouse maintained under natural conditions of temperature and photoperiod. *Can. J. Zool.* **57**, 2022–2027.

Gaston, S., and Menaker, M. (1968). Pineal function: the biological clock in the sparrow? *Science* **160**, 1125–1127.

Gavrilov, V. M., and Dolnik, V. R. (1974). Bioenergetika i regulyatsiya poslebrachnoi i postyuvenalnoi linek u zyablikov (*Fringilla coelebs coelebs* L.). *Trudy Zool. Inst. Akad. Nauk SSSR* **55**, 14–61.

George, J. C. (1980). Structure and physiology of posterior lobe hormones. *In* "Avian Endocrinology" (A. Epple, and M. H. Stetson, eds), pp. 85–115. Academic Press, New York.

Ghosh, A. (1962). A comparative study of histochemistry of the avian adrenals. *Gen. Comp. Endocrinol. Suppl.* **1**, 75–80.

Gilbert, A. B. (1971). The endocrine ovary in reproduction. *In* "Physiology and Biochemistry of the Domestic Fowl" (D. J. Bell, and B. M. Freeman, eds), Vol. 3, pp. 1450–1568. Academic Press, New York.

Gledhill, B., and Follett, B. K. (1976). Diurnal variation and the episodic release of plasma gonadotropins in Japanese quail during a photoperiodically induced gonadal cycle. *J. Endocrinol.* **71**, 245–247.

Goldsmith, A. R. (1982a). The Australian black swan (*Cygnus atratus*): prolactin and gonadotropin secretion during breeding including incubation. *Gen. Comp. Endocrinol.* **46**, 458–463.

Goldsmith, A. R. (1982b). Plasma concentrations of prolactin during incubation and parental feeding throughout repeated breeding cycles in canaries (*Serinus canarius*). *J. Endocrinol.* **94**, 51–59.

Goldsmith, A. R. (1983). Prolactin in avian reproductive cycles. *In* "Hormones and Behaviour in Higher Vertebrates" (J. Balthazart, E. Pröve, and R. Gilles, eds), pp. 375–387. Springer-Verlag, Berlin.

Goldsmith, A. R. (1991). Prolactin and avian reproductive strategies. *In* "Acta XX Congressus Internationalis Ornithologici" (B. Bell, ed.), pp. 2063–2071. New Zealand Ornithological Congress Trust Board, Wellington.

Goldsmith, A. R., and Follett, B. K. (1980). Anterior pituitary hormones. *In* "Avian Endocrinology" (A. Epple, and M. H. Stetson, eds), pp. 147–165. Academic Press, New York.

Goldsmith, A. R., and Follett, B. K. (1983). Avian LH and FSH: comparison of several radioimmunoassays. *Gen. Comp. Endocrinol.* **50**, 24–55.

Goldsmith, A. R., and Hall, M. (1980). Prolactin concentrations in the pituitary and plasma of Japanese quail in relation to photoperiodically induced sexual maturation and egg laying. *Gen. Comp. Endocrinol.* **42**, 449–454.

Goldsmith, A. R., and Nicholls, T. J. (1984). Thyroxine induces photorefractoriness and stimulates prolactin secretion in European starlings (*Sturnus vulgaris*). *J. Endocrinol.* **101**, R1–R3.

Goldsmith, A. R., Edwards, C., Koprucu, M., and Silver, R. (1981). Concentrations of prolactin and luteinizing hormone in plasma of doves in relation to incubation and development of the crop gland. *J. Endocrinol.* **90**, 437–443.

Gorbman, A., Dickhoff, W. W., Vigna, S. R., Clark, N. B., and Ralph, C. L. (1983). "Comparative Endocrinology". Wiley, New York.

Gorman, M. L. (1977). Sexual behavior and plasma androgen concentrations in the male eider duck (*Somateria mollissima*). *J. Reprod. Fertil.* **49**, 225–230.

Gorman, M. L., and Milne, H. (1971). Seasonal changes in the adrenal steroid tissue of the common eider *Somateria mollissima* and its relation to organic metabolism in normal and oil-polluted birds. *Ibis* **113**, 218–228.

Gratto-Trevor, C. L., Fivizzani, A. J., Oring, L. W., and Cooke, F. (1990a). Seasonal changes in gonadal steroids of a monogamous versus a polyandrous shorebird. *Gen. Comp. Endocrinol.* **80**, 407–418.

Gratto-Trevor, C. L., Oring, L. W., and Fivizzani, A. J. (1990b). The role of prolactin in parental care in a monogamous and polyandrous shorebird. *Auk* **107**, 718–729.

Greely, F., and Meyer, R. K. (1953). Seasonal variations in testes-stimulating activity of male pheasant pituitary glands. *Auk* **70**, 350–358.

Green, J. D. (1951). The comparative anatomy of the hypophysis, with special reference to its blood supply and innervation. *Amer. J. Anat.* **88**, 225–311.

Greenspan, F. S., Li, C. H., Simpson, M. E., and Evans, H. M. (1949). Bioassay of hypophysial growth hormone: the tibial test. *Endocrinology* **45**, 455–463.

Groothuis, T., and Meeuwissen, G. (1992). The influence of testosterone on the development and fixation of the form of displays in two age classes of young black-headed gulls. *Anim. Behav.* **43**, 189–208.

Groscolas, R. (1982). Modifications metaboliques et hormonale en relation avec le jeune prolongé, la reproduction et la mué chez le manchot emperor (*Aptenodytes forsteri*). PhD Thésis, Dijon University, Dijon.

Groscolas, R., and Leloup, J. (1986). The endocrine control of reproduction and molt in male and female emperor (*Aptenodytes forsteri*) and Adelie (*Pygoscelis adeliae*) penguins. II. Annual changes in plasma levels of thyroxine and tri-iodothyronine. *Gen. Comp. Endocrinol.* **63**, 264–274.

Groscolas, R., Jallageas, M., Goldsmith, A. R., Leloup, J., and Assenmacher, I. (1985). Changes in plasma LH, gonadal and thyroid hormones in breeding and molting emperor penguins. *In* "Current Trends in Comparative Endocrinology" (B. Lofts, and W. N. Holmes, eds), pp. 261–264. University of Hong Kong Press, Hong Kong.

Groscolas, R., Jallageas, M., Goldsmith, A. R., and Assenmacher, I. (1986). The endocrine control of reproduction and molt in male and female emperor (*Aptenodytes forsteri*) and adelie (*Pygoscelis adeliae*) penguins. I. Annual changes in plasma levels of gonadal steroids and LH. *Gen. Comp. Endocrinol.* **62**, 43–53.

Groscolas, R., Jallageas, M., Leloup, J., and Goldsmith, A. R. (1988). The endocrine control of reproduction in male and female emperor penguins (*Aptenodytes forsteri*). *In* "Acta XIX Int. Ornithol. Cong." (H. Ouellet, ed.), pp. 1692–1701. University of Ottawa Press, Ottawa.

Guhl, A. M. (1964). Psychophysiological interrelationships in the social behavior of chickens. *Psychol. Bull.* **61**, 277–285.

Gwinner, E. (1975). Circadian and circannual rhythms in birds. *In* "Avian Biology" (D. S. Farner, and J. R. King, eds), Vol. 5, pp. 221–288. Academic Press, New York.

Gwinner, E. (1981). Circannual systems. *In* "Handbook of Behavioral Neurobiology 4. Biological Rhythms" (J. Aschof, ed.), pp. 391–410. Plenum Press, New York.

Gwinner, E. (1986). "Circannual Rhythms". Springer-Verlag, Berlin.

Gwinner, E., Wozniak, J., and Dittami, J. (1981). The role of the pineal organ in the control of annual rhythms in birds. *In* "The Pineal Organ: Photobiology – Biochronometry – Endocrinology" (A. Oksche, and P. Pévet, eds), pp. 99–121, Elsevier/North-Holland Biomedical Press, Amsterdam.

Haack, D. W., Abel, J. H. Jr, and Rhees, R. W. (1972). Zonation in the adrenal of the duck: effect of osmotic stress. *Cytobiologie* **5**, 247–264.

Haase, E. (1975). Zur Wirkung jahreszeitlicher und sozialer Faktoren auf Hoden und Hormonspiegel von Haustauben. *Verh. Deut. Zool. Ges.* **1975**, 137.

Haase, E., and Sharp, P. J. (1975). Annual cycle of plasma luteinizing hormone concentrations in wild mallard drakes. *J. exp. Zool.* **194**, 553–558.

Haase, E., Paulke, E., and Sharp, P. J. (1976). Effects of seasonal and social factors on testicular activity and hormone levels in domestic pigeons. *J. exp. Zool.* **197**, 81–96.

Haase, E., Sharp, P. J., and Paulke, E. (1982). The effects of castration on the seasonal pattern of plasma LH concentrations in wild mallard drakes. *Gen. Comp. Endocrinol.* **46**, 113–115.

Haecker, V. (1926). Über jahresperiodik Veranderungen und klimatisch bedingte Verschiedenheit der Vogelschilddrüse. *Schweiz Med. Wochenschr.* **56**, 337–341.

Hall, B. K. (1968). The annual interrenal tissue cycle within the adrenal gland of Eastern rosella, *Platycercus eximus*, (Aves: Psittaciformes). *Austral. J. Zoology* **16**, 609–617.

Hall, M. R. (1986). Plasma concentrations of prolactin during the breeding cycle in the Cape gannet (*Sula capensis*): a foot incubator. *Gen. Comp. Endocrinol.* **64**, 112–121.

Hall, M. R. (1987). External stimuli affecting incubation behavior and prolactin secretion in the duck (*Anas platyrhynchos*). *Horm. Behav.* **21**, 269–287.

Hall, M. R., and Goldsmith, A. R. (1983). Factors effecting prolactin secretion during breeding and incubation in the domestic duck (*Anas platyrhynchos*). *Gen. Comp. Endocrinol.* **49**, 270–276.

Hall, M. R., Gwinner, E., and Bloesch, M. (1987). Annual cycles in moult, body mass, luteinizing hormones, prolactin, and gonadal steroids during the development of sexual maturity in the white stork (*Ciconia ciconia*). *J. Zool. Lond.* **211**. 467–486.

Haller, B. (1898). Untersuchungen über die Hypophyse und die Infundibularorgane. *Morphol. Jahrbuch* **25**, 31–115.

Halse, S. A. (1985). Gonadal cycles and levels of luteinizing hormone in wild spur-winged geese, *Plectropterus gambensis*. *J. Zool. Lond.* **205**, 335–355.

Hamner, W. M. (1964). Circadian control of photoperiodism in the house finch demonstrated by interrupted-night experiments. *Nature* **203**, 1400–1401.

Hamner, W. M. (1966a). Photoperiodic control of the annual testicular cycle in the house finch *Carpodacus mexicanus*. *Gen. Comp. Endocrinol.* **7**, 224–233.

Hamner, W. H. (1966b). The photorefractory period of the house finch. *Ecology* **49**, 212–227.

Hannover, A. (1824). "Recherches Microscopiques sur le Système Nerveux". Paris.

Harding, C. F. (1981). Social modulation of circulating hormone levels in the male. *Am. Zool.* **21**, 223–232.

Harding, C. F. (1983). Hormonal influences on avian aggressive behavior. *In* "Hormones and Aggressive Behavior" (B. Svare, ed.), pp. 435–467. Plenum Press, New York.

Harding, C. F., and Follett, B. K. (1979). Hormone changes triggered by aggression in a natural population of blackbirds. *Science* **203**, 918–920.

Harding, C. F., Walters, M. J., Collado, D., and Sheridan, K. (1988). Hormonal specificity and activation of social behavior in male red-winged blackbirds. *Horm. Behav.* **22**, 402–418.

Harris, M. P. (1969). Food as a factor controlling the breeding of *Puffinus Iherminieri*. *Ibis* **111**, 139–156.

Hartman, F. A., and Albertin, R. H. (1951). A preliminary study of the avian adrenal. *Auk* **68**, 202–209.

Hartman, F. A., Knouff, R. A., McNutt, A. W., and Carver, J. E. (1947). Chromaffin patterns in bird adrenals. *Anat. Rec.* **97**, 211–222.

Harvey, S., Phillips, J. G., Rees, A., and Hall, T. R. (1984). Stress and adrenal function. *J. Exp. Zool.* **232**, 633–645.

Hartwig, H. G. (1987). Structure and function of retinal and extraretinal photoreceptive organs. *In* "Comparative Physiology of Environmental Adaptations" (P. Pévet, ed.), Vol. 3, pp. 45–55. Karger, Basel.

Hartwig, H. G., and Oksche, A. (1981). Photoneuroendocrine cells and systems: a concept revisited. *In* "The Pineal Organ: Photobiology – Biochronometry – Endocrinology" (A. Oksche, and P. Pévet, eds), pp. 49–59. Elsevier/North Holland Biomedical Press, Amsterdam.

Hartwig, H. G., and Oksche, A. (1982). Neurobiological aspects of extraretinal photoreceptive systems: structure and function. *Experentia* **38**, 991–996.

Harvey, S., and Scanes, C. G. (1977). Purification and radioimmunoassay of chicken growth hormone. *J. Endocrinol.* **73**, 321–329.

Harvey, S., Scanes, C. G., Chadwick, A., and Bolton, N. J. (1978). The effect of thyrotropin-releasing hormone (TRH) and somatostatin (GHRIH) on growth hormone and prolactin secretion *in vitro* and *in vivo* in the domestic fowl (*Gallus domesticus*). *Neuroendocrinology* **26**, 249–260.

Harvey, S., Scanes, C. G., and Sharp, P. J. (1982). Annual cycle of plasma concentrations of growth hormone in red grouse (*Lagopus lagopus scoticus*). *Gen. Comp. Endocrinol.* **48**, 411–414.

Heald, P. J. and MacLachlan, P. M. (1964). The isolation of phosphovitin from plasma of oestrogen-treated immature pullets. *Biochem. J.* **92**, 51–55.

Hector, J. A. L. (1988). Reproductive endocrinology of albatrosses. *In* "Acta XIX Int. Ornithol. Cong." (H. Ouellet, ed.), pp. 1702–1709. University of Ottawa Press, Ottawa.

Hector, J. A. L., and Goldsmith, A. R. (1985). The role of prolactin during incubation: comparative studies of three *Diomedia* albatrosses. *Gen. Comp. Endocrinol.* **60**, 236–243.

Hector, J. A. L., Goldsmith, A. R., and Follett, B. K. (1984). Albatrosses: prolactin secretion during incubation and chick rearing. *Gen. Comp. Endocrinol.* **53**, 464.

Hector, J. A. L., Croxall, J. P., and Follett, B. K. (1986a). Reproductive endocrinology of the wandering albatross, *Diomedia exulans*, in relation to biennial breeding and deferred sexual maturity. *Ibis* **128**, 9–22.

Hector, J. A. L., Follett, B. K., and Prince, P. A. (1986b). Reproductive endocrinology of the black-browed albatross, *Diomedia melanophris*, and the grey-headed albatross, *D. chrysostoma*. *J. Zool. Lond.* **208**, 237–253.

Hegner, R. E., and Wingfield, J. C. (1986a). Behavioral and endocrine correlates of multiple brooding in the semi-colonial house sparrow, *Passer domesticus*. I. Males. *Horm. Behav.* **20**, 294–312.

Hegner, R. E., and Wingfield, J. C. (1986b). Behavioral and endocrine correlates of multiple brooding in the semi-colonial house sparrow, *Passer domesticus*. II. Females. *Horm. Behav.* **20**. 313–326.

Hegner, R. E., and Wingfield, J. C. (1987). Effects of experimental manipulation of testosterone levels on parental investment and breeding success in male house sparrows. *Auk* **104**, 462–469.

Hepp, G. R., Connolly, P., Kennamer, R. A., and Harvey, W. F. N. (1991). Wood duck hatch date: relationship to pairing chronology, plasma luteinizing hormone and steroid hormones during autumn and winter. *Horm. Behav.* **25**, 242–257.

Hiatt, E. S., Goldsmith, A. R., and Farner, D. S. (1987). Plasma levels of prolactin and gonadotropins during the reproductive cycle of white-crowned sparrows (*Zonotrichia leucophrys*). *Auk* **104**, 208–217.

Hinde, R. A. (1965). Interactions of internal and external factors in integration of canary reproduction. *In* "Sex and Behaviour" (F. Beach, ed.), pp. 381–415. Wiley, New York.

Hinde, R. A., and Steel, E. (1978). The influence of day length and male vocalizations on the oestrogen dependent behavior of female canaries and budgerigars, with discussion of data from other species. *In* "Advances in the Study of Behavior" (J. S. Rosenblatt, R. A. Hinde, C. Beer, and M.-C. Busnel, eds), Vol. 8, pp. 39–73. Academic Press, New York.

Hinde, R. A., Steel, E., and Follett, B. K. (1974). Non-ovary mediated effect of photoperiod on oestrogen-induced nest building. *J. Reprod. Fertil.* **40**, 383–399.

Hissa, R., Saarela, S., Balthazart, J., and Etches, R. J. (1983). Annual variation in concentrations of circulating hormones in capercaillie (*Tetrao urogallus*). *Gen. Comp. Endocrinol.* **51**, 183–190.

Höhn, E. O. (1947). Sexual behavior and seasonal changes in the gonad and adrenals of the mallard. *Proc. Zool. Soc. Lond.* **117**, 281–304.

Höhn, E. O. (1950). Physiology of the thyroid gland in birds. *Ibis* **92**, 464–473.

Hön, E. O. (1959). Prolactin in the cowbird's pituitary in relation to avian brood parasitism. *Nature* **184**, 2030.

Höhn, E. O., and Braun, C. E. (1977). Seasonal thyroid gland histophysiology and weight in white-tailed ptarmigan. *Auk* **94**, 544–551.

Höhn, E. O., and Cheng, S. C. (1967). Gonadal hormones in Wilson's phalarope (*Steganopus tricolor*) and other birds in relation to plumage and sex behavior. *Gen. Comp. Endocrinol.* **8**, 1–11.

Holberton, R. L. (1991). Fall migration in the Dark-eyed Junco: Internal and external factors influencing migratory behavior in a short-distance migrant. PhD Thesis, State University of New York at Albany.

Holmes, W. N., and Cronshaw, J. (1980). Adrenal cortex: structure and function. *In* "Avian Endocrinology" (A. Epple, and M. H. Stetson, eds), pp. 271–300. Academic Press, New York.

Holmes, W. N. and Phillips, J. G. (1976). The adrenal cortex of birds. *In* "General, Comparative and Clinical Endocrinology of the Adrenal Cortex" (I. Chester-Jones, and I. W. Henderson, eds), pp. 293–420. Academic Press, New York.

Homma, K., Ohta, M., and Sakakibara, Y. (1980). Surface and deep photoreceptors in photoperiodism in birds. *In* "Biological Rhythms and Photoperiodism in Birds" (Y. Tanabe, K. Tanaka, and T. Ookawa, eds), pp. 149–156. Jap. Sci. Soc. Press, Tokyo and Springer-Verlag, Berlin.

Huang, E. S.-R., Kao, K. J., and Nalbandov, A. V. (1979). Synthesis of sex steroids by cellular components of chicken follicles. *Biol. Reprod.* **20**, 454–461.

Huber, G. C., and Crosby, E. C. (1929). The nuclei and fiber paths of the avian diencephalon with consideration of telencephalic and certain mesencephalic centers and connections. *J. Comp. Neurol.* **48**, 1–225.

Huibregste, W. H., Ungar, F., and Farner, D. S. (1973). Adrenal steroidogenesis in the white-crowned sparrow (*Zonotrichia leucophrys*). *Comp. Biochem. Physiol.* **44B**, 1051–1056.

Hunt, G. L. Jr, Wingfield, J. C., Newman, A. L., and Farner, D. S. (1980). Sex ratio of western gulls on Santa Barbara Island, California. *Auk* **97**, 474–479.

Hunt, G. L. Jr, Newman, A. L., Warner, M. H., Wingfield, J. C., and Kaiwi, J. (1984). Comparative behavior of male–female and female–female pairs among western gulls prior to egg-laying. *Condor* **86**, 157–162.

Hunter, J. (1771). "The Natural History of Human Teeth". Printed for J. Johnson, London.

Hunter, J. (1792). "Observations on Certain Parts of the Animal Oeconomy". 2nd edn. G. Nichol, and J. Johnson, London.

Huston, T. M. (1975). The effects of environmental temperature in fertility of the domestic fowl. *Poultry Sci.* **54**, 1180–1184.

Ieromnimon, V. (1977). Beobachtungen uber die Wirkung von Hormonen auf des Zugverhalten bei Rotkehlchen (*Erithacus rubecula*). I. Die Wirkung von Prolaktin (LtH = HPr) im Jahreszyklus. *Vogelwarte* **29**, 126–134.

Ieromnimon, V. (1978). Beobachtungen uber die Wirkung von Hormonen auf des Zugverhalten bei Rotkehlchen (*Erithacus rubecula*). II. Die Wirkung von Pharmaka. *Vogelwarte* **29**, 221–230.

Immelmann, K. (1963a). Drought adaptations in Australian desert birds. *In* "Proc. XIII Int. Ornithol. Cong." pp. 649–657, American Ornithologist's Union, Louisiana State University, Baton Rouge.

Immelmann, K. (1963b). Tierische Jahresperiodik in Ökologische Sicht. Ein Beitrag zum Zeitgeberproblem unter besonderer Berucksichtigung der Brut—und Mauserzeiten australischer. *Vogel. Zool. Jahrb. Syst.* **91**, 91–200.

Immelmann, K. (1967a). Periodische Vorange in der Fortpflanzung tierischer Organismen. *Stud. Gen.* **20**, 15–33.

Immelmann, K. (1967b). Untersuchungen zur Endogenen und Exogenen Steuerung der Jahresperiodik afrikanischer Vogel. *Verh. Deut. Zool. Ges. Heid.* **1967**, 340–357.

Immelmann, K. (1971). Ecological aspects of periodic reproduction. *In* "Avian Biology" (D. S. Farner, and J. R. King, eds), Vol. 1, pp. 341–389. Academic Press, New York.

Immelmann, K. (1973). Role of the environment in reproduction as a source of "predictive" information. *In* "Breeding Biology of Birds" (D. S. Farner, ed.), pp. 121–147. Nat. Acad. Sci. USA, Washington DC.

Ishii, S., and Farner, D. S. (1976). Binding of follicle-stimulating hormone by homogenates of testes of photostimulated white-crowned sparrows, *Zonotrichia leucophrys gambelii*. *Gen. Comp. Endocrinol.* **30**, 443–450.

Ishii, S., and Furuya, T. (1975). Effect of purified chicken gonadotropins on the chick testis. *Gen. Comp. Endocrinol.* **25**, 1–8.

Ishii, S., and Tsutsui, K. (1982). Hormonal control of aggressive behavior in male Japanese quail. *In* "Aspects of Avian Endocrinology" (C. G. Scanes, M. A. Ottinger, A. D. Kenny, J. Balthazart, J. Cronshaw, and I. Chester-Jones, eds), pp. 125–131. Grad. Stud. Texas Technical University.

Jallageas, M., and Assenmacher, I. (1979). Further evidence for reciprocal interactions between the annual sexual and thyroid cycles in male Pekin ducks. *Gen. Comp. Endocrinol.* **37**, 44–51.

Jallageas, M., and Assenmacher, I. (1985). Endocrine correlates of molt and reproduction functions in birds. *In* "Acta XVIII Cong. Int. Ornithol." (V. D. Ilyichev, and V. M. Gavrilov, eds), pp. 935–945, Nauka, Moscow.

Jallageas, M. Assenmacher, I., and Follett, B. K. (1974). Testosterone secretion and plasma luteinizing hormone concentration during a sexual cycle in the Pekin duck and after thyroxine treatment. *Gen. Comp. Endocrinol.* **23**, 472–475.

Jallageas, M., Tamisier, A., and Assenmacher, I. (1978). A comparative study of the annual cycles in sexual and thyroid function in male Pekin ducks (*Anas platyrhynchos*) and teal (*Anas crecca*). *Gen. Comp. Endocrinol.* **36**, 201–210.

John, T. M., George, J. C., and Scanes, C. G. (1983). Seasonal changes in circulating levels of luteinizing hormone and growth hormone in the migratory Canada goose. *Gen. Comp. Endocrinol.* **51**, 44–59.

Johns, J. E. (1964). Testosterone-induced nuptial feathers in phalaropes. *Condor* **66**, 449–455.

Johnson, A. L., and Tilly, J. L. (1990). Evidence for protein kinase C regulation of steroidogenesis and plasminogen activator activity in preovulatory follicles from the domestic hen. *In* "Endocrinology of Birds; Molecular to Behavioral" (M. Wada, S. Ishii, and C. G. Scanes, eds), pp. 69–81, Jap. Sci. Soc. Press, Tokyo, and Springer-Verlag, Berlin.

Johnson, A. L., Pang, P., and Scanes, C. G. (1984). Third International Symposium on Avian Endocrinology. *J. Exp. Zool.* **232**, 385–745.

Johnston, D. W. (1956a). The annual reproductive cycle of the California gull. I. Criteria of age and the testis cycle. *Condor* **58**, 134–162.

Johnston, D. W. (1956b). The annual reproductive cycle of the California gull. II. Histology of the female reproductive system. *Condor* **58**, 206–221.

Jones, P. J., and Ward, P. (1976). The level of reserve protein as the proximate factor controlling the timing of breeding and clutch size in the red-billed quelea, *Quelea quelea*. *Ibis* **118**, 547–574.

Joseph, M. M., and Meier, A. H. (1973). Daily rhythms of plasma corticosterone in the common pigeon (*Columba livia*). *Gen. Comp. Endocrinol.* **20**, 326–330.

Kappers, J. A. (1981). Evolution of pineal concepts. *In* "The Pineal Organ: Photobiology—Biochronometry—Endocrinology" (A. Oksche, and P. Pévet, eds), pp. 3–23, Elsevier/North Holland Biomedical Press, Amsterdam.

Karten, H. J., and Hodos, W. (1967). "A Stereotaxic Atlas and the Brain of the Pigeon". The Johns Hopkins University Press, Baltimore.

Keast, J. A. (1953). The moulting physiology of the silvereye (*Zosterops lateralis*) (Aves). *In* "Proc. XIV Int. Cong. Zool." Copenhagen p. 314.

Kendeigh, S. C., and Wallin, H. E. (1966). Seasonal and taxonomic differences in the size and activity of the thyroid glands in birds. *Ohio J. Sci.* **66**, 369–379.

Kerlan, J. T., and Jaffe, R. B. (1974). Plasma testosterone levels during the testicular cycle of the red-winged blackbird (*Aegelaius phoeniceus*). *Gen. Comp. Endocrinol.* **22**, 428–432.

Kerlan, J. T., Jaffe, R. B., and Payne, A. H. (1974). Sex steroid formation in gonadal tissue homogenates during the testicular cycle of the red-winged blackbird (*Aegelaius phoeniceus*). *Gen. Comp. Endocrinol.* **24**, 352–363.

Kerlan, J. T., Greenspan, J. M., Trabuco, E. C., Schultz, D. M., and Winslow, J. B. (1991). Effects of surgical deafening and exposure to continuous darkness on cloacal gland size in scotorefractory Japanese quail. *Horm. Behav.* **25**, 97–111.

Kern, M. D. (1970). Annual and steroid-induced changes in the reproductive system of the female white-crowned sparrow, *Zonotrichia leucophrys gambelii*. PhD Thesis, Washington State University, Pullman.

Kern, M. D. (1972). Seasonal changes in the reproductive system of the female white-crowned sparrow, *Zonotrichia leucophrys gambelii*. *Z. Zellforsch.* **126**, 297–319.

Kern, M. D., DeGraw, W. A., and King, J. R. (1972). Effects of gonadal hormones on blood composition of white-crowned sparrows. *Gen. Comp. Endocrinol.* **18**, 43–52.

Ketterson, E. D., Nolan, V. Jr, Wolf, L., and Goldsmith, A. R. (1990). Effect of sex, stage of reproduction, season and mate removal, on prolactin in dark-eyed juncos. *Condor* **92**, 922–930.

Ketterson, E. D., Nolan, V. Jr, Wolf, L., Ziegenfus, C., Dufty, A. M. Jr, Ball, G. F. and Johnsen, T. S. (1991). Testosterone and avian life histories: The effect of experimentally elevated testosterone on corticosterone and body mass in Dark-eyed Juncos. *Horm. Behav.* **25**, 489–503.

King, J. A., and Millar, R. P. (1982a). Structure of chicken hypothalamic luteinizing hormone-releasing hormone. I. Structural determination on partially purified material. *J. Biol. Chem.* **257**, 10,722–10,732.

King, J. A., and Millar, R. P. (1982b). Structure of chicken hypothalamic luteinizing hormone-releasing hormone. II. Isolation and characterization. *J. Biol. Chem.* **257**, 10,729–10,732.

King, J. R., Follett, B. K., Farner, D. S., and Morton, M. L. (1966). Annual gonadal cycles and pituitary gonadotropin in *Zonotrichia leucophrys gambelii*. *Condor* **68**, 476–487.

Kirkpatrick, C. M. (1959). Interrupted dark period: tests for refractoriness in bobwhite quail hens. *In* "Photoperiodism and Related Phenomena in Plants and Animals" (R. B. Withrow, ed.), pp. 751–758. Amer. Assoc. Adv. Sci. Publ. 55, Washington DC.

Klandorf, H., Stokkan, K.-A., and Sharp, P. J. (1982). Plasma thyroxine and triiodothyronine levels during development of photorefractoriness in willow ptarmigan (*Lagopus lagopus lagopus*) exposed to different photoperiods. *Gen. Comp. Endocrinol.* **47**, 64–69.

Klein, G. C., Moore, R. Y., and Reppert, S. M., eds (1991). "Suprachiasmatic Nucleus. The Mind's Clock". Oxford University Press, New York.

Kleinholz, L. H., and Rahn, H. (1939). The distribution of intermedin in the pars anterior of the chicken pituitary. *Nat. Acad. Sci. U.S.A.* **3**, 145–147.

Knouff, R. A., and Hartman, F. A. (1951). A microscopic study of the adrenal of the brown pelican. *Anat. Rec.* **109**, 161–188.

Kobayashi, H. (1952). Effects of hormonic steroids on molting and broodiness in the canary. *Annot. Zool. Jap.* **25**, 128–134.

Kobayashi, H., and Wada, M. (1973). Neuroendocrinology in birds. *In* "Avian Biology" (D. S. Farner, and J. R. King, eds), Vol. 3, pp. 287–347. Academic Press, New York.

Kobayashi, H, Gorbman, A., and Wolfson, H. A. (1960). Thyroidal utilization of radioiodine in the white-throated sparrow and weaver finch. *Endocrinology* **67**, 153–161.

Kobayashi, H., Matsui, T., and Ishii, S. (1970). Functional electron microscopy of the hypothalamic median eminence. *Int. Rev. Cytol.* **29**, 281–381.

Konishi, M., Emlen, S., Ricklefs, R., and Wingfield, J. C. (1989). Contributions of bird studies to biology. *Science* **246**, 465–472.

Korenbrot, C. C., Schomberg, D. W., and Erickson, C. J. (1974). Radioimmunoassay of plasma estradiol during the breeding cycle of ring doves (*Streptopelia risoria*). *Endocrinology* **94**, 1126–1132.

Korf, H. W., Sato, T., and Oksche, A. (1989). Reflections on the nature and connectivities of pinealoctyes. *In* "Advances in Pineal Research" (R. J. Reiter, and S. F. Pang, eds), Vol. 3, pp. 11–16. John Libbey, London.

Krabbe, K. H. (1952). "Studies on the Morphogenesis of the Brain in Birds". Ejnar Munksgaard, Copenhagen.

Krabbe, K. H. (1955). Development of the pineal organ and a rudimentary parietal eye in some birds. *J. Comp. Neurol.* **103**, 139–148.

Krishnaprasadan, T. N., Kotak, V. C., Sharp, P. J., Schmedemann, R., and Haase, E. (1988). Environmental and hormonal factors in seasonal breeding in free-living male Indian rose-ringed parakeets (*Psittacula krameri*). *Horm. Behav.* **22**, 488–496.

Kroodsma, D. E. (1976). Reproductive development in a female song bird: differential stimulation by quality of male song. *Science* **192**, 574–575.

Küchler, W. (1935). Jahresperiodik Veränderungen im Histologischen Bau der Vogelschilddrüse. *J. Orn.* **83**, 414–461.

Kuenzel, W. J., and van Tienhoven, A. (1982). Nomenclature and location of avian hypothalamic nuclei and associated circumventricular organs. *J. Comp. Neurol.* **206**, 293–313.

Kurotsu, T. (1935). Über den Nucleus magnocellularis periventricularis bei Reptilien und Vögeln. *Koninklijke Akad. Wetenschappen Amsterdam, Proc. Sec. Sci.* **38**, 784–797.

Lack, D. (1968). "Ecological Adaptations for Breeding in Birds". Chapman and Hall, London.

Lam, F., and Farner, D. S. (1976). The ultrastructure of the cells of Leydig in the white-crowned sparrow (*Zonotrichia leucophrys gambelii*) in relation to plasma levels of luteinizing hormone and testosterone. *Cell Tiss. Res.* **169**, 93–109.

Lanyon, W. E. (1957). The comparative biology of the meadowlarks (*Sturnella*) in Wisconsin. *Publ. Nuttall Ornithol. Club* **1**. Cambridge, Mass.

Lea, R. W., and Sharp, P. J. (1991). Effects of presence of squabs upon plasma concentrations of prolactin and LH and length of time of incubation in Ring Doves on "extended" incubatory patterns. *Horm. Behav.* **25**, 275–282.

Lea, R. W., Sharp, P. J., Klandorf, H., Harvey, S., Dunn, J. C., and Vowles, D. M. (1986). Seasonal changes in concentrations of plasma hormones in the male ring dove (*Streptopelia risoria*). *J. Endocrinol.* **108**, 385–391.

Lea, R. W., Talbot, R. T., and Sharp, P. J. (1991). Passive immunization against chicken vasoactive intestinal peptide suppresses plasma prolactin and crop sac development in incubating ring doves. *Horm. Behav.* **25**, 283–294.

Legendre, R., and Rakotondrainy, A. (1963). Variation de l'epaisseur de l'epithélium thyroidien en rapport avéc la colouration nuptiale chez le mâle du Plocéidé malgache, *Foudia madagascariensis. C. R. Acad. Sci. 246*, 1019–1021.

Lehrman, D. S. (1965). Interaction between internal and external environments in the regulation of the reproductive cycle of the ring dove. *In* "Sex and Behavior" (F. A. Beach, ed.), pp. 355–380. Wiley, New York.

Leshner, A. I. (1978). "An Introduction to Behavioral Endocrinology". Oxford University Press, Oxford.

Lewis, R. A. (1975) Reproductive biology of the white-crowned sparrow (*Zonotrichia leucophrys pugetensis* Grinnell). I. Temporal organization of reproductive and associated cycles. *Condor* **77**, 46–59.

Lewis, R. A., and Farner, D. S. (1973). Temperature modulation of photoperiodically induced vernal phenomena in white-crowned sparrows (*Zonotrichia leucophrys*). *Condor* **75**, 279–286.

Lewis, R. A., and Orcutt, F. S. Jr. (1971). Social behavior and avian sexual cycles. *Scientia* **106**, 447–472.

Ligon, J. D. (1974). Green cones of the Pinyon pine stimulate late summer breeding in Pinyon Jays. *Nature* **250**, 80–83.

Lincoln, G. A., Racey, P. A., Sharp, P. J., and Klandorf, H. (1980). Endocrine changes associated with spring and autumn sexuality of the rook, *Corvus frugilegus*. *J. Zool. Lond.* **190**, 137–153.

Lissano, M. E., and Kennamer, J. E. (1977). Seasonal variations in plasma testosterone level in male eastern wild turkeys. *J. Wildl. Manag.* **41**, 184–188.

Ljunggren, L. (1968). Seasonal studies of wood pigeon populations. I. Body weight, feed habits, liver and thyroid activity. *Viltrevy* **5**, 435–504.

Ljunggren, L. (1969a). Seasonal studies of wood pigeon populations. II. Gonads, crop glands, adrenals and the hypothalamo-hypophysial systems. *Viltrevy* **6**, 41–126.

Ljunggren, L. (1969b). "Studies on Seasonal Activity in Pigeons With Aspects on the Role of Biogenic Monoamines in the Endocrine System". Gleerup, Lund.

Lloyd, J. A. (1965). Seasonal changes of the incubation patch in the starling. *Condor* **67**, 67–72.

Lofts, B. (1964). Evidence of an autonomous rhythm in an equitorial bird (*Quelea*). *Nature* **201**, 523–524.

Lofts, B., and Coombs, C. J. F. (1965). Photoperiodism and the testicular refractory period in the mallard. *J. Zool. Lond.* **146**, 44–54.

Lofts, B., and Murton, R. K. (1968). Photoperiodic and physiological adaptations regulating avian breeding cycles and their ecological significance. *J. Zool. Lond.* **155**, 327–394.

Lofts, B., and Murton, R. K. (1973). Reproduction in birds. *In* "Avian Biology" (D. S. Farner, and J. R. King, eds), Vol. 3, pp. 1–107. Academic Press, New York.

Lofts, B., Murton, R. K., and Westwood, N. J. (1966). Gonadal cycles and the evolution of breeding seasons in British Columbidae. *J. Zool. Lond.* **150**, 249–272.

Logan, C. A., and Carlin, C. A. (1991). Testosterone stimulates reproductive behavior during autumn in mockingbirds (*Mimus polyglottos*). *Horm. Behav.* **25**, 229–241.

Lorenzen, L. C., and Farner, D. S. (1964). An annual cycle in the interrenal tissue of the adrenal gland of the white-crowned sparrow, *Zonotrichia leucophrys gambelii*. *Gen. Comp. Endocrinol.* **4**, 253–263.

Lumia, A. R. (1972). The relationship among testosterone, conditioned aggression and dominance in male pigeons. *Horm. Behav.* **3**, 227–286.

MacKenzie, D. S. (1981). *In vivo* thyroxine release in day old cockerels in response to acute stimulation by mammalian and avian pituitary hormones. *Poultry Sci.* **60**, 2136–2143.

Malacarne, V. (1782). Esposizione anatomica delle parti relative all'encephallo degli uccelli. *Mem. Soc. Ital. Verona* **1**, 747–767.

Malamed, S., Gibney, J. A., and Scanes, C. G. (1988). Immunogold identification of the somatotrophs of domestic fowl of different ages. *Cell Tiss. Res.* **251**, 581–585.

Marshall, A. J. (1949a). On the function of the interstitium of the testis. The sexual cycle of a wild bird, *Fulmarus glacialis*. *Quart. J. Roy. Microsp. Sci.* **90**, 265–280.

Marshall, A. J. (1949b). Weather factors and spermatogenesis in birds. *Proc. Zool. Soc.* **119**, 711–716.

Marshall, A. J. (1959). Internal and environmental control of breeding. *Ibis* **101**, 456–478.

Marshall, A. J. (1960). The environment, cyclical reproductive activity and behavior in birds. *Symp. Zool. Soc. Lond.* **2**, 53–67.

Marshall, A. J., and Coombs, C. J. F. (1957). The interaction of environmental internal and behavioral factors in the rook, *Corvus f. frugilegus* L. *Proc. Zool. Soc. Lond.* **128**, 545–589.

Marshall, A. J., and Serventy, D. L. (1959). Experimental demonstrations of an internal rhythm of reproduction in a transequitorial migrant (the short-tailed shearwater, *Puffinus tenuirostris*). *Nature* **184**, 1704–1705.

Massa, R., Cresti, L., and Martini, L. (1977). Metabolism of testosterone in the pituitary gland and in the central nervous system of the European starling (*Sturnus vulgaris*). *J. Endocrinol.* **75**, 347–354.

Massa, R., Bottoni, L., Lucini, V., and McNamee, M. (1982). Testosterone metabolism in neuroendocrine tissues of male Japanese quail during photoinduced sexual maturation. *Bull. Zool.* **49**, 37–44.

Massa, R., Bottoni, L., and Lucini, V. (1983). Brain testosterone metabolism and sexual behavior in birds. *In* "Hormone and Behavior in Higher Vertebrates" (J. Balthazart, E. Pröve, and R. Gilles, eds), pp. 230–236. Springer-Verlag, Berlin.

Matsuo, S., Vitums, A., King, J. R., and Farner, D. S. (1969). Light microscope studies of the cytology of the adenohypophysis of the white-crowned sparrow, *Zonotrichia leucophrys gambelii*. *Z. Zellforsch. Mikrosk. Anat.* **95**, 143–176.

Matt, K. S. (1980). Sensitivity of the hypothalamic pituitary axis to negative feedback in a photoperiodic species. *In* "Functional Correlates of Hormone Receptors in Reproduction" (V. B. Mahesh, T. G. Muldoon, B. B. Saxena, and W. A. Sadler, eds.), pp. 499–501. Elsevier, N. Holland.

Matt, K. S., and Farner, D. S. (1979). The development of photorefractoriness: an endocrine phenomenon. *Amer. Zool.* **19**, 968.

Mattocks, P. W. Jr, Farner, D. S., and Follett, B. K. (1976). The annual cycle in luteinizing hormone in the plasma of intact and castrated white-crowned sparrows, *Zonotrichia leucophrys gambelii*. *Gen. Comp. Endocrinol.* **30**, 156–161.

Maung, Z. W., and Follett, B. K. (1977). Effects of chicken and ovine luteinizing hormone on androgen release and cyclic AMP production by isolated cells from quail testis. *Gen. Comp. Endocrinol.* **33**, 242–253.

Mays, N. A., Vleck, C. M., and Dawson, J. (1991). Plasma luteinizing hormone, steroid hormones, behavioral role and nest stage in coöperatively breeding Harris' Hawks (*Parabuteo unicinctus*). *Auk* **108**, 619–637.

McCarthy, C. P. (1979). Response of white-crowned sparrows (*Zonotrichia leucophrys gambelii*) to synthetic luteinizing hormone-releasing hormone: changes in plasma irLH. MS Thesis, University of Washington.

McCreery, B. R., and Farner, D. S. (1979). Progesterone in male white-crowned sparrows, *Zonotrichia leucophrys gambelii*. *Gen. Comp. Endocrinol.* **37**, 1–5.

McNaughton, F. J., Dawson, A., and Goldsmith, A. R. (1992). Puberty in birds: the reproductive system of starlings does not respond to short days until birds are fully grown. *J. Endocrinol.* **132**, 411–417.

McNeilly, A. S., Etches, R. J., and Friesen, H. G. (1978). A heterologous radioimmunoassay for avian prolactin: application to measurement of prolactin in the turkey. *Acta Endocrinol.* (*Copenh*) **89**, 60–69.

Meier, A. H., and Dusseau, J. W. (1968). Prolactin and the photoperiodic gonadal response in several avian species. *Physiol. Zool.* **41**, 95–103.

Meier, A. H., and Ferrell, B. R. (1978). Avian endocrinology. *In* "Chemical Zoology" (H. Florkin, H. Scheer, and A. Brush, eds), Vol. 10, pp. 213–271. Academic Press, New York.

Meier, A. H., Farner, D. S., and King, J. R. (1965). A possible endocrine basis for migratory behavior in the white-crowned sparrow, *Zonotrichia leucophrys gambelii*. *Anim. Behav.* **13**, 453–465.

Meier, A. H., Burns, J. T., and Dusseau, J. W. (1969). Seasonal variations in the diurnal rhythm of pituitary prolactin content in the white-crowned sparrow. *Gen. Comp. Endocrinol.* **12**, 282–289.

Meier, A. H., Ferrell, B. R., and Miller, L. J. (1980). Circadian components of the circannual mechanism in the white-throated sparrow. *In* "Acta XVII Int. Ornithol. Cong." (R. Nöhring, ed.), pp. 458–462. Deutsche Ornithologen-Gesellschaft.

Meijer, T., and Schwabl, H. (1989). Hormonal patterns in breeding and non-breeding kestrels *Falco tinnunculus*: field and laboratory studies. *Gen. Comp. Endocrinol.* **74**, 148–160.

Meijer, T., Daan, S., and Hall, M. (1990). Family planning in the kestrel (*Falco tinnunculus*): the proximate control of covariation of laying date and clutch size. *Behaviour* **114**, 117–136.

Menaker, M. (1982). The search for principles of physiological organization in vertebrate circadian systems. *In* "Vertebrate Circadian Systems" (J. Aschoff, S. Daan, and G. Groos, eds), pp. 1–12. Springer-Verlag, Berlin.

Menaker, M., and Keats, H. (1968). Extraretinal light perception in the sparrow. II. Photoperiodic stimulation of testes growth. *Proc. Natl. Acad. Sci. U.S.A.* **60**, 146–151.

Menaker, M., and Oksche, A. (1974). The avian pineal organ. *In* "Avian Biology" (D. S. Farner, and J. R. King, eds), Vol. 4, pp. 80–119. Academic Press, New York.

Menaker, M., Takahashi, J. S., and Eskin, A. (1978). The physiology of Circadian pacemakers. *Ann. Rev. Physiol.* **40**, 501–526.

Mench, J. A., and Ottinger, M. A. (1991). Behavioral and hormonal correlates of social dominance in stable and disrupted groups of male domestic fowl. *Horm. Behav.* **25**, 112–122.

Mialhe-Voloss C., and Benoit, J. (1954). L'intermédine dans l'hypophyse et l'hypothalamus du canard. *C. R. Soc. Biol.* **148**, 56–59.

Mikai, S.-I., Kurosu, T., and Farner, D. S. (1975). Light and electron-microscopic studies on the secretory cytology of the adenohypophysis of the Japanese quail, *Coturnix coturnix japonica*. *Cell Tiss. Res.* **159**, 147–165.

Mikami, S.-I. (1958). The cytological significance of regional patterns in the adenohypohysis of the fowl. *J. Fac. Agric. Iwate Univ.* **3**, 473–545.

Mikami, S.-I. (1969). Morphological studies of the avian adenohypohysis related to its function. *Gunma Symp. Endocrinol.* **6**, 151–170.

Mikami, S.-I. (1986). Immunocytochemistry of the avian hypothalamus and adenohypophysis. *Int. Rev. Cytol.* **103**, 189–248.

Mikami, S.-I., Vitums, A., and Farner, D. S. (1969). Electron microscopic studies on the adenohypophysis of the white-crowned sparrow, *Zonotrichia leucophrys gambelii*. *Z. Zellforsch.* **97**, 1–29.

Mikami, S.-I., Farner, D. S., and Lewis, R. A. (1973a). The prolactin cell of the white-crowned sparrow, *Zonotrichia leucophrys pugetensis*. *Z. Zellforsch.* **138**, 455–474.

Mikami, S.-I., Hashikawa, T., and Farner, D. S. (1973b). Cytodifferentiation of the adenohypophysis of the domestic fowl. *Z. Zellforsch.* **138**, 299–314.

Mikami, S.-I., Takagi, T., and Farner, D. S. (1980). Cytological differentiation of the interrenal tissue of the Japanese quail, *Coturnix coturnix*. *Cell Tiss. Res.* **208**, 353–370.

Mikami, S.-I., Homma, K., and Wada, M. (1983). "Avian Endocrinology: Environmental and Ecological Perspectives". Jap. Sci. Soc. Press, Tokyo, and Springer-Verlag, Berlin.

Miller, A. H. (1949). Potentiality for testicular recrudescence during the annual refractory period of the golden-crowned sparrow. *Science* **109**, 546.

Miller, A. H. (1954). The occurrence and maintenance of the refractory period in crowned sparrows. *Condor* **56**, 13–20.

Miller, D. S. (1939). A study of the physiology of the sparrow thyroid. *J. exp. Zool.* **80**, 259–286.

Miyamoto, K., Hasegawa, Y., Nomura, M., Igarashi, M., Kangawa, K., and Matsuo, H. (1984). Identification of the second gonadotropin-releasing hormone in chicken hypothalamus: evidence that gonadotropin secretion is probably controlled by two distinct gonadotropin-releasing hormones in avian species. *Proc. Natl. Acad. Sci. U.S.A.* **81**, 3874–3878.

Moens, L., and Coesseus, R. (1970). Seasonal variations in the adrenal cortex cells of the house sparrow *Passer domesticus* with special reference to a possible zonation. *Gen. Comp. Endocrinol.* **15**, 95–100.

Moore, M. C. (1980). Habitat structure in relation to population density and timing of breeding in prairie warblers. *Wilson Bull.* **92**, 177–187.

Moore, M. C. (1982). Hormonal responses of free-living male white-crowned sparrows to experimental manipulation of female sexual behavior. *Horm. Behav.* **16**, 323–329.

Moore, M. C. (1983). Effect of female sexual displays on the endocrine physiology and behavior of male white-crowned sparrows, *Zonotrichia leucophrys gambelii*. *J. Zool. Lond.* **199**, 137–148.

Moore, M. C. (1984). Changes in territorial defense produced by changes in circulating levels of testosterone. A possible hormonal basis for mate-guarding behavior in white-crowned sparrows. *Behaviour* **88**, 215–226.

Morgan, E. L., and Mraz, F. R. (1970). Effects of dietary iodine on the thyroidal uptake and elimination of I-131 in *Coturnix* and bobwhite quail. *Poultry Sci.* **49**, 161–164.

Morita, Y. (1966). Absence of electrical activity of the pigeon's pineal organ in response to light. *Experentia* **22**, 402.

Morita, Y. (1975). Direct photosensory activity of the pineal. *In* "Brain–Endocrine Interactions II: The Ventricular System in Neuroendocrine Mechanisms" (M. M. Knigge, D. E. Scott, H. Kobayashi, and S. Ishii, eds), pp. 376–387. Karger, Basel.

Morton, M. L., Pereya, M. E., and Baptista, L. (1985). Photoperiodically-induced ovarian growth in the white-crowned sparrow (*Zonotrichia leucophrys gambelii*) and its augmentation by song. *Comp. Biochem. Physiol.* **80A**, 93–97.

Morton, M. L., Pereya, L. E., Burns, D. M., and Allan, N. (1990). Seasonal and age related changes in plasma testosterone in mountain white-crowned sparrows. *Condor* **92**, 166–173.

Müller, J. (1929). Die Nebebbieren von *Gallus domesticus* und *Columba livia domestica*. *Z. Mikrosk. Anat. Forschung* **17**, 303–352.

Müller, W. (1871). Über Entwicklung und Bau der Hypophysis und des Processus infundibuli cerebri. *Jenaische Zeitschr. Med. Naturwiss.* **6**, 354–425.

Murton, R. K., and Westwood, N. J. (1977). "Avian Breeding Cycles". Clarendon Press, Oxford.

Myers, S. A., Millam, J. R., and El-Halawani, M. E. (1989). Plasma LH and prolactin levels during the reproductive cycle of the cockatiel (*Nymphicus hollandicus*). *Gen. Comp. Endocrinol.* **73**, 85–91.

Nakamura, T. (1957). The seasonal changes of thyroid gland and gonads in the Japanese tree sparrow. *Mem. Fac. Lib. Arts Educ.* **8**, 151–155.

Nakamura, T. (1958). Seasonal changes of thyroid gland and gonads of the Japanese tree sparrow on the thyroid activity in seasons based on the histology. *Misc. Rep. Yamashina Inst. Ornithol. Zool.* **12**, 494–496.

Narasimhacharya, A. V. R. L., Kotak, V. C. and Sharp, P. J. (1988). Environmental and hormonal interactions in the regulation of seasonal breeding in free-living male Indian weaver bird (*Ploceus phillipinus*). *J. Zool. Lond.* **215**, 239–248.

Narula, L., and Saxena, R. N. (1981). Variations in hypothalamic LHRH, pituitary and plasma LH in relation to testis cycle of the male Indian weaver bird, *Ploceus phillipinus*. *Ind. J. Exp. Biol.* **19**, 995–997.

Newton, I. (1964). The breeding biology of the chaffinch. *Bird Study* **11**, 47–68.

Newton, I. (1973). "Finches". Taplinger, New York.

Nice, M. M. (1943). "Studies in the Life History of the Song Sparrow". Vol. 2. The Behavior of the Song Sparrow and Other Passerines. Dover, New York.

Nicholls, T. J., and Storey, C. R. (1976). The effects of castration on plasma levels of LH in photosensitive and photorefractory canaries (*Serinus canarius*). *Gen. Comp. Endocrinol.* **29**, 170–174.

Nicholls, T. J., Goldsmith, A. R., and Dawson, A. (1984). Photorefractoriness in European starlings: associated hypothalamic changes and involvement of thyroid hormones and prolactin. *J. exp. Zool.* **232**, 567–572.

Nicholls, T. J., Goldsmith, A. R., and Dawson, A. (1988). Photorefractoriness in birds and comparison with mammals. *Physiol. Rev.* **68**, 133–176.

Nicoll, C. S. (1975). Radioimmunoassay and radioreceptor assays for prolactin and growth hormone: a critical appraisal. *Amer. Zool.* **15**, 881–903.

Nir, I., Yam, D., and Perek, M. (1975). Effects of stress on the corticosterone content of the blood plasma and adrenal gland of intact and bursectomized *Gallus domesticus*. *Poultry Sci.* **54**, 2101–2110.

Noble, G. K., and Wurm, M. (1940). The effects of hormones on the breeding cycle of the laughing gull. *Anat Rec.* **78**, Suppl. 50.

Noce, T., Ando, H., Ueda, T., Kubokawa, K., Higashinakagawa, T., and Ishii, S. (1989). Molecular cloning and nucleotide sequence analysis of the putative cDNA for the precursor molecule of the chicken LH-ß subunit. *J. Mol. Endocrinol.* **3**, 22–30.

Novikov, B. G. (1955). Eksperimetalnyi analizpolovykh i sezonnytch razlichii v reaktsii gonad ptits na cvet. *Akad. Nuak. Ukrainskoi SSR* **12**, 37–53.

Nussbaum, M. (1905). Inneresekretion und Nerveneinfluss. *Merkel-Bonnet Ergebnisse* **15**, 39–89.

Oakeson, B. B., and Lilley, B. R. (1960). Annual cycle of thyroid histology in two races of white-crowned sparrow. *Anat. Rec.* **136**, 41–57.

Obussier, H. (1948). Über die Grössenbeziehungen der Hypophyse und ihrer Teile bei Säugetieren und Vögeln. *Roux' Arch. Entwicklungsmechanik Organe* **143**, 181–274.

O'Connell, M. E., Reboulleau, C., Feder, H. H., and Silver, R. (1981). Social interactions and androgen levels in birds. I. Female characteristics associated with increased plasma androgen levels in the male ring dove (*Streptopelia risoria*). *Gen. Comp. Endocrinol.* **44**, 454–463.

O'Conner, R. J. (1978). Next box insulation and the timing of laying in the Wytham Woods population of great tits, *Parus major*. *Ibis* **120**, 534–537.

Oehmke, H. J. (1968). Regionale Strukturunterschiede im Nucleus infundibularis der Vögel (Passeriformes). *Z. Zellforsch.* **92**, 406–421.

Oehmke, H. J. (1969). Topographische Verteilung der Monoaminfloreszenz im Zwischenhim-Hypophysensystem von *Carduelis carduelis* und *Anas platyrhynchos*. *Z. Zellforsch.* **101**, 266–284.

Oehmke, H. J. (1971). Vergleichende neurohistologische Studien am Nucleus infundibularis einiger australischer vögel, *Z. Zellforsch.* **122**, 122–138.

Oehmke, H. J., Priedkalns, J., Vaupel-von Harnack, M., and Oksche, A. (1969). Fluoreszensz- und elektronenmikroskopische Untersuchungen am Zwischenhirn-Hypophysensystem von *Passer domesticus*. *Z. Sellforsch.* **95**, 109–133.

Oksche, A. (1968). Zur Frage extraretinaler Photorezeptoren im Pinealorgan der Vögel. *Arch. Anat. Histol. Embryol.* **51**, 497–507.

Oksche, A. (1971). Sensory and glandular elements of the pineal organ. *In* "The Pineal Gland" (G. E. Wolstenholme, and J. Knight, eds), pp. 127–146. Churchill Livingstone, Edinburgh.

Oksche, A. (1989). Pineal complex—the "third" or "first" eye of vertebrates?: A conceptual analysis. *Biomed. Res.* **10**, suppl. 3, 187–194.

Oksche, A., and Farner, D. S. (1974). Neurohistological studies of the hypothalamo-hypophysial system of *Zonotrichia leucophrys gambelii*. *Adv. Anat., Embryol. Cell Biol.* **48**, 1–136.

Oksche, A., and Kirschstein, H. (1969). Elektronenmikroskopische Untersuchungen am Pinealorgan von *Passer domesticus*. *Z. Zellforsch. Mikrosk. Anat.* **102**, 214–241.

Oksche, A., and Vaupel-von Harnack, M. (1965). Vergleichende Elektronmikroskopische Studien am Pinealorgan. *Prog. Brain. Res.* **10**, 237–258.

Oksche, A., Farner, D. S., Serventy, D. L., Wolffe, F., and Nicholls, C. A. (1963). The hypothalamo-hypophyseal neurosecretory system of the zebra finch, *Taenopygia castanotis*. *Z. Zellforsch. Mikrosk. Anat.* **58**, 846–914.

Oliver, J., and Baylé, J. D. (1982). Brain photoreceptors for the photo-induced testicular response in birds. *Experentia* **38**, 1021–1029.

Olson, D. M., Shimada, K., and Etches, R. J. (1986). Prostaglandin concentrations in peripheral plasma and ovarian and uterine plasma and tissue in relation to oviposition in hens. *Biol. Reprod.* **35**, 1140–1146.

O'Malley, B. W., McGuire, W. L., Kohler, P. O., and Korenman, S. G. (1969). Studies of the mechanism of steroid hormone regulation of synthesis of specific proteins. *Rec. Prog. Horm. Res.* **25**, 105–160.

van Oordt, G. J., and Junge, G. C. A. (1930). Die hormonale Wirkung des Hodens auf Federkleid und Farbe des Schnabels und der Fusse bei der Lachmöwe (*Larus ridibundus* L.). *Zool. Anz.* **91**, 1–7.

van Oordt, G. J., and Junge, G. C. A. (1933). Die hormonale Wirkung der Gonaden auf Sommerund Prachtkleid. I. Der Einfluss der Kastration bei mannlichen Lachmöwen (*Larus ridibundus* L). *Roux Arch. Entw.* **128**, 166–180.

Orban, F. (1929). Recherches cytologiques sur une sécrétion de l'epithélium séminal et l'évolution de la cellule interstitielle. *Arch. Biol.* **39**, 61–138.

Orians, G. H. (1969). On the evolution of mating systems in birds and mammals. *Amer. Nat.* **103**, 589–603.

Oring, L. W., Fivizzani, A. J., El-Halawani, M. E., and Goldsmith, A. R. (1986a). Seasonal changes in prolactin and luteinizing hormone in the polyandrous spotted sandpiper, *Actitis macularia*. *Gen. Comp. Endocrinol.* **62**, 394–403.

Oring, L. W., Fivizzani, A. J., and El-Halawani, M. E. (1986b). Changes in prolactin associated with laying and hatch in the spotted sandpiper. *Auk* **103**, 820–822.

Oring, L. W., Able, K. P., Anderson, D. W., Baptista, L. F., Barlow, J. C., Gaunt, A. S., Gill, F. B., and Wingfield, J. C. (1988a). Guidelines for use of wild birds in research. *Auk* **105** (supplement), 1a–41a.

Oring, L. W., Fivizzani, A. J., Colwell, M. A., and El-Halawani, M. E. (1988b). Hormonal changes associated with natural and manipulated incubation in the sex-role reversed Wilson's phalarope. *Gen. Comp. Endocrinol.* **72**, 247–256.

Oring, L. W., Fivizzani, A. J., and El-Halawani, M. E. (1989). Testosterone-induced inhibition of incubation in the spotted sandpiper (*Actitis macularia*). *Horm. Behav.* **23**, 412–423.

Ozon, R. (1972a). Estrogens in fishes, amphibians, reptiles and birds. *In* "Steroids in Non-mammalian Vertebrates" (D. R. Idler, ed.), pp. 390–413. Academic Press, New York.

Ozon, R. (1972b). Androgens in fishes, amphibians, reptiles and birds. *In* "Steroids in Non-mammalian Vertebrates" (D. R. Idler, ed.), pp. 329–389. Academic Press, New York.

Palmer, J. F. (1837). "The Works of John Hunter". Longman, Rees, Orme, Brown, Green and Longman, London, 4 vols.

Palmer, J. F. (1841). "The Complete Works of John Hunter". Haswell, Barrington, and Haswell Philadelphia, 4 vols.

Pandha, S. K., and Thapliyal, J. P. (1964). Effect of thyroidectomy upon the testes of Indian spotted Munia, *Lonchura punctulata*. *Naturwissenschaften* **51**, 102.

Papkoff, H., Licht, P., Bona-Gallo, A., MacKenzie, D. S., Oelofsen, W., and Oosthuizen, M. J. (1982). Biochemical and immunological characterization of pituitary hormones from the ostrich (*Struthio camelus*). *Gen. Comp. Endocrinol.* **48**, 181–195.

Pathak, V. K., and Chandola, A. (1982a). Thyroidal involvement in the development of migratory disposition in red-headed bunting (*Emberiza bruniceps*). *Horm. Behav.* **16**, 46–58.

Pathak, V. K., and Chandola, A. (1982b). Seasonal variations in extrathyroidal conversion of thyroxine to triiodothyronine and migratory disposition in red-headed bunting. *Gen. Comp. Endocrinol.* **47**, 433–439.

Pathak, V. K., and Chandola, A. (1983). Seasonal variations in circulating thyroxine and triiodothyronine concentrations in spotted munia *Lonchura punctulata*. *Gen. Comp. Endocrinol.* **50**, 201–204.

Paulke, E., and Haase, E. (1978). A comparison of seasonal changes in the concentrations of androgens in the peripheral blood of wild and domestic ducks. *Gen. Comp. Endocrinol.* **34**, 381–390.

Payne, R. B. (1969). Breeding seasons and reproductive physiology of tricolored blackbirds. *Univ. Calif. Publ. Zool.* **90**, 1–137.

Payne, R. B. (1972). Mechanisms of control of molt. *In* "Avian Biology" (D. S. Farner, and J. R. King, eds), Vol. 2, pp. 103–155. Academic Press, New York.

Payne, R. B., and Landolt, M. (1970). Thyroid histology in the annual cycle, breeding, and molt of tricolored blackbirds (*Agelaius tricolor*). *Condor* **72**, 445–451.

Pearce, R. B., Cronshaw, J., and Holmes, W. N. (1979). Structural changes in the interrenal tissue of the duck (*Anas platyrhynchos*) following adenohypohysectomy and treatment *in vivo* and *in vitro* with corticotropin. *Cell Tiss. Res.* **196**, 429–447.

Péczely, P. (1976). Étude circannuelle de la fonction corticosurrénalienne chez les espéces de passereaux migrants et non-migrants. *Gen. Comp. Endocrinol.* **30**, 1–11.

Péczely, P. (1986). Hormonal regulation of moulting in black-headed gulls. *In* "Acta XIX Cong. Int. Ornithol" (H. Ouellet, ed.), pp. 1710–1721. University of Ottawa Press, Ottawa.

Péczely, P., and Pethes, G. (1979). Alterations in plasma sexual steroid concentrations in the collared dove (*Streptopelia decaocto*) during the sexual maturation and reproduction cycle. *Acta Physiol. Acad. Sci. Hung.* **54**, 161–170.

Péczely, P., and Pethes, G. (1980). Plasma corticosterone, thyroxine, and triiodothyronine level in the collared dove (*Streptopelia decaocto*) during the reproductive cycle. *Acta Physiol.* **56**, 421–431.

Péczely, P., and Pethes, G. (1982). Seasonal cycle of gonadal, thyroid, and adrenocortical function in the rook, (*Corvus frugilegus*). *Acta Physiol. Acad. Sci. Hung.* **59**, 59–73.

Pedersen, H. C. (1989). Effects of exogenous prolactin on parental behavior in free-living female willow ptarmigan. *Anim. Behav.* **38**, 926–934.

Perrins, C. M. (1970). The timing of bird's breeding seasons. *Ibis* **112**, 242–255.

Perrins, C. M. (1973). Some effects of temperature on breeding in the great tit and Manx shearwater. *J. Reprod. Fertil. Suppl.* **19**, 163–173.

Pethes, G., Scanes, C. G., and Rudas, P. (1979). Effect of synthetic thyrotropin-releasing hormone on the circulating growth hormone concentration in cold and heat-stressed ducks. *Acta Vet. Acad. Sci. Hung.* **27**, 175–177.

Petra, P. H., and Schiller, H. S. (1977). Sex steroid binding protein in the plasma of *Macaca nemestrina*. *J. Steroid Biochem.* **8**, 655–661.

Pittendrigh, C. S., and Daan, S. (1976). A functional analysis of circadian pacemakers in nocturnal rodents: V. Pacemaker structure: A clock for all seasons. *J. Comp. Physiol.* **106A**, 333–355.

Proudman, J. A., and Corkoran, D. H. (1981). Turkey prolactin; purification by isotachophoresis and partial characterization. *Biol. Reprod.* **25**, 375–384.

Putzig, P. (1937). Von der Beziehung des Zugablaufs zum Inkretdrüsensystem. *Der Vögelzug* **8**, 116–130.

Quay, W. B. (1965). Histological structure and cytology of the pineal organ in birds and mammals. *Prog. Brain Res.* **10**, 49–86.

Quay, W. B., and Renzoni, A. (1963). Studio comparativo e superimentale sulla struttura e citologia della epifisi nei Passeriformes. *Riv. Biol.* **66**, 363–407.

Quay, W. B., and Renzoni, (1966a). I rapporti diencefalici e la variabilita nella duplice struttura del complesso epifisario degli uccelli. *Riv. Biol.* **40**, 9–75.

Quay, W. B., and Renzoni, A. (1966b). Observazioni sulle cellule neurosecrenenti commissuro-epifisarie degli uccelli. *Riv. Biol.* **59**, 239–266.

Rahn, H., and Painter, B. T. (1941). A comparative histology of the bird pituitary. *Anat. Rec.* **79**, 297–306.

Raitt, R. J. (1968). Annual cycle of adrenal and thyroid glands in Gambel quail of southern New Mexico. *Condor* **70**, 366–372.

Ralph, C. L. (1970). Structure and alleged functions of avian pineals. *Amer. Zool.* **10**, 217–235.

Ralph, C. L. (1976). Correlations of melatonin content in pineal gland, blood, and brain of some birds and mammals. *Amer. Zool.* **16**, 35–43.

Ralph, C. L., and Dawson, D. C. (1968). Failure of the pineal body of two species of birds (*Coturnix coturnix japonica* and *Passer domesticus*) to show electrical responses to illumination. *Experentia* **24**, 147–148.

Ralph, C. L., and Lane, K. B. (1969). Morphology of the pineal body of wild house sparrows (*Passer domesticus*) in relation to reproduction and age. *Can. J. Zool.* **47**, 1205–1208.

Ralph, C. L., Hedlund, L., and Murphy, W. A. (1967). Diurnal cycles of melatonin in bird pineal bodies. *Comp. Biochem. Physiol.* **22**, 591–599.

Ramenofsky, M. (1982). Endogenous plasma hormones and agonistic behavior in male Japanese quail, *Coturnix coturnix*. PhD Thesis, University of Washington, Seattle.

Ramenofsky, M. (1984). Agonistic behavior and endogenous plasma hormones in male Japanese quail. *Anim. Behav.* **32**, 698–708.

Ramsey, S. M., Goldsmith, A. R., and Silver, R. (1985). Stimulus requirements for prolactin and LH secretion in incubating ring doves. *Gen. Comp. Endocrinol.* **59**, 246–256.

Rankin, M. A. (1991). Endocrine effects on migration. *Amer. Zool.* **31**, 217–230.

Rattner, B. A., Siles, L., and Scanes, C. G. (1982). Oviposition and the plasma concentrations of LH, progesterone and corticosterone in bob white quail (*Colinus virginianus*). *J. Reprod. Fertil.* **66**, 147–155.

Rehder, N. B., Bird, D. M. and Laguë, P. C. (1986). Variations in plasma corticosterone, estrone, estradiol-17ß, and progesterone concentrations with forced renesting, molt, and body weight of captive female American kestrels. *Gen. Comp. Endocrinol.* **62**, 386–393.

Rehndahl, H. (1924). Embryologische und morphologische Studien über das Zwischenhirn beim Huhn. *Acta Zool.* **5**, 243–344.

Renzoni, A. (1965). Ancora sull'epifisi degli uccelli. *Boll. Zool.* **32**, 743–749.

Renzoni, A. (1968). Osservazione comparative sull'epifisi degle Strigiformi ed Ordini affini. *Arch. Ital. Anat. Embriol.* **73**, 321–336.

Riddle, O., and Honeywell, H. E. (1924). Studies on the physiology of reproduction in birds. XVIII. Effects of onset of cold weather on blood sugar and ovulation rate in pigeons. *Amer. J. Physiol.* **67**, 337–345.

Riddle, O. R., Bates, W., and Lahr, E. L. (1935). Prolactin induces broodiness in fowl. *Amer. J. Physiol.* **111**, 352–360.

Riley, G. M. (1936). Light regulation of sexual activity in the male sparrow (*Passer domesticus*). *Proc. Soc. Exp. Biol. Med.* **34**, 331–332.

Riley, G. M., and Witschi, E. (1938). Comparative effects of light stimulation and administration of gonadotropic hormones on female sparrows. *Endocrinology* **23**, 618–624.

Rissman, E. F., and Wingfield, J. C. (1984). Hormonal correlates of polyandry in the spotted sandpiper, *Actitis macularia*. *Gen. Comp. Endocrinol.* **56**, 401–405.

Robinson, G. G., and Warner, D. W. (1964). Some effects of prolactin on reproductive behavior in the brown headed cowbird (*Molothrus ater*). *Auk* **81**, 315–325.

Röhss, M., and Silverin, B. (1983). Seasonal variation in the ultrastructure of the Leydig cells and plasma levels of luteinizing hormone and steroid hormones in juvenile and adult male great tits, *Parus major*. *Ornis Scand.* **14**, 202–212.

Rohwer, S. (1981). The evolution of reliable and unreliable badges of fighting ability. *Amer. Zool.* **22**, 531–546.

Rohwer, S., and Wingfield, J. C. (1981). A field study of social dominance, plasma levels of luteinizing hormone and steroid hormones in wintering Harris' sparrows. *Z. Tierpsychol.* **57**, 173–183.

Roudneva, L. M. (1970). Contribution a l'analyse du mécanisme de la phase réfractaire des gonades chez les oiseaux. *Annal. Endocrinol. Paris*, **31**, 1065–1069.

Rowan, W. (1925). Relation of light to bird migration and developmental changes. *Nature* **115**, 494–495.

Rowan, W. (1926). On photoperiodism, reproductive periodicity, and the annual migrations of birds and certain fishes. *Proc. Boston Soc. Nat. Hist.* **38**, 147–189.

Runfeldt, S., and Wingfield, J. C. (1985). Experimentally prolonged sexual activity in female sparrows delays termination of reproductive activity in their untreated mates. *Anim. Behav.* **33**, 403–410.

Rutledge, J. T. (1974). The circannual rhythm of reproduction in the male starling (*Sturnus vulgaris*). *Int. J. Chronobiol.* **42**, 120–121.

Sailaja, R., Kotak, V. C., Sharp, P.J., Schmedemann, R., and Haase, E. (1988). Environmental dietary, and hormonal factors in the regulation of seasonal breeding in free-living female Indian rose-ringed parakeets (*Psittacula krameri*). *Horm. Behav.* **22**, 518–527.

Sakai, H., and Ishii, S. (1980). Isolation and characterization of chicken follicle-stimulating hormone. *Gen. Comp. Endocrinol.* **42**, 1–8.

Sakai, H., and Ishii, S. (1985). A homologous radioimmunoassay for avian FSH. *In* "Current Trends in Comparative Endocrinology" (B. Lofts, and W. N. Holmes, eds), pp. 195–197. University of Hong Press, Hong Kong.

Sakai, H., and Ishii, S. (1986). Annual cycles of gonadotropins and sex steroids in Japanese common pheasants, *Phasianus colchicus versicolor*. *Gen. Comp. Endocrinol.* **63**, 275–283.

Salomonsen, F. (1939). Moults and sequences of plumages in the rock ptarmigan (*Lagopus mutus*, Montin). *Vidensk. Medd. Dansk Naturhist. Foren.* **103**, 1–491.

Sandor, T., and Idler, D.R. (1972). Steroid methodology. *In.* "Steroids in Non-mammalian Vertebrates" (D. R. Idler, ed.), pp. 6–36. Academic Press, New York.

Sansum, E. L., and King, J. R. (1976). Long-term effects of constant photoperiods on testicular cycles of white-crowned sparrows (*Zonotrichia leucophrys gambelii*). *Physiol. Zool.* **49**, 407–416.

Sarkar, A., and Ghosh, A. (1964). Cytological and cytochemical studies on the reproductive cycle of the subtropical male house sparrow. *La Cellule* **65**, 111–126.

Saxena, R. N., and Narula, L. (1981). Seasonal and light induced changes in the pineal LHRH of the male Indian weaver bird, *Ploceus phillipinus* (L). *Ind. J. Exp. Biol.* **19**, 993–994.

Saxena, R. N., Malhotra, L., Kant, R., and Barreja, P. K. (1979). Effect of pinealectomy and seasonal changes on pineal antigonadotropic activity of male Indian weaver bird, *Ploceus phillipinus*. *Indian J. exp. Biol.* **17**, 732–735.

Scanes, C. G., and Harvey, S. (1981). Growth hormone and prolactin in avian species. *Life Sci.* **28**, 2895–2902.

Scanes, C. G., Cheeseman, P., Phillips, J. G., and Follett, B. K. (1974). Seasonal and age variation of circulating immunoreactive luteinizing hormone in captive herring gulls, *Larus argentatus*. *J. Zool. Lond.* **174**, 369–375.

Scanes, C. G., Bolton, N. J., and Chadwick, A. (1975). Purification and properties of an avian prolactin. *Gen. Comp. Endocrinol.* **27**, 371–379.

Scanes, C. G., Godden, P. M. M., and Sharp, P. J. (1977). An homologous radioimmunoassay for chicken follicle-stimulating hormone: observations on the ovulatory cycle. *J. Endocrinol.* **73**, 473–481.

Scanes, C. G., Jallageas, M., and Assenmacher, I. (1980). Seasonal variations in the circulating concentrations of growth hormone in male Pekin ducks (*Anas platyrhynchos*) and teal (*Anas crecca*); correlations with thyroidal function. *Gen. Comp. Endocrinol.* **41**, 76–89.

Scanes, C. G., Lauterio, T.J., and Buonomo, F.C. (1983). Annual, developmental, and diurnal cycles of pituitary hormone secretion. *In* "Avian Endocrinology: Environmental and Ecological Perspectives" (S.-I. Mikami, K. Homma, and M. Wada, eds), pp. 307–326. Jap. Sci. Soc. Press, Tokyo, and Springer-Verlag, Berlin.

Schildmacher, H. (1937). Histologische Untersuchungen an Vögelhypophysen. *J. Orn.* **85**, 587–592.

Schildmacher, H. (1938). Hoden und Schilddruse des Gartenrotschwazes *Phoenicurus ph. phoenicurus*

(L.) under dem Einfluss zusatzlicher Belichtung im Herbst und Winter. *Biol. Zentral.* **58**, 464–472.

Schleussner, G., and Gwinner, E. (1988). Photoperiodic time measurement during the termination of photorefractoriness in the starling (*Sturnus vulgaris*). *Gen. Comp. Endocrinol.* **75**, 54–61.

Schleussner, G., Dittami, J., and Gwinner, E. (1985). Testosterone implants affect molt in male European starlings, *Sturnus vulgaris*. *Physiol. Zool.* **58**, 597–604.

Schlinger, B. A. (1987). Plasma androgens and aggressiveness in captive winter white-throated sparrows (*Zonotrichia albicollis*). *Horm. Behav.* **21**, 203–210.

Schlinger, B. A., Fivizzani, A. J., and Callard, G. V. (1989). Aromatase, 5α- and 5ß-reductase in brain, pituitary and skin of the sex-role reversed Wilson's phalarope. *J. Endocrinol.* **122**, 573–581.

Schmidt, L. G., Bradshaw, S. D., and Follett, B. K. (1991). Plasma levels of luteinizing hormone and androgens in relation to age and breeding status among cooperatively breeding Australian magpies (*Gymnorhina tibicen*, Latham). *Gen. Comp. Endocrinol*, **83**, 48–55.

Schoech, S. J., Mumme, R. L., and Moore, M. C. (1991). Reproductive endocrinology and mechanisms of breeding inhibition in cooperatively breeding Florida scrub jays (*Aphelocoma c. coerulescens*). *Condor* **93**, 354–362.

Schuurman, T. (1980). Hormone correlates of agonistic behavior in adult male rats. *Prog. Brain Res.* **53**, 415–420.

Schwab, R. G. (1971). Circannian testicular periodicity in the European starling in the absence of photoperiodic change. *In* "Biochronometry" (M. Menaker, ed.), pp. 428–447. Natl. Acad. Sci., Washington DC.

Schwabl, H., and Kriner, E. (1991). Territorial aggression and song of male European robins (*Erithacus rubecula*) in autumn and spring: effects of antiandrogen treatment. *Horm. Behav.* **25**, 180–194.

Schwabl, H., Wingfield, J. C., and Farner, D. S. (1980). Seasonal variations in plasma levels of luteinizing hormone and steroid hormones in the European blackbird, *Turdus merula*. *Vogelwarte* **30**, 283–294.

Schwabl, H., Ramenofsky, M., Schwabl-Benzinger, I., Farner, D. S., and Wingfield, J. C. (1988). Social status, circulating levels of hormones and competition for food in winter flocks of the white-throated sparrow. *Behaviour* **107**, 107–121.

Searcy, W. A., and Wingfield, J. C. (1980). The effects of androgen and antiandrogen on dominance and aggressiveness in male redwinged blackbirds. *Horm. Behav.* **14**, 126–135.

Searcy, W. A., McArthur, P. D., Peters, S. S., and Marler, P. (1981). Responses of male song and swamp sparrows to neighbor, stranger, and self songs. *Behaviour* **77**, 152–163.

Selander, R. K. (1960). Failure of estrogen and prolactin treatment to induce brood patch formation in brown-headed cowbirds. *Condor* **62**, 65.

Selander, R. K., and Kiuch, L. L. (1963). Hormonal control and development of the incubation patch in icterids, with notes on the behavior of cowbirds. *Condor* **65**, 73–90.

Selander, R. K., and Yang, S. Y. (1966). Behavioral responses of brown-headed cowbirds to nests and eggs. *Auk* **83**, 207–232.

Selinger, H. E., and Bermant, G. (1967). Hormonal control of aggressive behavior in Japanese Quail (*Coturnix coturnix japonica*). *Behaviour* **28**, 225–268.

Sharp, P. J. (1980). The endocrine control of ovulation in birds. *In* "Acta Int. Ornithol. Cong." (R. Nöhring, ed.), pp. 245–248. Deutschen Ornithologen-Gesellschaft, Berlin.

Sharp, P. J., and Massa, R. (1980). Conversion of progesterone to 5α- and 5ß-reduced metabolites in the brain of the hen and its potential role in the induction of the pre-ovulatory release of luteinizing hormone. *J. Endocrinol.* **86**, 459–464.

Sharp, P. J., and Moss, R. (1977). The effects of castration on concentrations of LH in the plasma of photorefractory red grouse (*Lagopus lagopus scoticus*). *Gen. Comp. Endocrinol.* **32**, 289–293.

Sharp, P. J. and Moss, R. (1981). A comparison of the response of captive willow ptarmigan (*Lagopus*

*lagopus lagopus*) red grouse (*L.l. scoticus*) and hybrids to increasing day length with observations on the modifying effects of nutrition and crowding in red grouse. *Gen. Comp. Endocrinol.* **45**, 181–188.

Sharp, P. J., Massa, R., Bottoni, L., Lucini, V., Lea, R. W., Dunn, I. C., and Trocchi, V. (1986). Photoperiodic and endocrine control of seasonal breeding in grey partridge (*Perdix perdix*). *J. Zool. Lond.* **209**, 187–200.

Sherwood, N. M., Wingfield, J. C., Ball, G. F., and Dufty, A. M. (1988). Identity of gonadotropin-releasing hormone in passerine birds: comparison of GnRH in song sparrow (*Melospiza melodia*) and starling (*Sturnus vulgaris*) with five vertebrate GnRH's. *Gen. Comp. Endocrinol.* **69**, 341–351.

Shields, K. M., Yamamoto, J. T., and Millam, J. R. (1989). Reproductive behavior and LH levels of cockatiels (*Nymphicus hollandicus*) associated with photostimulation, nest box presentation and degree of mate access. *Horm. Behav.* **23**, 68–82.

Shimada, K. (1980). Rhythms of uterine activity during ovulatory cycles in the laying hen. *In* "Biological Rhythms in Birds: Neural and Endocrine Aspects" (Y. Tanabe, K. Tanaka, and T. Ookawa, eds), pp. 101–111. Jap. Sci. Soc. Press, Tokyo, and Springer-Verlag, Berlin.

Shimada, D., Zadworny, K., Sato, K., Seo, H., Murata, Y., and Matsui, N. (1987). Construction of cDNA library from the anterior pituitary of the chicken and isolation of prolactin cDNA. In "Proc. 1st Cong. Asia and Oceania Soc. Comp. Endocrinol." (E. Ohnishi, Y. Nagahama and H. Ishizaki, eds), pp. 51–52, Nagoya University Corporation, Nagoya.

Siegel, H. A. (1980). Physiological stress in birds. *BioScience* **30**, 529–534.

Silver, R. (1978). The parental behavior of ring doves. *Amer. Sci.* **66**, 209–215.

Silver, R., and Ball, G. F. (1989). Brain, hormone and behavior interactions in avian reproduction: Status and prospectus. *Condor* **91**, 966–978.

Silver, R., Feder, H. H., and Lehrman, D. S. (1973). Situational and hormonal determinants of courtship, aggressive and incubation behavior in male ring doves (*Streptopelia risoria*). *Horm. Behav.* **4**, 163–172.

Silverin, B. (1979). Activity of the adrenal glands in the pied flycatcher and its relation to testicular regression. *Gen. Comp. Endocrinol.* **38**, 161–171.

Silverin, B. (1980). Effects of long-acting testosterone treatment on free-living pied flycatchers *Ficedula hypoleuca*, during the breeding period. *Anim. Behav.* **28**, 906–912.

Silverin, B. (1982). Endocrine correlates of brood size in adult pied flycatchers, *Ficedula hypoleuca*. *Gen. Comp. Endocrinol.* **47**, 18–23.

Silverin, B. (1983a). Population endocrinology and gonadal activities of the male pied flycatcher (*Ficedula hypoleuca*). *In* "Avian Endocrinology: Environmental and Ecological Perspectives" (S.-I. Mikami, K. Homma, and M. Wada, eds), pp. 289–305. Jap. Sci. Soc. Press, Tokyo, and Springer-Verlag, Berlin.

Silverin, B. (1983b). Population endocrinology of the female pied flycatcher, *Ficedula hypoleuca*. *In* "Hormones and Behavior in Higher Vertebrates" (J. Balthazart, E. Pröve, and R. Gilles, eds), pp. 388–397. Springer-Verlag, Berlin.

Silverin, B. (1986). Corticosterone binding proteins and behavioral effects of high plasma levels of corticosterone during the breeding period in the pied flycatcher. *Gen. Comp. Endocrinol.* **64**, 67–74.

Silverin, B. (1988). Endocrine aspects of avian mating systems. *In* "Acta XIX Int. Ornithol. Cong." (H. Ouellet, ed.), pp. 1676–1684. University of Ottawa Press, Ottawa.

Silverin, B., and Deviche, P. (1991). Biochemical characterization and seasonal changes in the concentrations of testosterone-metabolizing enzymes in the European great tit (*Parus major*) brain. *Gen. Comp. Endocrinol.* **81**, 146–159.

Silverin, B., and Goldsmith, A. R. (1983). Reproductive endocrinology of free-living pied flycatchers (*Ficedula hypoleuca*): prolactin and FSH secretion in relation to incubation and clutch size. *J. Zool. Lond.* **200**, 119–130.

Silverin, B., and Goldsmith, A. R. (1984). The effects of modifying incubation on prolactin secretion in free-living pied flycatchers. *Gen. Comp. Endocrinol.* **55**, 239–244.

Silverin, B., and Goldsmith, A. R. (1990). Plasma prolactin concentrations in breeding pied flycatchers (*Ficedula hypoleuca*) with an experimentally prolonged brood. *Horm. Behav.* **24**, 104–113.

Silverin, B., and Wingfield, J. C. (1982). Patterns of breeding behavior and plasma levels of hormones in a free-living population of pied flycatchers, *Ficedula hypoleuca*. *J. Zool. Lond.* **198**, 117–129.

Silverin, B., Viebke, P. A., and Westin, J. (1986). Seasonal changes in plasma LH and gonadal steroids in free-living willow tits, *Parus montanus*. *Ornis Scand.* **17**, 230–236.

Silverin, B., Viebke, P. A., Westin, J., and Scanes, C. G. (1989). Seasonal changes in body weight fat depots, and plasma levels of thyroxine and growth hormone in free-living great tits (*Parus major*), and willow tits (*P. montanus*). *Gen. Comp. Endocrinol.* **73**, 404–416.

Singh, K. B. (1972). Studies on the hypothalamo-hypophysial complex in birds. PhD Thesis, Banaras Hindu University, Varanasi, India.

Singh, R. M., and Dominic, C. J. (1970). Disposition of the portal vessels of the avian pituitary in relation to the median eminence and the pars distalis. *Experentia* **26**, 962–964.

Singh, K. B. and Dominic, C. J. (1973). Hypophysial vascularization in the red munia, *Estrilda amandava* L. *Ind. J. Zool.* **1**, 129–134.

Singh, K. B., and Dominic, C. J. (1975). Anterior and posterior groups of portal vessels in the avian pituitary, incidence in forty nine species. *Arch. Anat. Microsc. Morphol. Exp.* **64**, 359–374.

Singh, S., and Chandola, A. (1981). Role of gonadal feedback in annual reproduction of the weaver bird: interaction with photoperiod. *Gen. Comp. Endocrinol.* **45**, 521–526.

Siopes, T. D., and Wilson, W. O. (1978). The effect of light intensity and duration of light on photorefractoriness and subsequent egg production of chukar partridge. *Biol. Reprod.* **18**, 155–159.

Smith, J. N. M., and Roff, D. A. (1980). Temporal spacing of broods, brood size, and parental care in song sparrows (*Melospiza melodia*). *Can. J. Zool.* **58**, 1007–1015.

Smith, J. P. (1979). Thyroid hormones and molt in two species of passerine birds. MS Thesis, University of Washington, Seattle.

Smith, J. P. (1982). Annual cycle of thyroid hormones in the plasma of white-crowned sparrows and house sparrows. *Condor* **84**, 160–167.

Snow, D.W. (1962). A field study of the black and white manakin in Trinidad. *Zool. N.Y.* **47**, 65–104.

Sossinka, R. (1980). Reproductive strategies of estrildid finches in different climate zones of the tropics: gonadal maturation. *In* "Acta Int. Ornithol. Cong." (R. Nöhring, ed.), pp. 493–500. Deutschen Ornithologen-Gesellschaft, Berlin.

Steelman, S., and Pohley, F. H. (1953). Assay of the follicle-stimulating hormone based on the augmentation with human chorionic gonadotropin. *Endocrinology* **53**, 604–616.

Steimer, T., and Hutchison, J. B. (1981). Metabolic control of the behavioral action of androgens in the ring dove brain: testosterone inactivation by 5ß-reduction. *Brain Res.* **209**, 189–204.

Stetson, M. H., Lewis, R. A., and Farner, D. S. (1973). Some effects of exogenous gonadotropins and prolactin on photostimulated and photorefractory white-crowned sparrows. *Gen. Comp. Endocrinol.* **21**, 424–430.

Stieve, H. (1919). Das Verhälfnis der Zwischenzellen zum generativen Anteil im Hoden der Dohle (*Corvus monedula*). *Arch Entwicklung. Organ.* **45**, 455–497.

Stieve, H. (1920). Entwicklung, Bau und Bedeutung der Keimdrüsenzwischenzellen. *Ergebnisse Anat. Entwicklung.* **23**, 1–249.

Stieve, H. (1921). Neue Untersuchungen über die Zwischenzellen. *Anat. Anz.* **54**, 63–76.

Stieve, H. (1926). Untersuchungen über die Wechselbeziehungen zwischen Gesamtkörper und Keimdrüsen. *V. Z. Mikrosk. Anat. Forsch.* **5**, 463–624.

Stockell-Hartree, A., and Cunningham, F. J. (1969). Purification of chicken pituitary follicle-stimulating hormone and luteinizing hormone. *J. Endocrinol.* **43**, 609–619.

Stokes, T. M., Leonard, C. M., and Nottebohm, F. (1974). The telencephalon, diencephalon, and mesencephalon of the canary, *Serinus canaria*, in stereotaxic coordinates. *J. Comp. Neurol.* **156**, 337–374.

Stokkan, K.-A., and Sharp, P. J. (1980a). Seasonal changes in the concentrations of plasma luteinizing hormone and testosterone in willow ptarmigan (*Lagopus lagopus lagopus*) with observations on the effects of permanent short days. *Gen. Comp. Endocrinol.* **40**, 109–115.

Stokkan, K.-A., and Sharp, P. J. (1980b). The roles of day length and the testes in the regulation of plasma LH levels in photosensitive and photorefractory willow ptarmigan (*Lagopus lagopus lagopus*) after suppression of photoinduced LH release with implants of testosterone. *Gen. Comp. Endocrinol.* **41**, 527–530.

Stokkan, K.-A., Sharp, P. J., and Moss, R. (1982). Development of photorefractoriness in willow ptarmigan (*Lagopus lagopus lagopus*) and red grouse (*L.l. scoticus*) exposed to different photoperiods. *Gen. Comp. Endocrinol.* **46**, 281–287.

Stokkan, K. A., Harvey, S., Klandorf, H., Unander, S., and Blix, A. S. (1985). Endocrine changes associated with fat deposition and mobilization in Svalbard ptarmigan (*Lagopus mutus hyperboreus*). *Gen. Comp. Endocrinol.* **58**, 76–80.

Stokkan, K.-A., Sharp, P. J., and Unander, S. (1986). The annual breeding cycle of the high arctic Svalbard ptarmigan (*Lagopus mutus hyperboreus*). *Gen. Comp. Endocrinol.* **61**, 446–451.

Stokkan, K.-A., Sharp, P. J., Dunn, I. C., and Lea, R. W. (1988). Endocrine changes in photostimulated willow ptarmigan (*Lagopus lagopus lagopus*) and Svalbard ptarmigan (*Lagopus mutus hyperboreus*). *Gen. Comp. Endocrinol.* **70**, 169–177.

Storey, C. R., and Nicholls, T. J. (1981). The effect of testosterone upon a photoperiodically-induced cycle of gonadotropin secretion in castrated canaries, *Serinus canarius*. *Gen. Comp. Endocrinol.* **43**, 527–531.

Storey, C. R., and Nicholls, T. J. (1982a). A photoperiodically-induced cycle of gonadotropin secretion in intact, hemi- and fully castrated male bullfinches, *Pyrrhula pyrrhula*. *Ibis* **124**, 55–60.

Storey, C. R., and Nicholls, T. J. (1982b). Low environmental temperature delays photoperiodic induction of avian testicular maturation and the onset of post-nuptial photorefractoriness. *Ibis* **124**, 172–174.

Storey, C. R., and Nicholls, T. J. (1983). Responses of photosensitive and photorefractory intact and castrated male canaries, *Serinus canarius*, to treatment with synthetic mammalian luteinizing hormone-releasing hormone. *Ibis* **125**, 228–234.

Storey, C. R., Nicholls, T. J., and Follett, B. K. (1980). Castration accelerates the rate of onset of photorefractoriness in the canary (*Serinus canarius*). *Gen. Comp. Endocrinol.* **42**, 315–319.

Stresemann, E., and Stresemann, V. (1966). Die Mauser der Vogel. *J. Ornithol.* **107**, 3–448.

Studnika, F. K. (1905). Die Parietalorgane. *In* "Oppel's Lehrbuch der Vergleichenden Mikroskischen Anatomie der Wirbeltiere" (V. G. Fischer, ed.), Jena.

Tanaka, T., Shui, R. P. C., Gout, P. W., Beer, C. T., Nobel, R. L., and Friesen, H. G. (1980). A new sensitive and specific biosassay for lactogenic hormones: measurement of prolactin and growth hormone in human serum. *J. Clin. Endocrinol. Metab.* **51**, 1058–1063.

Temple, S. A. (1974). Plasma testosterone titers during the annual reproductive cycle of starlings, *Sturnus vulgaris*. *Gen. Comp. Endocrinol.* **22**, 470–479.

Terkel, A. S., Moore, C. L., and Beer, C. G. (1976). The effects of testosterone on the rate of long calling vocalization in juvenile laughing gulls, *Larus atricilla*. *Horm. Behav.* **7**, 49–58.

Thapliyal, J. P. (1969). Thyroid in avian reproduction. *Gen. Comp. Endocrinol. Suppl.* **2**, 111–122.

Thapliyal, J. P., and Saxena, R. N. (1964). Absence of refractory period in the common weaver bird. *Condor* **66**, 199–208.

Thompson, A. L. (1950). Factors determining the breeding seasons of birds: an introductory review. *Ibis* **92**,173–184.

Threadgold, L. T. (1956a). The annual gonadal cycle of the male jackdaw, *Corvus monedula*. Qualitative aspects. *La Cellule* **68**, 19–42.

Threadgold, L. T. (1956b). The annual gonadal cycle of the male jackdaw, *Corvus monedula*. Quantitative aspects. *La Cellule* **68**, 45–54.

Threadgold, L. T. (1960). A study of the annual cycle of the house sparrow at various latitudes. *Condor* **62**, 190–201.

Thybusch, D. (1965). Jahresperiodik der Nebennieren bei der Stürmmöwe (*Larus canus* L.) unter Berückschtigung des Funktionszustandes von Schild- und Keimdrüse sowie des Körpergewichtes. *Z. Wissenschaft, Zool.* **173**, 72–89.

van Tienhoven, A. (1983). "Reproductive Physiology of Vertebrates". Cornell University Press, Ithaca.

van Tienhoven, A., and Juhász, L. P. (1962). The chicken telencephalon, diencephalon and mesencephalon in stereotaxic coordinates. *J. Comp. Neurol.* **118**, 185–197.

Tilney, F., and Warren, L. F. (1919). The morphology and evolutional significance of the pineal body. *Amer. Anat. Mem.* **9**, 1–257.

Tixier-Vidal, A. (1963). Histophysiologie de l'adénohypohyse des oiseaux. *In* "Cytologie de l'adénohypohyse" (J. Benoit, and C. De Lage, eds), pp. 255–274. CNRS. Paris.

Tixier-Vidal, A., and Assenmacher, I. (1963). Action de la métopirone sur la préhypophyse du canard mâle: Essai d'identification des cellules corticotropes. *C. R. Soc. Biol.* **157**, 1350–1354.

Tixier-Vidal, A., and Follett, B. K. (1973). The adenohypophysis. *In* "Avian Biology" (D. S. Farner, and J. R. King, eds), Vol. 3, pp. 110–182. Academic Press, New York.

Tixier-Vidal, A., Herlant, M., and Benoit, J. (1962). La préhypophyse du canard Pékin mâle au cours du cycle annuel. *Arch. Biol.* **73**, 319–367.

Tixier-Vidal, A., Follett, B. K., and Farner, D. S. (1968). The anterior pituitary of the Japanese quail, *Coturnix coturnix japonica*. The cytological effects of photoperiodic stimulation. *Z. Zellforsch. Mikrosk. Anat.* **92**, 610–635.

Tordoff, H. B., and Dawson, W. R. (1965). The influences of day length on reproductive timing in the red crossbill. *Condor* **67**, 416–422.

Tsutsui, K., and Ishii, S. (1981). Effects of sex steroids on aggressive behavior of adult male Japanese quail. *Gen. Comp. Endocrinol.* **44**, 480–486.

Tsuyoshi, H., and Wada, M. (1992). Termination of LH secretion in Japanese Quail due to high- and low-temperature cycles and short daily photoperiods. *Gen. Comp. Endocrinol.* **85**, 424–429.

Turek, F. W. (1972). Circadian involvement in termination of the refractory period in two sparrows. *Science* **178**, 1112–1113.

Turek, F. W. (1978). Diurnal rhythms and the seasonal reproductive cycle in birds. *In* "Environmental Endocrinology" (D. S. Farner, and I. Assenmacher, eds), pp. 144–152. Springer-Verlag, Berlin.

Turek, F. W., Desjardins, C., and Menaker, M. (1976). Antigonadal and progonadal effects of testosterone in male house sparrows. *Gen. Comp. Endocrinol.* **28**, 395–402.

Ueck, M. (1979). Innervation of the vertebrate pineal. *Prog. Brain Res.* **52**, 45–88.

Underwood, H., and Siopes, T. (1985). Melatonin rhythms in quail: regulation by photoperiod and circadian pacemakers. *J. Pineal Res.* **2**, 133–143.

Vaugien, L. (1948). Recherches biologiques et expérimentales sur le cycle reproducteur et la mue des oiseaux Passériformes. *Bull. Biol. France Belg.* **82**, 166–213.

Vaugien, L. (1954). Influence de l'obscuration temporaire sur la durée de la phase refractaire du cycle sexuel du moineau domestique. *Bull. Biol. France Belgique* **88**, 294–308.

Vitums, A., Mikami, S.-I., Oksche, A., and Farner, D. S. (1964). Vascularization of the hypothalamo-hypophysial complex in the white-crowned sparrow, *Zonotrichia leucophrys gambelii*. *Z. Zellforsch.* **64**, 541–569.

Vitums, A., Ono, K., Oksche, A., Farner, D. S., and King, J. R. (1966). The development of the hypophysial portal system in the white-crowned sparrow, *Zonotrichia leucophrys gambelii*. *Z. Zellforsch.* **73**, 335–366.

Vleck, C. M., Wingfield, J. C., and Farner, D. S. (1980). No temporal synergism of prolactin and corticosterone influencing reproduction in white-crowned sparrows. *Amer. Zool.* **20**, 899.

Vowles, D. M., Beazley, L., and Harwood, D. H. (1975). A stereotaxic atlas of the brain of the barbary dove (*Streptopelia risoria*). In "Neural and Endocrine Aspects of Behaviour in Birds" (P. Wright, ed.), pp. 351–359. Elsevier, Amsterdam.

Vyas, D. K., and Jacob, D. (1976). Seasonal study of the adrenal gland of some Indian avian species. *Acta Anat.* **95**, 518–528.

Wada, M., Hatanaka, F., Tsuyoshi, H., and Sonoda, Y. (1990a). Temperature modulation of photoperiodically induced LH secretion and its termination in Japanese Quail (*Coturnix coturnix japonica*). *Gen. Comp. Endocrinol.* **80**, 465–472.

Wada, M., Ishii, S., and Scanes, C. G. (1990b). "Endocrinology of Birds: Molecular to Behavioral". Jap. Sci. Soc. Press, Tokyo, and Springer-Verlag, Berlin.

Wada, M., Akimoto, R., and Tsuyoshi, H. (1992). Annual changes in levels of plasma LH and size of cloacal protrusion in Japanese Quail (*Coturnix coturnix japonica*) housed in outdoor cages under natural conditions. *Gen. Comp. Endocrinol.* **85**, 415–423.

Wagner, R. (1851). Mittheilung einer einfachen Methoden zu Versuchen über die Veränderungen thierischer Gewebe in morphologischer und chemischer Beziehung. *Arch. Physiol. Heilkunde* **10**, 521–528.

Ward, P. (1969). The annual cycle of the yellow-vented bulbul *Pyconotus goiavier* in a humid equatorial environment. *J. Zool. (Lond.)* **157**, 25–45.

Watson, A., and Parr, R. (1981). Hormone implants affecting territory size and aggressive and sexual behavior in red grouse. *Ornis. Scand.* **12**, 15–61.

Watzka, M. (1934). Physiologische Veränderungen der Schildrüse. *Z. Mikrosk. Anat. Forsch.* **36**, 67–86.

Wenn, R. U., Kamberti, I. A., Keyvanjah, M., and Johannes, A. (1977). Distribution of testosterone and estradiol binding globulin (TeBG) in higher vertebrates. *Endokrinologie* **69**, 151–156.

Wieslthier, A. S., and van Tienhoven, A. (1972). The effect of thyroidectomy on testicular size and on the photorefractory period in the starling (*Sturnus vulgaris*). *J. exp. Zool.* **179**, 331–338.

Williams, T. D., Dawson, A., Nicholls, T. J., and Goldsmith, A. R. (1987). Short days induce premature reproductive maturation in juvenile starlings, *Sturnus vulgaris*. *J. Reprod. Fertil.* **80**, 327–333.

Wilson, A. C., and Farner, D. S. (1960). The annual cycle in thyroid activity in white-crowned sparrows of eastern Washington. *Condor* **62**, 414–425.

Wilson, F. E. (1989). Extraocular control of photorefractoriness in American tree sparrows (*Spizella arborea*). *Biol. Reprod.* **41**, 111–116.

Wilson, F. E. (1990a). Extraocular control of seasonal reproduction in female tree sparrows (*Spizella arborea*). *Gen. Comp. Endocrinol.* **77**, 397–402.

Wilson, F. E. (1990b). On the recovery of photosensitivity in two passerine species, American tree sparrows (*Spizella arborea*), and Harris' sparrows (*Zonotrichia querula*). *Gen. Comp. Endocrinol.* **79**, 283–290.

Wilson, F. E., and Follett, B. K. (1974). Plasma and pituitary luteinizing hormone in intact and castrated tree sparrows (*Spizella arborea*) during a photoinduced gonadal cycle. *Gen. Comp. Endocrinol.* **23**, 82–93.

Wilson, F. E., and Follett, B. K. (1975). Corticosterone-induced gonadosuppression in photostimulated tree sparrows. *Life Sci.* **17**, 1451–1456.

Wilson, F. E., and Follett, B. K. (1978). Dissimilar effects of hemicastration on plasma LH and FSH in photostimulated tree sparrows (*Spizella arborea*). *Gen. Comp. Endocrinol.* **34**, 251–255.

Wilson, S. C., and Sharp, P.J. (1975). Episodic release of luteinizing hormone in the domestic fowl. *J. Endocrinol.* **64**, 77–86.

Wingfield, J. C. (1980a). Sex steroid binding proteins in vertebrate blood. *In* "Hormones, Evolution and Adaptation" (S. Ishii, T. Hirano, and M. Wada, eds), pp. 135–144. Jap. Sci. Soc. Press, Tokyo, and Springer-Verlag, Berlin.

Wingfield, J. C. (1980b). Fine temporal adjustment of reproductive functions. *In* "Avian Endocrinology" (A. Epple, and M. H. Stetson, eds), pp. 367–389. Academic Press, New York.

Wingfield, J. C. (1983). Environmental and endocrine control of reproduction: an ecological approach. *In* "Avian Endocrinology: Environmental and Ecological Aspects" (S.-I. Mikami, and M. Wada, eds), pp. 205–288. Jap. Sci. Soc. Press, Tokyo, and Springer-Verlag, Berlin.

Wingfield, J. C. (1984a). Environmental and endocrine control of reproduction in the song sparrow, *Melospiza melodia*. I. Temporal organization of the breeding cycle. *Gen. Comp. Endocrinol.* **56**, 406–416.

Wingfield, J. C. (1984b). Environmental and endocrine control of reproduction in the song sparrow, *Melospiza melodia*. II. Agonistic interactions as environmental information stimulating secretion of testosterone. *Gen. Comp. Endocrinol.* **56**, 417–424.

Wingfield, J. C. (1984c). Androgens and mating systems: testosterone-induced polygyny in normally monogamous birds. *Auk* **101**, 665–671.

Wingfield, J. C. (1984d). Influences of weather on reproduction. *J. Exp. Zool.* **232**, 589–594.

Wingfield, J. C. (1985a). Short-term changes in plasma levels of hormones during establishment and defense of a breeding territory in male song sparrows, *Melospiza melodia. Horm. Behav.* **19**, 174–187.

Wingfield, J. C. (1985b). Environmental and endocrine control of territorial behavior in birds. *In* "The Endocrine System and The Environment" (B. K. Follett, S. Ishii, and A. Chandola, eds), pp. 265–277. Jap. Sci. Soc. Press, Tokyo, and Springer-Verlag, Berlin.

Wingfield, J. C. (1985c). Influences of weather on reproductive function in male song sparrows, *Melospiza melodia. J. Zool. (Lond.)* **205**, 525–544.

Wingfield, J. C. (1985d). Influences of weather on reproductive function in female song sparrows, *Melospiza melodia. J. Zool. (Lond.)* **205**, 545–558.

Wingfield, J. C. (1988a). The challenge hypothesis: Interrelationships of testosterone and behavior. *In* "Acta Internationalis Ornithologici" (H. Ouellet, ed.), pp. 1685–1691. University of Ottawa Press, Ottawa.

Wingfield, J. C. (1988b). Changes in reproductive function of free-living birds in direct response to environmental perturbations. *In* "Processing of Environmental Information in Vertebrates" (M. H. Stetson, ed.), pp. 121–148. Springer-Verlag, Berlin.

Wingfield, J. C. (1990). Interrelationships of androgens, aggression and mating systems. *In* "Endocrinology of Birds—Molecular to Behavioral" (M. Wada, S. Ishii, and C. G. Scanes, eds), pp. 187–205. Jap. Sci. Soc. Press, Tokyo, and Springer-Verlag, Berlin.

Wingfield, J. C., and Farner, D. S. (1975). The determination of five steroids in avian plasma by radioimmunoassay and competitive protein binding. *Steroids* **26**, 311–327.

Wingfield, J. C. and Farner, D. S. (1976). Avian endocrinology—field investigations and methods. *Condor* **78**, 570–573.

Wingfield, J. C., and Farner, D. S. (1977). Zur endokrinologie einer brutenden population von *Zonotrichia leucophrys pugetensis. Vogelwarte* **29**, 25–32.

Wingfield, J. C., and Farner, D. S. (1978a). The endocrinology of a naturally breeding population of the white-crowned sparrow (*Zonotrichia leucophrys pugetensis*). *Physiol. Zool.* **51**, 188–205.

Wingfield, J. C., and Farner, D.S. (1978b). The annual cycle in plasma irLH nd steroid hormones in feral populations of the white-crowned sparrow, *Zonotrichia leucophrys gambelii. Biol. Reprod.* **19**, 1046–1056.

Wingfield, J. C., and Farner, D. S. (1979). Some endocrine correlates of renesting after loss of clutch or brood in the white-crowned sparrow, *Zonotrichia leucophrys gambelii. Gen. Comp. Endocrinol.* **38**, 322–331.

Wingfield, J. C., and Farner, D. S. (1980a). Environmental and endocrine control of seasonal reproduction in temperate zone birds. *Prog. Reprod. Biol.* **5**, 62–101.

Wingfield, J. C., and Farner, D. S. (1980b). Endocrinologic and reproductive states of bird populations under environmental stress. U.S.E.P.A. Report, contract No. cc699095.

Wingfield, J. C., and Farner, D. S. (1980c). Temporal aspects of the secretion of luteinizing hormone and androgen in the white-crowned sparrow, *Zonotrichia leucophrys. In* "Acta Cong. Int. Ornithol" (R. Nöhring, ed.), pp. 463–467. Deutschen Ornithologen-Gesellschaft, Berlin.

Wingfield, J. C., and Goldsmith, A. R. (1990). Plasma levels of prolactin and gonadal steroids in relation to multiple brooding and renesting in free-living populations of the song sparrow, *Melospiza melodia*. *Horm. Behav.* **24**, 89–103.

Wingfield, J. C., and Kenagy, G. J. (1991). Natural regulation of reproductive cycles. *In* "Vertebrate Endocrinology: Fundamentals and Biomedical Implications" (M. Schreibman, and R. E. Jones, eds), pp. 181–241. Academic Press, New York.

Wingfield, J. C., and Marler, P. R. (1988). Endocrine basis of communication: Reproduction and Aggression. *In* "The Physiology of Reproduction" (E. Knobil, and J. D. Neill, eds), pp. 1647–1677. Raven Press, New York.

Wingfield, J. C., and Moore, M. C. (1987). Hormonal, social, and environmental factors in the reproductive biology of free-living male birds. *In* "Psychobiology of Reproductive Behavior: An Evolutionary Perspective" (D. Crews, ed.), pp. 149–175. Prentice Hall, New Jersey.

Wingfield, J. C., and Ramenofsky, M. (1985). Hormonal and environmental control of aggression in birds. *In* "Neurobiology" (R. Gilles, and J. Balthazart, eds), pp. 92–104. Springer-Verlag, Berlin.

Wingfield, J. C., and Silverin, B. (1986). Effects of corticosterone on territorial behavior of free-living song sparrows, *Melospiza melodia*. *Horm. Behav.* **20**, 405–417.

Wingfield, J. C., and Wada, M. (1989). Male–male interactions increase both luteinizing hormone and testosterone in the song sparrow, *Melospiza melodia*: Specificity, time course and possible neural pathways. *J. Comp. Physiol.* **A 166**, 189–194.

Wingfield, J. C., Crim, J. W., Mattocks, P. W. Jr, and Farner, D. S. (1979). Responses of photosensitive and photorefractory white-crowned sparrows (*Zonotrichia leucophrys gambelii*) to synthetic mammalian luteinizing hormone releasing hormone (Syn-LH-RH). *Biol. Reprod.* **21**, 801–806.

Wingfield, J. C., Follett, B. K., Matt, K. S., and Farner, D.S. (1980a). Effect of day length on plasma FSH and LH in castrated and intact white-crowned sparrows. *Gen. Comp. Endocrinol.* **42**, 464–470.

Wingfield, J. C., Newman, A., Hunt, G. L. Jr, and Farner, D. S. (1980b). Androgen high in concentration in the blood of female western gulls, *Larus occidentalis wymani*. *Naturwissenschaften* **67**, S. 514.

Wingfield, J. C., Vleck, C. M., and Farner, D. S. (1981). Effect of day length and reproductive state on diel rhythms of luteinizing hormone levels in the plasma of white-crowned sparrows, *Zonotrichia leucophrys gambelii*. *J. Exp. Zool.* **217**, 261–264.

Wingfield, J. C., Newman, A., Hunt, G. L. Jr, and Farner, D. S. (1982a). Endocrine aspects of female–female pairing in the western gull (*Larus occidentalis wymani*). *Anim. Behav.* **30**, 9–22.

Wingfield, J. C., Smith, J. P., and Farner, D. S. (1982b). Endocrine responses of white-crowned sparrows to environmental stress. *Condor* **84**, 399–409.

Wingfield, J. C., Moore, M. C., and Farner, D. S. (1983). Endocrine responses to inclement weather in naturally breeding populations of white-crowned sparrows. *Auk* **100**, 56–62.

Wingfield, J. C., Matt, K. S., and Farner, D. S. (1984). Physiologic properties of steroid-hormone binding proteins in avian blood. *Gen. Comp. Endocrinol.* **53**, 281–292.

Wingfield, J. C., Ball, G. F., Dufty, A. M. Jr, Hegner, R. E., and Ramenofsky, M. (1987). Testosterone and aggression in birds: Tests of the "challenge hypothesis". *Amer. Sci.* **75**, 602–608.

Wingfield, J. C., Ronchi, E., Marler, C., and Goldsmith, A. R. (1989). Interactions of steroids and prolactin during the reproductive cycle of the song sparrow (*Melospiza melodia*). *Physiol. Zool.* **62**, 11–24.

Wingfield, J. C., Schwabl, H., and Mattocks, P. W. Jr (1990b). Endocrine mechanisms of migration. *In* "Bird Migration" (E. Gwinner, ed.), pp. 232–256. Springer-Verlag, Berlin.

Wingfield, J. C., Hegner, R. E., Dufty, A. M. Jr, and Ball, G. F. (1990a). The "challenge hypothesis": theoretical implications for patterns of testosterone secretion, mating systems, and breeding strategies. *Amer. Nat.* **136**, 829–845.

Wingfield, J. C., Hahn, T.P., Levin, R., and Honey, P. (1992). Environmental predictability and control of gonadal cycles in birds. *In* "Biology of the Chordate Testis" (H. Grier, and R. Cochran, eds), *J. exp. Zool.* **261**, 214–231.

Wingstrand, K. G. (1951). "The Structure and Development of the Avian Pituitary". Gleerup, Lund.

Wingstrand, K. G. (1954). The ontogeny of the neurosecretory system in chick embryos. *Publ. Staz. Zool. Napoli* **24**, 27–31.

Witschi, E. (1961). Sex and secondary sexual characters. *In* "Biology and Comparative Physiology of Birds" (A. J. Marshall, ed.), Vol. 2, pp. 115–168. Academic Press, New York.

Witt-Strømer, B., Ingritz, G., and Magnussen, L. (1956). Tidiga och sydliga hackninger av grasiska (*Carduelis flammea*) varen 1955. *Var Fagelvarld* **15**, 56–58.

Wittenberger, J. F. (1976). The ecological factors selecting for polygyny in altricial birds. *Amer. Nat.* **110**, 779–799.

Woitkewitsch, A. A. (1940a). On the inequality of the thyroid gland at various periods of moult in birds. *C. R. (Doklady) Acad. Sci. USSR*, **26**, 511–513.

Woitkewitsch, A. A. (1940b). Dependence of seasonal periodicity in gonadal changes on the thyroid gland in *Sturnus vulgaris. C. R. (Doklady) Acad. Sci. USSR*, **27**, 741–745.

Woitkewitsch, A. A., and Novikov, B. G. (1936). Sezonnye izmeneniya makotorykh endokrinykh organov i linka y *Passer domesticus. Tudy Inst. Eksp. Morf.* **5**, 331–341.

Wolfson, A. (1952). The occurrence and regulation of the refractory period in the gonadal and fat cycles of the junco. *J. exp. Zool.* **121**, 311–326.

Wolfson, A. (1966). Environmental and neuroendocrine regulation of the annual gonadal cycles and migratory behavior in birds. *Rec. Prog. Horm. Res.* **12**, 177–244.

Woods, J. W. (1957). The effects of acute stress and of ACTH upon ascorbic acid and lipid content of the adrenal glands of wild rats. *J. Physiol. Lond.* **135**, 390–399.

Wright, P. L., and Wright, M. H. (1944). The reproductive cycle of the male red-winged blackbird. *Condor* **46**, 46–59.

Yamashiro, D., Ho, C. L., and Li, C. H. (1984). Adrenocorticotropin 57. Synthesis and biological activity of ostrich and turkey hormones. *Int. J. Pep. Prot. Res.* **23**, 42–46.

Yokoyama, K., and Farner, D. S. (1976). Photoperiodic responses in bilaterally enucleated female white-crowned sparrows, *Zonotrichia leucophrys gambelii. Gen. Comp. Endocrinol.* **30**, 528–533.

Youngren, O. M., and Phillips, R. E. (1978). A stereotaxic atlas of the brain of the three-day old domestic chick. *J. Comp. Neurol.* **181**, 567–600.

Yu, J. Y. L., and Marquardt, R. R. (1973a). Effects of estradiol and testosterone on the immature female chicken (*Gallus domesticus*)—I. Quantitative changes in nucleic acids, proteins and lipids in liver. *Comp. Biochem. Physiol.* **46B**, 749–758.

Yu, J. Y. L., and Marquardt, R. R. (1973b). Interaction of estradiol and testosterone in regulation of growth and development of the chicken (*Gallus domesticus*) oviduct. *Comp. Biochem. Physiol.* **44B**, 769–778.

Yu, J. Y. L., and Marquardt, R. R. (1973c). Synergism of testosterone and estradiol in the development and function of the magnum from the immature chicken oviduct. *Endocrinology* **92**, 563–572.

Zweers, G.A. (1971). "A Stereotaxic Atlas of the Brainstem of the Mallard (*Anas platyrhynchos* L.)". Assen, Van Gorcum, Netherlands.

# INDEX TO BIRD NAMES

G



H



J



K



L



M

## Q



## R



## S



## T

# SUBJECT INDEX

## D

## Z